ACTIVATED CARBON

Activated Carbon

Harry Marsh
North Shields, United Kingdom

Francisco Rodríguez-Reinoso
Department of Inorganic Chemistry, University of Alicante, Spain

ELSEVIER

Amsterdam • Boston • Heidelberg • London • New York • Oxford
Paris • San Diego • San Francisco • Singapore • Sydney • Tokyo

Elsevier Ltd, The Boulevard, Langford Lane, Kidlington, Oxford OX5 1GB UK
Elsevier BV, Radarweg 29, PO Box 211, 1000 AE, Amsterdam, The Netherlands
Elsevier Inc., 525 B Street, Suite 1900, San Diego, CA 92101-4495, USA
Elsevier Ltd, 84 Theobald's Road, London, WC1Z 8RR, UK

First edition 2006

ISBN-13: 978-0-08-044463-5
ISBN-10: 0-08-044463-6

∞ The paper used in this publication meets the requirements of ANSI/NISO Z39.48-1992 (Permanence of Paper).

Typeset by Charon Tec Ltd, Chennai, India
www.charontec.com
Printed and Bound in Great Britain by CPI Antony Rowe, Eastbourne

This book is dedicated to our dear wives:
Audrey and Angela
for total support at all times

Contents

Preface

With its immense capacity for adsorption from gas and liquid phases, activated carbon is a unique material. It occupies a special place in terms of producing a clean environment involving water purification as well as separations and purification in the chemical and associated industries. In these roles, it exhibits a remarkable efficiency as the international production is a little more than half a million tonnes per year, with perhaps 2 million tonnes being in continuous use. This is equivalent to the allocation of 200 mg per person of the world population to be compared with the world use of fossil fuels of 2 tonnes per person of the world population.

Much is already written about activated carbon such that the Google search engine will offer over one million pages from the Internet. These pages are mostly descriptive and uncoordinated. Currently, there exists a need for a text which establishes a framework into which factual information can be located and explained. Further, carbon science has expanded rapidly in recent years with considerable relevance to an understanding of activated carbon. Again, these information are mostly uncoordinated and need to be assimilated into a comprehensive story. This assimilation is the purpose of this book.

An effective use of activated carbon requires a knowledge about the structure of its porosity, obtained from equilibrium data, namely the pore-size distributions of the microporosity in particular, of the pore-size distributions of the mesoporosity, of the composition of the carbon surfaces onto which adsorption occurs, and knowledge of the dynamics of adsorption to indicate its effectiveness in industrial use.

The authors experienced, during the course of writing, that two areas of major conflict exist within the scientific literature. Rather than simply describe these, attempts were made to resolve these two conflicts. First, the area of crystallography continuously presents the concept of the graphitic micro-crystallite as being the constituent of all carbons, the concept being carried over into the literature of Raman Spectroscopy. There is no evidence from adsorption studies that graphitic surfaces are part of activated carbons. In fact, the concept of the *crystallite* was flawed almost from the moment of its inception, several decades ago, and has blocked a realistic understanding of structure in both non-graphitizable and graphitizable carbons. The concept serves no purpose.

Second, despite what the advertisements may say, no instrument of adsorption exists (other than planometers) for the measurement of surface area. Such instruments measure adsorption capacity and a programmed microchip somewhere in the instrument converts from units of *grams adsorbed* to *meters squared per gram*, assuming that an atomic surface can be equated to a macro-surface, such as a mirror. This assumption has led to much non-productive debate. The characterization of activated carbon requires knowledge of pore volumes (e.g. $0.35\, cm^3\, g^{-1}$) but such numbers are less dramatic to the human ear than the sound of $1000\, m^2\, g^{-1}$. The concept of "surface area" is here to stay but its "image" should not be analyzed too deeply. This is the opinion of this book.

Central to activated carbon is the activation process which enhances the original porosity in a porous carbon. Activation uses carbon dioxide, steam, zinc chloride, phosphoric acid and hydroxides of alkali metals, each with its own activation chemistry. The story of what happens to a molecule of carbon dioxide after entering the porosity of carbon at 800 °C leading to the eventual emergence of less than two molecules of carbon monoxide is fascinating and talks about "atomic ballet".

This book, essentially, elaborates upon the origins of activated carbons, their structure and manufacture as well as their effectiveness for purification and separation.

This book points to the general direction of applications which is a subject area so immense that several volumes would be needed to summarize. However, searching the appropriate scientific journals, available electronically and with efficient computers, rapidly provides several thousands of applications.

Enjoy this book.

Harry Marsh
Francisco Rodríguez-Reinoso

Acknowledgments

The authors acknowledge the following Ph.D. students (University of Alicante) whose ideas and results are contained within this book.

Angel Linares Solano, Carlos Moreno Castilla, Luis Zurita Herrera, Antonia Martín Pozuelo, Francisco Javier Martínez Vilchez, Concepción Salinas Martínez de Lecea, Carmen Berenguer Melero, José Rivera Utrilla, José Miguel Martín Martínez, Miguel Molina Sabio, Antonio Guerrero Ruiz, Isadora Bautista Toledo, Inmaculada Rodríguez Ramos, Rosa Torregrosa Maciá, Julian Garrido Segovia, Celia Ana Prado Burguete, Itziar Pérez Lledó, A. Cesar Orgiles Barceló, Miguel Angel Martínez Sánchez, Antonio Sepúlveda Escribano, Miguel Angel Muñecas Vidal, Ana María Gutierrez León, Mª Teresa González Planelles, Fernando Coloma Pacual, Angeles Carmen Pastor Amorós, Carlota Gómez de Salazar Catarineu, Joaquín Silvestre Albero, Ana Huidobro Pahissa and Cristina Almansa Carrascosa.

The authors express deep appreciation for the assistance provided by the following in the initial readings of the chapters of this book.
Juan de Dios López-González, Miguel Molina-Sabio and Antonio Sepúlveda-Escribano.

The following Ph.D. students (University of Alicante) contributed to the preparation of figures of this book:

Javier Ruiz Martínez, Juan Manuel Juarez Galán, Yoshiteru Nakagawa, Enrique Vicente Ramos Fernández, Jose Manuel Ramos Fernández, Alejandro Rodríguez Guerrero, Segundo Antonio Sánchez Martínez, Juan Carlos Serrano Ruiz, Rachel Azzi Rios, Noelia Rojo Calderón,

H.M. thanks the support of the University of Alicante through the *Programa Senior*, requested

by Professor Rodríguez-Reinoso,
from September 2003 to January 2005.

CHAPTER 1

Introduction to the Scope of the Text

1.1 Activated Carbon

Question: *What is activated carbon?*
Answer: Activated carbon is porosity (space) enclosed by carbon atoms.

Question: *What is the size and shape of this porosity?*
Answer: Porosity has the size of molecules and is probably slit-shaped.

Question: *What are activated carbons used for?*
Answer: They are used for purification of water and of air and separation of gas mixtures.

Question: *What is activation?*
Answer: Activation is selective gasification of carbon atoms (thermal activation). Activation involves the use of phosphoric acid (chemical activation).

Question: *What are activated carbons made from?*
Answer: From hard woods, coconut shell, fruit stones, coals and synthetic macromolecular systems.

Question: *Can any natural organic material (NOM) be used to make activated carbon?*
Answer: No. Although NOM carbonize to porous carbon, only few provide commercially attractive activated carbons.

Question: *Are all activated carbons very similar to each other?*
Answer: No. There are several hundred commercial activated carbons available, with different sizes of porosity with specific applications.

Question: *Can different activated carbons be made from one source?*
Answer: Yes. Figure 1.1 contains isotherms of seven activated carbons from olive stones showing significant differences in pore sizes and pore contents.

Question: *What are isotherms?*
Answer: Please read Chapter 4.

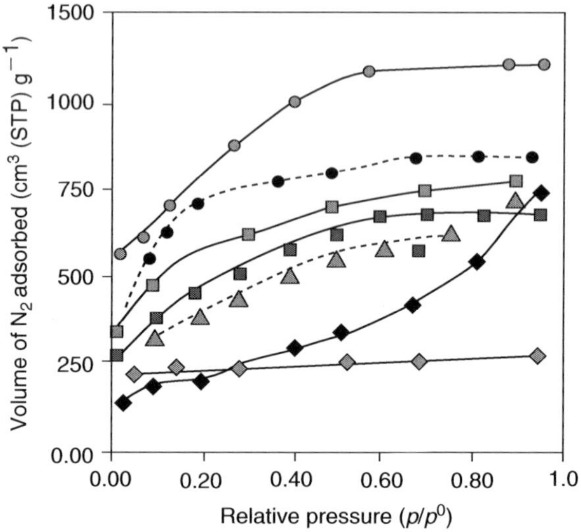

Figure 1.1. Adsorption isotherms of nitrogen at 77 K for seven activated carbons from a common precursor (olive stones), showing how the shape of the isotherm and extents of adsorption vary according to the method of preparation (adapted from Rodríquez-Reinoso F. The role of carbon materials in heterogeneous catalysis. *Carbon* 1998;36(3):159–175).

Question: *Is there a common structure in activated carbons?*
Answer: Yes, they are all assemblies of defective graphene layers.

Question: *Can activated carbons be described as amorphous materials?*
Answer: Emphatically, no.

Question: *Is shape and size of porosity all important for activated carbons?*
Answer: No. The chemistry of surfaces of porosity (functionality) has also to be considered.

1.1.1 Talking About Porosity

This text is a comprehensive overview of those issues within carbon science and technology which are considered to be essential to an understanding of the origins, preparation, characterization and applications of activated carbon. Carbon materials and activated carbons in particular are complex materials the study of which is not restricted to a single discipline but to several disciplines from geology to crystallography, to chemistry to theoretical physics, to manufacturing technology and environmental controls, and including aspects of economics and marketing. A good working knowledge of structural characterization (i.e. what is inside an activated carbon), of gasification kinetics (how to make an activated carbon) and of adsorption theory (a very effective way of understanding the porosity of carbons) is vital for those with interests and obligations toward industrial applications of activated carbon.

Activated carbon is a unique material because of the way it is filled with "holes" (voids, spaces, sites, pores, *whatever*) the size of molecules. What is special about these holes? It is that, although they are spaces of zero electron density, these pores possess intense van der Waals forces (from the near proximity of carbon atoms) and these are responsible for the adsorption process.

Throughout this book, reference is made to the characterizations or definitions used to describe this porosity. Thus, porosity in carbons is placed into four categories, not for convenience but out of necessity. The size of the pore determines how adsorption takes place in a pore be it narrow and wide microporosity <2.0 nm, mesoporosity 2.0–50 nm and macroporosity >50 nm. Chapter 1 introduces the reader through the several areas of relevant expertise.

1.2 Activated Carbon (Origins): Chapter 2

It is true to say that the entire world's flora is capable of providing a porous carbon on carbonization in an inert atmosphere. However, the casual production of activated carbon, capable of performing with efficiency the required industrial applications, does not happen. The commercial activated carbons on the market today are the result of continuous and intensive research and development toward optimization of application. The economics and availability of parent materials are as important (may be more so) as extents of available internal pore volumes (surface areas) associated with the right kind of porosity and surface chemistry. This means, of course, that a potential user of an active carbon should be as well acquainted with the capabilities of his purchase as is the producer of the activated carbon. And this means education and that means having available a text or a monograph devoted to the subject, and that means this book.

The availability of activated carbon for industrial use has much to do with accessing resources, renewing resources and processing to rigid specifications to control specific industrial applications. Only a handful of resources are used for activated carbon production, including coals of several rank, peat (not quite a coal) as well as woods, fruit stones and nutshells, as with coconut shells, as well as some synthetic organic polymers. A key element is the reliability and constancy of the resource. Manufacturing processes are so finely tuned that variations in the quality of the resource are unacceptable. Of the other resources which frequently come up for consideration, there are banana skins, straw, woodcuttings, casings from coffee beans and many other organic waste materials. Collection from separated areas, transportation, bulk availability and seasonal variations in quality and availability are reasons why these resources are not used.

Activated carbon is a member of a family of carbons ranging from carbon blacks to nuclear graphites, from carbon fibres and composites to electrode graphites, and many more. All come from organic parent sources but with different carbonization and manufacturing processes. As in a family, these carbon artifacts are related to each other. This text attempts to identify the most dominant of the carbon artifacts and to demonstrate their interrelationships. Mechanisms of carbonization are different, so creating the need to study the family of carbons in its completeness. Coal substances exhibit porosities ranging from microporosity to macroporosity, and the close similarities between porous carbons and coals are identified to advantage within the text.

The wide range of carbon artifacts which is available has a common ancestry, that is single-crystal graphite. All carbon forms (except the mineral diamond) are related to the graphite lattice in some way with each form of carbon representing one of an infinite number of assembled defective graphene layers, some very defective indeed. This text attempts to integrate these structural forms into a common pattern.

Carbons are prepared from parent organic material *via* carbonizations in the solid, the liquid and the gas phases. Mechanisms involved are significantly different. A solid-phase carbonization always proceeds within the solid phase with all structural changes involving atom removal and replacement within a solid lattice that remains essentially rigid throughout the process. No bulk movement of material occurs within the solid. Liquid-phase carbonizations are very different indeed. It is not a case of the liquid at some stage solidifying to give a structure as disorganized as the preceding liquid. In fact the opposite is true. The solids which emerge from a carbonizing liquid (coal-tar and petroleum pitches) are extremely well-organized and constitute the *graphitizable* carbons (also called *anisotropic* carbons), that is those carbons which on heating to beyond 2000 °C are capable of producing three-dimensional X-ray diffraction lines (*hkl*) of the graphite lattice.

Behind this story of liquid-phase carbonizations is the concept of growth of aromatic, nematic discotic liquid crystals which is one of the most exciting stories of carbon science. In comparison, carbonizations in the solid phase cannot permit any internal recrystallization (not quite true – there is *catalytic graphitization*), resultant structures being quite disordered but never disorganized and constitute what are termed the *non-graphitizable* carbons (also called *isotropic* carbons). That is, on heating these carbons to 2000 °C and beyond no (*hkl*) lines appear in X-ray diffractograms. Gas-phase carbonizations produce the carbon blacks, graphitizable, but to a limited extent only, in view of their dimensions of micrometers, and the pyrolytic carbons, highly graphitizable materials, produced by growth involving deposition of carbon atoms from the gas phase (methane or benzene). These pyrolytic (deposited) carbons have a role in controlling the diameters of entrances to porosities of some activated carbons, and for the carbon–lithium battery.

Of overwhelming importance is the structure (the arrangements of carbon atoms in space) within carbon materials, and in particular of activated carbon. Dominantly these originate from non-fusing parent materials and hence fall into the category of *non-graphitizable* carbons which exhibit considerable structural disorder, but are never amorphous. To describe in any meaningful and accurate way the structure within a disordered material is a challenge which still persists for *non-graphitizable* carbon materials. The very nature of this disorder creates the porosity upon which the properties of activated carbon depend. This disorder is central, totally, to what we need to know about activated carbon, and is therefore most critically assessed and debated within this book. This debate therefore has to enter the controversy over the existence (or otherwise) of the *carbon crystallite*, often reported in the literature as being graphitic, and being an integral part of carbon structure. This leads to assessments of how laser Raman microspectroscopy has a role to play in looking at this problem.

1.3 Porosity in Carbons (Modeling): Chapter 3

It cannot be over-emphasized that an understanding (best guess) of the structure of a microporous carbon, although difficult to achieve, is crucial as a prerequisite to any meaningful mental image of the shapes, sizes and continuities of porosities in carbons. In no way is it possible to have accurate modeling of these porosities without a close association with structural elements. The epicenter of an adsorption process, that is the containment of a molecule within (inside) the porosity, is the adsorption potential or intensity of the dispersion forces (van der Waals forces); this being dependent on distances between carbon atoms (not necessarily directly bonded to each other) and the specific bonding arrangements between the carbon atoms and heteroatoms which constitute is called a "surface".

Modeling of porosity is a serious challenge to the imagination and must relate to an understanding of what changes are possible and what are impossible within a solid phase. This text summarizes and comments on the models of porosity presented in the literature, making note of their useful points and rejecting what are misleading aspects. The aim of this review is to arrive at meaningful models to facilitate future applications of activated carbons and the theories supporting these processes. Some models may make crystallographic sense but are unacceptable when testing them against adsorption requirements. The two subjects must always to be considered together.

A significant hindrance to modeling of porosity in carbons, particularly microporosity, is the problem of scale (the scale of molecules) because a carbon particle, 1 mm^3 in size, has about 10^{11} pore entrances per 1 mm^2 of surface. Is it possible to model this?

The adsorbed state of a gas (e.g. nitrogen) within a microporous network can perhaps be considered as encapsulating the molecules of nitrogen in the gas phase by carbon atoms, such is the intimacy of the process. Another way of considering the adsorbed state is to say that there are resemblances to having a mixture of two gases which are immiscible (granted this is an impossibility) and separated from each other by thicknesses of only one or two molecules. The inter-twining and inter-dependency of one phase upon the other can be envisaged from this intimate relationship.

It is imperative to be sure that this intimate relationship between the porosity and structure is acceptable and realistic enough to be applied to theories of adsorption. This approach can be expanded upon as follows. Nitrogen gas, at 273 K and 0.1 MPa pressure, has a density of 0.00125 $g\,cm^{-3}$. When the fugacity of the gas is increased to 100 MPa the density increases to 1.25 $g\,cm^{-3}$ being a value comparable to the density of an active carbon. A "high-speed" (time lapse) photograph of the nitrogen molecules (frozen in time) would indicate the relative arrangement of the nitrogen molecules and carbon atoms with spaces between them, these not being randomly arranged but with groups of carbon atoms (a defective graphene layer) being associated with groups of nitrogen molecules (as adsorbed in a slit-like pore). Figure 1.2 is an attempt to model this situation using balloons to represent the nitrogen atoms. Always with modeling, some aspects are very appropriate and other aspects much less so.

The model of Figure 1.2, almost incidentally, suggests that the spaces between the balloons (carbon atoms) are more elongated (slit-like) than round. These spaces can now be renamed *pore entrances*.

When further considering this balloon model, the nitrogen atoms can be replaced by carbon atoms (or clusters of carbon atoms) which are of course bonded to each other chemically to

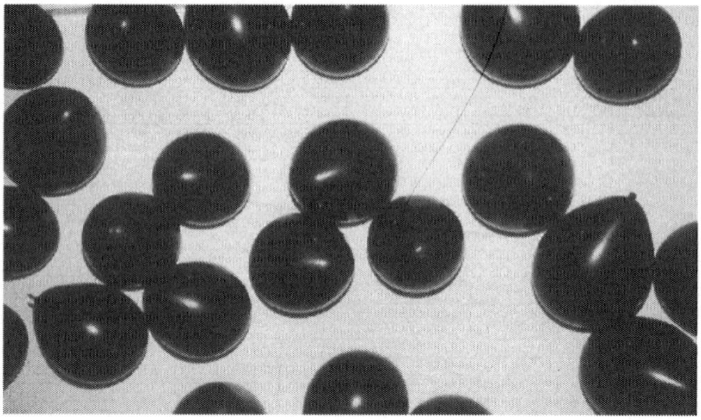

Figure 1.2. The use of balloons to demonstrate the structure of porosity and the shape of pore entrances and microsites.

create the integrity of carbon material. The model now becomes a little more sophisticated because the carbon atoms will not be randomly arranged but will be bonded to each other linearly and in defective cyclic arrangements of small size, and, most importantly, all of this bonding occurs in three dimensions to give the carbon strength and rigidity.

It is almost like the use of a glove to keep hands warm. The hand fills the space inside the glove closely because the glove has the shape of the hand. In a microporous carbon, the shapes adopted by the adsorbate molecules (in three dimensions) follow closely the shape of the microporosity. Microporous carbons from different parent sources produce different arrangements of carbon atoms leading to different structures of microporosity.

The simple approach of inverting in the computer Figure 1.2 provides Figure 1.3 in which the porosity shows up as black areas. An average dimension between the carbon atoms is about 2.0 nm. This model, although limited in vision as with most models, illustrates two aspects of porosity in carbons, namely (i) the variation in size of the adsorption sites and (ii) the reality that no two adsorption sites are identical.

Structure of porosity can be further imagined by making use of a two-dimensional maize as in Figure 1.4. This model (with its limitations) illustrates (a) that all positions of adsorption are interconnected and (b) the complexity of porosity in activated carbon when 1 cm^3 of an active carbon contains about 10^{20} of such adsorption spaces (sites). Further, the adsorption potential of an adsorption site changes when a neighboring site is occupied. During an adsorption process, the entire porosity is in a state of constant change.

Two other aspects of the adsorption process need to be introduced at this stage. First, an adsorbed state is not one of permanency. At the position of equilibrium between the gas phase (adsorptive) and the adsorbed phase (adsorbate) each molecule at a site of adsorption is replaced about every 10^{-12} s by another molecule. Equilibrium is a very dynamic state. Second, as each site of adsorption is different from any other site of adsorption, during an adsorption process, the sites of adsorption are occupied smoothly and progressively in order of decreasing energy of adsorption, wherever that site is within the carbon particle.

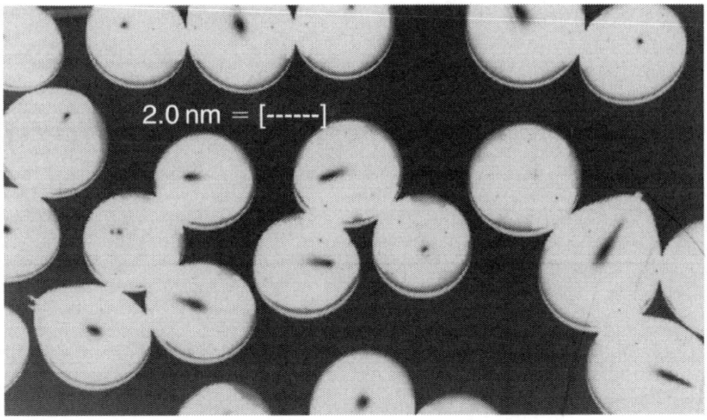

Figure 1.3. The use of balloons to illustrate the porous networks and the multivariant shapes of adsorption sites.

Figure 1.4. The use of a two-dimensional maize to illustrate networks of porosity and the multivariant shapes of adsorption sites within activated carbon.

Sites of equal (or almost equal) adsorption potential need not be physically close to each other.

Chapter 3 summarizes 27 proposed models for carbon structure, ranging from pen-and-paper models to advanced computer designed. The review demonstrates that there exists a gap between computer modeling and the most basic of properties of activated carbon as advanced from adsorption studies. There is much to debate and discuss in considerations of these models.

1.4 Characterization of Porosity: Chapter 4

As activated carbon has so many applications, it is imperative that a detailed knowledge should be available of the nature of the porosity within carbons, and those factors which control extents of adsorption and the strengths of these adsorptions. Although the models of Chapter 3 provide a visual image of porosity in carbons, predicting acceptable values of densities, they are as yet not sufficiently refined to predict pore-size distributions, mean pore sizes, surface compositions of the porosity, and to distinguish between carbons of different origins and different treatments.

There is no doubt that the most efficient experimental approach to obtain the above information about adsorption is to use the adsorption method itself. The experimental isotherm is packed with information and it becomes the task of the surface scientist to decode the isotherm and translate adsorption data into structural data. This book carefully takes the reader through the basic principles of adsorption theory pointing out some of the pitfalls which can beset the unwary. Carbon materials can be very unsympathetic when it comes to giving up its structural secrets to even the most supportive of investigators.

Although the complete equilibrium isotherm provides structural information of the entire carbon, for many applications adsorption is from dilute mixtures and for this that part of the adsorption isotherm obtained at low relative pressures (concentrations) $<0.0001\ p/p^0$, are extremely relevant. The other methods of characterization of activated carbon are all consistent within themselves and are complementary. Surface chemistry and dynamics of adsorption are of equal importance.

There is a limitation though in that methods of structural analysis of carbons have little to say about the arrangements of carbon atom within the graphene layer (with its wide range of contained defects). The adsorption process, *via* calculated enthalpies of adsorption and of immersion, does provide insights into the energetics (and hence defects) within the graphene layers of the carbon surfaces.

1.5 Activation Processes (Thermal or Physical): Chapter 5

Activated carbon does not just happen; it has to be synthesized. The porosities of a carbon, as initially prepared by carbonization, are not sufficiently developed for most applications and some amelioration is a prerequisite step. This is done in several ways involving creation of further porosity, widening of existing porosity, modifications to the surfaces of porosities and also modifying the carbonization process itself. All activation reactions are heterogeneous use is made of either carbon dioxide or steam or mixtures of these two gases. The classical chemical kinetics of these gasification reactions are applicable and are set out in Chapter 5.

Carbon atoms can be removed from within porous carbons by gasification using carbon dioxide or water vapor, usually at 800–900 °C. The reaction equation is simple enough, e.g. $CO_2 + \text{carbon} \rightarrow 2CO$, but the overall kinetics and topographical features (reaction anisotropy) contain additional detail which is relevant to activation processes. Activation by carbon dioxide and steam produce carbons with different porosities. Although discussions of the relevant importance of inhibition by chemisorbed oxygen and hydrogen provide some insight into this aspect, a final explanation, including molecular dynamics, is still some distance away.

In simple terms, thermal activation is a process of selective gasification (removal) of individual carbon atoms. Not all carbon atoms are of equal reactivity.

1.6 Activation Processes (Chemical): Chapter 6

In addition to the main processes of activation by carbon dioxide or steam, three other techniques of chemical activation are used, involving co-carbonization with (a) zinc chloride, (b) phosphoric acid and (c) with potassium hydroxide. Mechanisms for these activations are all different, with zinc chloride promoting the extraction of water molecules from the lignocellulosic structures of parent materials, and phosphoric acid combining chemically within the lignocellulosic structures. There is no selective removal of carbon atoms as during physical activation and carbonization yields are improved. The mechanisms by which potassium hydroxide activates an existing carbon are more complex and involve the disintegration (almost explosively) of structure following intercalation as well as some gasification by the oxygen of the hydroxide. The presence of oxygen is not essential (but may be helpful) to this form of activation. The details, in chemical terms, of these processes of chemical activation of carbons are elaborated upon in Chapter 6.

1.7 SEM and TEM Images of Structures in Activated Carbon: Chapter 7

It is often beneficial to move away from the theoretical equation and the computer-generated model to the scanning electron microscope and the transmission electron microscope to reveal actual structures with activated carbon. Chapter 7 contains a selection of such micrographs which provide an overall impression that activated carbon is not altogether that homogeneous and perfect in the way that pore sizes are distributed. Such micrographs reveal a heterogeneity of structure.

1.8 Applications of Activated Carbon: Chapter 8

1.8.1 Introduction

Charcoals (active or porous carbons) are as old as history itself and would be known to Hippocrates (Figure 1.5), the father of medicine. The most early of recorded applications include their use as a medicine to relieve digestion problems which continue today in the removal of overdoses of drugs from stomachs (as in television hospital dramas).

Figure 1.5. An image of Hippocrates, an early protagonist in the use of charcoal.

Figure 1.6. The structural formula of mustard gas, 1,1-thiobis(2-chloroethane).

Figure 1.7. Fireworks have given pleasure to millions, charcoal being an essential ingredient.

By far the most dramatic application of charcoal was in World War I with its use in gas masks for the protection of soldiers against chlorine, phosgene and mustard gas in trench warfare. Although the number of soldiers who lost their lives from gas warfare exceeded many tens of thousands, the number who survived making use of the gas mask was much higher. The gas mask (respirator) essentially is a cylinder filled with a charcoal (activation was still to come) prepared by the carbonization of coconut shell. It was reported at that time that these respirators were more effective against chlorine (Cl_2) and phosgene ($COCl_2$) than with mustard gas ($Cl—CH_2—CH_2—S—CH_2—CH_2—Cl$, 1,1-thiobis(2-chloroethane)) (Figure 1.6). This molecule, because of its size and shape, would be adsorbed more slowly than the smaller molecules of chlorine and phosgene.

Two other major applications of charcoal include its use in fireworks (Figure 1.7) and in gunpowder. The former has given immense pleasure to millions of people, but gunpowder does not have this reputation. Its use in ballistics must have resulted in the death of millions.

1.8.2 Adsorptions from Aqueous Solutions

Activated carbon, by comparison, is the workhorse of the porous carbons. Chapter 8 initially examines the chemistry and mechanisms of adsorption of inorganic and organic compounds from aqueous solution, and then moves on to summarize the most important of gas- and liquid-phase applications.

In industrial applications, adsorption takes place from mixtures of adsorptives (solute and solvent), this being much more complicated that single-component adsorptions. Adsorption from the liquid phase (mainly water) is more common than from the gas phase, the most common pollutants in water being phenols and NOM. Adsorption of inorganic compounds from aqueous solutions is somewhat controversial, when it appears that the presence of surface oxygen functionality is essential, the carbon acting as an ion-exchange material. Discussions of acidic and basic characteristics of graphene layers enter into the debate. There is the viewpoint that in the narrowest of pores, the adsorption potential is sufficiently high to attract and retain ions from the solution. As ions are usually hydrated, and as the experimental evidence supports the ion-exchange mechanism, Chapter 8 expresses doubts about the *ion-adsorption* theory.

Adsorption of organic molecules from aqueous solution is equally complicated. As many of the organic adsorptives have functionality (e.g. phenols) then pH of the solution and the chemical nature of the carbon surface are as important as in adsorption of inorganic adsorptives.

The literature relevant to adsorption from aqueous solutions is extensive and confusing. A dominant reason for this is that many of the studies are not comprehensive enough, meaning that an individual study must look, without exception, at all of the variables which influence extents of such adsorptions. The same variables apply, more or less equally, to adsorptions of both inorganic and organic solutes.

Each study, to be undertaken, must include the variables listed below. If not, then interpretations (conclusions) may not be conclusive enough to further clarify mechanisms.

> *A dominant aim of such studies is to maximize adsorption of the adsorptives (solutes), from solutions as existing within the environment, as efficiently as possible using activated carbons of known characteristics.*

The variables, within an adsorption system (dominantly an aqueous solution) controlling extents of adsorption include:

1. Volume of narrow micropores, <0.7 nm (V_{NM} – $cm^3 g^{-1}$), including the pore-size distributions.
2. Volume of wider micropores, 0.7–2.0 nm (V_{WM} – $cm^3 g^{-1}$), including the pore-size distributions.
3. Volume in mesopores, >2.0 and <50 nm (V_{ME} – $cm^3 g^{-1}$), including the pore-size distributions.

4. *Note*: The parameter "surface area" is a crude parameter and fails to quantify, adequately, the capacity of the all-important micropores and their size distributions.
5. The amphoteric nature of the carbon surface as measured by pH_{PZC} and pH_{IEP}.
6. The controlled additions of surface oxygen complexes from known sources, e.g. nitric acid, H_2O_2, etc., with controlled removal of the complexes, thermally or by reaction with hydrogen at temperatures $>750\,^{\circ}C$.
7. The effects of additions of surface oxygen complexes upon pore volumes and pore-size distributions (pore dimensions can be reduced).
8. The characterization of surface oxygen complexes by diffuse reflectance infrared spectroscopy (DRIFT) and temperature programmed desorption (TPD).
9. The pH of the solution, together with an understanding of the variation of speciation with changing pH, associated with knowledge of values of pK_A and pK_B.
10. Ionic strength (natural and industrial water, often, is a mixture of solutes).
11. Temperature.

A list of applications is almost endless and Chapter 8 emphasizes the major applications, including separation of gas mixtures and methane storage. Water treatment is of critical importance. The use of carbons in catalytic processes is reviewed. A look into the future concludes that active carbon will continue to be used for the foreseeable future with further growth in the market.

1.9 Production of Activated Carbon and Reference Material: Chapter 9

Activated carbons are more than a laboratory curiosity. They are working materials. This text leads the reader into the detail of commercial activation processes and the dedication of selected procedures for the preparation of carbons with specific applications. Carbonization and activation furnaces differ in constructions leading to different products requiring expertise in quality control. Dominantly, activated carbons are used in gas- and liquid-phase applications where quite different types of porosity are needed to optimize procedures. Although not excessively expensive, say \$1.0–10.00 per kg, the importance of regeneration is acknowledged.

Both the industry which supplies activated carbons and the many agencies which use them have devised test procedures as measures of quality control. Many of these are of an empirical nature relating to specific applications. These standard procedures are abstracted in this text to facilitate reference and use. A problem within the carbon industry and research laboratory is one of standardization of nomenclature. IUPAC published recommendations for terminology to be used for descriptions of carbon materials (1995). This has been updated and is also contained within Chapter 9. A list of references for further reading of material not used directly in this text, but of relevance overall to activated carbons, is also included. Finally, selected biographies of those researchers who pioneered understandings of adsorption processes, of structure in carbons and of industrial commercial applications, are included in Chapter 9.

CHAPTER 2

Activated Carbon (Origins)

It is the porosity within activated carbon which imparts their dominant characteristics of adsorption. Activated carbons are prepared from a wide range of carbon types (the family of carbons) of different structures. As porosity and structure are so intimately related, an understanding of the structures in this range of carbons together with a description of their origins and preparation are prerequisites to the effective use of activated carbons.

2.1 Carbon Materials

From a structural point of view, carbon materials, collectively, form a family of carbons. Yet, at the same time, each individual carbon possesses specific properties. Each carbon has a unique identity. This chapter considers this family of carbons in terms of related structures and how these affect porosity because the origins, extents and characteristics of porosity are intimate functions of structure within each carbon. Activated carbons are a major part of this family of carbons. *The specific objective here is to present descriptions of structures of carbons and to assess how these structures control the characteristics of adsorption in the porosity of carbons.* From this base of knowledge, the processes of adsorption and activation as described in Chapters 4–6 will be better understood.

With regard to the broad overview of the family of carbons, several comprehensive reviews are available providing relevant supportive information:

Baker (1992) reviewed activated carbon for the Kirk-Othmer Encyclopedia of Chemical Technology, and has a US Patent (1998).
Bandosz *et al.* (2003) review, in depth and historically, molecular models of porous carbons describing known forms of carbon.
Bansal *et al.* (1988) discuss active carbon.
Bourrat (1997) reviews structures in carbons and carbon artifacts.
Burchell (1999) summarizes the use of carbon materials, including porous carbons, for advanced modern technologies.
Cheremisinoff and Ellerbusch (1980) have a handbook on carbon adsorption.
Donnet and Voet (1976) describe the physics and chemistry of carbon black.
Jankowska (1991) discusses active carbon in book format.

Marsh *et al.* (1997, 2000) have published texts on carbon technology and science.
Mattson and Mark (1971) discuss activated carbons in terms of surface chemistry and adsorption from solution.
McGuire and Suffet (1980a, b) published two volumes on the use of activated carbons for adsorption of organics from the aqueous phase.
McGuire and Suffet (1983) edit a symposium on water treatment by granular-activated carbon.
Patrick (1995) reviews porosities in carbons.
Rodríguez-Reinoso (2002) reviewed the production and applications of activated carbon.
Rouquerol *et al.* (1999) describe the principles, methodology and applications of adsorption by powders and porous solids.
Ruthven (1984) outlines the principles of adsorption and adsorption processes mainly with industrial applications in mind.
Yang (2003) provides a comprehensive source of knowledge for all commercial and new sorbent materials.
Yasuda *et al.* (2003), in *Carbon Alloys*, provide 34 chapters describing studies of carbon structure from the point of view of incorporation of heteroatoms into their structures.

This chapter emphasizes that (apart from diamond, Figure 2.1) structures in all known carbon forms are considered as a continuous decrease in the degree of order from the single-crystal hexagonal graphite (Figure 2.2) to the most disordered (there is never total disorder) of the porous carbons and the glassy carbons with their closed (inaccessible) porosity (see Pikunic, 2001; Chapter 3). Within hexagonal graphite, the layers of hexagonal arrangements of carbon atoms are described as *graphene layers*. These layers do not lie immediately above and below each other but are displaced to form an ABABAB sequence. The density of hexagonal graphite is ~2.25 g cm^{-3}. The distance between the layers is 0.335 nm (335 pm) and the distance between two bonded carbon atoms is 0.142 nm (142 pm). Within the graphitic layers, the bonding is trigonal sp^2-hybrid σ-bonds with delocalized π-bonds within the layers. The interlayer spacing of 0.335 nm, being larger than the 0.142 nm of the C—C bond, indicates no chemical bonding between the layers, the forces of attraction being limited to van der Waals forces. This distance is between the values of 0.153 and 0.132 nm for Csp^3—Csp^3 bonding as in ethane, and Csp^2=Csp^2 bonding as in ethene. Resonance considerations indicate that the C to C bonding within a graphite layer has about one-third double-bond character. Such a graphite structure is referred to as AB graphite.

Diamond has a more dense structure than graphite with a density of 3.52 g cm^{-3} indicating, overall, a much closer packing of carbon atoms. The bond distances within diamond are 0.154 nm (for zinc blende structure) and 0.152 nm (for the Lonsdaleite structure). The hardness of diamond is associated with the close strong carbon–carbon bonds. The distance of the carbon–carbon within the graphite layer is 0.142 pm indicating a stronger bond than in diamond. The absence of any equivalent of a graphene layer in diamond excludes any possibility of creating a porous diamond. The defective *micro-graphene layer* is totally central to the structure of activated carbon.

There is always interest in the possible conversions of graphite to diamond. Graphite is the thermodynamically more stable phase with an enthalpy of transition of diamond to

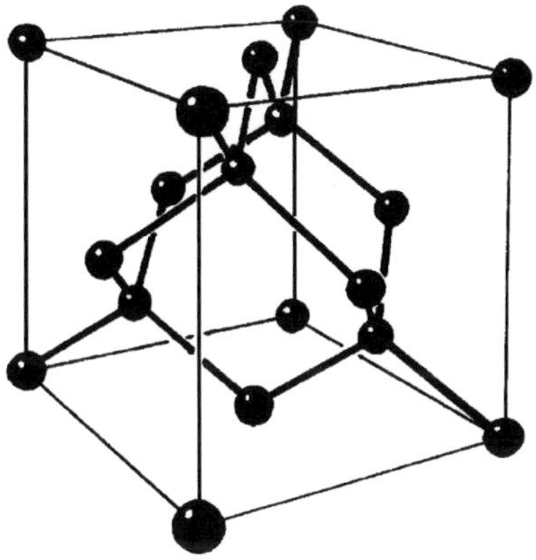

Figure 2.1. A model of the cubic unit cell of diamond. The internal carbon atoms are bonded to three other carbon atoms, with sp^3-symmetry, as in methane.

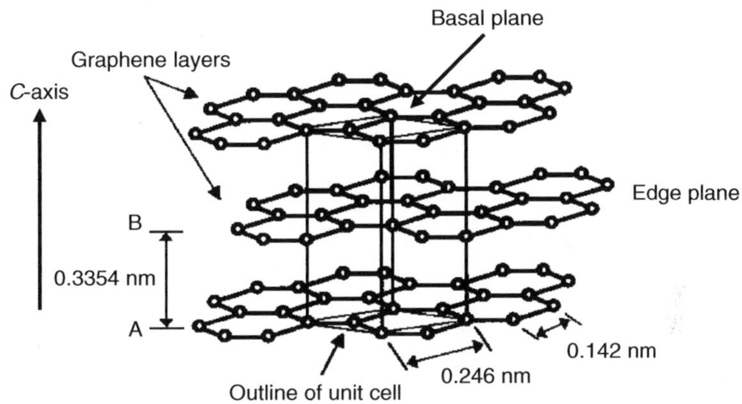

Figure 2.2. The structure of hexagonal graphite, with trigonal planar bonding within the graphene layers.

graphite of $+2.90\,\text{kJ mol}^{-1}$ at 298 K. The heating of diamond in an inert atmosphere to about 1200 K will accelerate its transformation to graphite. Diamond, like graphite and other forms of carbon, is oxidized to oxides of carbon when heated in air. Bundy *et al.* (1996) have produced a phase and transitional diagram for carbon and this is shown in Figure 2.3. It is relevant to have some mention of diamond because of suggestions that in activated carbons the carbon atoms, which contain the inherent porosity (the term pore walls may be too restrictive), may contain some sp^3-bonding.

Figures 2.4 and 2.5 are photographs of two of the early researchers. Kekulé (1858) first considered the ring structure of benzene (published in 1865). Bernal (1924) confirmed and refined previous suggestions for the structure of single-crystal graphite (Hull, 1917).

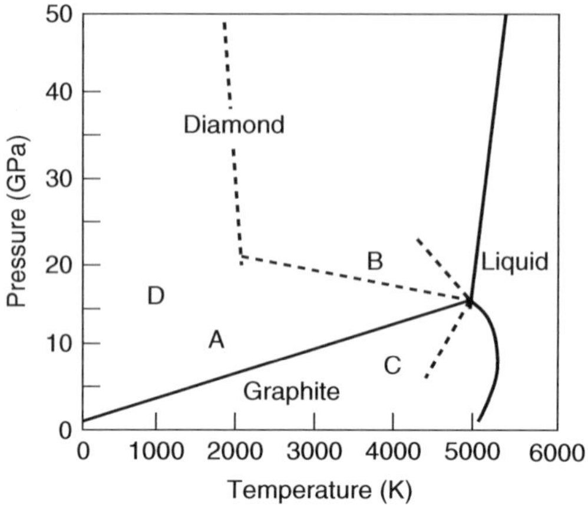

Figure 2.3. The phase diagram for carbon. Solid lines represent equilibrium phase boundaries: Position A: commercial synthesis of diamond from graphite by catalysts; B: P/T threshold of very fast (<1 ms) transformation of diamond to graphite; C: P/T threshold of very fast transformation of graphite to diamond and D: single-crystal hexagonal graphite transforms to retrievable hexagonal-type diamond (Bundy, 1996).

Figure 2.4. Photograph of Prof. FA Kekulé.

Figure 2.5. Photograph of Prof. JA Bernal.

2.2 Parent Materials for Carbons

Quite diverse source (parent) materials are currently used in active carbon production. Structures produced by their carbonization (heat treatment in an inert atmosphere (700–1000 °C) may be considered as segments of graphene sheets (not a three-dimensional segment) of different sizes reaching nanometer dimensions and degree of perfection (holes, non-planarity, linear carbon atoms (dangling bonds) and the presence of heteroatoms), bonded together in an infinite number of ways. These segments can be considered as being cut from a single graphene sheet using a pair of "atomic scissors". Segments can, in one sense, be more like polycyclic aromatic compounds of various complexity bonding into and forming part of a three-dimensional network of carbon atoms including carbon atoms with sp^3-bonding, but, and this is emphasized most strongly, not existing in isolation and so not manifestly identifiable as such.

Although structures in carbons can be envisaged (imagined) from this extremely simple but quite realistic point of view, how to describe them in words or mathematically (i.e. a quantitative characterization) is a much more difficult task; it may even be impossible. The most ordered carbons are single-crystal graphite, the pyrolytic carbons and the polycrystalline graphites with well-ordered stacking of graphene sheets (the layers of the model of Figure 2.2). Progressively, carbons exist in which the graphene layers abandon the ABABAB stacking, they become smaller and more defective, until eventually, the existence of identifiable graphene layers is almost lost, as happens with the glassy carbons (Pikunic *et al.*, 2001).

(Glassy carbons are isotropic, made from resins under pressure, are impermeable to gases, are of low helium density (with porosity closed to helium atoms) and have a high coefficient of elasticity.)

The random bonding together of these small defective polycyclic groups of carbon atoms, with linear carbon atoms, creates space (imperfect packing) called porosity. Figure 2.6(a–f) are fringe-image high-resolution electron microphotographs of carbons of decreasing extents of graphitization and illustrate the existence of a continuous and progressive reduction in structural order within the family of carbons. However, methods of preparation of these carbons are not continuous, but follow specific and different procedures. These methods of preparation form the basis of much of carbon chemistry.

The different arrangements of segments or groups of carbon atoms confer on carbons their extremely wide range of properties (i.e. physical, chemical and mechanical) as a result of which carbon materials have an extensive range of industrial applications. This represents a major feature of carbon materials. In fact, no other element has this facility of so controlling properties in terms of diverse structures.

2.3 Carbon Nomenclature

The characterization of members of the family of carbons is described in terms of the limits to their structural order. On the one hand, there are carbons which are graphitic (but not necessarily of single-crystal quality); that is, they exhibit X-ray diffraction (XRD) lines (*hkl*) of three-dimensional graphite or become graphitic on heating to beyond 2000 °C. These are the so-called *graphitizable*, *anisotropic* carbons in which graphene layers show considerable planarity and stacking. On the other hand, there are carbons which are always non-graphitic, that is of never exhibiting three-dimensional XRD lines even after heating to well beyond 2000 °C. These are the so-called *non-graphitizable*, *isotropic* carbons and all of the porous carbons fall into this category.

These terms of *graphitizable* and *non-graphitizable* carbons were introduced by Rosalind Elsie Franklin (1950, 1951), a pioneer researcher into structures of coals and carbons, as

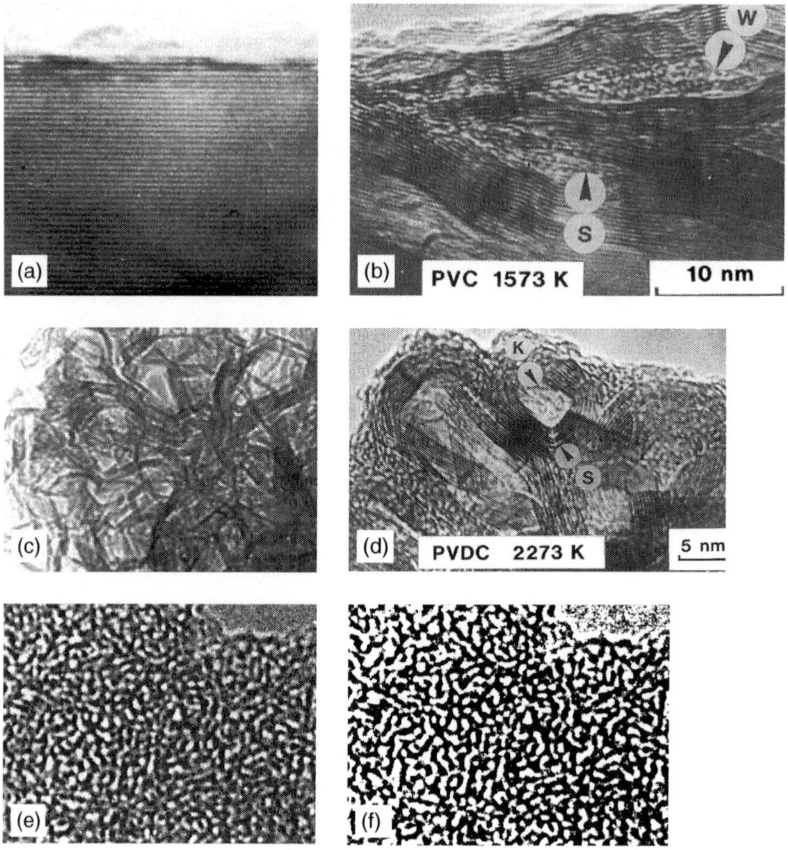

Figure 2.6. High-resolution fringe-image electron micrographs of structure in different carbons. (a) Single-crystal graphite showing parallelism of layers. (b) A partially graphitized carbon from PVC (1573 K). (c) A non-graphitizable carbon from PVDC (2973 K). (d) A non-graphitizable carbon from PVDC (2273 K). (e) A non-graphitizable carbon from PVDC (1073 K). (f) Fringe images of the non-graphitizable carbon from PVDC (1073 K) with computer manipulation to indicate the inability to decide which is carbon and which is porosity (Marsh, unpublished results).

well as the DNA double helix. The models she created to describe these two forms of carbon are shown in Figure 2.7 and a photograph of Dr. Franklin is shown in Figure 2.8.

Although these drawings (models) of Franklin (1950, 1951) are in two dimensions, they have proven to be remarkably accurate over the years since their publication. It is the parallelism of the layer planes (graphene layers) which facilitates graphitization. By contrast, the non-graphitizable carbons do not have this long-range parallelism of graphene layers.

The model of Franklin (1950) for non-graphitizable carbon is remarkably accurate. If this model is re-designed into three dimensions, then the models of Figures 3.9, 3.16 and 3.17

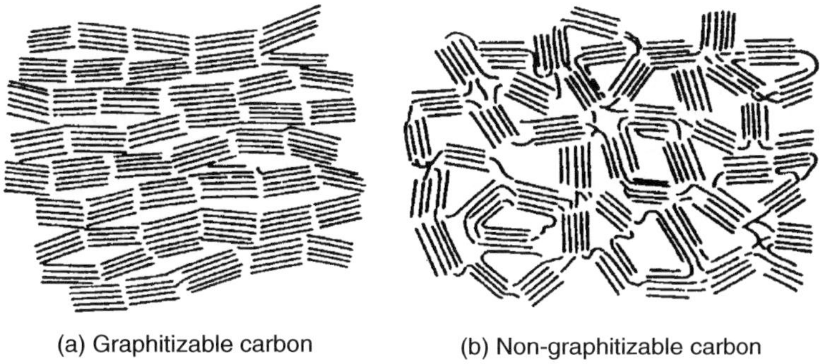

Figure 2.7. Drawings to illustrate the essential differences between (a) graphitizable and (b) non-graphitizable carbons (Franklin, 1950, 1951).

Figure 2.8. Photograph of Rosalind Elsie Franklin (1920–1958).

(Chapter 3) are created. The short-range parallelism of her model (distances of a few nanometers) only needs to be replaced by non-parallel, non-planar structural units and there is created an acceptable structure for porosity in activated carbon. Figure 2.7(b) indicates the origin of porosity.

This short-range structure of non-planar units became bonded into the carbon material during its formation in the carbonization stage. This short-range order cannot be removed by heat treatment to 2000 °C because so many bonds would need to be broken simultaneously as to create impossibly high activation energies. But, there appear to exist in these isotropic carbons zones (or space elements) where some form of imperfect parallelism (after heat treatment to 2973 K) which is capable of producing fringe images in the diffraction mode of transmission electron microscopy (TEM) (Figure 2.6(c)). It is relevant to note that this incipient parallelism is too weak and too limited in extent, in carbons of heat treatment temperature (HTT) <2000 °C, to be resolved by XRD into (*hkl*) lines. However, carbons exist in all of the intermediate stages of restricted graphitizability, some having major commercial use like carbon anodes (aluminum production) and metallurgical coke for use in blast furnaces and metallurgical foundries. Such carbons, on heat treatment to 2000 °C and beyond, produce a limited number of (*hkl*) XRD lines (often bands rather than lines) indicating limited extents of graphitization.

The terms *soft* and *hard*, very frequently found in the literature, are really not that helpful being based on early working experiences of their physical properties. These terms should be avoided and relegated to the past literature. To use these imprecise physical terms loses sight of the basis of their structures. For example, a baked carbon anode, as used in the aluminum industry, is an extremely hard material, but is graphitizable. The isotropic carbons in their industrial use are selected for their hardness; otherwise they would easily degrade to dust. An isotropic carbon, of low HTT of about 600 °C, may be quite soft, depending. The terms should be avoided.

The terms *anisotropic* and *isotropic* form the basis of the characterization of carbons and are, if anything, under-used. This terminology is based on their examination in an optical microscope which uses reflected plane polarized light, preferably with a retarder plate, from a polished surface of carbon. As graphite is an anisotropic solid, with different structural characteristics along the *x*-, *y*- and *z*-axes, the quality of the polarized light reflected from the polished surface changes as the carbon specimen is rotated on the microscope stage. This is seen as progressive changes in black and white areas or in changes in color from blues to reds in these areas (the phenomenon exhibited is termed pleochroism or *optical texture*) with the retarder plate in place in the optical microscope. The more graphitic is the carbon the larger are the sizes of the anisotropic areas (Figures 2.9 and 2.10) the

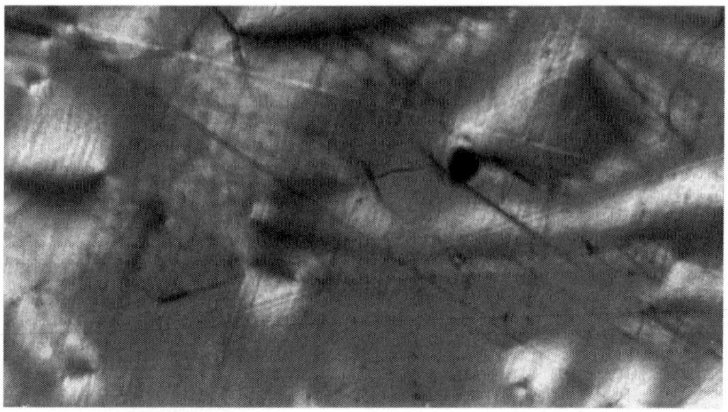

Figure 2.9. Optical micrograph of polished surfaces of an anisotropic carbon of large optical texture, domains, sizes of >20 μm. The size of the image of the micrograph is 30 × 60 μm.

larger sizes being termed domains (>20 μm in size) and the smaller sizes being termed mosaics (<5 μm in size), respectively.

In addition, it is possible to measure, quantitatively, percentages of polarized light reflected from these polished surfaces to provide *reflectance* data. Such reflectance measurements (%) increase as the degree of "aromaticity" increases in the carbon, indicating higher degrees of graphitizability (Table 2.1). The reflectance is also a function of the position of the graphene layer relative to the plane of polarized light used by the microscope. Values of minimum and maximum reflection separate as the degree of graphitization increases, providing the function of bireflectance. The properties of reflectance and bireflectance, although used routinely by the coal industry, could be used with profit by

Figure 2.10. Optical micrograph of polished surfaces of an anisotropic carbon of small optical texture composed of mosaics, sizes of <5 μm. The size of the image of the micrograph is 30 × 60 μm.

Table 2.1. Variation of specific reflectance (%) of polarized light from polished surfaces of coals of increasing rank and coal properties.

Rank stages	wt% carbon content	% volatile matter (daf)	% moisture *in situ*	% vitrinite reflectance (random/maximum)
Wood	50	>65	–	–/–
Peat	60	>60	75	0.20/0.20
Brown coal	71	52	30	0.40/0.42
Sub-bituminous	80	40	5	0.60/0.63
High volatile bituminous coal	86	31	3	0.97/1.03
Medium volatile bituminous coal	90	22	<1	1.47/1.58
Low volatile bituminous coal	91	14	1	1.85/2.63
Semi-anthracite	92	8	1	2.65/2.83
Anthracite	95	2	2	6.55/7

daf: dry and ash free.

the carbon and graphite industries, and be more meaningful than crystallite size determinations (Sections 2.10. and 2.12).

On heating a graphitizable carbon to beyond 2000 °C, XRD lines become sharper and more and more three-dimensional lines appear as parallelism of the graphene layers progresses. An emphasis is critical for the fact that parallelism and enhanced stacking order of graphene layers (graphitization) is *not* the result of movement of entire graphene layers within the solid carbon phase. Rather, it occurs by the independent movement of individual carbon atoms to positions of minimum energy, both within an individual graphene layer and relative to those graphene layers, both above and below.

Isotropic materials have properties which are independent of the direction of examination, x-, y- or z-direction. As such, they appear with a purple color in the polarized light optical microscope and this color does not change on rotation of the stage of the microscope. Thus, reflectance and bireflectance are not relevant properties for isotropic carbons. The porous, isotropic carbons are easily identified and distinguished from anisotropic carbons by this technique.

Another term, which also has to be avoided, is the use of the word *turbostratic* (adjective). This term entered the early literature of carbon materials before any realistic knowledge of structure in all carbons was available. The term *turbostratic*, in one sense, begs the question about structure by assuming, in the first place, that there are identifiable layers of carbon atoms (in their more perfect form they are called graphene layers) and that these layers exist with some parallelism but are not in an ABABAB sequence or any other sequence. Instead, the assumption is made that they are randomly orientated relative to each other, like playing cards thrown one by one on to a table. This definition, if it has any meaning at all, indicates that all carbons, apart from AB graphites, are turbostratic including both the graphitizable anisotropic (but not graphitic) and non-graphitizable isotropic carbons. Such a purposeless definition can hardly be helpful and it is a pity that the carbon literature uses the term indiscriminately. It is also interesting to note that the IUPAC recommended terminology for the description of carbon as a solid (Fitzer *et al.*, 1995) omits this definition.

2.4 The Element Carbon

A question now is why should the element carbon be so versatile in the way it can organize itself. In the periodic table, the element carbon occupies quite a unique position, atomic number of six with four outer electrons capable of multi-bonding (sp-linear, sp^2-in-ring, and sp^3, as in diamond) (Figure 2.11). It is of low atomic weight (12.011 amu). It is an element of prehistoric discovery and is ubiquitous being found in meteorites and in abundance in suns, stars, comets and the atmospheres of most of the planets. That is, it is an element, an atom, of significant stability. The atmosphere of the planet Mars contains 96.2 vol% of carbon dioxide. It is an element of immense versatility, combining with hydrogen, oxygen, nitrogen and other elements to form at least 10 million compounds (organic). Without carbon, all life forms, that we know of, would not exist. This branch of chemistry, that is organic chemistry, does have relevance here because of a geological connotation. So why does geology have to enter into discussions of carbon materials? It is because activated carbon is manufactured from woods, nutshells, and coals of various

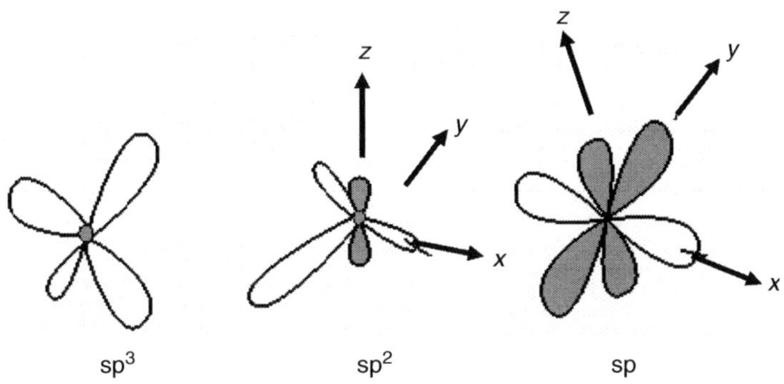

Figure 2.11. The four outer electrons of carbon atoms are capable of multi-bonding as tetrahedral sp^3-hybrid, trigonal planar sp^2, linear (180°) sp.

ranks (age or level of maturity along the coalification path), all being organic in nature and all the result of millions of years of floral evolution. Knowing the origins of the parent sources of activated carbons must be part of our overall knowledge of activated carbons. Coals of different ranks are used in activated carbon manufacture. Differences between coals of different ranks have to be recognized.

It is about 2000 million years ago (2×10^9) since life forms on the landmasses of planet Earth made their first appearances. Initially, the life forms were of a primitive nature and belonged to the world of plants (algae). The climatic conditions of the Earth were vastly different from what we know today, the landmasses (continents) were in different locations relative to each other on the surface of the planet with a slow movement of continents (tectonic) through what are now considered to be the tropical zones (tropics). This was the time when growth of organic material took place on a scale never seen since on this planet. The vast growth, death and preservation of algae in aqueous environments, followed by millions of years of metamorphosis, lead to the creation of the petroleum deposits of today on a scale commensurate with the size of the Earth.

Incidentally, the appearance of plants (flora) with their ability for photosynthesis involving the release of oxygen into the atmosphere, totally transformed the reduced state of the Earth's constituent minerals into an oxidized state (e.g. red iron oxide). More so, however, the appearance of plants (oxygen releasing) opened the pathways for the evolution of fauna (the animal kingdom) including, ultimately man, who was destined to use activated carbon. The variation of the composition of the Earth's atmosphere during geological time is shown in Figure 2.12. The time from the Earth's formation is about 5,000,000,000 (5×10^9) years when its atmosphere was composed principally of nitrogen and carbon dioxide associated with volcanic activity. The creation of life, initially as flora, resulted in a progressive reduction of carbon dioxide, the carbon forming part of the plant material and oxygen entered the atmosphere, as it is known today.

It was about 500 million years ago (5×10^8) that flora invaded the landmasses to create tropical rain forests of unimaginable size and extent (Figure 2.13). Almost all of these trees would have grown, died and rotted away in the same location. However, a minute

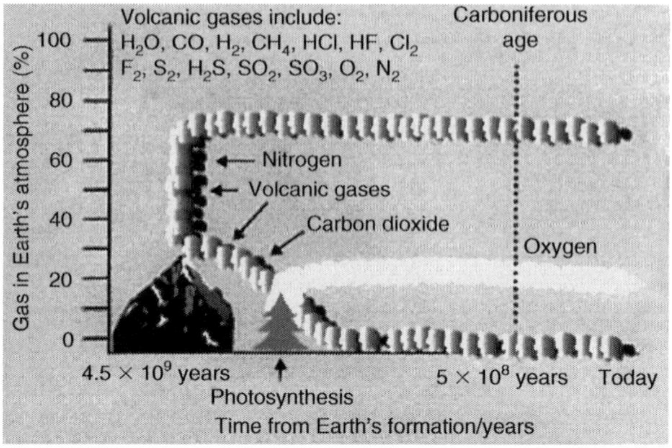

Figure 2.12. The variation of composition of the earth's atmosphere during the period of its evolution (Marsh, unpublished results).

Figure 2.13. Schematic of a carbonaceous tropical rain forest – precursor to coals.

proportion of these trees, still a massive amount of wood, when subject perhaps to violent tectonic disturbances (earthquakes and volcanic eruptions), was buried in the ground or submerged in salt or fresh water. Under these conditions, processes were initiated leading to the preservation of this organic material. Thus, the process of coalification began as a result of which, these millions of years later, the material known, as coal, is available to man (Figure 2.14). The amount of coal so deposited can be roughly estimated as about 10^{12} tonnes. These coals and their derivatives are used as the basis for many of the activated carbons of industrial use. This is some of the relevance of floral evolution to this chapter.

Figure 2.14 is a photograph of a specimen of bituminous coal as extracted from a mine. Horizontal layering can be seen within the coal which is the result of the tectonic pressures applied to the coal over geological time. This layering imparts an anisotropic property of the coal with higher reflectance and bireflectance from the coal material parallel to layering as compared with vertical sections. This anisotropy is preserved in the higher-rank semi-anthracites and anthracites.

Figure 2.14. Photograph of bituminous coal (fusible) as extracted from a coalmine. Note the horizontal striations within the specimen.

It is to be noted that, by far, most of the carbon element on this planet is in the form of carbonates and of carbon dioxide. In passing, it may be mentioned that diamond may result from the cracking of carbon monoxide, in geological time, within subterraneous conditions of high temperature and pressure.

2.5 Carbons with Industrial Applications

Although this book is dominantly concerned with porosities in activated carbons, it is helpful to list other carbon forms which have crucial industrial roles, as summarized below (Table 2.2).

Prior to about 1880, the carbon forms available included lampblack which was the only available writing material and obtained from smoking flames. Porous carbons commonly obtained by carbonizing wood, known as charcoals, had uses as a medicine (e.g. stomach problems), as a deodorant (soap and showers were non-existent) and in gunpowder (ballistics) and in pyrotechnics. World War II, where gas warfare made its appearance, demanded extensive use of activated carbon in respirators the effective use of which was a matter of life and death.

In the period up to about 1940, there was a significant expansion of the carbon black industry associated with the growth in size and number of newspapers and printing in general. This period of history saw increasing numbers of petrol-driven vehicles on the roads, all requiring pneumatic tyres which incorporated carbon blacks to promote wear resistance. The demand for iron and steel products accelerated the number and size of blast furnaces which consume metallurgical coke produced from the carbonization of coal in cokeries, a by-product being coal-tar pitch. The petrol-driven vehicles created an expanded petroleum industry which refined crude petroleum leaving behind viscous barrel residues which were carbonized to solid cokes in the delayed coker so putting on to the market a variety of delayed cokes, known as shot cokes, sponge (regular) cokes and needle cokes. This period also saw the rapid growth of a graphite-making industry, via the

Table 2.2. The range of industrial carbon forms.

Activated carbons (some with specialty applications)
Activated carbon fibers
Activated carbon cloth
Activated carbon felt
Carbon molecular sieves
Carbon aerogels, cryogels and zerogels
Carbon as catalyst support
Carbon black
Carbon composites – ceramic
Carbon composites – metal
Carbon fiber/carbon composites
Carbon fiber/resin composites
Carbon fibers based on PAN
Carbon fibers-based pitch (anisotropic fibers)
Carbon fibers, porous and activated
Carbon filaments and whiskers
Carbon films
Carbon lithium-ion battery
Diamond films
Elastic carbon
Fullerenes
Glassy carbon
Graphite/oxide refractories
Intercalation compounds
Mesocarbon microbeads
Nanorods (nano-fibers) without a central cavity
Nanotubes with open and closed ends
Prosthetics
Pyrolytic carbon

Acheson process, the electrodes being made from a blend of delayed coke (the needle coke quality) and coal-tar pitch. Further, this period of time also saw the rapid growth of the aluminum-making industry using the Héroult–Hall electrochemical cell with its carbonaceous anodes made from "regular" delayed coke and coal-tar pitch as the binder.

The period from 1940 saw a further continuous development in the use of carbon materials, their use becoming more and more sophisticated. Carbon fibers made their appearance, based on polyacrylonitrile (PAN) and pitch materials. Certain of these fibers could be selectively oxidized to produce porous-activated fibers. The fibers, when woven into three-dimensional units (prepregs), which were impregnated with a resin or a suitable pitch and then carbonized/graphitized, became carbon–carbon composites with a wide range of structural applications. Currently, the range of carbon types which have been produced is extensive and growing. Many no doubt will remain as interesting products but, without applications, others will not enter the commercial world. Table 2.2 contains this range.

Derbyshire *et al.* (2001) reviewed in depth the roles adopted for carbon materials, including the use of activated carbons, in improving environmental conditions of water and the atmosphere as well as other specialty applications.

Table 2.3. Definitions of porosity in carbons.

Porosity within a porous solid is space which is accessible to molecules from the gas/vapor and liquid phases.
The adsorbent is the solid porous material, in this case a carbon.
The adsorbate is the gas or vapor, or solute from solution, which is adsorbed within the adsorbent.
The adsorptive is the vapor, gas or solute which will eventually be adsorbed as an adsorbate.
The adsorption process is when an adsorbate enters into the porosity of the adsorbent.
Micropores have entrances dimensions <2.0 nm.
Mesopores have entrance dimensions between 2 and 50 nm.
Macropores have entrance dimensions >50 nm.

2.6 Preparation of Carbons in Solid Phase

2.6.1 Introduction

Carbon forms, although from a structural point of view representing a progressive and continuous series of change, individually have quite specific methods of preparation and these have to be considered to demonstrate the several mechanisms by which carbon forms are created. Always, carbon materials result from the *carbonization* of organic materials; that is, their heat treatment in an inert atmosphere to a required temperature resulting in increases in carbon contents and decreases in contents of heteroatoms. Such carbonizations occur in the solid, the liquid and the gas phases.

As this book is dominantly concerned with porosity in carbons, current definitions of the terminology of the subject area are given in Table 2.3. Later chapters expand fully on these definitions. The literature is not totally consistent with definitions and terminology, and it is unlikely that consistency will happen in the immediate future.

Some porosity exists which is inaccessible (closed) to helium but accessible to lithium.
Some porosity exists which is inaccessible (closed) to nitrogen but accessible to carbon dioxide.

Note 1: As entrances to porosities have not easily defined geometric shape, the use of such terms as pore radius, pore diameter, etc. can be misleading. These terms are still to be found in the literature and will continue to be used. However, care has to be exercised with a too literal interpretation.

Note 2: The nomenclature assigned to pore dimensions is one which has been inherited over past decades. In the literature, the use of the term *nanoporosity* is appearing to distinguish it from other porosities. This is where some confusion now arises. The porosities of concern to adsorption processes are the micro- and mesoporosities, with dimensions of <2.0 nm and between 2.0 and 50 nm respectively; that is, both have dimensions of nanometers. The terms micro and meso, as such, are essentially only a name and have no significance beyond that. In recent times, interest has centered on porosities in carbons with dimensions <1.0 nm and which are responsible for the phenomena of activated diffusion (Chapter 4) and uptake of lithium as for the lithium-ion battery (the so-called nanoporosity). The literature also refers to ultra-microporosity, of suggested dimensions <0.7 nm as well as super-microporosity assigned to microporosity with dimensions nearer to the limit of 2.0 nm, where three or four layers of adsorbate can exist in the micropore. There is a need to emphasize that adsorption processes in porosity <0.7 nm are distinct in the way that the intense dispersion forces, which operate in such confined volumes, influence the physical *state* of the adsorbed phase.

It has to be said at the onset that the numerical values attached to these definitions of porosity do not have the precision as is attached to a weight or a volume. Rather, a pore is defined according to the way it adsorbs an adsorbate molecule and that is a function of the size and polarity of the adsorbate molecule as well as the size of the porosity and surface polarity within the adsorbent.

The above discussions have stressed that porosity in this family of carbons essentially is that space where carbon material (atoms and heteroatoms) is absent. If the structure or relative arrangements of carbon atoms is changed in some way then automatically a new set of porous properties will be created. It therefore follows that a knowledge of structure assists with knowledge of the nature of porosity in that carbon. This aspect was mentioned earlier when graphitizable and non-graphitizable carbons were introduced.

2.6.2 Solid-phase Carbonizations

In solid-phase carbonizations, the parent material, almost always being a macromolecular system, naturally occurring or synthetic, essentially "decomposes" as the HTT increases, the process being accompanied by the evolution of gases and liquids of low molecular weight resulting from the decomposition processes. Consequently, the resultant carbon is a "pseudomorph" of the parent material, having more or less the same original shape but it is now of a lower bulk density (Figures 2.15 and 2.16). A solid-phase carbonization process is a progressive decomposition which will cease when the heat treatment is stopped. Increasing further the HTT results in the formation of progressively more stable intermediate structures. The structure in a carbon of a defined HTT can be considered as being "frozen" at that HTT. During the carbonization process, when the parent macromolecular system is decomposing, the remaining carbon atoms of the macromolecular network move

Figure 2.15. Photograph of stacked wood trunks prior to carbonization.

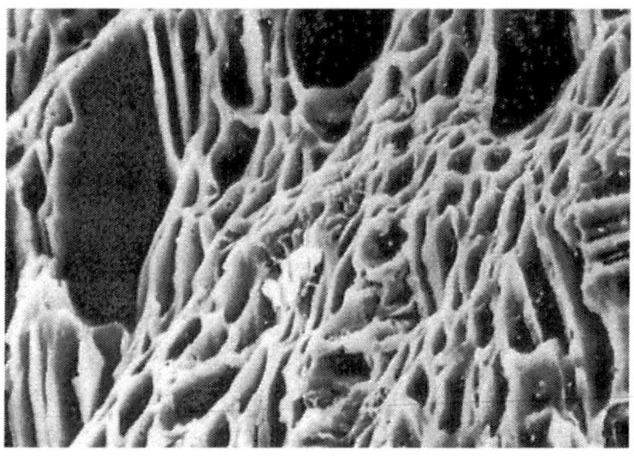

Figure 2.16. Scanning electron microscopy (SEM) photograph of carbonized wood (100 × 70 μm) showing that the structure of the wood has not liquefied (Marsh, unpublished results).

short distances (probably <1 nm) within the network to positions of greater stability (e.g. formation of six-membered ring systems), eventually creating a network of carbon atoms (with residual hydrogen bonded to it) which constitutes structure in such carbons. Different parent substances (e.g. a brown coal or coconut shell) decompose in their own distinctive ways to produce a specific type of carbon. The spaces (of atomic dimensions) vacated by the evolution of the heteroatoms, the migration of carbon atoms and their co-bonding to form the network of carbon atoms constitute porosity, each porous carbon having its own specific porous characteristics. Graphitizable carbons are never produced in solid-phase carbonizations. From the industrial point of view, lignin–cellulose precursors and coals dominate carbonizations in the solid state. The carbons produced are the *non-graphitizable, isotropic* type with considerable microporosity.

A question which now arises is how best to consider structure in such materials. The idea was introduced above that one approach is to consider such carbons to be composed of small segments of a graphene layer (defective micro-graphene layer), containing say about 20–30 carbon atoms, the layers being non-linear, enclosing five- and seven-membered ring systems with considerable strain. Such fragments of carbon material bond together in a continuous way to create a stable "macromolecular" network of carbon atoms (it could be called macromolecular) imparting physical strength (as with glassy carbons; Pikunic, 2001).

> *A microporous carbon, with a surface area of 1000 $m^2\,g^{-1}$, adsorbs nitrogen at 0.1 MPa pressure and 77 K to the extent of 10 mmol g^{-1}. One gram of carbon is equivalent to **80** mmol of carbon atoms indicating that one adsorption site (pore) uses about **8** carbon atoms for the adsorption of one nitrogen molecule. This small number indicates that with such a carbon (1000 $m^2\,g^{-1}$) all (or almost all) of the carbon atoms of its structure make up an adsorption site; that is, they are surface atoms. This feature must enter into any modeling of porosity which is undertaken (see Chapter 3).*

This proportion of carbon atoms to nitrogen molecules, for example, or the dilution of the nitrogen molecule with the carbon atoms is shown in Figure 2.17 (oversimplified). The

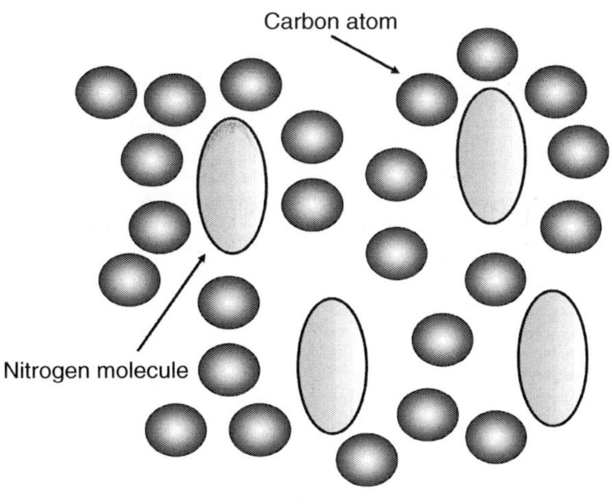

Figure 2.17. This model shows the ratio of nitrogen molecules to carbon atoms for a carbon of 1000 $m^2\,g^{-1}$, filled with adsorbed nitrogen at 77 K.

three-dimensional aspect is not considered. If the model diagram (Figure 2.7(b)) of Franklin (1951) is re-considered, then the walls, containing the cages, must be composed of no more than two defective micro-graphene layers, not four or five as in the Franklin model. This is an important consideration.

2.6.2.1 Variation of surface area with HTT

The carbonization process essentially is the conversion, by progressive heating, of a three-dimensional organic macromolecular system (e.g. coals, woods, nutshells, etc.) to a three-dimensional "macro-atomic" network of carbon atoms. As small molecules, such as water, methanol and carbon dioxide, are eliminated from the organic system, so the resultant free radicals (dangling bonds whatever) are removed by movement of atoms over short atomic distances to create an intermediate stable phase of higher carbon content. It is emphasized again that this process does not produce a continuous solid phase but one in which spaces of nanometer dimensions result, that is a network of porosity is produced. As the carbonization temperature continues to rise so the extent of this space network increases as well as its dimensions, length and breadth, to a maximum. With further increases in HTT (above about 800 °C), accessibility to adsorbate gases decreases as cross-linking of carbon atoms reduces space between atoms and essentially closes off porosity to an adsorptive gas.

These effects are seen in Figures 2.18 and 2.19 for the carbonization of two polymeric, non-fusing solids which produce carbon (isotropic) of the same shape and size (about) as the parent material. The polyvinylidene chloride (PVDC) yields a carbon of surface area of about 1200 $m^2 g^{-1}$ at 1300 °C, whereas the carbon from polyfurfuryl alcohol (PFA) does not have this temperature stability and maximizes at about 750 °C giving a surface

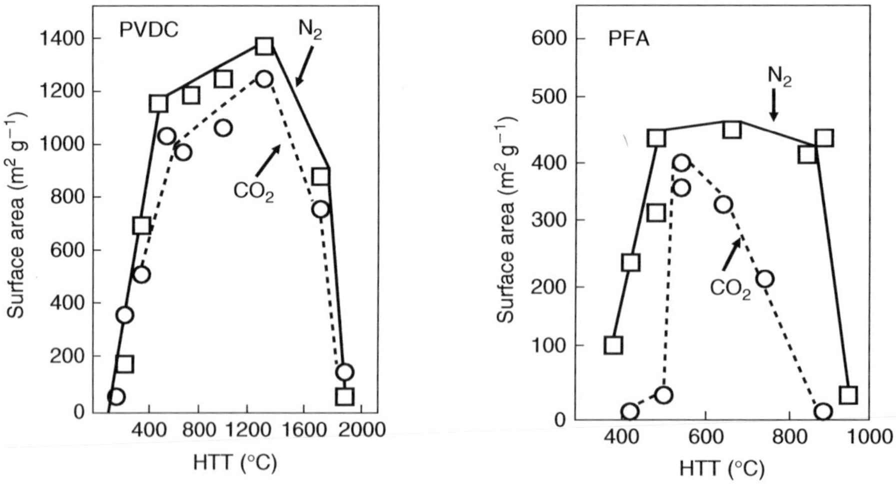

Figure 2.18 (left) and 2.19 (right). The variation of surface area ($m^2 g^{-1}$) with HTT (°C) for the carbonization of PVDC and PFA. The solid line is for adsorption of nitrogen at 77 K and the dotted line for carbon dioxide at 195 K (Marsh and Wynne-Jones, 1964).

area of 450 $m^2 g^{-1}$. Each point of a carbonization process essentially is a point of stability. The carbonization process does not continue once a final HTT has been reached. The new carbon lattice is in a *quasi-equilibrium* state.

2.6.2.2 Variation of free spin concentration (ESR) with HTT

Whereas adsorption techniques are informative about extents of surface generation, they provide little information about the composition (structure) of the surfaces. On the other hand, the techniques of electron spin resonance (ESR) and magnetic susceptibility, by measuring concentrations of free radicals within a carbon, and the properties of these free electrons offer information of changes within the surfaces with increasing HTT. These techniques were pioneered and developed extensively by Mrozowski (1979).

The ESR studies of Lewis and Singer (1981), of formation of a graphitizable carbon, noted that for polyvinyl chloride (PVC) a fusible parent material, a maximum of about 15×10^{19} spin g^{-1} occurred at an HTT of 750 °C (decreasing to 1.5×10^{19} spin g^{-1} at 1000 °C (Figure 2.20). For carbons from non-fusible PVDC, a first maximum of 38×10^{19} spin g^{-1} occurred at 450 °C, followed by a second maximum of 20×10^{19} spin g^{-1} at 900 °C (5×10^{19} spin g^{-1} at 1000 °C) (Figure 2.21).

Generally, free spin concentrations in the graphitizable carbons were lower than for the non-graphitizable carbons, because the larger graphene layers of the former facilitated electron pairing within the layer. The lack of facility for pairing in the non-graphitizable carbons is indicative of the absence of such large graphene layers. The maximum in free spin concentration does not occur at the same HTT for all carbons, indicating that variations exist in the structure/composition of the micro-graphene (defective) layers. The maximum in free spin concentration is at about 880 °C for the PVDC carbon compared

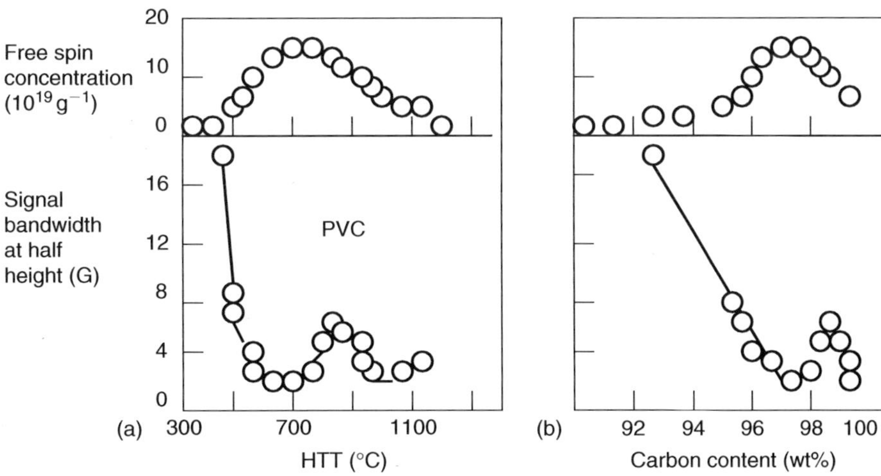

Figure 2.20. The variation of free spin concentration (unpaired electrons) ($n_s g^{-1}$) and signal bandwidth at half-peak height (G) with (a) HTT (°C) and (b) carbon content (wt%) for carbon from PVC (fusible and graphitizable).

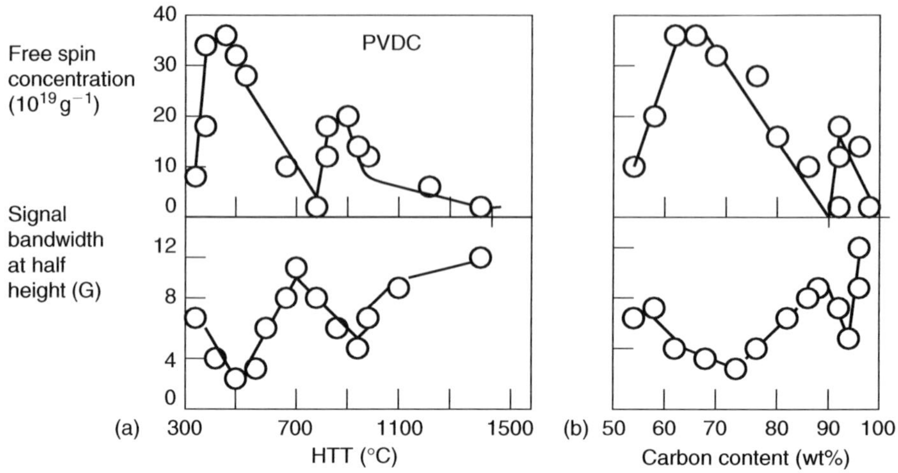

Figure 2.21. The variation of free spin concentration (unpaired electrons) (n_s g^{-1}) and signal bandwidth at half-peak height (G) with (a) HTT (°C) and (b) carbon content (wt%) for carbon from PVDC (non-fusible and non-graphitizable) (Lewis and Singer, 1981).

Table 2.4. Spin resonance properties of three activated carbons (Manivannan *et al.*, 1999).

Carbon	Parent material	Specific surface area/m^2 g^{-1}	Number of spins g^{-1}		Ratio (N_a:N_b)
			Before evacuation (N_b)	After evacuation (N_a)	
GX203	Coconut	1000	1.8×10^{17}	6.9×10^{18}	38
P1400	Wood	1150	5.8×10^{17}	7.7×10^{19}	133
Med 50	Coconut	2000	1.8×10^{16}	1.2×10^{19}	667

with about 720 °C for the PVC carbon indicating the higher stability of the PVDC carbon structure.

Using this ESR approach, Manivannan *et al.* (1999) arrived at important and relevant conclusions. They examined three activated carbons, made from coconut and a wood, of known surface areas, supplied by PICA (USA Inc) and measured their free spin concentrations. Measurements were made both in air and in evacuated specimen tubes. Results are summarized in Table 2.4.

Figures 2.22 and 2.23 show the influence of the presence and adsorption of paramagnetic oxygen by the carbon on the ESR signal. On evacuation to remove the oxygen, the signal intensity increases from N_b to N_a. (spins g^{-1}) and the width of the signal decreases indicating an increase in the number of measurable free spins.

As 1.0 g of these carbons contains 5×10^{22} carbon atoms, then there is one free spin (dangling bond) (based on N_a) for every 7200 carbon atoms (GX203), 650 (P1400) and

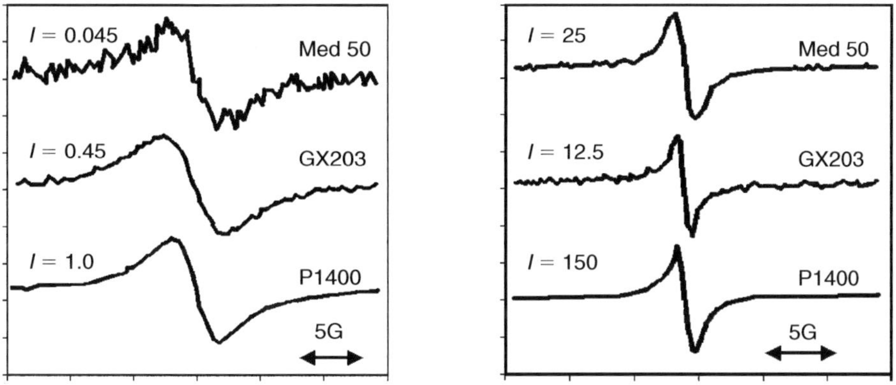

Figures 2.22 and 2.23. (Left) Room temperature ESR spectra of the as-received carbons. The intensity factor gives the relative intensity of the ESR signal. (Right) Room temperature spectra of the three carbons after evacuation to remove physically adsorbed oxygen. The intensity factor is relative to the ESR spectra of the non-evacuated P1400 carbon (Manivannan *et al.*, 1999).

4200 (Med 50) carbon atoms, not that many relatively. This suggests an overall stability of the carbon lattice if it is assumed that a dangling bond can be equated to a free spin.

Parts of the spectra of these free spins were broadened beyond measurement in the presence of the paramagnetic oxygen molecule (a well-established phenomenon). The free spins are equated to "dangling bonds" where a carbon atom(s) has been unable to move into a more stable position within the carbon lattice network and the orbitals of the carbon atom are not complete. Using a stable free radical compound as a standard, spin concentrations were calculated. For the porous carbon from wood, P1400, with a surface area of $1150\,m^2\,g^{-1}$, the number of free spins for the evacuated sample was $7.7 \times 10^{19}\,g^{-1}$.

Table 2.4 further indicates that the interactions of oxygen with the free spins were not the same for all three carbons. The ratio of N_a:N_b varied from 38 to 667 indicating a significant difference in the ability of the oxygen to reach and interact with the free spins. The implication is that parts of the structure of the activated carbons are shielded from the oxygen by other parts of the carbon. For example, using the concept of pore walls, if a pore wall is composed of three or more layers, then the middle (central) layers cannot be reached by the oxygen. Carbons of lower surface areas are more likely to show this effect, as with GX203. The high surface area carbon, Med 50, shows a maximum accessibility to the oxygen.

The location of these dangling bonds can be of relevance when considering the nature of surfaces of porosity. Whether or not a "dangling-bond" location may be a site of high adsorption energy in an adsorption process (by a contribution to the adsorption potential within an adsorption site) remains a matter for future discussion. Thus, the technique of ESR gives further insight into structure within a microporous-activated carbon.

The above considerations of the number of carbon atoms at surfaces of adsorption sites lead to a conclusion that the surfaces of some of these porous carbons are not dominated by "dangling-bond" atoms. A conclusion is that should a "dangling bond" contribute to the adsorption potential, its effect upon an adsorption isotherm will be at very low relative pressures (e.g. <0.0001 p/p^0). Mrozowski (1979) had previously noted that the concentration of free radicals within an HTT range of carbons increased to a maximum at about 800 °C to subsequently fall to zero at HTTs >1000°C. A further conclusion is that the "macromolecular system" of carbon structure is dominated (perhaps exclusively) by interconnected defective micro-graphene layers with little in the way of "side-chain" carbon atoms. Carbon atoms with sp^3-bonding may act as bridges to join the micro-graphene layers into a continuity, so promoting electron pairing.

A further consideration emerges from these experiments with ESR. What is intriguing in this process of development of a space network (that is porosity) is its *continuity*, where every part of the space within this network is connected to every other part. The concept of *connectivity* of each adsorption site is all important. This creation of connectivity of network space (porosity) appears to be an intrinsic part of the carbonization process. If it had not happened then the spaces, resulting from the elimination of the small molecules and the creation of the carbon atom network, may have been isolated and inaccessible, and microporous carbon would not exist. Incidentally, for the glassy carbons which are prepared by carbonizations of resins under pressure, the release of small molecules of volatilization is prevented and are retained so closing, within itself, the network porosity. This process is not identical for all organic macromolecular materials which can be carbonized. Each parent material (an organic macromolecular system), biologically distinct, provides its own specific carbons with a characteristic porosity. As indicated in the definitions above, porous carbons contain a range of pores conveniently described, according to their behavior to various adsorbates, as being micro- or mesoporous, with most, in fact, containing all of these pore sizes.

This raises the question, to be discussed more fully later, as to whether or not all of these pore sizes occur within a single interconnected porous network or whether there are several networks, quite independent of each other, for the other sizes of porosity.

The porosity of a carbon can be enlarged by chemical gasification reactions in which such gases as hydrogen, water vapor, oxygen or carbon dioxide enter the porosity and gasify the surface atoms of the porosity removing carbon atoms as carbon dioxide, carbon monoxide, hydrogen and methane. This is the process of one method (thermal or physical) of activation.

2.6.2.3 Activated carbon from coal

Coal is an organic sedimentary rock consisting of a complex mixture of substances, organic and mineral, derived from plant debris deposited where it grew, and from similar sources deposited in a place different from where it grew, including spores and aquatic plants. *Peat* is the material first formed from this organic material when there is a restricted supply of oxygen. The type of original plant input, the availability of nutrients, climatic conditions, level of the water table and pH determine the type of peat that is created. Peats include large trees, herbaceous shrubs, grasses, aquatic plants and micro-organisms (Figure 2.13). For a coal to form from this organic detritus, it must be buried and preserved by a process termed *coalification* and the extent of coalification (peats to anthracites) determines the *rank* of the coal. Peats are normally found near to the surface and are therefore easily mined. Their moisture content is high as well as their mineral matter, so pre-treatments are often necessary prior to their use in equipment for activated carbon production.

The coalification transformation of plant material, over millions of years, occurs in two stages, involving an initial *biochemical* degradation followed by *physico-chemical* degradation. The former involves oxidation of plant material and destruction by fungi, insects and aerobic bacteria where oxygen is present, and by anaerobic bacteria below ground. Much of the softer tissues of the plants are removed by these processes. Physico-chemical coalification results from burial and the overburden allows for increases in temperature and pressure from the Earth's crust, tectonic movement, such as folding and mountain formation, and movement of magma (lava) either vertically upwards (dyke) or horizontally parallel to the coal seams (sill). Coalification involves the removal of water and carbon dioxide, a reduction in oxygen and hydrogen contents, an increase in carbon content, and the formation, entrapment (adsorption) of methane and its release (firedamp). The names given to coals of increasing rank (increasing carbon content) are shown in Table 2.1.

The rank range of coal offers a microporous system the early study of which revealed much of what is known today about microporous systems, such as their structure and characterization. A brief summary of coal properties is indeed useful in the context of this book, not only because of the surface science involved but also because a significant proportion of industrially prepared activated carbon uses coal as a source material.

Coals are unique of the parent materials used for production of activated carbons in that they possess an inherent microporosity which can be taken advantage of without any carbonization treatments. Other parent materials, such as woods, nutshells, etc., need to be carbonized, initially.

Coals are buried in the ground to extents of about 10^{12} tonnes (recoverable resources). Coals have been studied extensively both from an investigative point of view and also from an application point of view. About 5×10^9 tonnes of coal are utilized annually, worldwide.

The availability of studies of structure in coals, their formation (coalification) and their carbonization behavior, assisted significantly understandings of activated carbon, are discussed further below. To a first approximation, the processes of coalification (geological) and carbonization (chemical) have broad similarities leading to *carbonaceous* materials. The enthalpies and entropies of activation of the two processes are very similar and the laws of the Arrhenius equation apply indicating how temperature can be compensated for by time. However, there is a difference of some importance. During coalification, because the coal-forming material is often trapped underneath an overburden of impermeable rock, the volatile materials cannot escape and many are trapped within the coal-forming material. This entrapment and the non-compressibility of the gas molecule in these conditions results in porosity being "frozen" into the coal structure (somewhat akin to the formation of glassy carbons). Thus, the pathways of coalification and carbonization diverge but not significantly. The low temperatures of coalification, in general not $>350\,^{\circ}$C, are compensated by long reactions times (geological) and the effects of pressure arising from the overburden (being buried) and of tectonic displacements (compressive mountain folding). As coalification involves densification, the effects of pressure would be to support coalification.

The porosity of coals has received considerable attention with several journals devoted to the material (*Coal Age, Coal Geology, Energy and Fuel, Fuel, Fuel Processing Technology*). There are three main reasons for this interest: **One** is an assessment of the quantities of methane adsorbed within coals for reasons of safety in mines and the use of coal-bed methane as a fuel; **Two** is that all types of coal (rank) are used for the making of activated carbons; **Three** is the need for intensive study to obtain liquid fuels from coals to replace fuels from traditional petroleum sources, an international program of research promoted by the fuel crises of the 1970s and 1980s. For coal solvolysis, selected solvents were required to penetrate the macromolecular structure of coal via its porosities requiring that porosities had to be studied in the first place.

Carbonization and activation of coal resources for modern usage complete the process of coalification initiated anything from 300×10^6 to 15×10^6 years ago. It is somewhat unfortunate that the scientific worlds of coal and activated carbon have developed rather independently of each other. They have so much in common, both starting from the organic materials which constitute wood, grass, seeds, etc. Resultant coals, from an adsorption point of view, are not straightforward materials to study and exhibit many of the deep-seated problems which beset the adsorption chemists of today who work with microporous-activated carbon.

2.6.2.4 Coal rank

A brief but relevant scan of the dominant porosities of coals is a worthwhile exercise as it supplements studies of activated carbon prepared from all ranks of coals.

Extents of *coalification* (coal rank) are usually measured in terms of carbon and volatile contents as well as reflectance measurements, the latter being a direct indicator of extents of aromaticity in the coal (Table 2.1). Coalification (rank) has proceeded continuously without discontinuities in terms of analytical data. Some physical properties, however, have not developed continuously. Fluidity (plasticity), an important industrial property, rises and falls with increasing coal rank. On heating coals in an inert atmosphere, fluidity increases from zero at 81 wt% carbon, reaching a maximum at about 87 wt% carbon, falling to zero at about 91 wt% carbon (Figure 2.24).

Figure 2.24 is a plot of fluidity (measured as dial divisions per minute (ddpm) of the Gieseler equipment) against the carbon content of the coal (wt% dry mineral matter free basis (dmmf)). Considerations of fluidity are a matter of considerable importance in activated carbon production. A coal which fuses (melts), such as a bituminous coal, on carbonization requires a different pre-treatment from a coal which does not melt (brown coals and anthracites).

The measurement of the fusibility (melting) of a coal (it becomes plastic rather than liquid on heating) is a difficult operation and use is made of a Gieseler plastometer shown in diagrammatic form in Figure 2.25. The plastometer heats the coal in an inert atmosphere in a crucible in which is located a stirrer, the energy supplied to which is kept constant. As the coal becomes fluid, the speed of rotation of the stirrer increases to a point of maximum

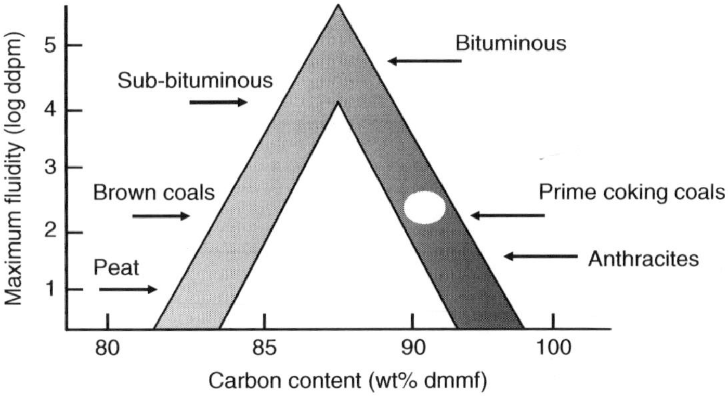

Figure 2.24. The variation of maximum fluidity (log ddpm) of a rank range of coals with carbon content (wt% (dmmf)) of the coals.

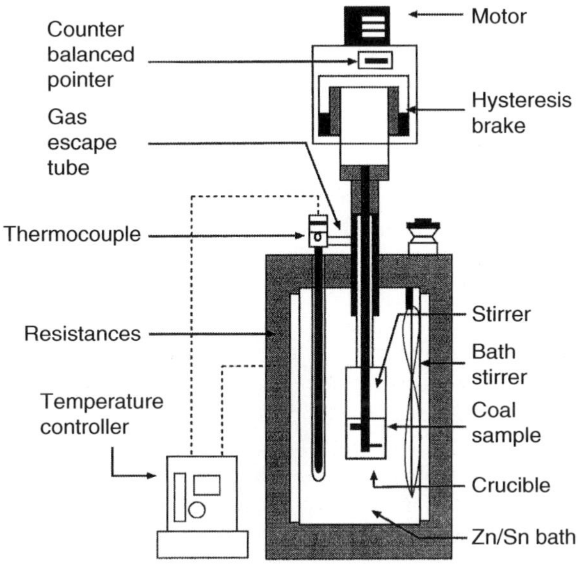

Figure 2.25. Simplified Gieseler plastometer.

fluidity. As the coking (solidification) process takes over, then the speed of rotation reduces to zero.

Figure 2.26 is a typical plot of fluidity, measured as ddpm (quite an arbitrary unit, but one which is established within the industry). Bituminous coals, used for active carbon production, have to be tested in this way. The Gieseler plastometer has proved to be a reliable instrument over many years for measurements of coal fluidity. There is one disadvantage, for which there is no answer, and that is the tedious operation of cleaning the stirrer and crucible after each testing.

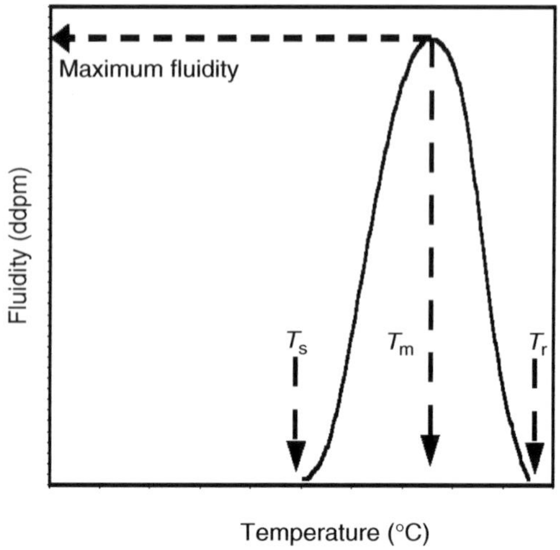

Figure 2.26. A typical fluidity curve. T_s: initial softening temperature and T_r: solidification temperature (Marsh *et al.*, 1997).

Coals are not homogeneous but are made up with macerals, these being recognizable components (by optical microscopy) within the coal derived from specific plant components. Coals may possess different quantities of macerals so accounting for small differences in rank. The exinite group of macerals exhibit maximum fluidity, the vitrinites have an intermediate position and the fusinites (like a wood charcoal) are non-fusible.

2.6.2.5 Microporosity in coal

It is necessary to consider the macromolecular structure of the vitrinite component (the principal maceral component) and the chemical changes which occur during coalification. The initial chemistry of conversions of parent vegetable materials (buried and free of oxygen) produces oxygen cross-linkage, ether-like, within the material. Over geological time, the slow (very slow) removal of oxygen allows hydrogen atoms of the macromolecular structure to be attached to the increasing aromaticity of this macromolecular structure. Thus, the covalent bonding within the lignites and brown coals becomes replaced by weaker hydrogen bonding within the more aromatic sub-bituminous and bituminous coals, resulting in the fluidity of these coals.

As coalification proceeds in geological time with loss of hydrogen (as methane), the hydrogen bonding within bituminous coals is replaced by carbon–carbon covalent bonds causing the fusibility of the bituminous coals to decrease to zero as found in the semi-anthracites and anthracites (Figure 2.24).

On heating the fusible bituminous coals, structures soften (melt, become fluid) with the release of volatile material from within the fused system so causing swelling. As a result, the initial microporosity is lost. Hence, to effectively use such coals for activated carbon production, this fusibility has to be destroyed. Oxidation in air, at temperatures 100–200 °C, effectively will do this.

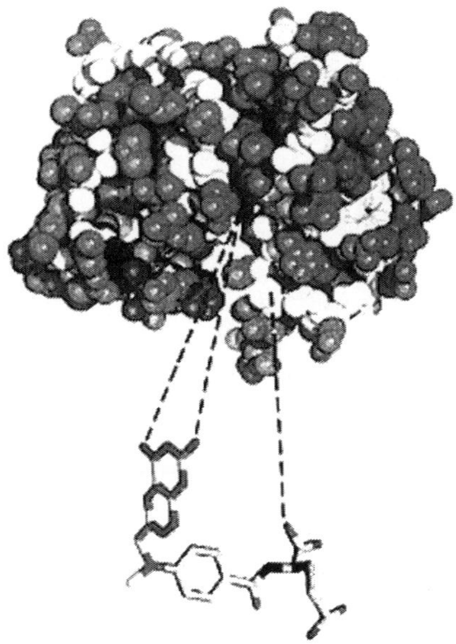

Figure 2.27. A computer-generated three-dimensional model of the macromolecular structure of vitrinite in coal showing porosity between coal substance (Marsh, unpublished results).

A computer-generated three-dimensional model of a "possible" macromolecule of coal is shown in Figure 2.27. It is a structure-less three-dimensional cage-like aggregate of atoms (C, H, O, N, S) with a water density of about $1.3\,g\,cm^{-3}$. The model shows that the sizes of spaces within the macromolecular system approximate to those of the atoms and demonstrates the presence of microporosity within the coals.

Figure 2.28 is a more open, two-dimensional model, Shinn (1984), indicating three relevant properties, namely (i) the continuous bonding within the three-dimensional network, (ii) the porosity (space) between the molecular (atomic) components and (iii) the presence of surface functionality (O, S, N) within the porosity. These three properties are to be found in activated carbon.

The surface chemistry of coals has been extensively studied, not only because of the use of coals as source for active carbons but also for the processes of methane extraction, carbonization, of gasification and of liquefaction. Measuring the surface properties of coals encompasses most of the problems encountered by experimentalists in studies of activated carbon.

For example, Figure 2.29 is a plot of the variation of the surface area ($m^2\,g^{-1}$) of coals against carbon content (a measure of coal rank) using two adsorbates, namely carbon dioxide at 298 K and nitrogen at 77 K. It is seen that the results fall into two distinct bands with extents of adsorption of carbon dioxide lying well above the results for adsorption of nitrogen. Whereas extents of adsorption of carbon dioxide reach a *minimum* at about 80 wt% carbon, extents of nitrogen *maximize* at about 75 wt% of carbon.

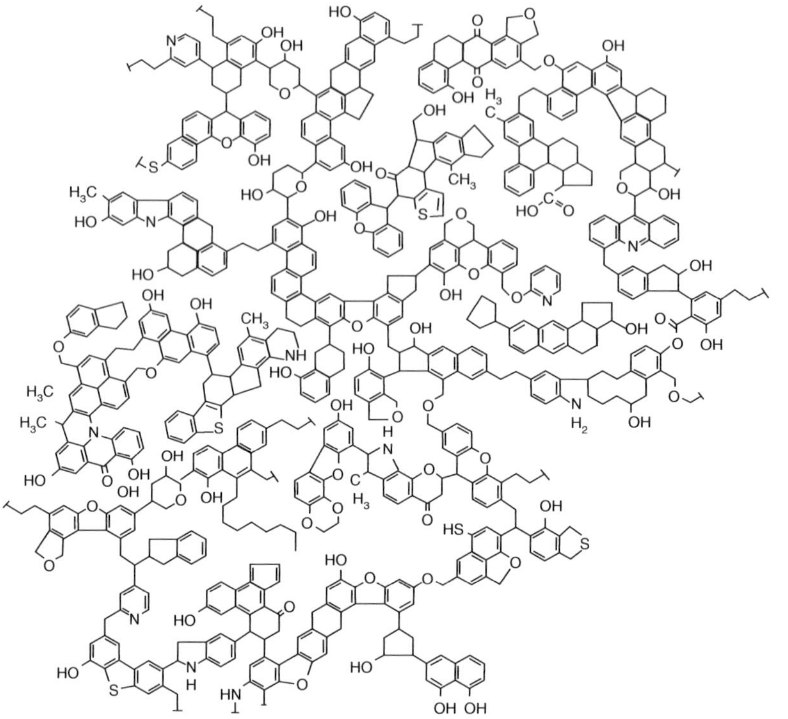

Figure 2.28. A two-dimensional model for the macromolecular structure of coal illustrating (i) the network, (ii) surface functionality and (iii) porosity (Shinn, 1984).

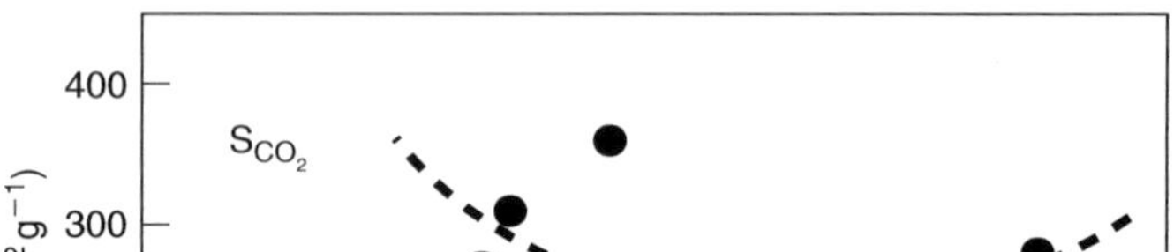

Figure 2.29. The variation of surface area of a rank range of coals. The upper set of data is of surface areas determined from the adsorption of carbon dioxide at 298 K, based on the Dubinin–Radushkevich (DR) equation. The lower set of data is of surface areas determined from the adsorption of nitrogen at 77 K using the BET equation (see Chapter 4) (Marsh and Rodríguez-Reinoso, 2000). daf: dry and ash free.

The restricted adsorption of nitrogen at 77 K is the result of the activation energy of adsorption being too high (>40 kJ mol^{-1}) to allow adsorption into narrow microporosity (dimensions of <~0.5 nm) at 77 K at measurable rates, giving equilibrium times of weeks or even years (see Chapter 4 for further details). It is also noted that the experimental data fall into bands rather than on one line, simply because coals of equal carbon content are not necessarily of the same rank. Reflectance/bireflectance data are often considered a better measure of rank than carbon or volatile contents.

Porosities in coals exhibit properties which can be related to those of activated carbons. Figures 2.30–2.32 need to be considered. Figure 2.30 shows the variation of total porosity

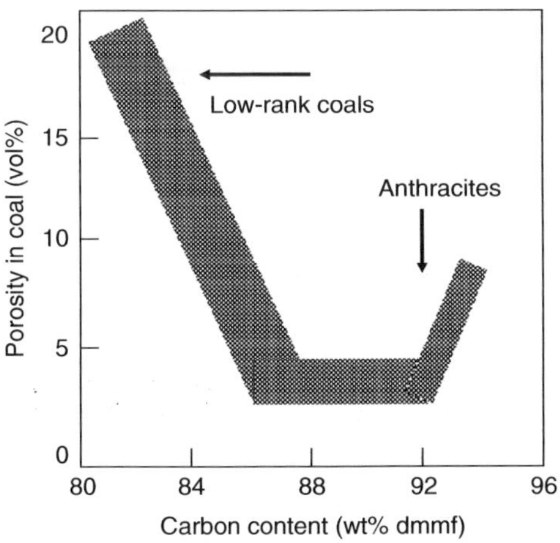

Figure 2.30. The variation of total porosity (vol%) with coal rank.

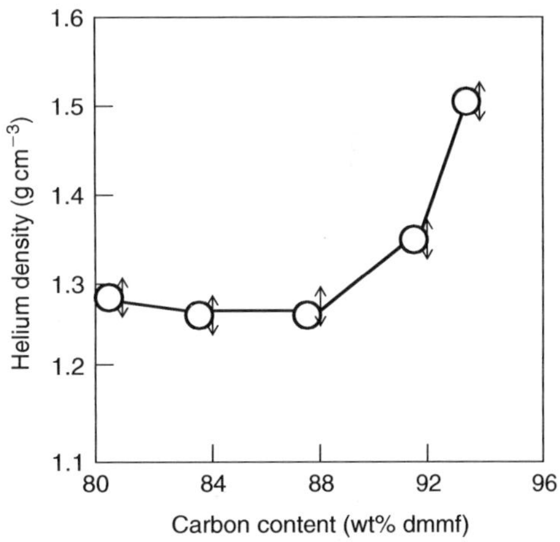

Figure 2.31. The variation of helium density (g cm^{-3}) with coal rank (Toda, 1972).

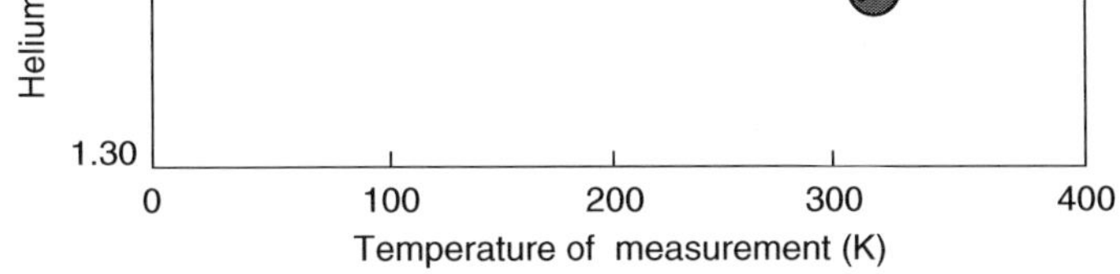

Figure 2.32. The variation of helium density (g cm^{-3}) of an anthracite with temperature (K) of measurement. Apparently, the helium density decreases with increasing temperature of measurement.

(vol%) with carbon content (wt%) for all coals. The coals of lowest rank possess about 20 vol% but this decreases to a minimum of about 3 vol% for bituminous coals (90 wt% carbon) rising to about 12 vol% for the anthracites (94 wt% carbon). However, beyond the point of minimum porosity, the anthracites are associated with the further formation of microporosity which arises from the generation of *coal-bed methane* which is initially trapped within the coal structure but whose subsequent release leaves the vacant site as microporosity.

Figure 2.31 shows the variation of helium density ($g\,cm^{-3}$) with coal rank for all coals. Helium, as a gas, has the advantage of being able to penetrate into porosities and yet not to be adsorbed (This is not strictly true and adsorption of helium can be detected but this effect does not radically alter the major conclusions).

The helium density, initially being about $1.45\,g\,cm^{-3}$ for brown coals (64 wt% carbon), decreases to about $1.25\,g\,cm^{-3}$ for coals of 85 wt% carbon to rise to $1.55\,g\,cm^{-3}$ for the anthracites. Clearly, porosity must have been generated in these coals into which the helium cannot penetrate, with entrance diameters of narrow microporosity. The continuous removal of hydrogen (as coal-bed methane) in the later stages of coalification results in the generation of narrow microporosity to allow access of the helium (Figure 2.31).

There is however some adsorption of helium (probably in the narrow microporosity) as demonstrated by consideration of Figure 2.32. The effect of increasing temperature of measurement on helium densities is illustrated in Figure 2.32. The decreasing values of density are indicative of adsorption of helium at the lowest temperatures. That densities do not increase, with increasing temperature, suggest that there is no restriction of helium penetration into porosity because of activated diffusion effects. Should this occur, then increasing temperature would enhance penetration, so providing for higher values of density.

The surface functionalities present within coal mean that the coals are more responsive to polar adsorbates than activated carbons without surface treatments. The liquefaction studies indicated that there exists heterogeneity of bonding within the macromolecular systems because not all of the coal could be liquefied relatively easily. Ultimately, all of the coals were solubilized when solvents of high-solubility parameters (dipole moment) were used.

Activated carbons are prepared from coals, those from the non-fusible low-rank coals and the anthracites not requiring pre-carbonization treatment. However, for the coals which do fuse, an oxidation pre-treatment is required to introduce carbon–oxygen cross-linking to the macromolecular system to prevent the softening and melting of the coal. Carbonization of bituminous coals does not develop porosity, the porosity of the low- and high-rank coals becoming further developed during carbonization.

2.7 Preparation of Carbons in Liquid Phase: Liquid Crystals

2.7.1 Introduction

Liquid-phase carbonizations, of aromatic, pitch-like materials, lead to the formation of graphitizable carbons which, essentially, are non-porous carbons. Hence, to make a porous-activated carbon from these materials requires a reaction which will force apart the stacked graphene-type layers. How this is done is explained in Chapter 6 which is concerned with chemical activation methods. But first, the mechanism for formation of graphitizable carbons has to be described.

In liquid-phase carbonizations, the mechanisms are completely different from those in the solid phase. It is via liquid-phase carbonizations (but not all liquid phases) that graphitizable forms of carbon result. How does this come about? The explanation takes us to a quite different subject area, that of anisotropic aromatic, discotic, nematic liquid crystals (called mesophase) formed as a result of growth and self-assembly of the constituent polycyclic aromatic molecules of the parent material. These usually are the highly aromatic coal-tar pitches, a liquid product from the making of metallurgical coke, from aromatic pitches synthesized by the petroleum industry as well as polycyclic aromatic model compounds.

2.7.2 Nematic Liquid Crystals

The discovery that the conversion of a coal-tar pitch to a graphitizable carbon is a significant milestone in the history of carbons. Up to 1961, when this story begins, the detail of a mechanism by which the aromatic moieties of the coal-tar pitch could orientate themselves parallel to each other was unclear.

The story begins when GH Taylor (a geologist with interests in coal deposits, photograph Figure 2.33) was asked to take samples of coal at distances of about 1 km from where a dyke had penetrated a coal seam. A dyke is a vertical penetration of magma (~1000 °C) through the seam resulting in a thermal gradient through the coal for about this 1 km. A photograph of the tip of a dyke through a coal is shown in Figure 2.34.

Brooks and Taylor (1968) reported changes in the microscopy of vitrinite (reflectance and bireflectance) with distance from the dyke. The vitrinite, at a distance of 1 km from the dyke, was unaltered and retained its anisotropy resulting from the overburden pressures.

Figure 2.33. Photograph of Geoffrey H. Taylor, who advanced the liquid crystal theory for structure formation in anisotropic carbons.

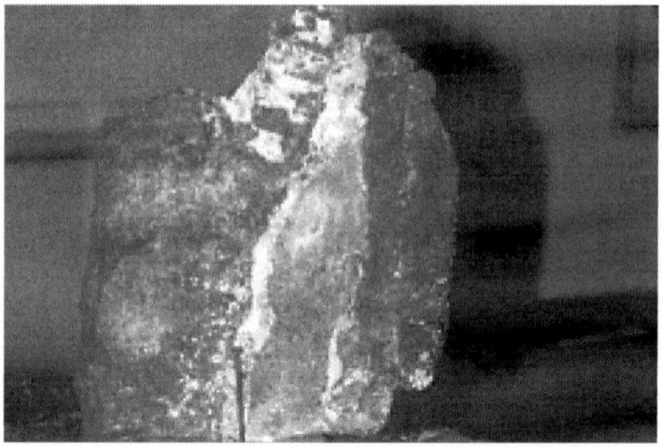

Figure 2.34. Photograph of the head of a dyke after penetrating through a coal seam. The cooled magma is in the center with metamorphosed coal adjacent on both sides.

As the dyke was approached, this anisotropy disappeared indicating that the vitrinite had been heated, had melted and then cooled down without further change. As the dyke was approached, small spheres, anisotropic in structure, of higher reflectivity than the matrix of vitrinite, were observed in the microscope. An optical micrograph of comparable spheres produced during the carbonization of anthracene is shown in Figure 2.35. On further approaching the dyke, the number of spheres increased as well as their size until they all coalesced into each other to create a mosaic structure as seen in metallurgical cokes (see Figure 2.10). In order to coalesce, the spheres must have been fluid. These observations could be reproduced in the laboratory using slow rates of heating and sampling over short temperature intervals. The conclusion from the analyses of the formation and structure of these spheres was that they represented a system of aromatic, nematic, discotic liquid crystals. The aromatic molecular constituents of the coal-tar pitch undergo condensations reactions to increase in size to about 1000 amu. At this size, such molecules (termed mesogens) self-assemble with parallelism and the normal forces of surface tension between two liquid phases establish the spherical shape.

Structure within a sphere of mesophase is modeled in Figure 2.36 where the lamellar polycyclic aromatic mesogens are stacked parallel to each other to create the anisotropy. On coalescence, these structures are maintained when the two adjoining structures merge into each other to create bulk anisotropy.

Figure 2.37(a–d) model the growth and self-assembly of mesogens during the formation of mesophase. The initial, relatively small polycyclic aromatic molecules (about 250 amu) (Figure 2.37(a)), undergo simultaneous decomposition and condensation reactions involving the elimination of small hydrocarbon molecules, the molecular weight of the constituent molecules eventually reaching about 1000 amu when they are called mesogens (Figure 2.37(b)).

At this stage, these molecules self-assemble on collision, that is they “stick” together initially via polar interactions (associated with their benzenoid structures), to create laminar stacking over distances of micro/millimeters. Initially, these will be small assemblies

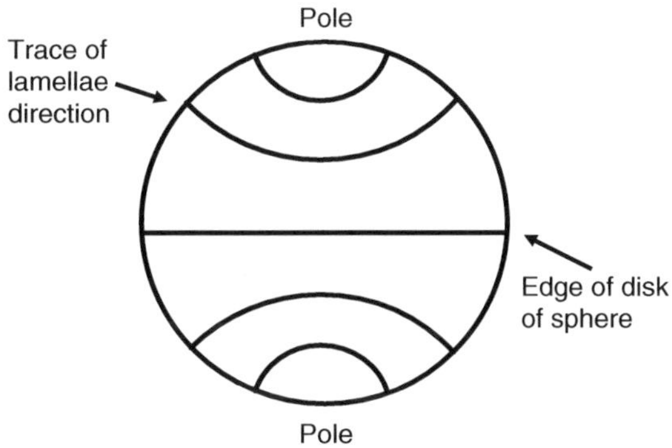

Figure 2.35, Figure 2.36. Optical micrograph of spheres of mesophase as developed during the carbonization of anthracene (left). Drawing of stacking arrangements of mesogens parallel to an equator of the sphere (right) (Marsh, unpublished results).

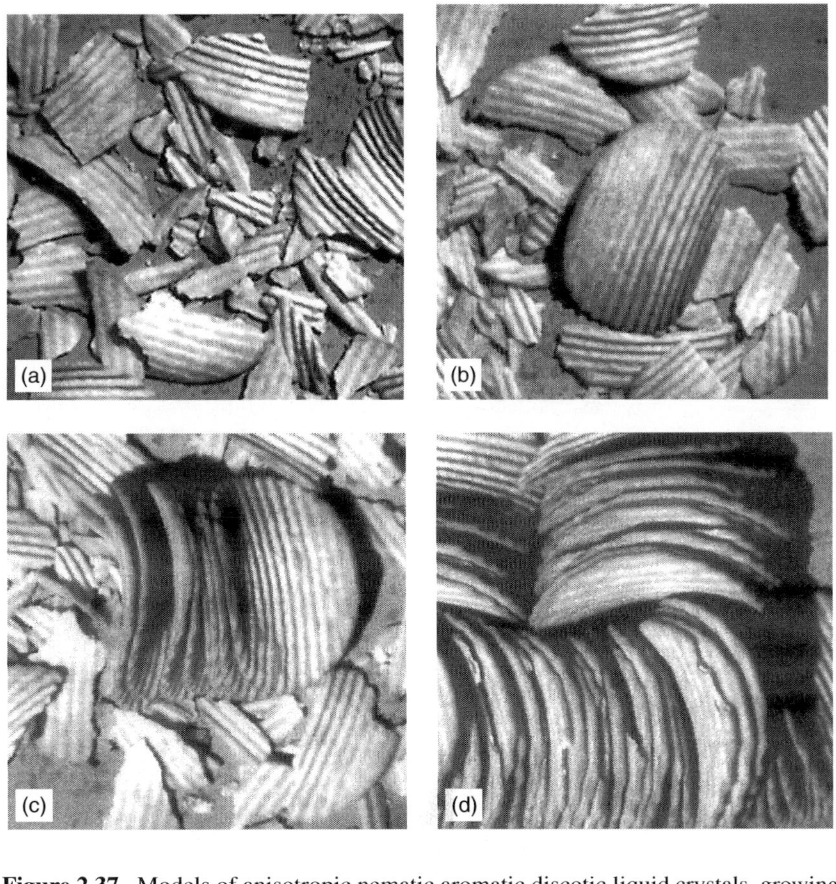

Figure 2.37. Models of anisotropic nematic aromatic discotic liquid crystals, growing from (a) an isotropic pitch to (d) an anisotropic carbon.

(Figure 2.37(c)). Eventually, all of the carbonization system converts to mesophase (Figure 2.37(d)), reading from top left to bottom right. This growth of mesophase is not a process of involving nucleation or crystallization or precipitation, but is one of self-assembly as is established in many biological systems.

The orientation of the constituent polycyclic aromatic molecules is maintained during the subsequent carbonization process with chemical cross-linking ensuring the stability of such a structure. With feedstocks of lower aromaticity, on heating, the perfection of growth and stacking of the polycyclic molecules diminishes and carbons become less and less graphitizable until they approach the structures of the non-graphitizable carbons of solid-state carbonizations. With increasing HTT, the helium densities of these graphitizable carbons increase to about 1.8–1.9 g cm^{-3} indicating the imperfect stacking of the mesogens as well as defects within the planes of the mesogens themselves.

The structures established in graphitizable carbon, via the liquid crystal mechanism, of parallelism and cross-linking of graphene layers, prevent the formation of porosity as seen in solid-phase carbonizations, this necessitating other procedures to be adopted for activation of such graphitizable carbons.

Following on from the explanations of Brooks and Taylor (1968), Mochida *et al.* (1994), Bonnamy (1999), Marsh *et al.* (1999) and Marsh (2001) (using petroleum pitch residues) reported on the kinetics of formation of liquid crystals. Granda *et al.* (2003) reviewed the pyrolysis of coal-tar pitch (derived from the carbonization of coal in cokeries) in terms of its chemical composition.

Figure 2.38 is a diagram which attempts to include the principal stages involved in the formation of graphitizable carbons (Marsh, 1991). The left-hand side of Figure 2.38 sets out the stacking of the mesogens in the mesophase, these mesogens having buckled structures with significant defects, vacancies, etc. within the graphene-like layer. With increasing HTT, the planarity increases and the defect density within the layers decreases. But then, extensive cross-linking occurs across the layers resulting in further buckling, until at about 2000 K, the planarity of the layers becomes established leading to the production of (*hkl*) diffraction lines. At the same time, residual hydrogen is desorbed.

2.8 Preparation of Carbons in Gas Phase

2.8.1 Introduction

Gas-phase carbonizations need careful control of the feedstocks being used. Carbonaceous products range from the lustrous, mirror-like **pyrolytic carbons** to a sooty mass. The pyrolytic carbons are considered first. The feedstock can be methane, or propane or benzene, but of dominant importance is that in the carbonization process (cracking or pyrolysis) the gaseous material must be at a low relative pressure usually brought about by dilution with helium. The fragments of the pyrolyses (radicals) from the parent feedstock interact with a suitable substrate and, by a mechanism involving carbon atom movement, the hexagonal lamellar structure of graphite is established. Highly orientated pyrolytic graphite (HOPG) has some of the most perfect of crystalline arrangements within the family of carbons. With a helium density of HOPG of 2.23 g cm^{-3}, porosity is absent.

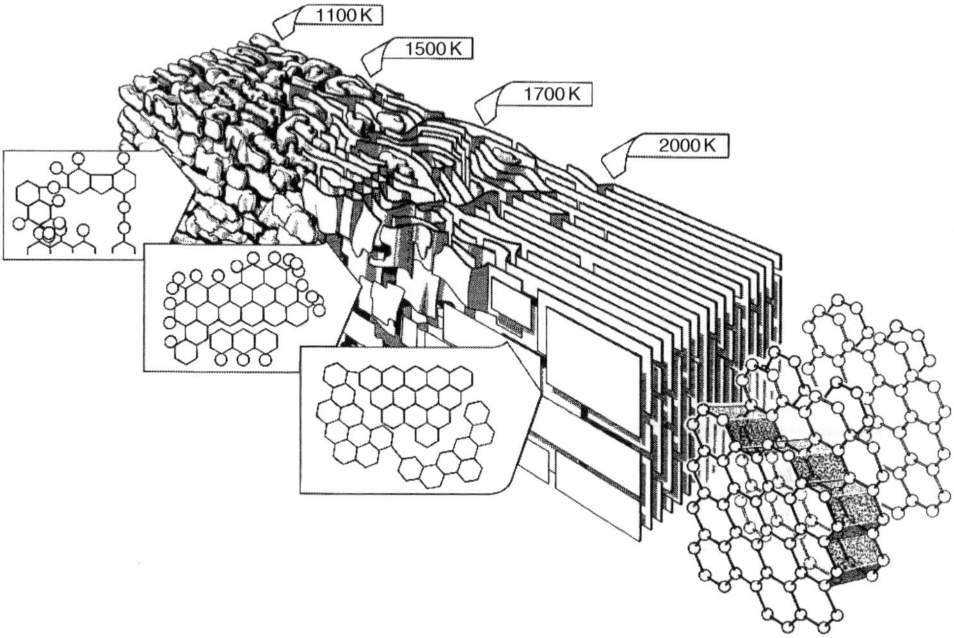

Figure 2.38. This diagram comprehensively models the structural changes which occur during the heat treatment of a graphitizable carbon, HTT 1100–3300 K (Marsh, 1991).

If a catalytic surface is present in such a pyrolysing system, for example nickel, then the carbon atoms formed at the surface interact in such a way as to produce **graphitic tubes or fibers** of diameters of some micrometers, and, if the conditions are appropriate, to produce **carbon whiskers**, diameters of <1.0 mm, of low packing density.

2.8.2 Carbon Black

Carbon black production is a significant modern industry. Carbon blacks are formed by the incomplete combustion or cracking of hydrocarbon gases and vapors derived from petroleum sources (Taylor, 1997; Sebok and Taylor, 2001) and are classified as *furnace blacks* (<50 nm diameter), *acetylene blacks* (40–55 nm diameter) and *thermal blacks* (100–500 nm diameter). Manufacturers of carbon blacks produce at least 70 products designed for specific applications.

Donnet and Voet (1976) report major differences in surface areas of carbon blacks as determined by TEM (S_{TEM}) and by adsorption of nitrogen at 77 K (S_{BET}). For example, Carbolac 1 $S_{TEM} = 264$, $S_{BET} = 972\,m^2\,g^{-1}$; Mogul A $S_{TEM} = 82$, $S_{BET} = 244\,m^2\,g^{-1}$. Differences between the two surface area values are due to (a) the presence of microporosity with the carbon black particle and (b) to capillary condensation of the nitrogen adsorbate (77 K) occurring between the carbon black particles which are close packed. On the other hand, for Spheron 6, values of S_{TEM} and S_{BET} both equal to $106\,m^2\,g^{-1}$, making this carbon black (and its graphitized product *Graphon*) suitable for adsorption studies of carbon surfaces. In studies of the use of carbon black as a metal catalyst support, in this case platinum, Prado-Burguete *et al.* (1991) report a N_2-BET (Brunauer–Emmett–Teller) surface area of $1570\,m^2\,g^{-1}$.

The main uses of carbon blacks are as a reinforcing agent in rubber-based goods, such as tyres, tubes, conveyor belts and cables. Carbon blacks form the basis of printing inks, ranging from the early days of handwriting to the manual typewriter with its ink ribbon, to the modern laser printer with its cartridge of toner (carbon black). Industrially, the vast market of newspaper, magazine and book production is dependent on carbon black as the ink source. Carbon blacks are to be found in paints, coatings, lacquers, plastics, fibers, ceramics and enamels. The list can be extended to include leather finishes, dry-cell batteries, carbon electrodes and brushes, as well as conductive rubbers and plastics, electromagnetic shielding and high-temperature insulating materials. Modifications are made to surface properties of carbon blacks (oxidation treatments) to promote the use of water-based inks and coatings which are preferable to the use of solvent-based inks (so avoiding health risks associated with some solvents).

There are *four* fundamental properties of carbon blacks which determine how these materials can be used with optimization. **One**: Fineness and particle size distribution (Figure 2.39) influence blackness and tint. **Two**: Structure within the carbon black particle and the aggregation of the particles (also called structure, Figure 2.39) influence dispersibility and electrical conductivity. **Three**: Porosity and pore-size distributions influence viscosity and coverage requirements. **Four**: The presence of surface functionality influences wettability, viscosity and electrical conductivity. In carbon black usage, it is their external surfaces, with oxygen functionality in some cases, which dominate their "sorption" properties, as distinct from the internal microporosity as found in activated carbon. Carbon blacks are supplied by manufacturers both as powders and as pellets.

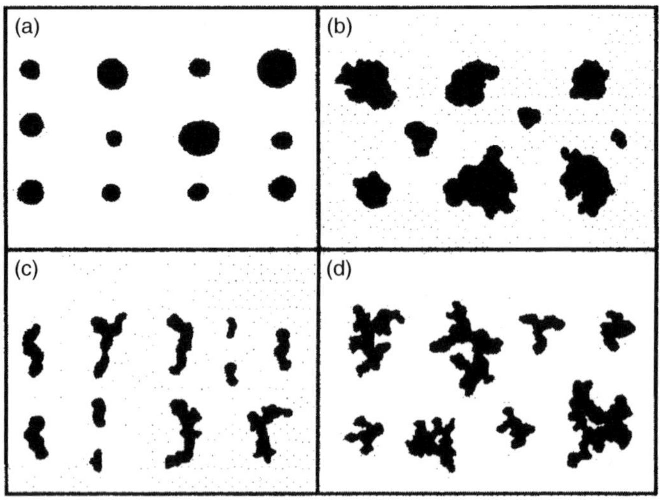

Figure 2.39. Silhouettes of particles of carbon black morphologies (a) spherical and individual, (b) agglomerated into approximately spherical shapes, (c) agglomerated linearly and (d) agglomerated into linear but branched systems (Taylor, 1997).

Carbon blacks must not be considered as being soots and possess relatively low volatile contents ranging from about 0.5 to about 10.5 wt%, with values of about 1.0 wt% for the majority of carbon blacks. When handled appropriately, they are not considered to be a health hazard, as are low-temperature soots which contain solvent-extractable material (carcinogenic). However, inhalation of such small particles and skin contact should be avoided, absolutely.

Whereas the pyrolytic carbons are formed by movement and bonding of individual carbon atoms or small groups of atoms at a surface, carbon blacks are formed by reactions in a way which is not that different from pyrolyses in the liquid phase. The small molecules originating from the pyrolysis of petroleum vapor at about 1100 °C, as well as other source gases, are further pyrolyzed into cyclic molecules leading to polycyclic aromatic molecules which form droplets. Further growth of these polycyclic molecules follows the pattern of the formation of mesogens in bulk liquid. Here, mesogens are formed in the liquid droplets and it is a characteristic of mesogens (nematic, aromatic, discotic liquid crystals) that a position of minimum energy is when they are placed parallel to a surface.

Hence, a carbon black has a layered structure, the layers being parallel to the surface. Sufficient order is established in this way to ensure some degree of graphitizability, a severe limitation being the smallness of the carbon black particle, see Figure 2.40(a, b).

These models, derived from fringe-imaging TEM, as drawn by Heinenreich *et al.* (1968), indicate the layering of the embryonic graphene layers parallel to the external surface of the carbon black particle (Figure 2.40(a)) and the way it graphitizes (with clearly seen flat-surfaced facets) as in Figure 2.40(b).

2.8.3 Fullerenes, Nanotubes and Necklaces

As mentioned above, pyrolyses in the gas phase can produce a sooty mass. Within this mass, however, there were discovered carbon (graphitic) forms of nanometer size. Recent years have seen the presentation of new graphitic (or defective graphitic) forms of the graphene layer whereby it is transformed into a spherical shape to create the fullerene

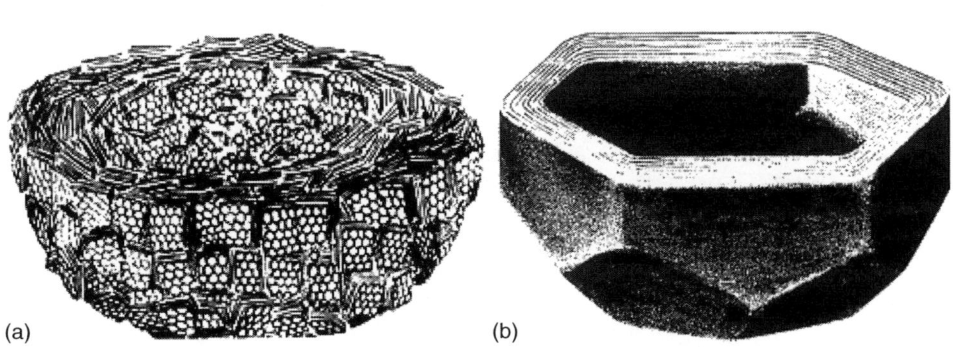

Figure 2.40. (a) Models of structure in a thermal carbon black as prepared and (b) after graphitization (Heinenreich *et al.*, 1968).

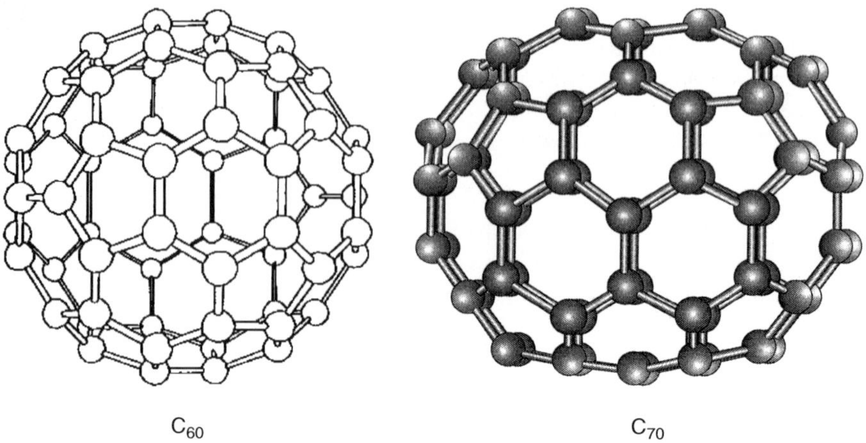

Figure 2.41. Structures of the non-porous cage-like structures of carbon, the C_{60} and C_{70} fullerene (adapted from Dresselhaus *et al.*, 1995).

family (C_{60}, etc.), made up of five-, six- and seven-membered carbon rings (Harris, 2003; Figure 2.41). These are of nano-size and although the center of the sphere can be considered as a pore, this is inaccessible to normal adsorbates under normal conditions. During their syntheses, fullerenes may incorporate other elements.

The revelation of spherical graphite caused considerable excitement in the carbon community at that time, eventually leading to the award of a Nobel Prize. Predictions of unique properties, catalytic or otherwise, were published almost on a daily basis. Millions of dollars of research money were invested and several hundred research papers were published across a wide range of literature. But, research into developments has been slow with no major breakthroughs at the time of writing. An exciting possibility is the blending of two nano-technologies to create efficient inhibitors of HIV enzymes and assisting cancer treatments.

There is, however, more to this story. The fullerenes were essentially a stepping stone in the realization that pyrolyses of carbonaceous gases, under different conditions, would provide a totally new family of carbon structures, now known as nanotubes. These take different forms, such as nano-horns, necklaces, bamboo and sausage formats, and others which will no doubt be found in future literature. These materials have a potential for applications because they are graphitic and filamentous and as such have an ability to transmit heat and electricity, over and above the mechanical properties of a near-perfect graphitic filament (fiber).

Nanotubes can be produced by several, often quite diverse methods, such as arc discharge, pyrolysis of hydrocarbons over catalysts and laser evaporation, even in explosions. Much of recent research into nano-structures originate in Japan and a review of work carried out there has been made available by Inagaki *et al.* (2004).

Nanotubes belong to the modern world of nano-materials and offer exciting possibilities to chemical physicists and physicists with interests in electronic structures. A question of relevance here is whether or not there is relevance to activated carbon. The answer is quite positive. Whereas the nanotubes and their variations lend themselves easily to structural analyses by TEM, the detail of structure in activated carbon is still elusive. It is a central theme in this book that the "walls" of slit-shaped porosity are not flat/smooth in a conventional sense, but are curve/distorted to accommodate stresses within the defective graphene layers. The various curved structures of the nanotubes may possibly exist in activated carbon, the imagination having to be restrained a little. Further, the ends of nanotubes are exposed edge sites, armchair or zigzag (see Chapter 5) and hence offer a direct method of monitoring gasification, which is a dominant method of activation of carbons. And lastly, as a nanotube offers an isolated pore, micro- or mesopore, without a pore-size distribution, the possibility of studying adsorption mechanisms can be simplified. For example, the dynamics of hydrogen and nitrogen were studied by Ötvös *et al.* (2004) who report as to how equilibrium and dynamic sorption properties of hydrogen and nitrogen are influenced by oxidative treatment. The introduction of surface functionality to surfaces of activated carbons is important to the adsorption of polar molecules. The changes which occur to nanotubes and nano-horns on heat treatment to 2400 °C provide a remarkable insight into possible rearrangements of structure when single-walled nanotubes are transformed into multi-walled nanotubes (Yudasaka *et al.*, 2003).

Then there are the nanotubes, some with closed ends, as well as nanorods (Qian *et al.*, 2003; Yudasaka *et al.*, 2003). These are made up of rolled graphene sheets, with five-membered rings to accommodate the strains involved. The internal diameters of these nanotubes range from $\approx$0.7 nm into the mesopore range $>$2.0 nm for multi-walled nanotubes (Figure 2.42(a)).

Further examples of nano-shapes are given in Figure 2.42(b–d). Figure 2.42(b) is a nano-coil from acetylene vapor deposition on cobalt nano-particles (Lu *et al.*, 2004a). Figure 2.42(c, d) are micrographs of carbon necklaces prepared using a thermal plasma technology (Okuno *et al.*, 2004). Carbon nano-forms, with a bamboo technology, have been prepared in the controlled explosion of picric acid (Lu *et al.*, 2004b). Carbon nano-horn formation is reported by Sano *et al.* (2004).

Fiala (2000) discusses the problem of the growth of curvature in graphitic materials (fullerenes) as distinct from planar growth, as in hexagonal graphite. The driving force toward curvature and eventual closure is associated with the energy released by the elimination of edge dangling bonds formed by rupture of bonds within the parent material that is being carbonized. This energy release can be brought about by formation of five- or seven-membered rings. As the π-bonds deviate from planarity, the carbon atoms are rehybridized and a π-orbital is no longer of π-character and the σ-orbitals no longer contain all of the σ-character. The strain energy involved makes these curved structures less stable than hexagonal graphite.

The question to be asked is why curved structures are formed when they are less stable than hexagonal graphite. The answer lies within the kinetics of the processes. The end product of the reaction is not always the most stable, but the one which is formed the fastest. Rates of reaction are slower for reactions of higher activation energy. Another factor to consider is the similarity between starting structures and the final product. During pyrolysis in both the solid and the gas phase, intermediates are unstable collections of carbon atoms with incomplete bonding arrangements. What happens is opportunism. The curved structures, whether incomplete as in the solid phase or complete as in the formation

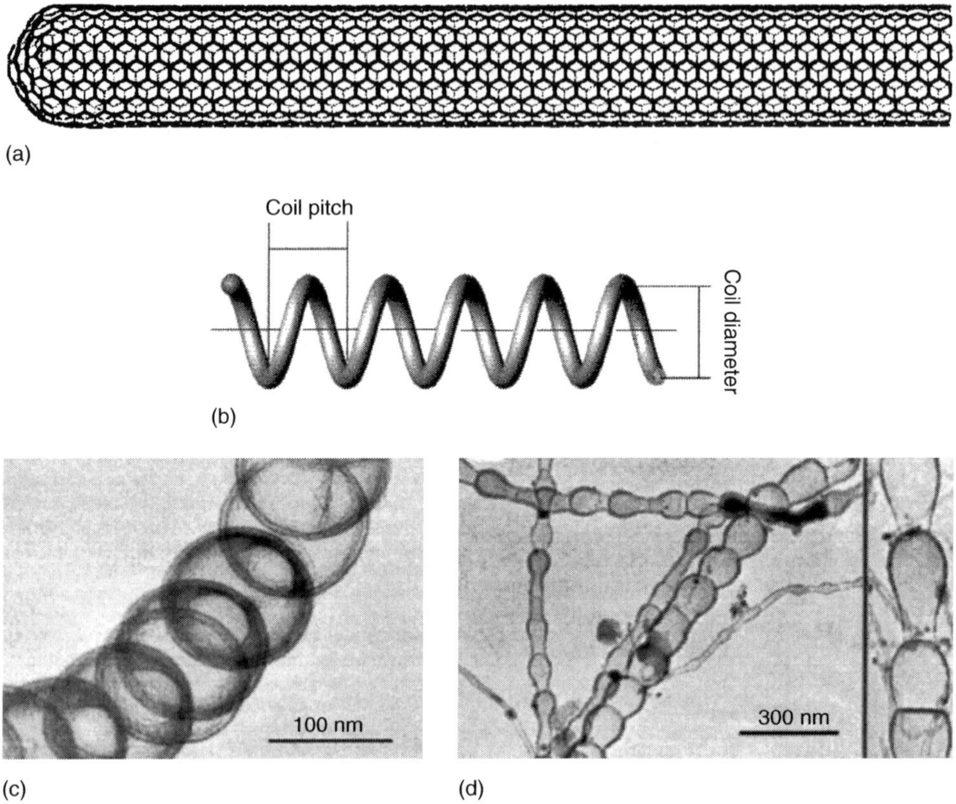

Figure 2.42 (a) Structure of a single-walled, open-ended nanotube with entrance diameter of about 1.0 nm. The wall is composed of a rolled graphene sheet, the closed end containing five-membered rings to accommodate the stresses involved. (b) A schematic of a coiled nanotube showing coil diameter and pitch, prepared by chemical vapor deposition of acetylene on cobalt nano-particles (Lu *et al.*, 2004a). See also Lu *et al.* (2004b). (c, d) Transmission electron micrograph of carbon necklaces prepared in a thermal plasma generated between graphite electrodes (Okuno *et al.*, 2004).

of fullerene-type structures and nanotubes, are formed and established before the graphite arrangements are possible. The graphite layers (graphene layers) have no carbon atoms available to tie up the dangling bonds at their edges, a process which is possible on closure of spherical shapes. These considerations are relevant to discussions of the validity of the graphitic crystallite, as discussed in Section 2.7.2.

2.9 Structures within Carbons

2.9.1 Bonding and Structure

Without doubt, the element carbon exists in a multiplicity of structural forms, more than any other element of the periodic table. These structural forms exhibit a very wide range of properties, both physical and chemical as well as mechanical, thus ensuring a wide range of industrial application, all of which are structurally dependent.

These properties are a direct result of the bonding properties of the carbon atom with its four outer electrons. The most stable form (allotrope) of carbon is that of hexagonal graphite (Figure 2.2). All other forms of carbon, when heat treated, attempt to move toward this structure of minimum free energy and maximum stability. When successful, as with the anisotropic carbons, this process is called *graphitization* (Oberlin, 1989). During graphitization, increases in structural order are brought about by movement of carbon atoms across surfaces and edges of graphene layers to sites of reduced energy. *There does not involve movement, within the solid phase, of entire graphene layers, perfect or otherwise.* If the parent carbon, although being graphitizable, possesses significant structural defects, then the activation energy for their removal, on heating, is so high as to make structural changes impossible. As a result, the graphitization process is limited, and the product remains as a *polycrystalline graphite*. This *kinetic trapping*, as it is called, is always present in parent materials and the realization of graphitic structures of single-crystal quality from carbon forms is never attained. Figure 2.6(b) is a high-resolution, fringe-imaging, electron micrograph of polycrystalline graphite, from graphitizable PVC, showing the extremes of convolution of the graphene layers which can never be removed on heat treatment to temperatures well above 2000 °C. Graphite occurs naturally, ranging from single-crystal graphite (Ticonderoga) to the seams of less-perfect graphitic material (as with Ceylon graphite). Although carbonaceous diamond, as a jewel, is intensely attractive, it reverts to the more mundane graphitic form on heating (Figure 2.3).

The positions of the carbon atoms relative to each other in solid carbons dominate the properties of carbon materials. The close packing of carbon atoms in the sp^3-mode, with a high bonding energy, creates the isotropic cubic diamond (density 3.50–3.53 gcm^{-3}) with its extreme hardness and low elasticity. Diamond has the maximum number of atoms per unit volume of known solids. Diamond materials have no porosity and are impermeable to the helium atom (see Figure 2.1). On the other hand, with carbon atoms bonding together in the sp^2-mode, there is created anisotropic graphite (density 2.23 gcm^{-3}) made up of stacked layers (graphene layers) composed of carbon atoms organized into a hexagonal arrangement (Figure 2.2). The sp^2-bonding within the graphene layers confers a high tensile strength within the layer, this forming the basis of strength in those carbon fibers which have a graphitic structure. However, the bonding between the graphene layers is of a *van der Waals* nature and is extremely weak resulting in a considerable ease of displacement (sliding) of the layers relative to each other. Hence, there is the use of graphitic materials as lubricants and polishing materials. This structure of Figure 2.2 confers the properties of electrical and thermal conductivity, and tensile strength in such materials, of dominating importance to carbon/carbon composites as used in the braking systems of aircraft, for example.

Some care is needed when considering Figure 2.2. This model gives the impression that there exists considerable space between the graphene layers which constitute the structure of hexagonal graphite. In fact, this is not correct and is a problem associated with the drawing of models. As the use of space-filling models does impart opacity to the model, it becomes difficult to see the relative positions of atoms. The π-bonds of the bonding within the graphene layers project above and below the layers effectively filling this space.

There is relevance of these considerations to the processes of activation of carbons (discussed in Chapters 5 and 6). A series of compounds, known as *intercalation compounds*, exists in which ions and molecules are located between the graphene layers of carbon materials, principally graphitic structures, but not necessarily so. These reactions, leading to the formation of intercalation compounds, involve electron transfer processes. For example, electron transfer from the graphene layer forms a bromine anion (Br^-), or electron transfer to the graphene layer forms a potassium cation (K^+). Ferric chloride also forms an intercalation compound.

These *intercalates* are not randomly distributed between the graphene layers but have well-defined positions of maximum stability related to their interactions and the electronic state of the graphene layers. The penetration between the graphene layers of these intercalates results in the expansion of the graphite lattice (swelling) with the formation of the so-called *intercalation* compounds (Tanaike and Inagaki, 1999). These intercalation compounds are not thermally stable and react easily with water vapor.

This process of intercalation must always be distinguished from a process of physical adsorption into porosity when no electron transfer takes place between the adsorbate and the adsorbent.

Structure within the non-graphitizable isotropic porous carbon is extremely difficult to elucidate, even impossible, with any degree of precision. The experimental techniques, which are available in structural analyses, all require very considerable mathematical support with interpretations and in the end, the suggested structures are best estimates only. In fact, probably the best method (but by no means being totally informative) is that of physical adsorption of gases and vapors whereby it can be said that if porosity is a function of structure then structure is a function of porosity.

A wide range of analytical methods, physical, chemical and mechanical exist for the characterization of carbons (Bandosz *et al.*, 2003; Yasuda *et al.*, 2003). These are summarized in Table 2.5.

> *The structure of the surface of porosity must be a reflection of structure within the carbon. If porosity is a function of structure, then structure is a function of porosity.*

2.10 The Non-validity of the "Crystallite" (Graphitic Microcrystallite) Concept

2.10.1 Introduction

The literature of carbon materials repeatedly refers to the *crystallite* and to the *crystallite size*, with its graphitic connotations, in analyses of structure within activated carbon (based on XRD data). Crystallite size is described in terms of a crystallite height (L_c) and width (L_a) (nm). If in fact these graphitic crystallites exist within a carbon, then their surfaces must be the surfaces of porosity and are therefore extremely relevant to all aspects of adsorption behavior. So, what justification is there, if any, for their existence? This section concludes by expressing a total denial as to the existence and the usefulness of this concept and sets out the reasons why.

Table 2.5. Techniques available for the characterization of carbons.

Method	Information provided
Boehm titration	Surface oxygen functionality
Computer simulation	pK_a of functional groups
Electron energy-loss spectroscopy	Bonding states and band gaps
Electron spin resonance or electron paramagnetic resonance	Free spins and dangling bonds
Electrophoresis	Separation – analysis
Flow adsorption	Average polarity
Fourier transform infrared spectroscopy	Surface group functionality
Gas adsorption	Surface areas and energetics, pore-size distributions
Immersion calorimetry	Surface areas and energetics
Inverse gas chromatography	Average acidity
Magneto-resistance	Electronic properties
Nuclear magnetic resonance	Molecular structure, atom groups
Potentiometric titration	pK_a of functional groups
Scanning electron microscopy	Surface characterizations
Scanning tunneling microscopy	Atomic scale structure
Small angle X-ray scattering	Total surface area including closed porosity, pore sizes
Temperature programmed desorption	Surface functionalities mainly for oxygen
Transmission electron microscopy	Fringe imaging, diffraction, light and dark field
X-ray absorption fine structure	Atomic distances and coordination chemistry
X-ray diffraction (wide-angle X-ray diffraction)	Crystallite sizes, carbon–carbon distribution functions
X-ray photoelectron spectroscopy	Identification of surface functional groups

It is necessary to look back at the earlier literature to see how the concept of the "crystallite" came about and how restrictive qualifications upon its use and interpretation have been ignored by later workers, to the detriment of carbon science. A useful survey, with references, is that of Short and Walker Jr (1963).

The story of the *crystallite* began when Scherrer (1918) derived an expression which connected the edge dimension of a cubic crystal to determined XRD line-broadening, as in the Scherrer equation:

$$L = K\lambda/B\cos\theta$$

where L is a crystallite dimension (nm), B is the line-broadening (radians), λ is the wavelength of the X-radiation (nm), θ is the Bragg angle and K is a constant (of the order of about 1, known as the Scherrer constant). It was considered at that time that line-broadening phenomena resulted from a decrease in crystallite size, below the micrometer level. No other effects were considered.

Experimentally, the line-broadening effect was observed in the diffractograms as shown in Figure 2.43(a–d). For the coal, C518 (sub-bituminous) a broad band based on the *(002)* diffraction line of graphite is visible. On heating C698, also a sub-bituminous coal, to

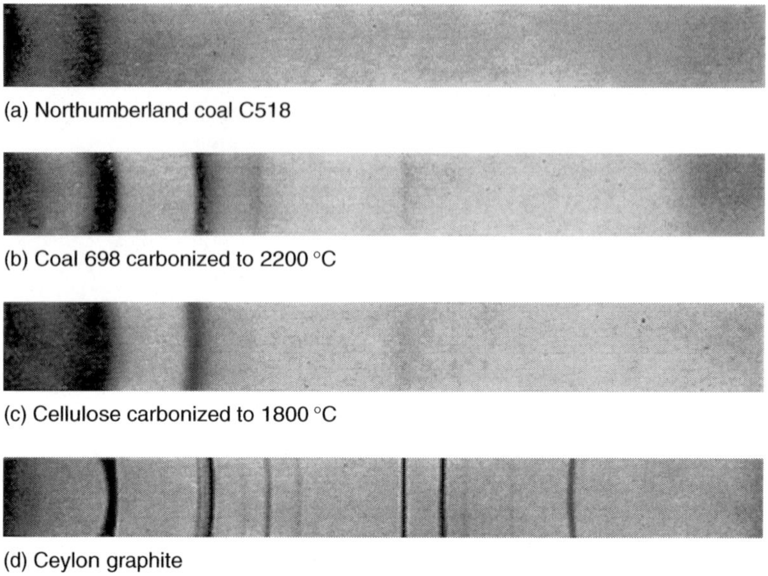

Figure 2.43. Copy of original of XRD, on film, obtained from (a) UK coal (92.6 wt% C), (b) UK coal HTT 2200 ° C (98.7 wt% C), (c) cellulose HTT 1800 ° C (99.6 wt% C) and (d) Ceylon natural graphite, showing line-broadening of the (*002*) and (*100*) bands (Blayden *et al.*, 1944).

2200 °C, additional bands appear based on the (*100*), (*004*) and (*008*) graphite diffraction lines (but not three dimensional).

The cellulose carbon (HTT 1800 °C) was intermediate between the coals, and the last diffractogram (Figure 2.43(d)) is from crystal quality naturally occurring Ceylon graphite. From these diffractograms, density profiles were obtained as the upper curve of Figure 2.44. From this upper profile was subtracted the background irradiation component to give the lower curve, from which the widths of the peaks were measured at half their intensity, and the data then used in the Scherrer equation to evaluate the crystallite parameters.

As reported by Ergun (1967), Ramdohr, independently, in Germany in 1928 was commenting that cokes, similarly, produced line-broadening effects and interpreted these effects as being due to a structure within the coke composed of single-crystalline units of graphite of variable grain size. In 1936, Marshall studied coal to coke conversions and commented that there was no graphite in the cokes, although there were similarities in the X-ray data. In 1951, Mackowsky (an eminent coal petrologist of the time) reported that graphite does not, and cannot occur in coal and coke. In 1952, Stach (also an eminent coal petrologist of the period) concurred with Mackowsky that graphitic material was not present in coals, the broadening being due to imperfect anisotropic stacking of aromatic molecular moieties (Stach *et al.*, 1982).

Warren (1941) and later Biscoe and Warren (1940, 1942) observed that carbon blacks gave broadened XRD based on the diffraction pattern of graphite. As, at that time, no

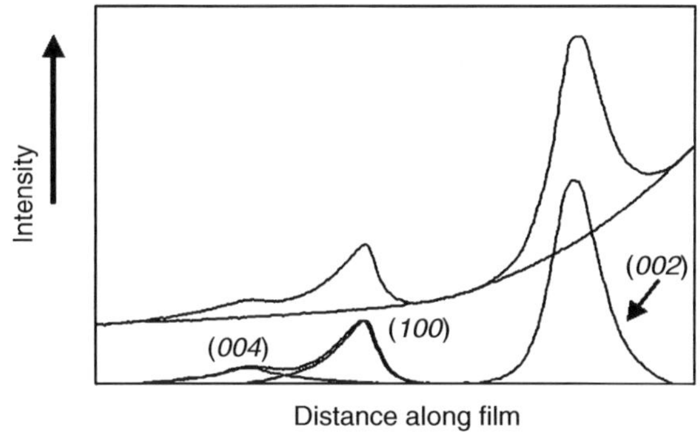

Figure 2.44. Profile of the diffraction bands from a carbon (upper curve) with the background subtracted (lower curve). Crystallite sizes are calculated from the half-peak width (radians) of the (*002*) and (*100*) bands of the lower curve (Blayden *et al.*, 1944).

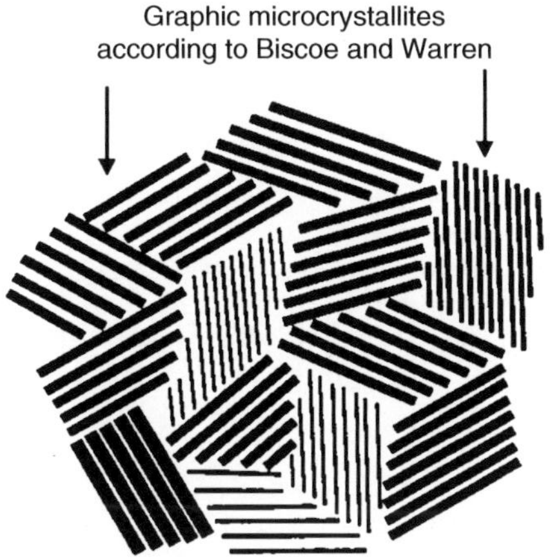

Figure 2.45. The model for structure in a carbon black using the graphitic microcrystallite theory (Biscoe and Warren, 1942).

detail of structure within these materials was available, recourse was made to the graphitic microcrystallite theory resulting in the model of Figure 2.45 for structure in a carbon black. The inappropriateness, these days, of this model is seen when Figure 2.45 is compared with Figures 2.39 and 2.40.

2.10.2 Graphitic Microcrystallites in Coal?

Difficulties with the acceptance of the graphitic microcrystallite model become more apparent when a rank series of coals was considered as in Figure 2.46 or when coals (representative of other organic sources of carbons) are heat treated and the changes of crystallite sizes measured against the HTT, as in Figures 2.47 and 2.48 (Blayden *et al.*, 1944). The graphitic

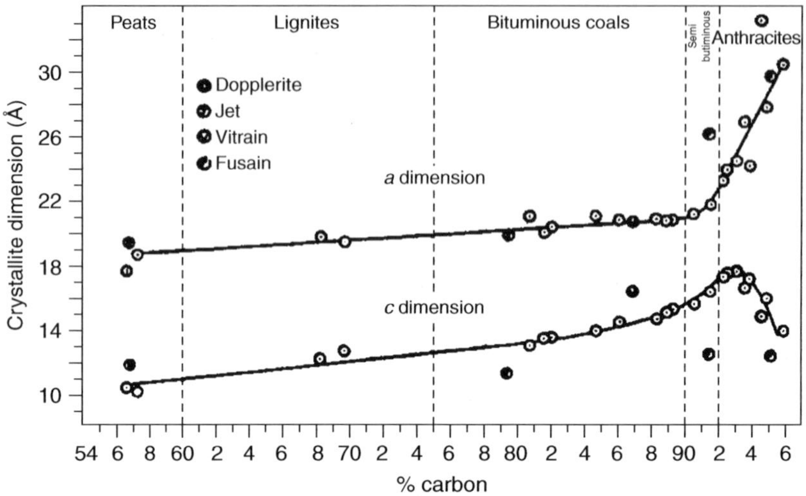

Figure 2.46. The variation of crystallite dimensions, L_c and L_a (Å), with carbon content of a rank range of coals. Reproduced without change from the original publication (Blayden *et al.*, 1944) (10 Å = 1.0 nm).

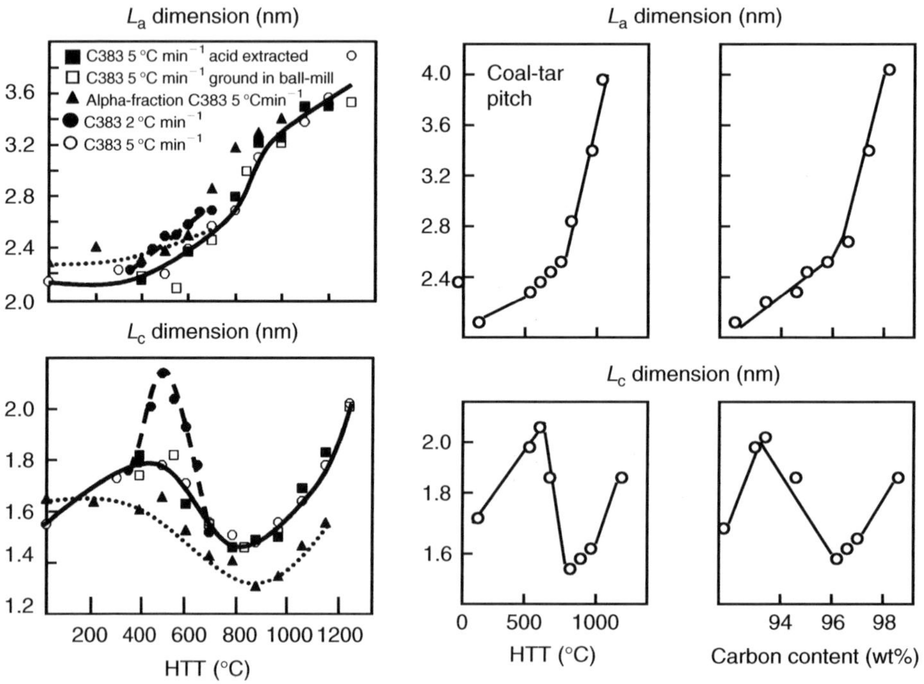

Figures 2.47 and 2.48. The variation of crystallite dimensions, L_c and L_a (nm), with HTT (°C) and carbon content (wt% C) of coals and a coal-tar pitch (Blayden *et al.*, 1944).

microcrystallite theory of structure could not explain these changes. The problem was that the diffractograms for carbon blacks, carbons of low HTT (<800 °C) and coals were so essentially similar. Could the graphitic microcrystallite theory explain it all or were there different explanations for different materials?

Figure 2.46 is a plot of crystallite size (L_a, L_c) against coal rank, as expressed in wt% of carbon content. These crystallite dimensions, as determined from X-ray line-broadening data, show a progressive increase with carbon content of the coals until about 93 wt% carbon, when the L_c dimensions decrease sharply but the L_a dimension increases sharply. It is difficult to explain the decrease in height associated with an increase in width of the crystallite bundle. Above 93 wt% carbon is the region of the anthracites which show significant properties of graphitizability, a property which appears to contradict the decreases in the L_c value.

Figures 2.47 and 2.48 show a similar phenomenon for the carbonization of coals and of a coal-tar pitch where L_c values, initially, are seen to increase with increasing HTT. They then decrease, sometimes to values below those of the starting materials, and then rise again at about 94 wt% carbon (for the coal-tar pitch). The phenomenon of a very pronounced maximum in the L_c versus HTT curve occurs when the carbonization system passes through a liquid (plastic or fluid) phase and is associated with formation of the ordered nematic liquid crystals of the mesophase. However, as indicated in Figure 2.38, any increase in HTT results in cross-linking between the mesogens and this, in turn, introduces buckling, twisting and bending in the micro-graphene layers (in other words, increases the strain energy within the layers). The outcome of this, in terms of XRD, is to increase the width of (*002*) band which is interpreted as a decrease in L_c values.

This example is to illustrate that a single determination of crystallite sizes is of extremely limited value as no distinction is possible between contributions from size and contributions from strain. The unchanging graphitic microcrystallite model cannot explain these phenomena.

The technique of ESR (Section 2.6.2.2) provides information concerning variation of the formation and removal of unpaired electrons, with HTT, during a carbonization process (Figures 2.20 and 2.21). It is apparent that significant structural changes occur within the structural units of the carbon which are responsible for the line-broadening effects. These units cannot be inert graphitic microcrystallites. *Blayden et al. (1944) made the comment that it is pressing the graphitic microcrystallite theory too far to regard structure in coals and carbons (cokes) as consisting of minute graphitic crystallites.*

2.10.3 Causes of Line-broadening in XRD

Earlier warnings, about the misuse of the graphitic crystallite theory, were apparently ignored. Ruland (1965), Ergun (1967, 1968, 1970) and Maire and Méring (1970) opposed the use of the graphitic crystallite theory to describe structure in carbons, and stressed that defects, such as cross-linking between the aromatic planes, dislocations and tetrahedrally bonded carbon, would contribute to diffraction broadening. Nagle and Thrower (1973) emphasized the importance of strain in crystallite size determinations in graphite. These authors studied a polycrystalline graphite, using the *(002)* and *(004)* peaks, and estimated that strain accounts for around 12% of the width of the *(002)* peak. The sample was then ground to <38 mm size and re-examined when it was found that strain accounted for only 4% of the observed broadening. This result confirmed the importance of strain in crystallite size determinations and the danger of analyzing only one diffraction peak, when using the Scherrer equation (see also Thrower, 2004).

The causes of line-broadening phenomena in carbons can therefore be summarized as follows:

1. Instrumental broadening caused by using an X-ray beam of finite size and apertures of finite size. Instrumental broadening can easily be corrected by using a known single crystal as a standard.
2. Strain within the lattice, causing an interlayer distance to vary, as in curvature.
3. Smallness of size of perfect crystallites (without strain).
4. Smallness of size associated with strain within the perfect lattice.
5. Smallness of size associated with imperfect layering of defective structural units, that is the defective micro-graphene layer (and not graphitic microcrystallites).

When studying carbonaceous materials using XRD procedures, e.g. coals, cokes, carbons from whatever source, even graphites, it is impossible to know which factors are creating the broadening effect. A resolution to the problem would be the co-analysis of diffraction peaks, such as the *(002)* *(004)* and *(008)*, but as carbons usually provide the *(002)* peak only, such further analysis is impossible.

Line-broadening is now explained, for coals and non-graphitic carbons (usually of HTT > 1200 °C), as resulting from structures made up of associations, roughly parallel, of hydrocarbon (polycyclic) moieties (for the macromolecular structure of coal) and of quite defective, non-planar but roughly parallel associations of carbon atoms as summarized in Figure 2.38, and termed the defective micro-graphene layers. That is, there is no planarity, only strain and defects.

Further, a plot of crystallite height, L_c (nm), against HTT, for all carbons, is essentially a broad band from which it can be seen that many carbons of quite different origins and structure have the same half-peak width and hence the same apparent crystallite sizes. A carbon prepared in the liquid phase, from mesogen components, being highly graphitizable but of HTT of about 700 °C, has a diffraction band indistinguishable from a porous isotropic carbon of the same HTT. It is now appreciated that structure in these two carbons is so different as to demand that the concept of the graphitic microcrystallite be most severely restricted in its use. Thus, as an analytical tool, because the "crystallite" cannot distinguish between carbons of minimum and maximum graphitizability, *why continue to use it*?

On the other hand, the concept of the graphitic crystallite and its assumed existence within all carbons have stimulated crystallographers to fine-tune their analyses and interpretations of line-broadening phenomena. The Japanese Society for the Promotion of Science (JSPS, 1963) standardized procedures for the calculation of crystallite sizes (Shiraishi, 2003).

A theoretical study of X-ray scattering intensities is based on the crystallite concept (Fujimoto, 2003). The author based his calculations on diffractions of an AB graphite and a *turbostratic* carbon (a term which is not recommended but which exists in the literature). No further details are given of the origins and preparation of the carbons, and the use of the term "turbostratic" sheds little light on the carbon structure which is being modeled. Calculations are based on the benzene–pyrene and benzene–coronene models using Debye and Warren–Bodenstein equations (Fujimoto, 2003). Calculated values of L_c, L_a and d_{002} (nm) are quoted to six significant figures. A specification for a standard procedure of XRD measurements on carbon materials has been published by Iwashita *et al.* (2004).

In addition, it is noted in the literature that although crystallite sizes are evaluated in many publications (in one case to an accuracy of 0.0001 nm), no subsequent use appears to be made of these numbers. In the optimum of circumstances, all that can be said is that one carbon, in a given series, appears to have a more organized structure than a closely related carbon. A comparison of half-peak widths is sufficient to make such comparisons, as there is little evidence that different carbons of the same half-peak width can have different meaningful crystallite sizes. To compare different carbons, from different parent materials, is fraught with danger. Current knowledge of carbon chemistry cannot seriously support the concept of the crystallite.

A further possibility can be considered in analyses of line-broadening phenomena. Considering again the model of non-graphitizable carbon of Franklin (1950) of Figure 2.7(b), its microporosity (extent of) could relate to the thickness of the carbon material (pore wall) which separates (individualizes) the microporosity. This material contributes to line-broadening, the thinner the pore wall, the broader the peak. Thus, there is a possibility (to be explored further) that line-broadening in porous carbons may relate more to porosity than pore wall carbon.

It is impossible to make structural analyses of carbons, independent of line-broadening experimentation, to support the theory of the graphitic microcrystallite. The use of high-resolution fringe-imaging TEM (Oberlin, 1989) is a most valuable tool for graphitizable carbons of HTT above about 550 °C indicating the basic parallelism of the constituent embryonic graphene layers. However, for the non-graphitizable isotropic carbons this technique has serious problems of interpretation. Certainly, it has not been used extensively for the identification of the graphitic microcrystallite, but rather for resolution of porosity in carbons (see Figure 2.6(e, f)). Moreover, as indicated above, the solubilization of coals never left a residue of the insoluble graphitic microcrystallite.

Sadly, this graphitic microcrystallite concept appears to have been passed down over the years (perhaps uncritically) from publication to publication. Now, it requires a careful reassessment to enable the subject of structure in non-graphitizable carbons to move on.

Clearly, if the crystallite based on the graphite hexagonal lattice is an essential component of all carbons, then it must be present in the activated porous carbons which are so relevant to this book. This implies that the surfaces of the porosity of micro and mesoporous carbons must be the surfaces of the graphitic microcrystallites (if they exist). This aspect is not discussed in the literature. The evidence which comes from adsorption studies of carbons of increasing HTT to a maximum of >2000 °C would never support this concept. The quest to know the structure of the adsorbing surfaces of micro and mesoporosity must be pursued, as it is fundamental to the adsorption process itself. Structural assessments are also important to the processes of activation as described in Chapters 5 and 6.

The earlier literature also uses the term "apparent crystallite size" indicating that the sizes are indicative of some level of order of carbon atoms within the carbon. And that is about as far as the concept should be developed. Carbons of quite low HTT, say 500 °C, exhibit line-broadening phenomena. In fact, all carbons do this. There exists no discussion as to how a graphitic microcrystallite can appear in such a low-temperature material. It cannot.

2.10.4 The Graphitic Microcrystallite Theory: Conclusions

1. The concept adds nothing to an understanding of carbonization processes.
2. The concept adds nothing to facilitate distinguishing between carbons.
3. The concept has nothing to say about any non-crystalline component of a carbon.
4. The concept has nothing to say about the structure of surfaces of porosity.
5. The concept has nothing to say about how microporous carbons respond to increasing HTT.
6. The concept has no relevance to processes of physical adsorption, isotherm interpretation as well as processes of activation (physical and chemical).

The concerns about the validity of the graphitic microcrystallite theory are summarized in Figure 2.49. This model shows the approximate parallelism of micro-graphene layers with variable interlayer distances (a cause of line-broadening), the small dimensions of the scattering units, ~20 nm in the *C*-direction (a cause of line-broadening), the lack of planarity in the micro-graphene layers (a cause of line-broadening), and that part of the carbon which may not be able to contribute to any form of coherent scattering of the X-ray beam.

2.11 Raman Microspectroscopy in Structural Analyses: An Assessment

2.11.1 Introduction to the Use of RMS

*An analysis of the literature of RMS has aided considerably in the search for an acceptable structural model of isotropic microporous carbons. Raman spectra of carbons contain two bands of major interest, namely the **D**-band (~1350 cm^{-1}) and the **G**-band (~1580 cm^{-1}). The **D**-band is identified with **D**efective or **D**isorganized carbon, and the **G**-band with **G**raphitic carbon.*

*The origins of the **D**-band, in terms of possible defective structures in carbons, remain quite obscure. Similarly, the origins of the **G**-band cannot be identified with graphitic carbon.*

Certainly, there is no evidence from RMS of the existence of the graphitic microcrystallite even though this concept appears firmly to be rooted in much of the RMS literature. It would appear that this concept has simply been carried over into the RMS literature from wide-angle X-ray diffraction (WAXD) data without too much critical analysis.

The following account of some RMS studies clearly sets out a clearer picture of structure in these microporous carbons.

Figure 2.49. A model which summarizes concerns about the validity of the graphitic microcrystallite theory as applied to XRD studies. The existence of very limited parallelism between graphene layers accounts for the X-ray scattering. The concept of graphitic microcrystallites is not needed to explain this scattering.

Chemists have tended to use the word ***DISORDERED*** *whereas physicists use the word* ***DEFECTIVE*** *when it comes to describing the* ***D****-band of RMS. The term "disordered" conjures in the mind an assembly of carbon atoms with little in the way of relationships to each other in space, certainly not with the spatial order of a graphene layer, even a defective graphene layer. However, for the physicist and Raman spectroscopy belongs to the physicist, the word to use is "defective" and by this is meant defective arrays of carbon atoms in spatial arrangements of hexagons, that is in defective graphene layers. These graphene layers can be defective in so many ways that their contribution to RMS is significant and ongoing. It would appear that the* ***G****-band could originate from a graphitic graphene layer. There is little in the literature to indicate how RMS spectra can originate from the so-called disordered carbons meaning disorganized and defective (containing defects relative to a graphene layer).*

2.11.2 Incorrect Interpretations of RMS Data

WAXD (or XRD) is an extremely blunt tool for structural (three dimensional) analyses of isotropic carbons. RMS is an appropriate tool for studying, *specifically*, chemical bonding in the crystalline solid state, both well and poorly organized (Kakihana and Osada, 2003). This technique has been used for some time to study structure in carbons and the changes induced by heat treatment (Lespade *et al.*, 1984; Cuesta *et al.*, 1994) to the extent of becoming a routine analytical tool. The availability of nanotubes (Figure 2.42) in many structural forms has assisted with interpretations of RMS spectra (Dresselhaus *et al.*, 2002). RMS is a spectroscopic tool used to study vibrational modes in graphene layers whatever their size,

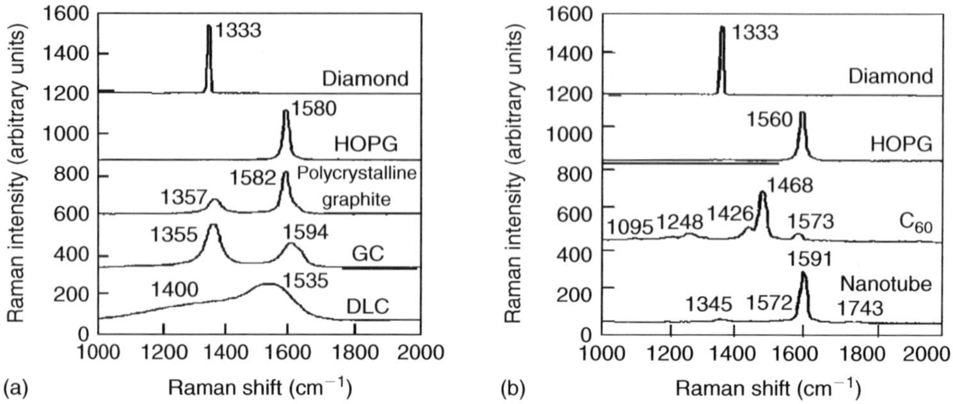

Figure 2.50. RMS of a series of carbons including (i) diamond, (ii) highly ordered pyrolytic graphite (HOPG), (iii) polycrystalline graphite, (iv) glassy carbon (GC), (v) diamond-like carbon (DLC), (vi) C_{60} fullerene and (vii) nanotube (Kakihana and Osada, 2003).

shape and composition. RMS is not a structural tool in the sense that the vibrational modes of one graphene layer are not influenced by the proximity of a second layer.

Unfortunately, for studies of carbons, many of the RMS publications have assumed, uncritically, the existence of the graphitic microcrystallites. There is no independent justification for the assumption. This section therefore assesses the contributions which RMS can make to further understandings of carbon materials, in particular the isotropic porous-activated carbons of which surface structure is of major importance in adsorption studies.

The hexagonal graphite structure has two Raman-active vibrational modes corresponding to in-plane vibrational modes of carbon atoms one of which involves the stretching of the sp^2—C—C bond, this E_{2g2} mode having a frequency of 1582 cm^{-1} (**G**-band) (**G** implies graphitic; Figure 2.50). For the carbons dominantly of low HTT, both anisotropic and isotropic, Raman spectra show an additional band at about 1350 cm^{-1} (**D**-band) (**D** implies defective). The **D**-band is assigned to defects within the carbon material relative to the graphene layer.

Laser Raman spectroscopy is a specialist subject area for the physicist. The experimental technique is relatively easy to carry out although care has to be taken with the intensity of, for example, the laser excitation beam. Results of selected studies reported in the literature are commented upon below to illustrate the need for some re-interpretations of Raman spectra.

An example of this problem can be seen in a study by Manivannan *et al.* (1999) who examined three porous carbons (by PICA) from coconut (surface area 1000 m^2 g^{-1}), from a wood (surface area 1150 m^2 g^{-1}) and a coconut shell (2000 m^2 g^{-1}). L_c values of graphitic microcrystallites are reported as 0.9, 1.0 and 1.5 nm, and L_a values as 3.9, 2.5 and 2.7 nm, respectively. However, these authors also interpreted their Raman spectra in terms of L_a values of 2.9, 3.1 and 2.6 nm of these graphitic microcrystallites and commented on the good agreement between the results. No comment is provided on any possible relationship between graphitic microcrystallite size and surface area. There does not appear to be any inter-dependency. No comment is provided as to how graphitic crystallites are generated within an isotropic carbon from a lignocellulosic parent material.

Values of L_c and L_a are a maximum for the commercial carbon of surface area of 2000 $m^2 g^{-1}$. High adsorption capacity has to be associated with a low ratio of carbon atoms to adsorbate molecules; that is, the pore walls have to have the minimum possible thickness. The above crystallite dimensions do not support this requirement.

Manivannan *et al.* (1999) also report on the presence of magnetic inorganic impurities of size too small to be detected by WAXD. Such impurities can have significance in terms of catalytic gasification (during physical activation) and in terms of catalytic graphitization (during carbonization processes).

This article assumes, however, without questioning, the concept that the presence of a **G**-band automatically means the existence of the graphitic microcrystallite made up of more than one graphene layer. It is this assumption which has to be debated. Again, there is no mention as to the stage, in the carbonization process, how and when these crystallites appear. Such a phenomenon does not fit well into the general understanding of formation of carbon materials. Manivannan *et al.* (1999) appear to assume that as the **G**-band of single-crystal graphite or pyrolytic graphite appears at the same frequency as the **G**-band of these isotropic porous carbons, then such graphitic structures, as found in pyrolytic carbons, must be present in such carbons. Within the scope of Raman spectroscopy, this may be a reasonable possibility but a complete analysis of this problem has to move into a wider context. The central question is whether or not quite different structures in carbons generate similar **G**-bands.

2.11.3 Definitive Studies Using Pitch (Dumont et al., *2002)*

An article by Dumont *et al.* (2002) appears to answer the question as to what are the origins of these **G**- and **D**-bands (that is how they develop from a parent feedstock), and is therefore well worth quoting at some length. Dumont *et al.* (2002) report that a carbonaceous material does not have to be graphitic to produce a **G**-band and conversely that a relatively well-ordered carbon could produce a **D**-band. An article by Dresselhaus *et al.* (2002) indicates the complexity (and therefore amount of contained information) within a Raman spectrum from single-walled nanotubes.

Dumont *et al.* (2002) studied the pyrolysis/carbonization of a synthetic pitch derived from naphthalene (ara24r) (Mochida *et al.*, 1990, 2000). Such pitches, as well as coal-tar pitches and some petroleum-based pitches, carbonize to yield highly anisotropic graphitizable carbons. This naphthalene pitch had been developed commercially by Mitsubishi Gas Chemical and Mitsubishi Oil (probably for pitch-based carbon fibers) with a narrow molecular weight distribution such that with increasing HTT polycondensation chemical reactions increased the molecular weights in the system so forming self-assembling molecules to form mesophase (anisotropic aromatic nematic discotic liquid crystals) (see Section 2.6.3 above) (Figures 2.35–2.38). By selecting a naphthalene pitch, the number of oligomers in the carbonization system was kept to a minimum in order to simplify the Raman spectra. In these experiments, the condensation reactions occurring within the parent naphthalene pitch can be followed by Raman spectroscopy extremely closely in a way that no other technique would find possible.

Dumont *et al.* (2002) noted that the spectrum for the ara24r pitch exhibited features in common with those obtained from the so-called disordered carbon materials. A **G**-band occurred around 1600 cm^{-1} and a broad complex band lay between around 1150 and 1500 cm^{-1} and, enclosing the usual wave numbers for the **D**-band. The **G**-band at 1600 cm^{-1} consisted of two components at 1609 and at 1580 cm^{-1} which corresponded well with the **G**-band of graphite. Dumont *et al.* (2002) explained the origins of the **G**-band and of the **D**-band (with associated peaks at 1609, 1248, 1288 and 1376 cm^{-1}) in terms of the structure of the mesophase pitch (Figures 2.51 and 2.52).

The ara24r pitch consists of naphthalene-based oligomers from the polycondensations of naphthalene, these being polycyclic hydrocarbons (benzenoid) rather than being graphitic. The oligomers are smaller in size and possess aryl–aryl C—C bonds, naphthenic rings and peripheral alkyl groups. These authors consider that there is similarity between the vibrational modes of the C—C bonds within the plane of these aromatic molecules (1570–1620 cm^{-1}) and the E_{2g2} mode for graphite at 1580 cm^{-1}. The two-component 1600 cm^{-1} band would be due to the different types of covalent C—C bonds within the molecules with their different lengths, strengths and therefore different vibrational frequencies. It is considered that the broader **D**-bands result from such large defects (distortions) within a carbon network containing six-membered rings being more benzenoid than graphitic.

On heating the ara24r pitch to 550 °C, little change is observed in the Raman spectrum with the peaks at 1250 and 1290 cm^{-1} decreasing slightly as the pitch is transformed into a carbon (coke). However, on heating to 750 °C major changes are observed. The peaks at 1250 and 1290 cm^{-1} disappear leading to a spectrum characteristic of the so-called disordered carbons with a broad **D**-band at 1350 cm^{-1} and a **G**-band at 1600 cm^{-1}. Heating beyond 750 °C is considered to result in the introduction of further disorder into these (embryonic) graphene layers which form part of the carbon network, an observation supported by high-resolution, fringe-imaging, TEM.

This research by Dumont et al. (2002) describes the development of Raman spectra from a parent pitch to a resultant coke. The RMS spectrum of the coke originates from within one (embryonic) graphene layer which most probably forms part of a cross-linked carbon network. The proposed existence of a "graphitic crystallite" composed of several graphene layers stacked parallel to each other is not required. Hence, the frequent use of the term "graphitic microcrystallite" within the RMS literature is simply an uncritical "carry-over" of words from WAXD studies and which unfortunately impart seriously misleading ideas.

2.11.4 Interpretations of RMS Data

Dresselhaus *et al.* (2002) (in the language of the physicist) state that the **G**-band in carbon single-walled nanotubes (1500–1605 cm^{-1}) is derived from the Raman-allowed optical mode E_{eg} of 2D graphite by two-zone folding of the 2D graphene Brouillon zone. The **G**-band is thus an intrinsic feature of carbon nanotubes that is closely related to vibrations

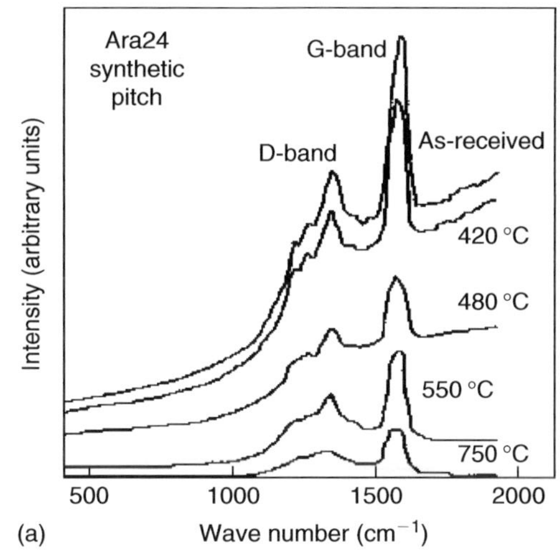

Figure 2.51. RMS of a series of carbons of increasing HTT, 420–750 °C, from ara24 synthetic pitch (Dumont *et al.*, 2002).

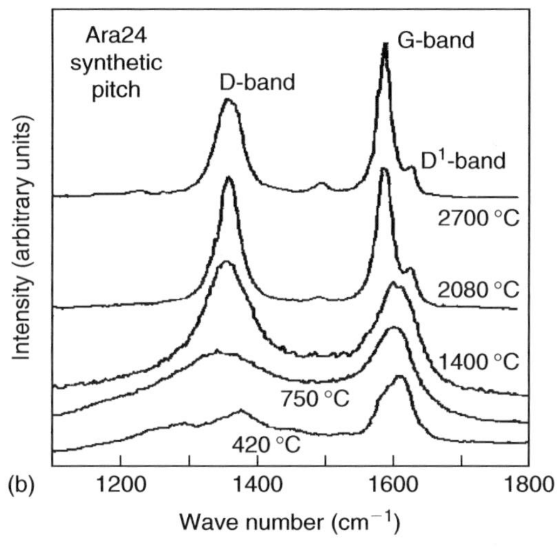

Figure 2.52. RMS of a series of carbons of increasing HTT, 420–2700 °C, from ara24 synthetic pitch within a C—C composite (Dumont *et al.*, 2002).

in all sp^2-carbon materials. These authors stress the importance of analysis of line shape of the **G**-band in terms of structure within the graphene layer. The **D**-band in graphite involves scattering from a defect which breaks the basic symmetry of the graphene sheet and is observed in sp^2-carbons containing vacancies, impurities or other symmetry-breaking defects (Figure 2.53; Dresselhaus *et al.*, 2002).

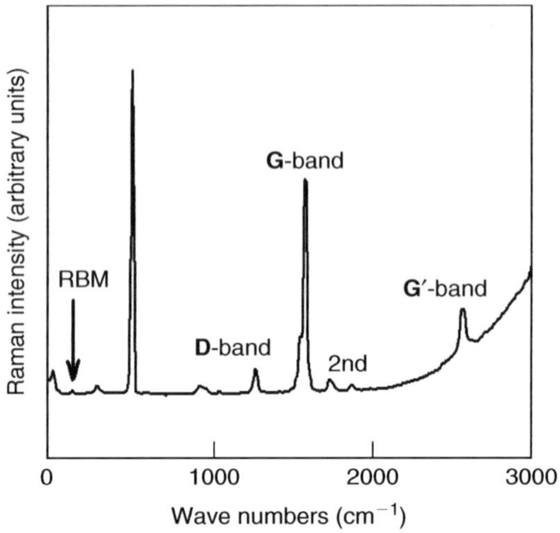

Figure 2.53. Raman spectrum from one nanotube taken over a broad frequency range using E_{laser} = 785 nm and showing the D, the G and the **G′**-band (associated with the **D**-band) (Dresselhaus *et al.*, 2002). RBM: radial breathing mode.

Qian *et al.* (2003) studied the Raman spectra of single and multi-walled carbon nanotubes and noted that, in mixtures of these two types, as the percentage of multi-walled carbon nanotubes increased, so the intensity of the **G**-band of the mixture decreased but the intensity of the **D**-band increased. This effect is attributed to the increased number of defects within the multi-walled carbon nanotubes as a result of accommodations to maintain these structures and in the kinetics of their formation (more difficult to obtain perfection during growth).

Another study of relevance to this discussion is that of Wakayama *et al.* (1999) who studied defects introduced into mesophase microcarbon beads (MCMB) (HTT 2800 °C) by mechanically grinding. These authors describe three Raman spectra, as in Figure 2.54, where the original material has a distinct **G**-band and a minimum **D**-band. After 0.5 h of grinding, the spectrum changed, the **D**-band having a higher intensity than the **G**-band, being indicative of the defects introduced by the grinding process. A third spectrum is that of the mesocarbon microbeads (HTT 1000 °C) where the **D**-band is much broader than the **G**-band of the ground graphite. This indicates a difference in the type of structural defects involved between the heat-treated carbon and ground graphite. A further consideration is that the ground graphite could have more uniform boundaries and less-defective hexagonal structure within the graphene layers than does the heat-treated carbon. The understanding from these two studies is that the **D**-band is associated with arrangements of carbon atoms in hexagonal format ranging from a benzenoid molecular type to defective graphene layers, the defects being lack of continuity in the layer, point defects, multiple vacancy defects, nitrogen or oxygen in the layer or at edges of layers, etc. Those publications which report the existence of graphitic crystallites, even quoting crystallite sizes, based on Raman spectra are projecting ideas from WAXD studies which are not considered to be acceptable any more. The available analytical techniques point to atoms

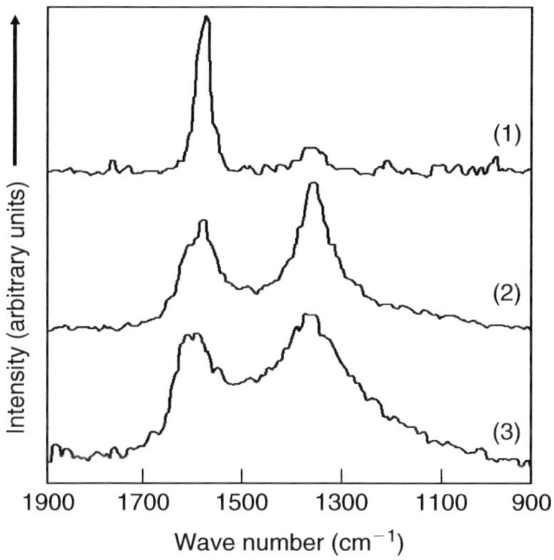

Figure 2.54. Raman spectra of (1) graphitized MCMB HTT 2800 °C, (2) MCMB ground for 0.5 h and (3) MCMB HTT 1000 °C (Wakayama *et al.*, 1999).

of carbon in the structure of microporous carbons being incorporated into ring systems of various levels of perfection. If carbon atoms are present outside of the ring systems, then their detection remains difficult, indeed.

What emerges from these discussions is that to use WAXD for isotropic carbons makes a minimum contribution to understandings of their structure. RMS, if not over-interpreted, tells us that, in a carbonization process, from the moment the parent macromolecular system (coal, plant materials) begins to turn black (all carbons are black), a **G**-band appears indicating that carbon atoms (but not all) from the decomposition process, when released of associated heteroatoms-like hydrogen, oxygen and nitrogen, arranged themselves in groups of hexagonal arrays (no doubt positions of minimum structural energy). This has been known for some time because transmission infrared (IR) spectroscopy indicates complete absorption of the IR at these temperatures, indicative of a multitude of absorption sites of an aromatic, benzenoid even graphitic nature.

These ideas are discussed further in Chapter 3 which examines the various models put forward for structure in microporous carbons.

2.12 Quantitative Reflectance Microscopy and Carbon Structure

The technique of reflectance optical microscopy has proven to be invaluable in the analyses and characterization of fossil fuels, in particular coal materials and petroleum-containing source rocks. The technique involves the measurement of the percentage of white polarized light reflected from a polished surface of a coal specimen using oil immersion and analyzed within the microscope. This percentage value gives a measure of the aromaticity of the carbonaceous matter which constitutes coal and rock materials. Table 2.1 contains specific reflectance values which increase from about 0.2% for the lowest rank of coal, a peat, to about 7.0% for an anthracite within a coalification series. The technique can be further developed to provide "bireflectance" data which measure extents of three-dimensional ordering of the graphene layers. Here, it is necessary to rotate the stage of the microscope through 180 ° to obtain maxima and minima values. Isotropic materials do not show bireflectance. This subject area is well developed for pleochroic minerals.

The coal industry is closely linked to the iron and steel industries where metallurgical coke reduces iron ore to iron in the blast furnace. The quality of the metallurgical coke is critical to the efficient running of the blast furnace. In turn, this quality is controlled by the specifications of the coal supplied to the cokerie (coke ovens) which blends several coals which are then carbonized to metallurgical coke. Coals are classified according to their carbon and volatile contents and fluidity properties. It is not uncommon to find that coals, which are otherwise indistinguishable, behave differently in the cokerie. The use of reflectance microscopy distinguishes between such coals.

The processes of coalification and carbonization are not that dissimilar and hence carbonizations should be amenable to studies by reflectance microscopy. However, the method appears to have been confined to the laboratories of coal petrologists with few exceptions. An early study as that of Jüntgen *et al.* (1988) who used carbonized bituminous coal and activated products from this coke (carbon). Reflectance values increased from about 1.2% (HTT 440 °C) to 6.3% (HTT 900 °C). The microscope technique measures reflectance values from a large number of areas of the polished section (computerized microscopy make this relatively easy) when a distribution of reflectances is found. This indicates that not all of the structural units (now called graphene layers) are identical but that there is a range of structural development. Further, with increasing HTT, this distribution changes from being symmetrical to being skewed with a higher proportion of graphene layers which are structurally less well developed. In a further series of experiments, Jüntgen *et al.* (1988) monitored the surfaces of their carbons in terms of activity for the oxidation of ammonia (NH_3) by oxides of nitrogen (NO_x). This activity (measured as a conversion percentage) reached a maximum at about 5% reflectance and then decreased sharply. This effect can be interpreted as the initial growth of aromaticity, from the constituents of the coal, into defective graphene layers, the number of defects maximizing with increasing HTT as the graphene layers become more graphite like (but not entirely graphitic).

The use of the optical microscope has therefore confirmed that the development of structure within a carbonized product in a progressive phenomenon, so supporting the anti-graphitic microcrystallite concepts. The technique indicates that the growth of the graphene layers is via defective graphene layers which become more perfect with increasing HTT. More, the evidence is complete that not all of the graphene layers have equal structural perfection; there is range of degrees of perfection.

Such information is relevant to considerations of the structure of surfaces in activated carbons and responsible for adsorption, as well as to considerations of gasification reactions in physical activation processes. Here, selective gasification is crucial to the development of porosity and these considerations provide clues as to how selective gasification occurs. Neither the use of XRD nor RMS is able to provide these information. In fact, the opposite may be true, that reflectances may assist with interpretations of Raman spectra.

A later use of this reflectance technique was to study chemical activation of wood and coals, by Jagtoyen *et al.* (1995). The full details of this study belong to Chapter 6, but sufficient to state here that, like Jüntgen *et al.* (1988), reflectances of carbons from the wood and coal increased with increasing HTT, reaching ~5.5% at 700 °C for white oak wood. The coals gave values of ~4.5% for the same temperature of 700 °C. Rather than make a full study of the optical properties of their carbons, these authors concentrated on how reflectances were influenced by the development of mesoporosity in their activated carbons.

Reflectance parameters are a function of the refractive index of the solid. The theoretics of reflectance lead to scattering phenomena and into the world of theoretical physics. This may be one reason why the creation of reflectance data is limited. Such a worry has not, however, restricted the use of X-ray and neutron scattering techniques, as outlined in Chapter 4. Accordingly, a comprehensive study of carbonizations and activations, as part of a program of structural analyses of active carbon, is long overdue.

2.13 What Is Carbon Structure? Concluding Comments

This chapter has described the origins of activated carbons. It is a long story, and can be linked directly the beginnings of planet Earth. The story includes the beginnings of life now known as flora and fauna (plants and animals). The development of the coal fields over at least 3×10^8 years and of the oil fields for even longer provide now, in the 21st century, the raw materials used for making isotropic and anisotropic carbons. The flora of today, woods, nutshells, etc. are also parent materials for active carbons. The origins of activated carbon are lost in prehistory, but this is always the case as can be found in metallurgy and medicine. Sufficient to say that the uses of activated carbons, as in those early days extending backwards for at least 5000 years from now, are not that different from their uses today, namely one of separation. What has changed, of course, is the extent of their use, at least 0.6 million tonnes per year worldwide, and the complexity of the many separation processes which constitute their applications.

The success and continuing use of activated carbons are a direct consequence of the uniqueness and versatility of activated carbons. To put it another way, there does not exist any other material which is such a powerful adsorbent and which, at the same time, can exhibit extreme specificity of application via variations in structure, these variations being relatively easy to bring about.

An activated carbon can be considered as molecular space (i.e. the space occupied by a molecule usually in the liquid phase), this space being contained within a three-dimensional network of carbon atoms arranged in layers composed of ring structures joined together somewhat imperfectly. This network is continuous in three dimensions with some layers being stacked, roughly parallel to each other, in groups of two, or three, probably not much more.

Such a unique and versatile material, naturally, has attracted the attention of a wide range of investigators, from water engineers, environmental groups, design engineers, to chemists, physicists, mathematicians, computer modelers, etc. There is a need to know the characteristics of a specific activated carbon which make it work so well. Proposed models, ranging from paper and pencil sketches to computer-designed suggestions, are compared and contrasted in Chapter 3. It is ironic to have to report that almost all of these models are some distance away from the reality of carbon porosity. And what is more, Chapter 3 concludes with models drawn for children, using simple software which can be down-loaded on to household computers, of a maze or labyrinth, which approximates closely, in two and three dimensions, to the characteristics of porosity in carbons.

Modeling has two main objectives. The first is an attempt to define the shape and size factors of the porosity; that is, the spaces of molecular size within the porous carbon. The second is to define the composition/structure of the carbonaceous pore wall material. A massive literature already exists discussing and defining porosity in immense detail. The second objective has still some way to go because of the difficulty of the task. High-resolution electron microscopy can look at pore walls edge-on, that is, with the electron beam passing parallel to pore walls. The elucidation of the detail of the arrangements of carbon atom bonding within the layer is still outstanding.

Appropriate words, to conclude this chapter, have been published from laboratories in France. Salver-Disma *et al.* (1999) demonstrated the versatility of carbon materials, in this case, a graphite, by converting it into microporous carbons. The conversion from porous carbon to graphite has never been achieved. The authors achieved this remarkable transition by selecting a polycrystalline graphite (F399) and subjecting it to mechanical grinding. Two types of grinding were used, namely a grinder that generates normal mechanical strain (shock interaction) and tangential mechanical strain (shear interaction). As reference carbons, a mesoporous non-graphitizable saccharose carbon (HTT 1000 and 2500 °C) and a graphitizable carbon (HTT 1000 °C) based on an anthracene coke were used. Effects of grinding were monitored by TEM and adsorption experiments.

The disordered (not amorphous) structure of the saccharose carbon (HTT1000 °C) was not affected by the grinding. The saccharose carbon (HTT 2500 °C) contained nano-sized textures of parallelism of layers (to call them graphitic is being a little too optimistic). The effect of shock type of grinding is to break these groups of layers, as shown in Figures 2.55 and 2.56. The graphitizable anthracene coke (HTT 1000 °C) has a lamellar structure,

Figure 2.55. Lattice-fringe images (from (*002*) diffraction band) for saccharose-based carbon (HTT 2500 °C) (Salver-Disma *et al.*, 1999).

but is not graphitic; that is, there is repetitive order between the layers (Figure 2.57). The effect of grinding (with associated mechanical pressure) was not to enhance the graphitizability as might have been expected, but to destroy the lamellar ordering. This was replaced by concentric units as in Figure 2.58 with the characteristics of microporosity. The graphite (commercial name F399), after being subject to a shear type of grinding, completely lost all traces of a graphitic structure, becoming indistinguishable from a microporous carbon, with a surface area of $500\,m^2\,g^{-1}$ (Figure 2.59).

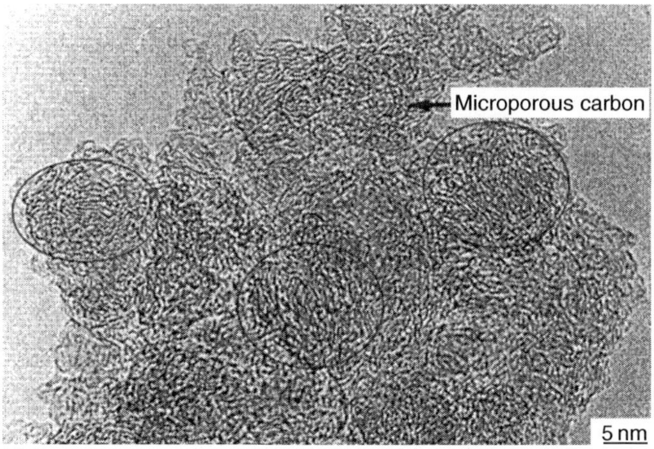

Figure 2.56. Lattice-fringe images (from (*002*) diffraction band) for saccharose-based carbon (HTT 2500 °C), ground using a shock-type interaction for 40 h (Salver-Disma *et al.*, 1999).

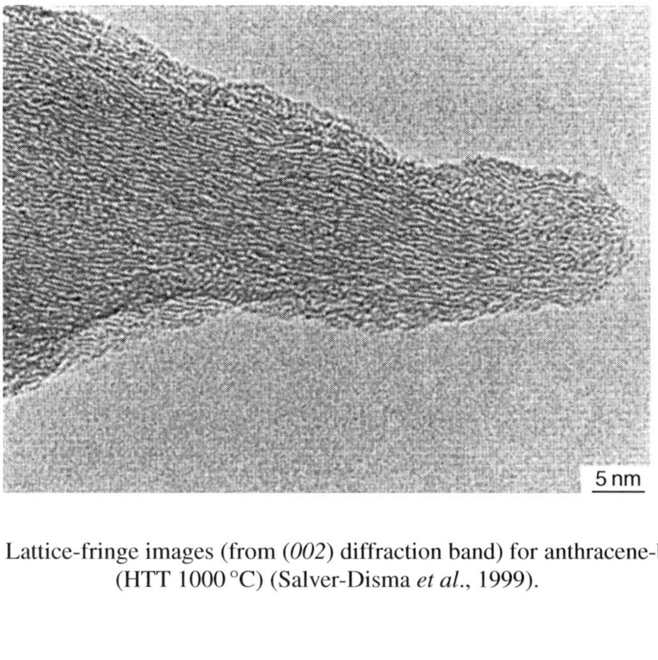

Figure 2.57. Lattice-fringe images (from (*002*) diffraction band) for anthracene-based carbon (HTT 1000 °C) (Salver-Disma *et al.*, 1999).

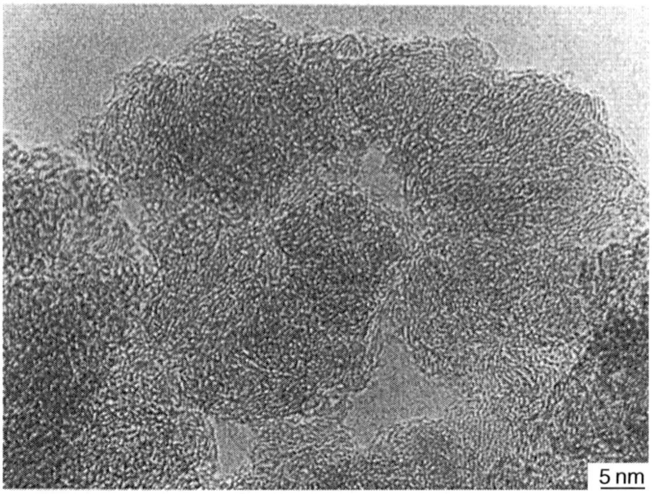

Figure 2.58. Lattice-fringe images (from (*002*) diffraction band) for anthracene-based carbon (HTT 1000 °C) after grinding in a shock-type interaction for 40 h (Salver-Disma *et al.*, 1999).

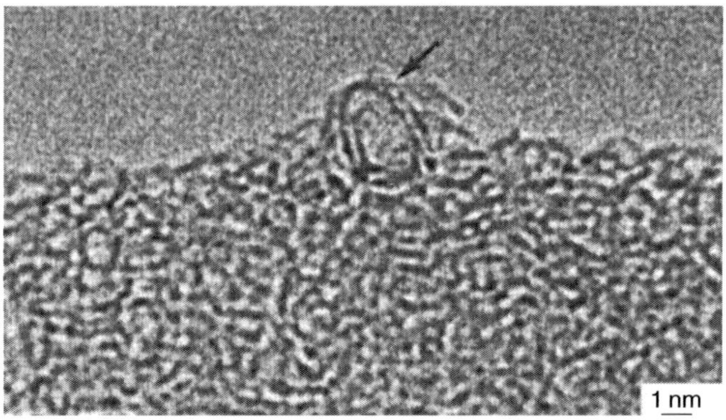

Figure 2.59. Lattice-fringe images (from (*002*) diffraction band) for graphite after grinding in a shear-type interaction for 40 h. The formation of microporous carbon is irreversible (Salver-Disma *et al.*, 1999).

Ii is not proposed that active carbons can be prepared from graphites on a commercial scale. This discussion does however indicate that if the graphite (graphene) layers are broken down to sufficiently small size, the edges are sufficiently reactive to combine chemically to form a network essentially similar to that of "traditional" activated carbon. Salver-Disma *et al.* (1999) in the course of their work, examining many TEM fringe images, observed that they did not find evidence for the *house of cards* model as postulated by Dahn *et al.* (1997) as discussed in Chapter 3, Section 3.4.8.

In Section 2.2, the suggestion was put forward that isotropic, non-graphitizable carbons might be considered as imperfect assemblies of segments (nanometer size) cut from a graphene layer using "atomic scissors". In essence, Salver-Disma et al. (1999) have done just this, not by using the refinement of atomic scissors but by using a brutal grinding technique. However, the results are the same.

To define the composition/structure of the carbonaceous pore wall material is a problem that will take a while to resolve. At the moment, there is no method offering direct visual imaging so interpretative techniques need to be used. It would appear that where interpretations of data are necessary, then there is always room for debate. The use of Raman spectroscopy, as indicated in Section 2.11, is a most promising tool. Beyssac *et al.* (2003) studied a wide range of carbons, from different origins, and gave suggestions for the way ahead. This article is thus appropriate to quote in the concluding pages of this chapter.

The carbons examined by Beyssac *et al.* (2003) included carbons from natural deposits (coals) with different geological histories as well as a saccharose-based carbon, anthracene-based cokes and an anthracite. The article reviews the applicability of Raman spectroscopy to the characterization of carbons bearing in mind the heterogeneity of structure which exists in such materials. Measurement of Raman spectra is not straightforward and sample preparation can be critical. The authors discuss the most accurate parameters which are available. About 90% of all spectra exhibit a D1-band position between 1345 and 1355 cm^{-1}. There is no significant evolution and the D1-band does not appear as a reliable parameter, its position being strongly dependent on the wavelength of the incident laser beam. The **G**-band position shows a larger range of values from 1575 to 1607 cm^{-1}. Raman spectroscopy provides information on the vibrational properties and is thus sensitive to carbon structure. For the natural samples, despite a large heterogeneity, information obtained at the 1 μm level are representative. For the synthetic carbons, these authors conclude that it is almost impossible to estimate an average degree of organization in such heterogeneous materials by using conventional Raman spectroscopy. The use of Raman area mode microspectroscopy is therefore a necessity.

Beyssac *et al.* (2003) are optimistic for the future of Raman spectroscopy making use, initially, of the D1:G1 ratios. The technique is complimentary to the use of TEM, as in Figures 2.55–2.59, the latter being the best method to image the sample structure and microtexture, but only semi-quantitatively, currently. It is Raman area microspectroscopy which examines the structure of carbon atoms within the graphene layers.

A "best guess" at the moment is that the defective micro-graphene layers is dominated by six-membered ring systems, under strain/stress which is accommodated by ring systems of other sizes.

As often happens, the world of commercial advertising can offer images of interest and relevance to those studying structures in solids. There is an example of this for the surface of activated carbon. The city of Alicante, Spain, celebrates the end of Winter and the beginning of a new Summer by 1 week of Fiestas, at the end of June, including massive fireworks displays and the burning of effigies (*Hogueras*). An advertisement for this event, *La Fiesta de Fogueres* including two *Bellesas* (women and girls in traditional costumes) is shown in Figures 2.60 and 2.61; the artist using a possible structure of a graphene layer (imperfect) as a background.

Figures 2.60 and 2.61. The artist who designed this advertisement for the *Barracas* of *La Fiesta de Fogueres* of Alicante, Spain, used as a background a multi-ring system as may be found in the graphene layers of an isotropic, non-graphitizable carbon (right). Antoni Gaudi I Cornet (1852–1926) may have been an influence with his ceramic mosaics.

2.14 Applications Related to Porous and Chemical Structure

2.14.1 The Uses of Porous Carbons

Microporous carbons have a *unique* structure which essentially is a three-dimensional network, or labyrinth of carbon atoms some in hexagonal arrangements and some as individual carbon atoms bonded together extremely closely but not close packed. This bonding arrangement results in space existing between the atoms to create an interconnecting three-dimensional passageway in which every space unit has a connection to all others within the carbon.

The dimensions of the passageway (width) are those of molecules, about 0.5 nm, such that any molecule which enters into this space is subject to intense dispersion forces from the carbon atoms (about eight per space) which make up the "surface" of the space. These forces can "trap" the molecule for periods of time much longer than on an open surface and so the phenomenon of physical adsorption of gases is generated.

The site, place or space where a molecule can be trapped (adsorbed) is called an adsorption site. These adsorption sites can be modified in terms of their size (widened or narrowed) and in terms of their chemical composition. The surface can be bonded to hydrogen, oxygen, chlorine, nitrogen, etc. to alter the polarity of the surface. Changing this polarity can enhance the adsorption process for polar molecules.

The porosity can be widened by gasification with water vapor, oxygen or carbon dioxide (hydrogen is also a gasifying agent for production of methane – effects on porosity appear not to have been examined). The parent materials may also be impregnated with zinc chloride, potassium hydroxide or phosphoric acid, these treatments improving adsorption characteristics, a process known as activation.

The sizes of the pores or the pore entrances can be finely adjusted to allow the passage of a smaller molecule, but not a larger molecule from a mixture of gases, a process known as molecular sieving. Unwanted pollutants in the atmosphere can be removed. By widening of the porosity, adsorption processes from solution (water treatment) can be carried out. In addition, catalyst particles can be supported within the porosity to promote required catalytic conversions. Surface oxygen complexes can promote the adsorption of polar molecules. When the latter are in aqueous solution, the polar water molecules can be adsorbed preferentially and may block entrances. Initial adjustments to pore dimensions are therefore needed. However, for activated carbon which is relatively free from surface oxygen complexes, and other surface polarities, the ability to adsorb preferentially from the atmosphere is not inhibited by the presence of water vapor.

Physically, these porous carbons can be available in granular form, as a powder and as a fiber, and can be made from nutshells, coals and woods as well as selected anisotropic carbons usually in fiber format, such as cloths and felts.

Regeneration of spent (used) carbons is part of the technology of the use of these carbons.

This very brief summary of the properties of activated carbons illustrates the versatility of these materials, this topic being expanded during the course of this book.

2.14.2 Take-up of Lithium into Carbons for Batteries

Worldwide and extensive investment in the carbon lithium-ion battery (LIB) has resulted in a plethora of publications to make this application an intensely studied form of carbon. It is also unique in the sense that the range of carbon structures studied for their suitability in lithium-ion batteries stretches from the most ordered of graphites (highly ordered pyrolytic graphites (HOPG)) to isotropic carbons of high surface area with a minimum of structural order. Such studies are relevant here because this complete range of carbons can be used to illustrate clearly the various porosities which exist within carbons as well as mechanisms of activation of carbons.

> *Before entering into the discussions of how lithium is incorporated into carbons, it has to mentioned, quite strongly, that the matter is not straightforward, one reason being that lithium is incorporated electrochemically, and as such, reaches into parts of carbon structures it might otherwise not reach.*

The interest in carbon for the LIB rests with its ability to "take-up" lithium to form a battery capable of being charged and discharged, reversibly and many times, with a capacity to make the battery commercially viable for modern electronic equipments, such as the mobile telephone. The most favored carbon material is graphitic carbon, both natural and synthetic. Here, it is considered that the lithium *intercalates* between the graphene layers in a reversible way and offers a maximum theoretical capacity of $372\,mAh\,g^{-1}$ corresponding to the intercalation compound C_6Li. However, carbons have been produced with extremely high capacities of about $950\,mAh\,g^{-1}$. This section examines how this may come about considering dominantly the mechanisms of uptake lithium by these carbons. Further details of the electrochemistry of these batteries are available (Flandrois and Simon, 1999; Endo *et al.*, 2000; Ko *et al.*, 2001; Béguin *et al.*, 2004).

Dahn *et al.* (1993), Xing *et al.* (1998) and Stevens and Dahn (2000a, b) conveniently summarized published capacities for a wide range of graphitizable and non-graphitizable

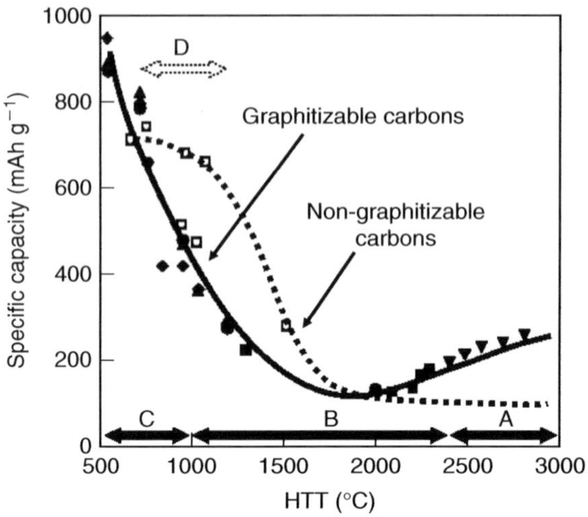

Figure 2.62. The variation of reversible specific capacity (mAh g^{-1}) with HTT (°C) for a series of graphitizable (solid line) and non-graphitizable carbons (dashed line) as used in lithium batteries (Dahn *et al.*, 1993; Liu *et al.*, 1996).

carbons plotting HTT (°C) of the carbons against the reversible specific capacity (mAh g^{-1}), this plot being Figure 2.62. Despite the major differences in structure between these two forms of carbon, their plots show major similarities, decreasing from 800–950 mAh g^{-1} at 500 °C to a minimum of about 150 mAh g^{-1} at about 2000 °C to increase to about 325 mAh g^{-1} at 2800 °C for the graphitizing carbons. Increasing HTT affects both the structure and porosity of these carbons.

For the graphitizable carbons (HTT 2800 °C), the mechanism of uptake by the lithium is clearly one of intercalation. Here, the lithium initially enters between every fourth layer (Stage 4), when this is complete between every third layer, etc. until the process is complete with the lithium occupying well-defined positions within the lattice which is unrestricted to adopt an AAAAAA configuration. This is called *staging* and the process is critically dependent on the graphitic carbon having the initial ABABAB configuration. With graphitizable carbons (not necessarily graphitic) of HTT about 2000 °C the specific capacity reaches a minimum, the electrochemical behavior indicating that lithium has intercalated but without staging; that is, the distribution of the lithium is random and is lesser in extent than when staging occurs. This in turn is an indication of the imperfect stacking of the graphene layers as existing at this HTT, itself a result of defects internally within the graphene layer and pinning between the layers so inhibiting the adoption of the AAAAAA-staged configuration.

The high capacities (950–450 mAh g^{-1}) of the graphitizing carbons, HTT 500–1000 °C, are more difficult to explain. Clearly, a mechanism of intercalation only is not possible. High-resolution TEM producing fringe images demonstrates that the parallelism of the graphene layers is far from perfect. Extensive bending/buckling is seen such that the packing density is low. There exist hexagonal arrangements of carbon atoms as suggested by Dumont *et al.* (2002) as well as carbon atoms which form a network of atoms, disordered and not completely space filling to leave residual porosity. Helium densities may not exceed about $1.8\,g\,cm^{-3}$ confirming these packing densities and the existence of a *residual porosity* (<0.5 nm dimensions) not accessible to the usual adsorbates, such as nitrogen and carbon dioxide. The uptake of lithium could be a combination of several phenomena including limited intercalation, the "adsorption" of lithium atoms, singly or in clusters, into this residual porosity as well as the attachment of ions to internal surfaces via electron transfer processes. There are ^{7}Li-NMR (nuclear magnetic resonance) studies which indicate the presence of such clusters (Nishizawa, 2003). With increasing HTT, the parallelism (stacking efficiency) of the imperfect graphene layers improves and extents of the residual porosity accordingly diminish so reducing the capacities to the minimum at an HTT of about 2000 °C.

The isotropic carbons are not that dissimilar in their electrochemical behavior. Their structure may have aspects somewhat similar to the graphitizable carbons. The same hexagonal groups of carbon atoms (not crystallites) as suggested by Dumont *et al.* (2002) could exist alongside the disorganized carbon atoms (not in rings). It can be considered as a variation on a theme and it would appear that the tighter bonding within the carbon network of the isotropic carbons (compared with the lighter bonding between layers in the anisotropic carbons) maintains a higher level of residual porosity. This maintains the higher capacity of the isotropic carbons. With increasing HTT to about 2800 °C, the isotropic carbons are unable to increase their capacities which remain at about $140\,mAh\,g^{-1}$. It is known that some porosity remains.

Stevens and Dahn (2000b), using an *in situ* electrochemical cell within small-angle X-ray scattering (SAXS) equipment, consider that they have indisputable evidence for the filling of porosity (probably <1.0 nm) by lithium or sodium at a potential near to that of the metallic lithium or sodium. At a higher cell voltage, the lithium or sodium is inserted between the graphene layers via an intercalation mechanism. Stevens and Dahn (2000a) also report success when using sodium and not lithium, having a sodium cell with a reversible capacity of $300\,mAh\,g^{-1}$.

The use of electron paramagnetic resonance (EPR or ESR) affords information about unpaired electrons within petroleum cokes. Zhecheva *et al.* (2002) found these in the cokes of HTT 480–800 °C (this is well-established information). These paramagnetic centeres correspond to aromatic π-radicals. For cokes of HTT 1000 °C, no EPR signal was detected. For cokes, with HTT 1400–2800 °C, the EPR signal originates from delocalized electrons. After the electrochemical reactions, for cokes of HTT <1000 °C, the EPR signal narrows with decreasing intensity caused by interactions between the localized centers and lithium or sodium. Further, a new narrow signal appears and assigned to lithium within the smallest of the microporosity. For the petroleum cokes of HTT >1000 °C, the number of delocalized electrons increases as a result of lithium intercalation.

Letellier *et al.* (2004) are the first to report experimentation in which an electrochemical cell is located within the magnet of NMR equipment, enabling the simultaneous collection of ^{7}Li-NMR and electrochemical data. The location of lithium, either intercalated or adsorbed (contained within cavities), was demonstrated.

An article by Ruike *et al.* (1994) summarized this question of *closed porosity* from the adsorption point of view. For decades, the use of SAXS has predicted surface area values for porous carbons in excess of values determined from interpretations of the adsorption isotherm. Overviews of SAXS techniques are available from Hoinkis (1997) and Nishikawa (2003). The SAXS technique measures all surfaces and the automatic deduction is that porous carbons contain surfaces access to which by a gaseous adsorbate is not possible (closed, inaccessible or hidden porosity are terms used). This is a correct deduction as far as traditional adsorbates go.

The use of helium density measurements now indicates that closed porosity is a relative term, closed to nitrogen but open to helium, and now there is the ultimate when porosity closed to helium is open to lithium.

The reason why the carbons of the highest specific reversible capacity are not suitable commercially is that as well as containing the smallest of micropores (<0.5 nm) they also contain mesoporosity (>20 nm diameter). The electrolyte of the cell unfortunately can enter into this mesoporosity and stays there so preventing the egress of the lithium. It is interesting to note in passing that the success of the lithium-ion battery is dependent on the existence and use of the so-called closed or hidden porosity which has little use in traditional adsorption applications.

The uniqueness and versatility of carbonaceous porous materials is demonstrated by Mukai *et al.* (2004) in their attempt to reduce the phenomenon of irreversibility of the LIB. As indicated above, irreversibility is associated with the formation of solid electrolyte films on surfaces of carbons by an irreversible reaction of lithium ions with the electrolytes. For the isotropic porous carbons (not amorphous carbons as quoted by Mukai *et al.*, 2004), the electrolyte film is formed preferentially in the entrances to the porosity (mainly microporosity). Should it be possible to prevent this deposition, then the irreversible component of battery performance could be reduced. It is established that increasing the heat treatment of carbons (normally beyond about 800 °C) decreases the pore dimensions, but at the same time there is reduction in volume of porosity which is available for lithium entry. Quite separately, Suzuki *et al.* (2003) report on the impossibility of bringing about a meaningful reduction in the irreversible component, maintaining the reversible component, by changing the porosity of the material. That is, an improvement automatically creates a deterioration. The use of an approach of carbon vapor deposition (as for pyrolytic carbons) has been tried whereby carbon is deposited in the entrances to the microporosity. There is no overall change to carbon structure. This method was successful but applications on an industrial scale are expensive.

Mukai *et al.* (2004) used a simple and ingenious approach to alleviate the problem. They added lithium carbonate, in carefully controlled amounts, to the parent material before carbonization. During carbonization, the lithium carbonate distills to the outer surfaces of the carbon accumulating in the entrances (or in this case the exits) of the microporosity. With this situation, the pore entrances are reduced sufficiently to reduce irreversibility significantly without altering in capacity within the micropore volume. A carbon based on sucrose was used as the parent material. Water was removed from the sucrose by heating to 453 K with a soak time of 10 h. Three methods of addition of lithium carbonate were used, namely (1) The lithium carbonate was added to the parent sucrose, (2) the lithium carbonate was added to the sucrose after an HTT at 453 K and (3) the lithium carbonate was added to the product after heating to 923 K. Samples were then heated 1273, 1373 or 1473 K. It was found that the mixing of lithium carbonate with the preheated sucrose not only reduced the irreversible component but also increased the reversible component. Thus, the lithium carbonate may not be an entirely inert additive but may have properties comparable to those of potassium hydroxide (Chapter 6: Activation Processes (Chemical)). The uniqueness and versatility of carbonaceous porous materials is thus demonstrated.

It is the porosity within activated carbons which imparts their dominant characteristics of adsorption. Activated carbons are prepared from a wide range of carbon types (the family of carbons) of different structures. As porosity and structure are intimately related, an understanding of the structures in this range of carbons together with a description of their origins and preparation are prerequisites to the effective use of activated carbons.

References

Baker FS. Highly microporous carbon derived from lignocellulose material. US Patent 5416056, 1998.

Baker FS, *et al. Activated Carbon. Kirk-Othmer Encyclopedia of Chemical Technology*, 4th Edition, Vol. 4. John Wiley & Sons Inc., New York, 1992, pp. 1016–1037.

Bandosz TJ, Biggs MJ, Gubbins KE, Hattori Y, Iiyama T, Kaneko K, Pikunic J, Thomson KT. Molecular models of porous carbons. In: *Chemistry and Physics of Carbon*, Vol. 28, Ed. Radovic LR. Marcel Dekker Inc., New York, 2003, pp. 42–228.

Bansal RC, Donnet J-B, Stoeckli F. *Active Carbon*, Marcel Dekker Inc., New York, 1988, p. 482.

Béguin F, Chevallier F, Vix C, Saadallah S, Rouzaud JN, Frackowiak E. A better understanding of the irreversible lithium insertion mechanisms in disordered carbons. *J Phys Chem Solids* 2004;65(2–3):211–217.

Bernal JD. The structure of graphite. *Proc Roy Soc* 1924;106A:749.

Beyssac O, Goffé B, Petitet JP, Froigneux E, Moreau M, Rouzaud JN. On the characterization of disordered and heterogeneous materials by Raman spectroscopy. *Spectrochimica Acta, Part A* 2003;59:2267–2276.

Biscoe J, Warren BE. An X-ray study of carbon black. Part I. Optical microscopy study. *J Appl Phys* 1940;11:1691–1705.

Biscoe J, Warren BE. An X-ray study of carbon black. *J Appl Phys* 1942;13:364–371.

Blayden HE, Gibson J, Riley HL. An X-ray study of the structure of coals, cokes and chars. In: *Proceedings Conference of Ultrafine Structure of Coals and Cokes*, BCURA, London, 1944, pp. 176–231.

Bonnamy S. Carbonization of various precursors. Effect of heating rate. Part II. Transmission electron microscopy and physicochemical studies. *Carbon* 1999;37(11):1707–1724.

Bourrat X. Structure in carbons and carbon artefacts. In: *Sciences of Carbon Materials*, Eds. Marsh H, Rodríguez-Reinoso F. Universidad de Alicante, Secretariado de Publications, Spain, 1997, 1–97.

Brooks JD, Taylor GH. The formation of some graphitizing carbons. In: *Chemistry and Physics of Carbon*, Vol. 4, Ed. Walker Jr PL. Marcel Dekker Inc., New York, 1968, pp. 243–285.

Bundy FP, Bassett WA, Weather MS, Hemley RJ, Mao HU, Goncharov AF. The pressure–temperature phase and transformation diagram for carbon, updated through 1994. *Carbon* 1996;34(2):141–153.

Burchell TD. *Carbon Materials for Advanced Technologies*, Pergamon, Amsterdam, 1999.

Cheremisinoff PA, Ellerbusch F. *Carbon Adsorption Handbook*, Ann Arbor Science Publishers Inc., Michigan, 1980, p. 1058.

Cuesta A, Dhamelincourt P, Laureyns J, Martinez-Alonso A, Tascon JMD. Raman microprobe studies on carbon materials. *Carbon* 1994;32(8):1523–1532.

Dahn JR, Sleigh AK, Shi H, Way BM, Weycanz WJ, Reimers JN, Zhong Q, von Sacken U. In: *Lithium Batteries New Materials and New Perspectives*, Ed. Pistoia G. Elsevier, Amsterdam, 1993, p. 1.

Dahn JR, Sleigh AK, Shi H, Reimers JN, Zhong Q, Way BM. Dependence of the electrochemical intercalation of lithium in carbons on the crysal structure of the carbon. *Electrochimica Acta* 1993;38(9):1179–1191.

Dahn JR, Xing W, Gao T. The "falling-card" for the structure of microporous carbons. *Carbon* 1997;35(6):825–830.

Derbyshire F, Jagtoyen M, Andrews R, Rao A, Martin-Gullón I, Gruike E. Carbon materials in environmental applications. In: *Chemistry and Physics of Carbon*, Vol. 27, Ed. Radovic LR. Marcel Dekker Inc., New York, 2001, pp. 21–66.

Donnet J-B, Voet A. *Carbon Black*. Marcel Dekker Inc., New York, 1976, p. 351.

Dresselhaus MS, Dresselhaus G, Eklund PC. *Science of Fullerenes and Nanotubes*. Academic Press, California, 1995, p. 66.

Dresselhaus MS, Dresselhaus G, Jorio A, Souza Filho AG, Saito R. Raman spectroscopy on isolated single wall carbon nanotubes. *Carbon* 2002;40(12):2043–2061.

Dumont M, Chollon G, Dourges MA, Pailler R, Bourrat X, Naslain R, Bruneel JL, Couzi M. Chemical, microstructural and thermal analyses of a naphthalene-derived mesophase pitch. *Carbon* 2002;40(9):1475–1486.

Endo M, Kim C, Nishimura K, Fujino T, Miyashita K. Recent developments of carbon materials for Li-ion batteries. *Carbon* 2000;38(2):183–197.

Ergun S. X-ray studies of carbon. In: *Chemistry and Physics of Carbon*, Vol. 3, Ed. Walker Jr PL. Marcel Dekker Inc., New York, 1967, pp. 211–288.

Ergun S. Structure in carbon. *Carbon* 1968;6(2):141–159.

Ergun S. X-ray scattering by very defective lattices. *Phys Rev* 1970;B1(8):3371–3380.

Fiala J. New structures in carbon-based materials. In: *Science of Carbon Materials*, Eds. Marsh H, Rodriguez-Reinoso F. Universidad de Alicante, Secretariado de Publications, Spain, 2000, pp. 99–115.

Fitzer E, Kochling K-H, Boehm HP, Marsh H. Recommended terminology for the description of carbon as a solid. *Pure Appl Chem* 1995;67(3):473–506.

Flandrois S, Simon B. Review: carbon materials for lithium-ion re-chargeable batteries. *Carbon* 1999;37(2):165–180.

Franklin RE. The interpretation of diffuse X-ray diagrams of carbon. *Acta Crystallogr* 1950;3:107–121.

Franklin RE. The structure of graphitic carbons. *Acta Crystallogr* 1951;4:252–261.

Fujimoto H. Theoretical X-ray scattering intensity of carbons with turbostratic stacking and AB stacking structures. *Carbon* 2003;41(8):1585–1592.

Granda M, Santamaria R, Menendez R. Coal-tar pitch, composition and pyrolysis behaviour. In: *Chemistry and Physics of Carbon*, Ed. Radovic LR. Marcel Dekker Inc., New York, 2003, pp. 263–330.

Harris PJF. Impact of the discovery of fullerenes on carbon science. In: *Chemistry and Physics of Carbon*, Vol. 28, Ed. Radovic LR. Marcel Dekker Inc., New York, 2003, pp. 1–39.

Heinenreich RD, Hess NM, Ban LL. A test objective for high-resolution electron microscopy. *Appl Crystallogr* 1968;1:1–19.

Hoinkis E. Small-angle scattering of neutrons and X-rays from carbons and graphites. In: *Chemistry and Physics of Carbon*, Vol. 25, Ed. Thrower PA. Marcel Dekker Inc., New York, 1997, 71–241.

Hull AW. A new method of X-ray crystal analysis. *Phys Rev* 1917;10:661–696.

Inagaki M, Kaneko K, Nishizawa T. Nanocarbons – recent research in Japan. *Carbon* 2004;42(8–9):1401–1417.

Iwashita N, Park CR, Fujimoto H, Shiraishi M, Inagaki M. Specification for a standard procedure of X-ray diffraction measurements on carbon materials. *Carbon* 2004;42(4):701–714.

Jagtoyen M, Derbyshire F, Rimmer S, Rathbone R. Relationship between reflectance and structure of high surface area carbons. *Fuel* 1995;74(4):610–614.

Jankowska H. *Active Carbon*. Ellis Horwood Limited, Chichester, England, 1991, p. 280.

Jüntgen H, Richter E, Kühl H. Catalytic activity of carbon catalysts for the reaction of NO_x with NH_3. *Fuel* 1988;67(6):775–780.

Kakihana M, Osada M. Raman spectroscopy as a characterization tool for carbon materials. In: *"Carbon Alloys": Novel Concepts to Develop Carbon Science and Technology*, Eds. Yusada E, Inagaki M, Kaneko K, Endo M, Oya A, Tanabe Y. Elsevier, Amsterdam, 2003, pp. 285–298.

Kekulé FA. Studies in aromatic compounds. *Annalen der Chemie* 1865;137:129–196 (in German).

Ko KS, Park CW, Yoon S-H, Oh SM. Preparation of Kevlar-derived carbon fibres and their anodic performances in Li-secondary batteries. *Carbon* 2001;39(11):1619–1625.

Lespade P, Marchand A, Couzi M, Cruege F. Characterisation de materiaux carbones par microspectrometrie Raman. *Carbon* 1984;22(4–5):375–385.

Letellier M, Chevallier F, Béguin F, Frackowiak E, Rouzaud JN. The first *in situ* ^{7}Li NMR study of the reversible lithium insertion mechanism in disorganised carbons. *J Phys Chem Solids* 2004;65:245–251.

Lewis IC, Singer LS. Electron spin resonance and the mechanism of carbonization. In: *Chemistry and Physics of Carbon*, Vol. 17, Eds. Thrower PA, Walker Jr PL. Marcel Dekker Inc., New York, 1981, pp. 1–88.

Liu Y, Xue JS, Zheng T, Dahn JR. Mechanism of lithium insertion in hard carbons prepared by pyrolysis of epoxy resins. *Carbon* 1996;34(2):193–200.

Lu M, Liu W-M, Guo X-Y, Li H-L. Coiled carbon nanotubes growth via reduced-pressure catalytic chemical vapour deposition. *Carbon* 2004a;42(4):805–811.

Lu Y, Zhu Z, Su D, Wang D, Liu Z, Schlögl R. Formation of bamboo-shape carbon nanotubes by controlled rapid decomposition of picric acid. *Carbon* 2004b;42(15):3199–3207.

Maire J, Méring J. Graphitization of soft carbons. In: *Chemistry and Physics of Carbon*, Vol. 6, Ed. Walker Jr PL. Marcel Dekker Inc., New York, 1970, pp. 125–190.

Manivannan A, Chirila M, Giles NC, Seehra MS. Microstructure, dangling bonds and impurities in activated carbons. *Carbon* 1999;37(11):1741–1747.

Marsh H, Wynne-Jones WFK. Surface properties of carbon: I. The effect of activated diffusion in the determination of surface area. *Carbon* 1964;1(3):269–279.

Marsh H. A tribute to Philip L Walker. *Carbon* 1991;29(6):703–704.

Marsh H, Heintz EA, Rodríguez-Reinoso F (Eds.). *Introduction to Carbon Technologies*, Universidad de Alicante, Secretariado de Publications, Spain, 1997, p. 669.

Marsh H, Martinez-Escandell M, Rodríguez-Reinoso F. Review: Semicokes from pitch pyrolyses: mechanisms and kinetics. *Carbon* 1999;37(3):363–390.

Marsh H, Rodríguez-Reinoso F (Eds.). *Science of Carbon Materials*. Universidad de Alicante, Secretariado de Publications, Spain, 2000, p. 673.

Marsh H. Carbon mesophase. *Encyclopedia of Materials: Science and Technology*. Elsevier Science Ltd, Amsterdam, 2001, pp. 926–932.

Mattson JS, Mark HB. *Activated Carbon*, Marcel Dekker Inc., New York, 1971, p. 237.

McGuire MJ, Suffet IH (Eds.). *Activated Carbon Adsorption of Organics from the Aqueous Phase*, Vol. 1. Ann Arbor Science Publishers Inc., Michigan, 1980a, p. 508.

McGuire MJ, Suffet IH (Eds.). *Activated Carbon Adsorption of Organics from the Aqueous Phase*, Vol. 2. Ann Arbor Science Publishers Inc., Michigan, 1980b, p. 589.

McGuire MJ, Suffet IH (Eds.). *Treatment of Water by Granular Activated Carbon, Advances in Chemistry Series, 202*. American Chemical Society, Washington, 1983, p 599.

Mochida I, Shimizu K, Korai Y, Otsuka H, Sakai Y, Fujiyama S. Preparation of mesophase pitch from aromatic hydrocarbons by aid of HF/BF_3. *Carbon* 1990;28(2–3):311–319.

Mochida I, Fujimoto K, Oyama T. Chemistry in the production and utilization of needle coke. In: *Chemistry and Physics of Carbon*, Vol. 24, Ed. Walker Jr PL. Marcel Dekker Inc., New York, 1994, pp. 111–212.

Mochida I, Korai Y, Ku CH, Watanabe F, Sakai Y. Chemistry of synthesis, structure, preparation and application of aromatic-derived mesophase pitch. *Carbon* 2000;38(2):305–328.

Mrozowski S. Specific heat anomalies and spin–spin interactions in carbons: A review. *J Low Temp Phys* 1979;35(3,4):231–298.

Mukai SR, Hasegawa T, Takagi M, Tamon H. Reduction of irreversible capacities of amorphous carbon materials for lithium ion battery anodes by Li_2CO_3 addition. *Carbon* 2004;42(4):837–842.

Nagle DC, Thrower PA. The importance of strain in crystallite size determinations. *Carbon* 1973;11(6):663–664.

Nishikawa K. Pore structure analyses of carbons by small-angle X-ray scattering. In: *"Carbon Alloys": Novel Concepts to Develop Carbon Science and Technology*, Eds. Yusada E, Inagaki M, Kaneko K, Endo M, Oya A, Tanabe Y. Elsevier, Amsterdam, 2003, pp. 175–188.

Nishizawa T. Basics of nuclear magnetic resonance and its application to carbon alloys. In: *"Carbon Alloys": Novel Concepts to Develop Carbon Science and Technology*, Eds. Yusada E, Inagaki M, Kaneko K, Endo M, Oya A, Tanabe Y. Elsevier, Amsterdam, 2003, pp. 299–318.

Oberlin A. High-resolution TEM studies of carbonization and graphitization. In: *Chemistry and Physics of Carbon*, Vol. 22, Ed. Thrower PA. Marcel Dekker Inc., New York, 1989, pp. 1–143.

Okuno H, Grivei E, Fabry F, Gruenberger TM, Gonzalez-Aguilar J, Palnichenko A, Fulcheri L, Probst N, Charlier J-C. Synthesis of carbon nanotubes and nano-necklaces by thermal plasma process. *Carbon* 2004;42(12–13):2543–2549.

Otvös Z, Onyestyák G, Valyon J, Kiricsi I, Kónya Z. Rees LVC. The dynamics of H_2 and N_2 sorption in carbon nanotubes. *App Patrick Surf Sci* 2004;238:73–76.

Patrick JW (Ed.). *Porosity in Carbons*. Edward Arnold, London, 1995, p. 331.

Pikunic J, Pellenq RJ-M, Thomson KT, Rouzaud J-N, Levitz P, Gubbins KE. Improved molecular models for porous carbons. *Stud Surf Sci Catal* 2001;132:647–652.

Prado-Burguete C, Linares-Solano A, Rodriguez-Reinoso F, Salinas-Martinez de Lecea. Effect of carbon support and mean Pt particle size on hydrogen chemisorption by carbon-supported Pt catalysts. *J Cat* 1991;128(2):397–404.

Qian W, Liu T, Wei F, Yuan H. Quantitative Raman characterization of the mixed samples of the single and multi-wall carbon nanotubes. *Carbon* 2003;41(9):1851–1856.

Rodríguez-Reinoso F. Production and applications of activated carbons. In: *Handbook of Porous Solids*, Eds. Schüth F, Sing KSW, Weitkamp J. Wiley-VCH, 2002, Gmbh, Weinham (Germany) pp. 1766–1827.

Rouquerol F, Rouquerol J, Sing K. *Adsorption by Powders and Porous Solids*. Academic Press, San Diego, 1999, p. 467.

Ruike M, Kasu T, Setoyama N, Suzuki T, Kaneko K. Inaccessible pore characterization of less-crystalline microporous solids. *J Phys Chem* 1994;98(38):9594–9600.

Ruland W. X-ray studies on the structure of graphitic carbons. *Acta Crystallogr* 1965;18:992–996.

Ruthven DM. *Principles of Adsorption and Adsorption Processes*. John Wiley & Sons Inc., New York, 1984.

Salver-Disma F, Tarascon JM, Clinard C, Rouzaud JN. Transmission electron microscopy studies on carbon materials prepared by mechanical milling. *Carbon* 1999;37(12):1941–1959.

Sano N, Kikuchi T, Wang H, Chhowalla M, Amaratunga GAJ. Carbon nanohorns hybridized with a metal-included nanocapsule. *Carbon* 2004;42(1):95–99.

Scherrer P. *Nachr, Gottingen Gesell* 1918;98.

Sebok EB, Taylor RL. *Carbon Blacks. Encyclopedia of Materials: Science and Technology*. Elsevier Science Ltd, Amsterdam, 2001, pp. 902–906.

Shinn JH. From coal to single-stage and two-stage products: a reactive model of coal structure. *Fuel* 1984;63(9):1187–1196.

Shiraishi M, Inagaki M. X-ray diffraction methods to study crystallite size and lattice constants of carbon materials. In: *"Carbon Alloys": Novel Concepts to Develop Carbon Science and Technology*, Eds. Yusada E, Inagaki M, Kaneko K, Endo M, Oya A, Tanabe Y. Elsevier, Amsterdam, 2003, pp. 161–173.

Short MA, Walker Jr PL. Measurement of interlayer spacings and crystallite sizes in turbostratic carbons. *Carbon* 1963;1(1):3–9.

Stach E, Mackowsky M-Th, Teichmüller M, Taylor GH, Chandra D, Teichmüller R. *Stach's Textbook of Coal Petrology*. Gebrüder Borntraeger, Berlin, 1982.

Stevens DA, Dahn JR. High capacity anode materials for rechargeable sodium-ion batteries. *J Electro Chem Soc* 2000a;147(4):1271–1273.

Stevens DA, Dahn JR. An *in situ* small-angle X-ray scattering study of sodium insertion into a nanoporous carbon anode material within an operating electrochemical cell. *J Electro Chem Soc* 2000b;147(12):4428–4431.

Suzuki T, Hasegawa T, Mukai SR, Hajimi T. A theoretical study on storage states of Li ions in carbon anodes of Li ion batteries using molecular orbital calculations. *Carbon* 2003;41(10):1933–1939.

Tanaike O, Inagaki M. Degradation of carbon materials by intercalation. *Carbon* 1999;37(11):1759–1769.

Taylor R. Carbon blacks: production, properties and applications. In: *Introduction to Carbon Technologies*, Eds. Marsh H, Heintz EA, Rodriguez-Reinoso F. Universidad de Alicante, Secretariado de Publications, 1997, pp. 167–210.

Toda Y. Densities of coals measured with various liquids. *Fuel* 1972;51(2):108–112.

Thrower PA. Editorial. *Carbon* 2004;42(4):699–700.

Wakayama H, Mizuno J, Fukushima Y, Nagano K, Fukunaga T, Mizutani U. Structural defects in mechanically ground graphite. *Carbon* 1999;37(6):947–952.

Warren BE. X-ray diffraction in random layer lattices. *Phys Rev* 1941;59:693–698.

Yang RT. *Adsorbents: Fundamentals and Applications*, John Wiley & Sons Inc., New Jersey, 2003.

Yasuda E, Inagaki M, Kaneko, Endo K, Oya A, Tanabe Y. Novel concepts to develop carbon science and technology. *Carbon Alloys*, Eds. Yusada E, Inagaki M, Kaneko K, Endo M, Oya A, Tanabe Y. Elsevier Science, Amsterdam, 2003.

Yudasaka M, Ichihashi T, Kasuya D, Kataura H, Iijima S. Structure changes of single-wall nanotubes and single-wall nanohorns caused by heat treatment. *Carbon* 2003;41(6):1273–1280.

Zhecheva E, Stoyanova R, Jiménez-Mateos JM, Alcántara R, Lavela P, Tirado JL. EPR study on petroleum cokes annealed at different temperatures and used in lithium and sodium batteries. *Carbon* 2002;40(13):2301–2306.

Xing W, Dunlap RA, Dahn. JR. Studies of lithium insertion in ball-milled sugar carbons. *J Electro Chem Soc* 1998;145(1):162–169.

CHAPTER 3

Porosity in Carbons: Modeling

Models designed to describe the microporous nature of carbons are compared and contrasted leading to an assessment of the requirements of a comprehensive model to account for the properties of microporous systems. No comprehensive model has, as yet, been created. The use of a model, based on a maze, provides insights. Another approach is the use of an informed imagination.

3.1 Introduction

Ruthven (1984) in his book *Principles of Adsorption and Adsorption Processes* states that activated carbon is made by the thermal decomposition of carbonaceous material followed by activation with carbon dioxide or steam, this process bringing about the removal of residual tarry material so opening pores. The activated carbon is said to consist of elementary crystallites of *graphite* stacked in random orientation, the porosity being the space between these *graphitic microcrystallites*. On the other hand, the American Society for Testing Materials (ASTM) is much more cautious and defines activated carbon as being a family of carbonaceous substances manufactured by processes that develop porosity. There is no mention of graphitic microcrystallites. A similar approach is adopted by Fitzer *et al.* for IUPAC (1995) where it is stated simply that an activated carbon is a porous carbon material, a char, which has been subjected to reaction with gases, sometimes with the addition of chemicals, e.g. $ZnCl_2$, before, during or after carbonization, in order to improve its adsorptive properties.

The literature which attempts to model structure in activated carbon and its consequential porosity is somewhat in disarray and is contradictory. The purpose of this chapter is an attempt to create a rational of this subject area. The concept of crystallites, microcrystallites, graphitic or otherwise, cannot be looked upon favorably and the use of this concept may not have been to the advantage of a realistic understanding of structure in activated carbon. In fact, it may have led to serious misunderstandings of structure in activated carbon. The complex nature of activated carbon is such that it requires the assessment and contrasting of many models, available in the one text, to assist in gaining an understanding, perhaps incomplete, of structure of porosity in carbons. The use of such words as honeycomb, sponge-like and labyrinth to describe porosity in activated carbon is quite appropriate, with care.

Chapter 2 has presented summaries of mechanisms of formation of carbons, the origins of microporosity, the structure of carbons, and the structure and composition of surfaces which make up microporosity. From these information, over recent years, many attempts have been made to create models, ranging from hand-drawn cartoons, to computer graphics and to computer simulations, all of which attempt to describe essential properties of these carbons. This is far from being an easy task because not only must the model meet the requirements of, say, the crystallographer but it must also meet the requirements of all the other disciplines (Table 3.1), in particular the surface chemist whose requirements are detailed and lengthy. In this respect, activated carbon is one of the most complex and unique of materials existing in the solid phase (there is no doubt about this).

An assessment of the literature indicates that some models meet the requirements of one discipline (e.g. crystallography) but which are quite unacceptable to the surface chemist

Table 3.1. Dominant properties of microporous carbons to be included in modeling.

1. The physical property of hardness.
2. The ultimate chemical analysis of contents (wt%) of hydrogen, carbon, nitrogen, sulfur with oxygen determined by difference.
3. The bulk, water, mercury and helium densities.
4. The existence of a porosity which is stable.
5. The ability to adsorb from gaseous and liquid phases.
6. The co-existence of micro- and mesoporosities.
7. The interconnectivity and tortuosity of networks of porosity.
8. Pore-size distribution.
9. The nature of an adsorption site; is it spherical, elliptical, slit shaped, tubular or whatever? An adsorption "site" is a volume element, or location, or place for adsorption.
10. The ability to predict the required ratio of carbon atoms (containing an adsorption site) to adsorbate molecules as determined from adsorption isotherms.
11. An ability to separate gases from mixtures of gases, that is to have molecular-sieve properties which are a function of the size and shape of the adsorbate gases and of the porosity.
12. The kinetics (dynamics) of the adsorption process which includes activated diffusion effects.
13. The energetics of the adsorption process including enthalpies and entropies of adsorption, and enthalpies of immersion (wetting).
14. The effect of HTT of carbons on isotherm shape, enthalpies of adsorption and adsorption capacities.
15. The ability of the carbon to act as a catalyst and as a catalyst support material.
16. The isotropic nature and non-graphitizability of microporous carbons.
17. The processes (physical and chemical) of activation of carbons.
18. Structure as revealed by XRD studies.
19. Structure as revealed by small-angle scattering of X-rays (SAXS) and of neutrons (SANS).
20. Diffuse reflectance infrared spectroscopy of surface complexes.
21. Raman microspectroscopy.
22. ESR spectroscopy.
23. X-ray photoelectron spectroscopy (XPS).
24. Electronic properties in particular as used in the lithium-ion battery.
25. To provide a facility for theoreticians to calculate adsorption potentials of adsorption sites.

(for example) who has to explain the many and varied properties of activated carbon from an adsorption point of view. A multi-discipline approach by the modeler is necessary to validate a model, not an easy task. It would appear that there are as many models as there are modelers. This chapter discusses several of these models pointing out (as with all models) their usefulness and their limitations.

3.2 Model Requirements

Carbon materials exhibit a wide range of properties and any modeling, which is undertaken, ultimately has to be able to explain and describe these various properties. Table 3.1 is a list of those dominant properties of isotropic microporous carbons that a model has to be able to satisfy, all of them, without exception.

> *Of immediate importance, a model must be such that its structure can be seen to originate from its parent material(s) by mechanisms which are understood and acceptable.*

3.3 Why Modeling?

Porosity in carbons essentially is molecular space, that is, it is space with the dimensions of atoms and molecules. Such porosity is not easy to visualize, in particular its heterogeneity of size, shape and capacity for adsorption (adsorption potential). Activated carbon is thus quite unlike the homogeneous porosity of the crystalline zeolites. The dispersion forces, which are contained within the porosity, are probably the most powerful within the subject area of surface chemistry. Molecular space is the space within the network of carbon atoms and between the carbon atoms. As such, this molecular space must reflect, totally, the porosities relative to each other, of the carbon atoms which constitute the surface of this space. An initial task is to model the structure of the carbon network and, from this structural model, the structure of porosity will automatically be created. The surface chemist, studying adsorptions by carbons, needs to know the nature of the surfaces which are being studied. Molecular space can only be considered as porosity when it is created by groups of carbon atoms in extremely close proximity to each other. An "open surface" is of quite limited interest from the point of view of adsorption processes.

Porosity is created during the carbonization process of an organic macromolecular precursor (such as cellulose and lignin). The removal of small molecules (the volatiles) brings about "distortions" within the structure of the original precursor at the same time as hydrogen and oxygen atoms are eliminated. The resultant "skeletal" structures become extremely unstable and carbon atoms within these structures will combine with other near-neighbor-carbon atoms to create positions of maximum stability (minimum energy) as far as the available stereochemistry will allow.

Structure is a function of heat treatment temperature (HTT) of the carbon and represents positions of quasi-equilibria. An increase in HTT moves the structure to a position of greater stability. A picture which emerged from the survey of Chapter 2 is of formation of rings of carbon atoms, five-, six- or seven-membered rings (or more) in non-planar layers, with some surfaces having six-membered ring systems and other areas being defective, relatively. Such non-planar layers do not exist in isolation but are part of a continuously interconnected three-dimensional network of such non-planar layers, some of the bonding (cross-linking) being of an sp^3 nature, single carbon atoms or linear atoms. This cross-linking maintains a maximum proximity to each other of the non-planar layers and at the same time imparts the physical hardness which is a characteristic of many microporous carbons (but not for KOH activated carbons).

This concept of a network of bonded together, non-planar layers of carbon atoms (with some single or linearly bonded carbon atoms) constituting the entire carbon specimen is fundamental to understanding porosity in such microporous carbons. This concept then demands that the molecular space (porosity) must equally constitute part of the entire carbon specimen. That is, there exists one continuous network of porosity and it is within this one porous network that there exist the micro- and mesoporosities. Macroporosity, >50 nm, is considered as being open surface, and assists only with transport of adsorptive into the interior of carbon particles, where adsorption takes place mainly in the micropores. There are implications to such a concept. A first consideration is whether or not distinct porous networks exist solely made up of micropores (0.3–2.0 nm) and of mesopores (2.0–50 nm). A second consideration is that an adsorption site, which behaves as a micropore could be adjacent to a larger molecular space which behaves as a mesopore, this occurring throughout the entire porous network (molecular space).

The term *site* is introduced to describe that volume element where one adsorbate molecule may reside. The term is also used in chemisorption studies to describe the location of a chemisorbed molecule, as for example, the site of a surface oxygen complex. There is a general acceptance that, for example, a micropore will accommodate many adsorbate molecules. Here, each adsorbate molecule occupies a specific (unique) adsorption site within the micropore. Basically, the concept of a *site* is identical for both chemisorption and physical adsorption, a difference being the residence times of the adsorbed molecule. With chemisorption, bond formation stabilizes the chemisorbed system: in physical adsorption, it is the adsorption potential at the location of the site which is the effective control.

Modeling in three dimensions requires computational treatments and, as yet, attempts to create a single model, which can explain everything, have not as yet been realized. These computational models are, however, becoming quite close to realism of the structures of porosity within activated carbon. It is acknowledged that it is difficult for a computer-orientated worker to be fully conversant with the detailed requirements of a surface chemist and vice versa. So, a procedure adopted here is to present the several models available in the literature and by explaining their advantages and disadvantages, to gain an integrated concept of the carbon atom network and of porosity (molecular space). Initially, as said above, some of the computer-generated models are very realistic but have some limitations which are explained. The various models available are discussed separately each receiving positive comment and negative comment (what the authors feel is missing from the models).

Sections 3.4.1–3.4.6 are of models used before the time of computers and from an historical point of view indicate the development of concepts. Sections 3.4.7–3.4.18 summarize computer modeling in recent years acknowledging the several approaches used and arriving at what can be considered to be acceptable models. Sections 3.4.19–3.4.21 are electron micrographs assisting structural analyses. Section 3.4.22 contains hand-drawn models to indicate the cage-like nature of porosity and its interconnectivity. Section 3.4.23 is an introduction to studies which use models to simulate experimental isotherms with Sections 3.4.24–3.4.27 discussing further modeling of porosity.

Restraints have to be introduced to the modeling programs to prevent structures being suggested which could not have reality. Density function theory (DFT) is used to equate calculated densities with known densities. Also, the use of small-angle X-ray scattering (SAXS), of small-angle neutron scattering (SANS) and electron diffraction to calculate radial distributions of carbon atoms (probabilities of pairs of carbon atom distances) associated with minimum energy designs have resulted in significant advances in the modeling of structures in carbon atom networks, and hence of the porosity networks and hence of composition of the surfaces (walls) of porosity, and hence the nature of adsorption sites.

3.4 Models to be Considered

3.4.1 The Drill-hole Models, Figure 3.1(a, b)

These models are about as simple a model as is possible to draw and consist essentially of filling a solid body with holes using an electric drill to make the holes which are circular (Figure 3.1(a) or slit shaped, Figure 3.1(b)). These can be in three dimensions and can be interconnected. They can be considered as a starting point and give some idea of internal accessibility and tortuosity.

However, these models have limitations in that they make no attempt to relate the shape and structure of the drill-holes (porosity) to the structure of the solid into which the holes are drilled. In reality, the drill holes need to have the dimensions of molecules. In activated carbon, most of the carbon atoms contribute to the phenomenon of adsorption. These models fail to meet all of the requirements of Table 3.1. But, a start has been made with the idea of holes in a solid.

3.4.2 The Branched-tree Model, Figure 3.2(a, b)

This is a somewhat more advanced model and demonstrates the multiplicity of size of porosity that may exist within a porous carbon from micro to macroporosity (Figure 3.2). It suggests the possibility of transport of material (usually an adsorbate gas) through the widest of pores (the trunk) to the narrowest of pores (branches and twigs).

Unfortunately, the model has limitations in that it does not follow that microporosity has to lead from mesoporosity. By being drawn in this way, it provides for misconceptions concerning the structure of the molecular space and of the carbon network. It also gives the impression that parts of the carbon possess no porosity, that is, those parts of the model (photograph) which do not contain branches. They fail to meet all of the requirements of Table 3.1. But, the model does indicate the concept of transportation and of interconnectivity.

3.4.3 The Norit Model, Figure 3.3(a, b)

This model gives the impression that porous carbons are made up of isolated quite planar (maybe graphitic) layers packed closely together and that the space in between these

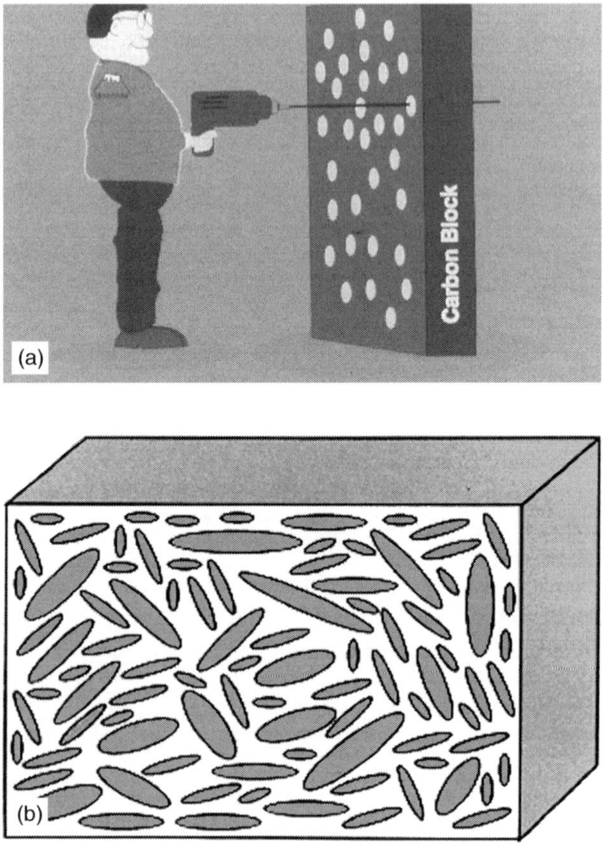

Figure 3.1. A model of porosity in activated carbon as being (a) isometric and (b) non-isometric drill holes in a solid.

Figure 3.2. The branched-tree model (a) direct and (b) inverted to illustrate connected types of porosity.

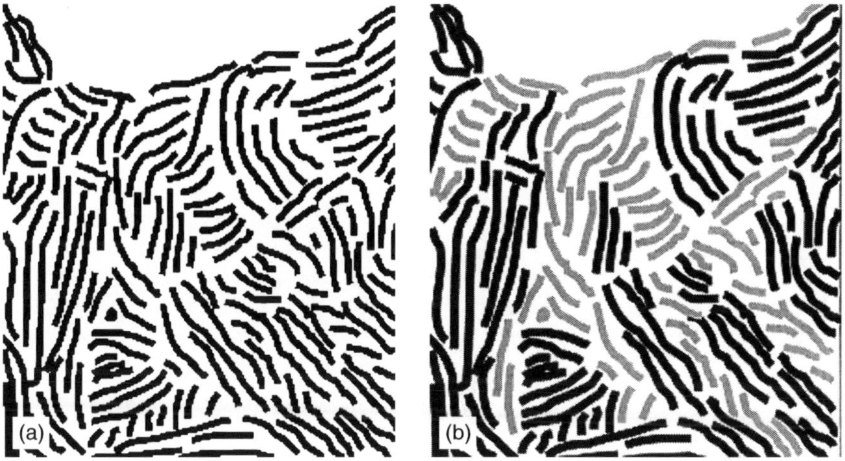

Figure. 3.3. (a) A Norit model indicating a structure in activated carbon made up of layers of carbon atoms (defective graphene layers) stacked so as to create porosity between them. (b) The process of activation removes totally, or parts of, these layers.

layers is the porosity (Figure 3.3(a)). This is correct as far as it goes. The model can be extended to indicate that the process of physical activation (Chapter 5) is the removal of all or part of these layers to create larger-sized spaces, microporosity (<2.0 nm) extending to mesoporosity (2–50 nm) between the remaining layers (Figure 3.3(b)). This is correct as far as it goes.

This model, however, gives no indication of arrangement of carbon atoms within a layer, defective or otherwise. The model does not indicate whether there is any cross-linking between the layers and also suggests a uniformity of size of molecular space (porosity). The concept that the layers have perfect planarity is also totally misleading. The process of physical activation, that is of enhancing the number of pores and their dimensions, by reaction with oxidizing gases, may not involve the removal of entire identifiable layers. The model raises the problem of explaining why one layer should be more chemically reactive than other layers. The structural analytical data (Table 3.1), using any diffraction technique, would not support such a homogeneous microcrystalline structure. However, the concept of selective removal of carbon atoms during the activation processes (see Chapter 5) is introduced.

3.4.4 Carbon from Benzene Hexachloride, Gibson et al. *(1946) and Riley (1947), Figure 3.4*

Gibson *et al.* (1946) and Riley (1947) were early researchers into the structure of carbons and coals. They attempted to make a graphitic carbon by reacting benzene hexachloride (C_6Cl_6) with molten sodium metal (a spectacular reaction). The idea was that the sodium would strip the chlorine from the benzene rings which would then combine together in a way which would create a graphene layer and then graphite. Unfortunately, the experiment

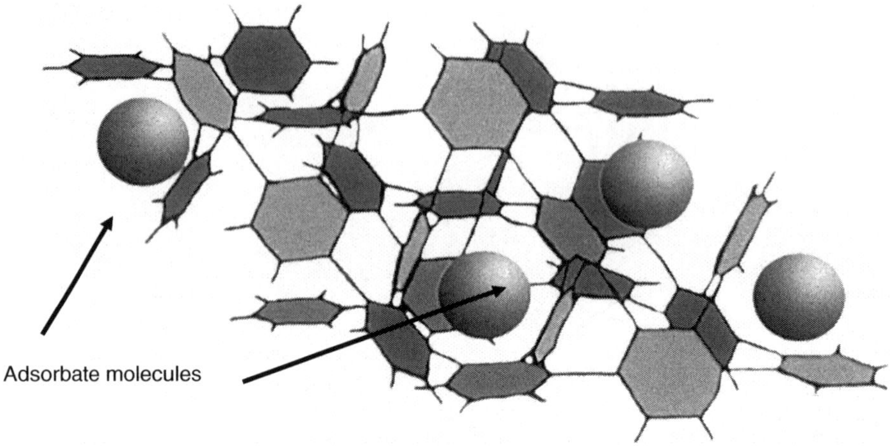

Figure 3.4. The polyphenylene model based on the concepts of Gibson *et al.* (1946) and Riley (1947).

did not work that way and the carbonaceous product was an extremely light (low-density) flocculent type of carbon which gave an extremely broad 002 diffraction band (X-ray diffraction, XRD) indicating a minimum of coherent diffraction. If adsorption techniques had been available at that time, this carbon would have exhibited a high adsorption capacity (Figure 3.4).

The model these authors proposed for this flocculent carbon is as Figure 3.4 and has some relevant aspects to current understandings in that the requirement of a "cage-like" structure is present. The model certainly describes a carbon of low density. In its own way, the model describes the close intimacy that exists between the structures within activated carbon and adsorption capacity. The model meets the requirement (No. 10) of Table 3.1. This is one of the first models of carbon structure ever to be published and makes use of small metal rods, sheets of a plastic material and some glue, to create the concept (incorrect) of an adsorption site being a cage without any means of ingress and egress for the adsorptive molecule. The detail of the carbon structure may be incorrect, but the fundamental concept is correct. In fact, if the polyphenylene unit is replaced by a defective micrographene layer, then the model becomes acceptable.

It is extremely doubtful, however, if such an ordered array of polyphenylene rings would be formed in such a violent reaction. The experiments appear not to have been repeated in later years and these carbons have not been examined by modern techniques. An important point to note is that the graphene layers (imperfect or otherwise) are not formed in such reactions nor are they formed in carbonizations of organic macromolecular material (e.g. coals and wood) by a process involving clusters of carbon atoms. The model of Gibson *et al.* (1946) and Riley (1947) possesses considerable strain energy and would never have formed with such extensive repetitive. The model was designed to elaborate upon possible carbon structures with little relevance to structure of porosity in carbons, but it has turned out to be useful toward an understanding of origins of porosity in carbons. The proposed model structure has never been confirmed. So near and yet so far.

Figure 3.5. (a) Photograph of potato chips to model the appearance of defective graphene layers of a surface of an activated carbon, to an adsorbate molecule and (b) slit-like possibility of shape of an adsorption site, as created between defective graphene layers.

3.4.5 Potato-chip Models, Figure 3.5(a, b)

These simple models (almost domestic) demonstrate how the shapes of spaces (porosity) between the potato chips (defective micro-graphene layers) are a direct function of the shapes of the potato chips themselves, that is porosity is dependent on the structure of the solid phase. The models illustrate that because no single potato chip is the same as a second potato chip then every molecular space (adsorption site for an adsorbate molecule) must differ from all others. The non-planarity of the potato chip is demonstrated as well as the fact that the surface of each potato chip is not homogeneous.

The irregular packing together of these potato chips permits the visualization of how the microporosities are generated as variations in distances between the layers (wedge-shaped porosity). It can also be appreciated from this model that if another brand of potato chips was used, of different shapes and sizes, then the porous characteristics would change and a different microporous carbon would be created. However, the principles governing the relationship between structure and porosity would be unchanged. Another relevant aspect to be seen in this model is that some of the defective micro-graphene layers present their surfaces (basal planes in graphite) while others present the edges of layers (prismatic edges of graphite).

With some imagination, the events of an adsorption process can be visualized. Figure 3.5(a) indicates that not all collisions of an adsorbate molecule with the carbon surface would lead automatically to adsorption. There could be elastic collisions on the surface, or mobility over the surface until a pore space is found, or there could be a direct entry of an adsorbate molecule into the porosity. Figure 3.5(b) is another perspective of the model and illustrates that if the carbon atoms were arranged more in layers than anything else, this would lead automatically to slit-shaped pores. These models allow for considerable development of the imagination.

As with all models, only some of the important properties of activated carbon are demonstrated; there are however limitations. The models do not indicate that structure in carbons is internally bonded together to give a hard material. In addition, there is probably too much uniformity with the potato chips. An improvement is to use a mixture of sizes of potato chips made by breaking some and mixing the different sizes together. Of course, the carbon chemistry is missing from the models but progress is made toward the demonstration of some important aspects of porosity in carbon.

3.4.6 Models of Kaneko et al. *(1992a, b), Figure 3.6(a–c)*

Kaneko *et al.* (1992a, b) used the graphitic microcrystallite theory of carbon structure, as introduced and commented upon in Chapter 2, to explain some of the adsorption properties of activated carbon. As will be discussed further in Chapter 4, some carbons exhibit apparent surface area in excess of $2000\,m^2\,g^{-1}$. Kaneko *et al.* (1992a) based calculations of surface areas on assemblies of graphitic microcrystallites. Figure 3.6(a) shows their calculations of the variation of surface area ($m^2\,g^{-1}$) of assemblies of graphite crystallites with size (l in nm) and thickness of the graphite crystallite wall (t in nm).

The structures of possible graphitic assemblies are in Figure 3.6(b) ranging in size from 56 to 212 carbon atoms (samples A–E). For a single molecule, adsorbing on both sides, such an assembly of graphite crystallites, made up of 56 carbon atoms, would generate a surface area (including adsorption on edges) of $5800\,m^2\,g^{-1}$ ($5.8\,km^2\,g^{-1}$). (*Note*: These surface areas are calculated geometric areas, and are not derived from adsorption data.) For an infinite graphite sheet (graphene layer) the calculated area for adsorption, on both sides, is $2630\,m^2\,g^{-1}$ (see Table 3.2).

Equation (3.1) was developed to relate surface area to the dimensions of the graphitic microcrystallite, which has the form:

$$a = \frac{2\{nt + w(n-1)\} + 4t}{\{nt + w(n-1)\}\rho t} \tag{3.1}$$

where a is the surface area, n is the number of graphitic crystallites in the model, t is the thickness of the micropore wall, w is the micropore width and ρ is the density of the graphitic microcrystallite. The variation of surface area with thickness of the walls of the graphitic assemblies is shown in Figure 3.6(a) showing that surface areas as high as $3600\,m^2\,g^{-1}$ are possible according to these calculations.

Kaneko *et al.* (1992a) make the case that these considerations can be used to explain the high surface areas of activated carbon. A three-layered structure of molecule type B can possess a surface area $>3000\,m^2\,g^{-1}$ assuming that adsorption occurs at the edges as well as on the surface and that the theoretical limitation of $2630\,m^2\,g^{-1}$ for a single layer of graphite need not be restrictive. (This argument is somewhat spurious.)

What Kaneko *et al.* (1992a) are doing is to make use of the exponential rise in surface area which occurs with diminution of particle size (comminution) as is found with the carbon blacks. Here, the large surface areas are not associated with internal porosity but with external surfaces of small spheres of carbon black with diameters of nanometers.

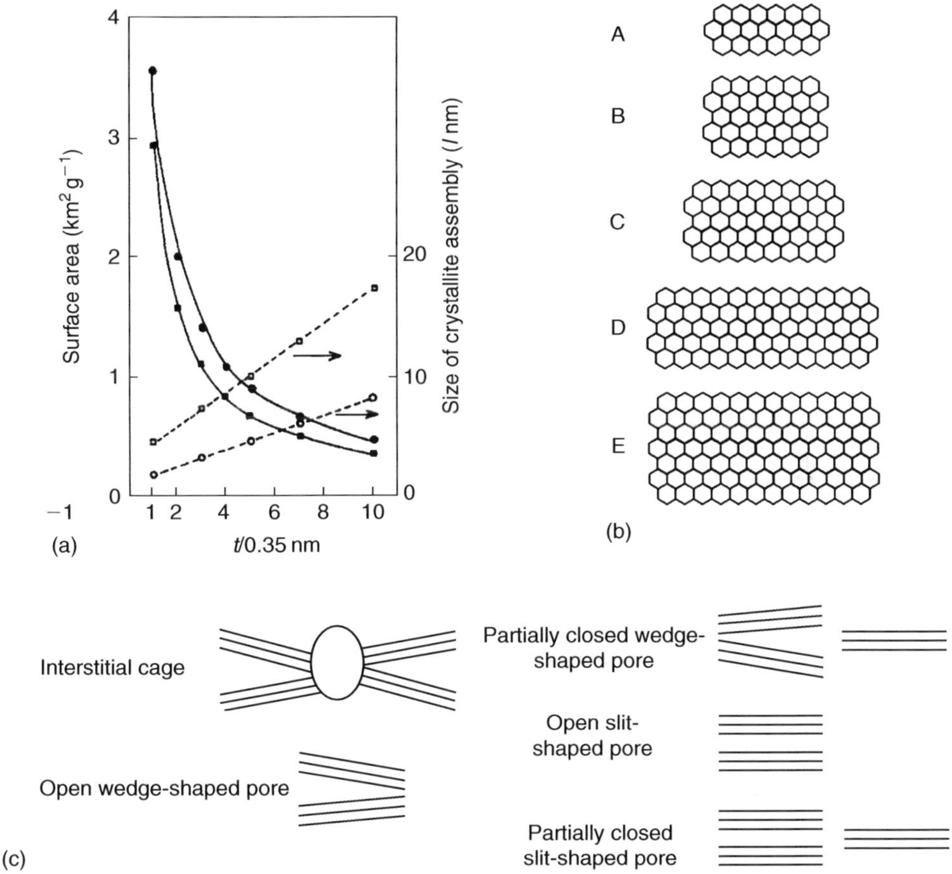

Figure 3.6. (a) Variation of surface area ($km^2 g^{-1}$) of assemblies of graphite crystallites with size (*l* in nm) and thickness of the graphite crystallite wall (*t* in nm) (Kaneko *et al.*, 1992a, b). *Note*: $1\,km^2 = 1000\,m^2$. (b) Molecules made up of carbon polyhexagons used to calculate surface areas (Kaneko *et al.*, 1992a). (c) Structures allocated for the various type of pore considered to exist in activated carbon (Kaneko *et al.*, 1992a).

Table 3.2. Surface areas of carbons ($m^2 g^{-1}$) composed of graphitic crystallites of various sizes.

Polyhexagonal molecule	Number of carbon atoms	Area of molecule (nm^2)	Surface area of carbon using one molecule ($m^2 g^{-1}$)	Surface area of carbon using three molecule ($m^2 g^{-1}$)
A	56	1.1×2.1	5800	3200
B	71	1.5×2.1	6000	3100
C	110	1.5×2.6	4700	2400
D	158	1.5×3.5	4400	2200
E	212	1.9×3.5	4100	1900

The surface area of a single graphite layer is $2630\,m^2 g^{-1}$.
Source: Kaneko *et al.* (1992a).

Kaneko *et al.* (1992b) further develop the use of the graphitic microcrystallite to explain the structures of different types of porosity which occur in carbons. Figure 3.6(c) describes how these authors envisage (a) an interstitial cage, (b) an open wedge-shaped pore, (c) a partially closed wedge-shaped pore, (d) a slit-shaped pore and (e) a partially closed slit-shaped pore. The graphitic crystallites are considered to be bonded together by sp^3 carbon atom linkages.

These calculations and considerations are of course correct in one way, but the question immediately arises as to whether or not such structures do actually exist in carbons; that is, is there a reality to this? The direct evidence (not models based on interpretations of, for example, XRD data) available indicates that such structures do not exist in carbons. There is no mechanism in the carbonization processes, at comparatively low temperatures of $<800\,°C$ for the creation of such structures. The discussions of Chapter 2 for formations of both isotropic and anisotropic carbons make this quite clear.

Thus, the dominant requirement of a model (Section 3.2) of being able to establish a mechanism of formation of the structure is not met. It is not conceivable that such large polycyclic aromatic molecules can be created within and exist, with stability, within activated carbon made, for example from the macromolecular lignin-based composition of coconut shell. The internal movement (internal recrystallization) of carbon atoms, requiring the breakage of all of the C—C bonds (one of the strongest bonds known in chemistry) could never occur. Further, the limited availability of sp^3-bonded carbon atoms, bonding the graphitic crystallites together, could never generate the mechanical strength of activated carbon.

However, the model indicates that surface areas of $>1000\,m^2\,g^{-1}$ and greater must be associated with adsorption processes in positions of molecular dimensions.

Apart from the rather unusual phenomenon of catalytic graphitization, graphitic structures never appear at temperatures below about 2500 °C. The crystallite theory for carbon structure has hindered a realistic understanding of what makes an activated carbon for too long a time.

3.4.7 Model of Ruike et al. *(1994), Figure 3.7(a, b)*

Ruike *et al.* (1994) studied microporous carbons using the techniques of XRD, small-angle X-ray scattering (SAXS), adsorption of nitrogen and density measurements. These authors assume that the XRD provides information on the *micro-graphitic* structure of the micropore walls of the micropores of activated carbon, that is the ordered solid parts taking their model from the previous articles by Kaneko *et al.* (1992a). They further assumed that SAXS leads to an average structural unit of pores and solids including both ordered *micro-graphites* and disordered *inter-micro-graphitic* parts. Referring to Figure 3.7(a), the XRD provided information of the size of the micro-graphitic assembly (say the top left-hand side group of three layers) while the SAXS provided information of the sizes of the assemblies of micro-graphitic parts, together with interconnecting carbon structures, that is, all of Figure 3.7(a). It was suggested that activation of the microporous carbon removed the interconnecting carbon separating the micro-graphitic parts followed by oxidation of carbon atoms at the edges of the micro-graphitic parts.

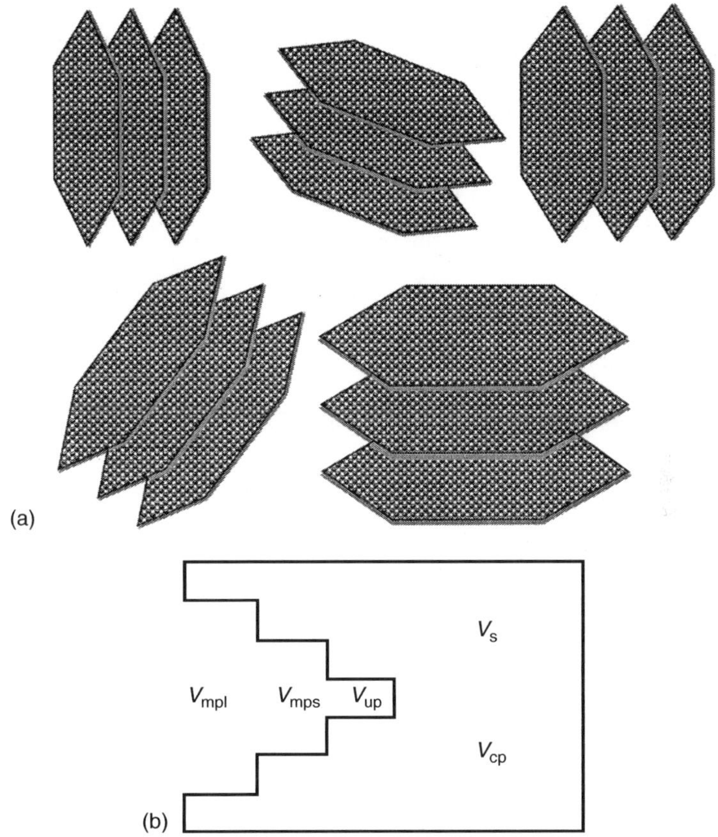

Figure 3.7. (a) The modeling of structure (based on the microcrystallite theory) and activation of microporous carbons (Ruike *et al.*, 1994). (b) The modeling of types of porosity in activated carbon, where V_s, V_{mpl}, V_{mps}, V_{up} and V_{cp} are volumes occupied by solid carbon, larger micropores, smaller micropores, ultrapores (not defined) and closed pores, respectively (Ruike *et al.*, 1994).

Density data were used to develop a new concept of inaccessible pores which were to include closed pores and pores accessible to helium at 303 K but not to nitrogen at 77 K. Surface areas as determined by SAXS were higher than surface areas determined from the nitrogen isotherm. This latter result is to be anticipated as SAXS measures the areas of interfaces between volume elements of high and low electron density, that is all porosity without relation to accessibility of any adsorbate or density medium which is used. The density data are used to create the description of porosity as shown in Figure 3.7(b) and which includes a volume (V_s) of solid carbon.

This model is firmly based on the acceptance of graphitic microcrystallites within the structure of these microporous carbons, as are the models of Kaneko *et al.* (1992a, b). The case against the acceptance of this type of model is as explained in Section 3.4.6. The conclusions are based on one interpretation of the diffraction data. They make no mention of the nature of the interconnecting carbon and assume (without saying so directly) that it has no role to play in the adsorption properties of activated carbon. The nature of *solid carbon* is not elaborated upon (only single-crystal graphite can be said to have no porosity) nor are the origins of closed porosity discussed.

3.4.8 Falling Card Model of Dahn et al. *(1997), Figure 3.8(a–c)*

The work of Dahn *et al.* (1997) is concerned with explaining the carbonization behavior of sugar, starch and oak wood. These authors needed to describe structure in such carbons and their dependent adsorption capacity and the effects of increasing HTT. These authors envisaged their structures in terms of carbon density and the existence of graphene platelets of different diameters, their alignment and the distribution of diameters of platelets, and possible curvature within the platelets. Pore sizes within the carbons were determined from SAXS and changes in porosity observed with increasing HTT of the carbons. Dahn *et al.* (1997) developed their falling card model (Figure 3.8(a–c)) to explain these changes which involved the reorganization of these graphene platelets in terms of complete platelet movement relative to each other. They found it possible to explain the pore-size changes determined experimentally in terms of the movements of these platelets and changes in their sizes, following increases in HTT.

Carbons as prepared from such materials as sugar and starch, because of their high oxygen contents, do not form graphitizable carbons; rather, they form isotropic carbons with some

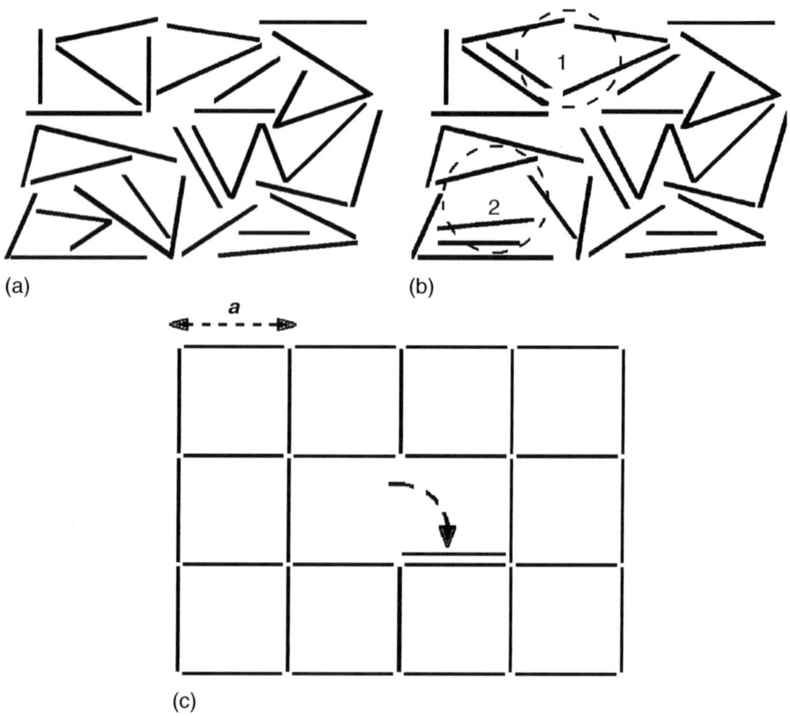

Figure 3.8. The falling card model (Dahn *et al.*, 1997).

porosity being formed (see Chapter 2). From current understanding of carbon structures, the existence of platelets within these carbons cannot be supported, the arguments of Section 3.4.6 applying. Further, as has been shown from graphitization studies, the possibility of considerable movement of an entire platelet within a solid matrix is virtually zero. The porosity within such carbons could never have graphene platelets for their surfaces because considerable experiences with published isotherms would clearly prevent the acceptance of such a model. The model is of little value in terms of explaining basic adsorption phenomena and the processes of activation. Although this model of falling cards makes a link between porosity and structure, its assumptions are considered to be unrealistic and coincidental.

It is worth noting that where phase changes occur in the solid phase, as with allotropic changes in tin (for example) atom movements occur only over dimensions of molecular size. The suggestions of Dahn *et al.* (1997) are for groups of carbon atoms moving as a single unit. This idea cannot be supported.

3.4.9 Glassy Carbon Model of Yoshida et al. *(1991), Figure 3.9*

Yoshida *et al.* (1991) created a model based on direct experimental observations of the structure of surfaces of a glassy carbon which had been heat treated in the range 1000–3000 °C. A Hitachi S-900, field-emission electron gun type (model name DEM) was used with a possible resolving power of about 2 nm. A series of micrographs were published for five samples, crushed to 2–3 mm size, all of which indicated a granular texture to the surface. These surfaces were therefore essentially fracture surfaces and at the magnifications used the structure of the carbon in the glassy carbon should be revealed. An analogy is the surface of broken concrete when the components of the aggregate will be exposed. The size of the granules averaged out at about 20 nm. From these observations, and from magneto-resistance measurements which indicated defects in layer planes, Yoshida *et al.* (1991) postulated a model illustrated in Figure 3.9. The model indicates what could possibly exist within a surface granule, the model basically being a cross-section.

This model supports the concept of porosity in activated carbon as being an assembly of cage-like space, of nanometer dimensions, the walls of the cages being the assembly of

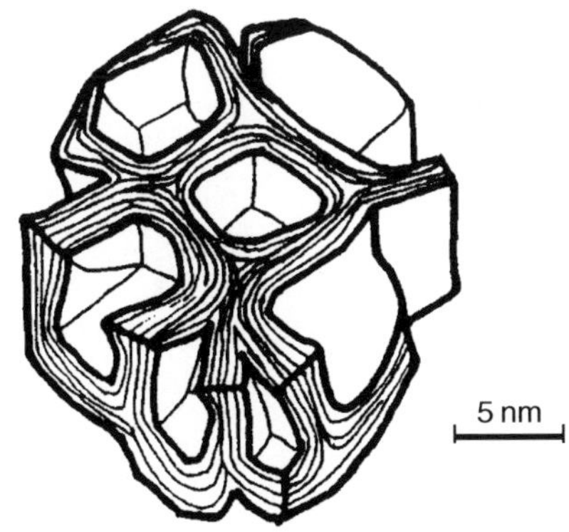

Figure 3.9. Model of assemblies of defective graphene layers within a glassy carbon to create porosity (Yoshida *et al.*, 1991).

carbon atoms in imperfect micro-graphene layer structures. The model of Figure 3.9 does not show interconnectivities between the pores but as this is a model of a high-temperature glassy carbon, of zero permeability to gases, the model is correct. The models of Figures 3.4 and 3.9 are not that dissimilar.

3.4.10 Model of Porous Microtexture of a Carbonaceous Particle of Oberlin et al. *(1980, 1999) and Oberlin (1989), Figure 3.10*

Oberlin *et al.* (1980) studied, intensively, structures within the family of carbons using mainly diffraction techniques of electron microscopy, especially the use of fringe imaging. The concept of the basic structural unit (BSU) was advanced (Figure 3.10(a)). This was considered to be the fundamental building block of structure in activated carbons. It was further postulated that these BSU assemble to form regions of local molecular orientation (LMO) as modeled in Figure 3.10(b). Finally, according to these authors, these bundles of BSU assemble to form an LMO resulting in the complex structures of Figure 3.10(c).

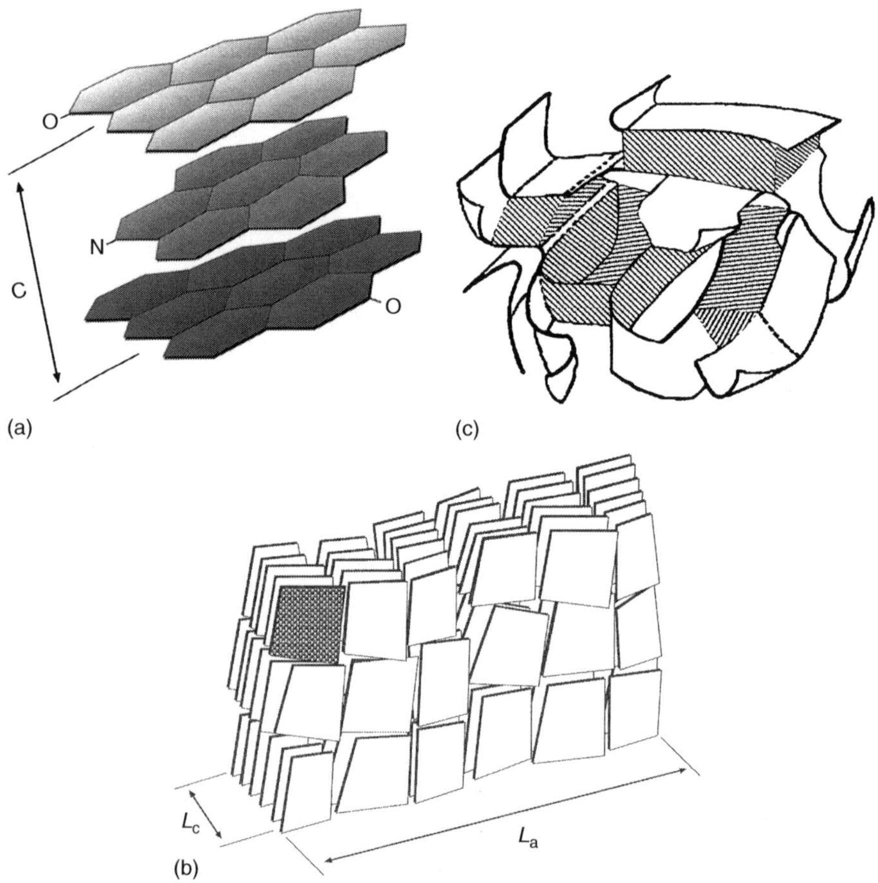

Figure 3.10. Model of porous microtexture (crumpled sheets of paper) of a carbonaceous particle, (a) BSU, (b) LMO and (c) assembly of LMO (Oberlin *et al.*, 1980).

Thus, these models resemble more the ideas of Kaneko *et al.* (1992a) than of Yoshida *et al.* (1991) in that they are based on the theory of structure of activated carbon as being made up of graphitic microcrystallites. Again, the problems associated with this model concept (Section 3.4.6) apply. The publications of Oberlin *et al.* (1980) and Oberlin (1989) are dominantly associated with graphitizable carbons, with an anisotropic structure and their model is more relevant in that situation. However, these publications refer to structure in carbons as resembling *crumpled paper*. This suggestion is helpful in that the porosity of activated carbons can be considered as resembling crumpled paper to facilitate inter-connectivity and transport of adsorbate gases. The thickness of the paper is probably no more than two or three carbon layers.

3.4.11 The Model of Virtual Porous Solids of Biggs and Agarwal (1992, 1994) and Biggs et al. *(2004a, b), Figure 3.11(a–c)*

The models of Sections 3.4.3–3.4.10 are concerned with volume elements of activated carbon of sizes of only about 1.0 nm^3. The models apparently are created more to support adsorption theory than structure in solids. The models of Biggs and Agarwal (1992, 1994) move from the single pore to bulk properties making use of computer programs.

How microporous carbons are used is controlled not only by their capacity to adsorb materials to acceptable extents and their ability to maintain retention times such that the adsorbate is not immediately eluted from the carbon, but also by the kinetics of the process, that is the rates at which the adsorbate can flow through the networks of porosity to all adsorption sites. When working with fluids (as distinct from gases) this aspect of flow or transport is of significance (see MacElroy *et al.*, 1999). Working toward this end, Biggs and Agarwal (1992, 1994) modeled structures in a 100 nm cube ($1 \times 10^6 nm^3$) of microporous carbon in such a way as to be able to perform molecular simulations of fluid movement within this 100 nm cube, named the virtual porous solid (VPS). This approach of Biggs and Agarwal (1992, 1994) is based on the concepts of Oberlin *et al.* (1980) whereby carbons have a repeating structure made up of BSUs composed of stacks of polyaromatic-like molecules, these assembling into regions of LMO, these LMO themselves forming complex associations which according to Oberlin *et al.* (1980) are structures normally associated with microporous carbons.

A note of caution is necessary here (as in Section 3.4.10) in that the formation of the BSU and the LMO leading to the complex associations of Oberlin *et al.* (1980) is a process of significant decrease in the entropy of the system and would require a "template" of some sort to create such a complex repetitive structure. It is therefore extremely unlikely that such formations occur particularly in microporous isotropic carbons, originating as they do from macromolecular organic parent materials.

Biggs and Agarwal (1992, 1994), copying this approach, designed what they termed basic building elements (BBE) for their modeling made up of polyaromatic-like molecules (polycyclic aromatic compounds such as phenanthrene and coronene) as described by some early work of Clar (1964). (*Note: Such compounds as phenanthrene and coronene would never survive, as such, the carbonization temperatures of 800 °C.*) Each cube, the BBE, within the cubic visualization contains these polyaromatic-like molecules, in 26 combinations

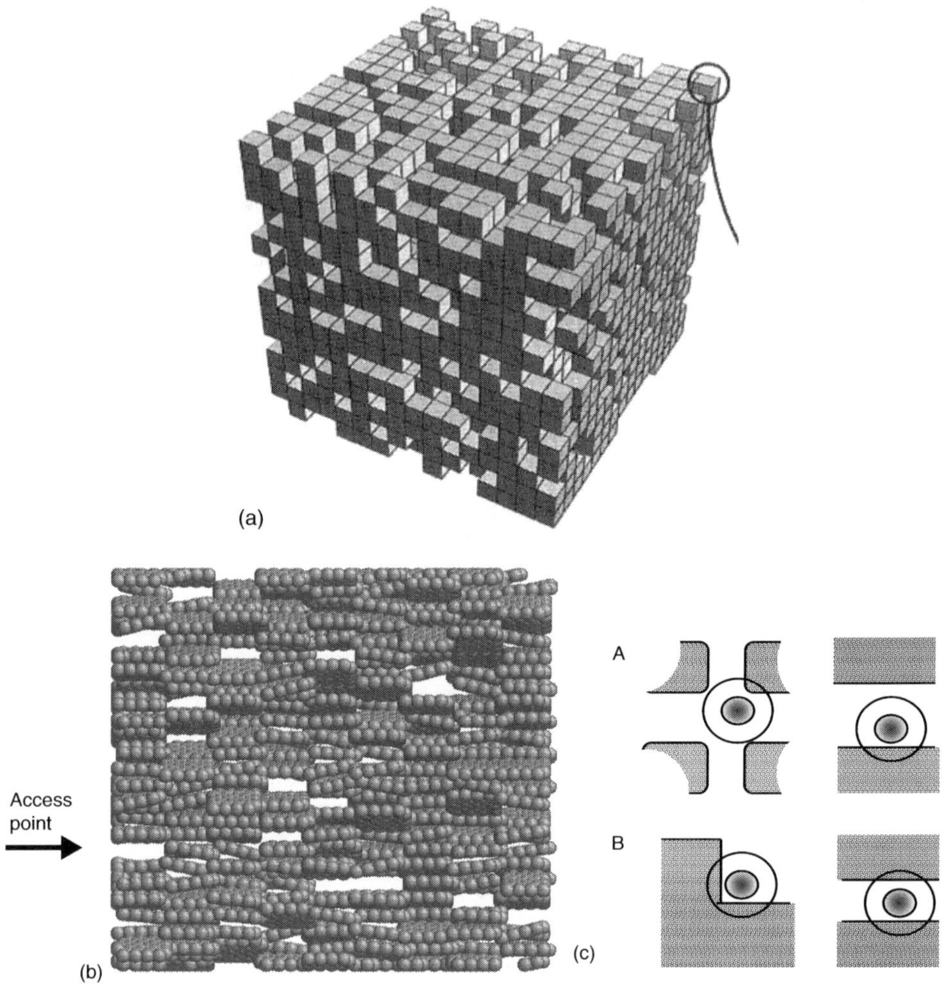

Figure 3.11. (a) The model of cubic visualization of virtual porous carbon solids (Biggs and Agarwal, 1992, 1994). (b) An isoenergy map of a finite microporous region of LMO showing the nature of microporosity and access points for the adsorbate molecules. For an activated carbon, this model is too anisotropic without sufficient porosity (Biggs *et al.*, 2004a, b). (c) Simulation of adsorption processes: (A) locally convex porosity and (B) locally concave porosity. The molecule is shown with its field of substantial interaction with the carbon (Biggs and Agarwal, 1992, 1994; Biggs *et al.*, 2004a, b).

(Figure 3.11(a)). Microporosity could possibly exist between the BSU of polycyclic aromatic molecules (better described today as defective graphene layers).

These models were generated to study self-diffusion of fluids within microporous carbons as a function of temperature and porosity. However, their immediate application to microporosity of activated carbons from traditional sources is not that obvious. From this point of view, the model has some misconceptions. The selection for the BBE of the model of polycyclic aromatic molecules seems unnecessary and these molecules, as such, can never exist in a microporous carbon that has an HTT of >800 °C. There is no evidence that in microporous carbons there exists a repeating structure as visualized with the LMO of the model.

In two later articles Biggs *et al.* (2004a, b) considered adsorption of a model nitrogen vapor on a range of complex nanoporous carbon structures (Figure 3.11(b)). This was simulated by grand canonical Monte Carlo (GCMC) simulation for a single sub-critical temperature above the bulk freezing point. Adsorption and desorption isotherms, heats of adsorption and three-dimensional singlet distribution functions (SDFs) were generated. Inspection of the SDFs revealed significant levels of solid-like adsorbate at saturation even in the most complex of the microporous solids considered. The presence of significant levels of solid-like adsorbate has implications for characterization of microporous solids where adsorbate density is used (e.g. determination of pore volume from loading). Detailed consideration of the SDF at different loadings for a model microporous solid indicates solid-like adsorbate forms at distributed points throughout the pore space at pressures dependent on the nature of the local porosity. Introduction of mesoporosity into the model causes hysteresis between the adsorption and desorption isotherms. Adsorption in the hysteresis loop occurs by a series of local condensation events. It appears as if the presence of adjacent microporosity and/or adsorbate within it affects the pressure at which these events occur. Reversal of the condensation during desorption occurs throughout the mesoporosity at a single pressure; this pressure is unaffected by the presence of adjacent microporosity or the adsorbate within it. The importance of pore shape is discussed further in Section 3.4.23.

It is also shown that the empirical concept of "pore size" is not consistent for describing adsorption in the complex solids considered here. A new concept is, therefore, proposed that seeks to account for the factors that affect local adsorption energy: local geometry, microtexture, surface atom density and surface chemistry.

Adsorption potential within a pore is a function not only of pore size but also of the position of the adsorbate molecule within the adsorption site (pore). Figure 3.11(c) illustrates this effect. For example, the adsorption site in Figure 3.11(c) in (A) with interactions between four walls could have the same adsorption potential as in a narrow micropore (classical considerations). This raises some quite fundamental issues. The adsorption isotherm is created by adsorption in adsorption sites (pores) of decreasing adsorption. This is normally interpreted as the filling of porosity of increasing size. In actual fact, there is no way of knowing if this is indeed the case. In an extreme consideration, the concepts of porosity may need some adjustment. It can be said that porosity is not all that important. Another way of characterizing carbons would be to use distributions of adsorption potentials, without consideration of structure of the adsorption site. The articles of Biggs *et al.* (2004a, b) present non-traditional approaches which are well worth following up.

3.4.12 Model of Porous Carbon of Segarra and Glandt (1994), Figure 3.12(a, b)

The model of Segarra and Glandt (1994), and those described subsequently, interpret structural analytical data using computer simulations of possible structures. There is a possible danger here, because of the capacity of computers to generate so many possible models, that severe restraints have to be imposed. For example, the models must suggest acceptable densities for the activated carbon, as well as carbon atom distribution values which are comparable to experimental values.

The computer visualization of a model of porous carbon of Segarra and Glandt (1994), Figure 3.12(a, b), was obtained using a Conical Monte Carlo program to move the relative positions of the disks of this model until a thermodynamic (minimum energy) equilibrium was obtained.

Once again, this model is based on the assumption of the graphitic microcrystallite as the foundation unit of structure in activated carbon (see Section 3.4.10 for concerns about this assumption). The disks are all of a cylindrical shape and composed of stacked graphene layers with variable thicknesses dependent on the number of graphene layers within the disk. A further assumption or approximation in the design of the model is that the carbon atoms of the surfaces are of uniform and continuous density.

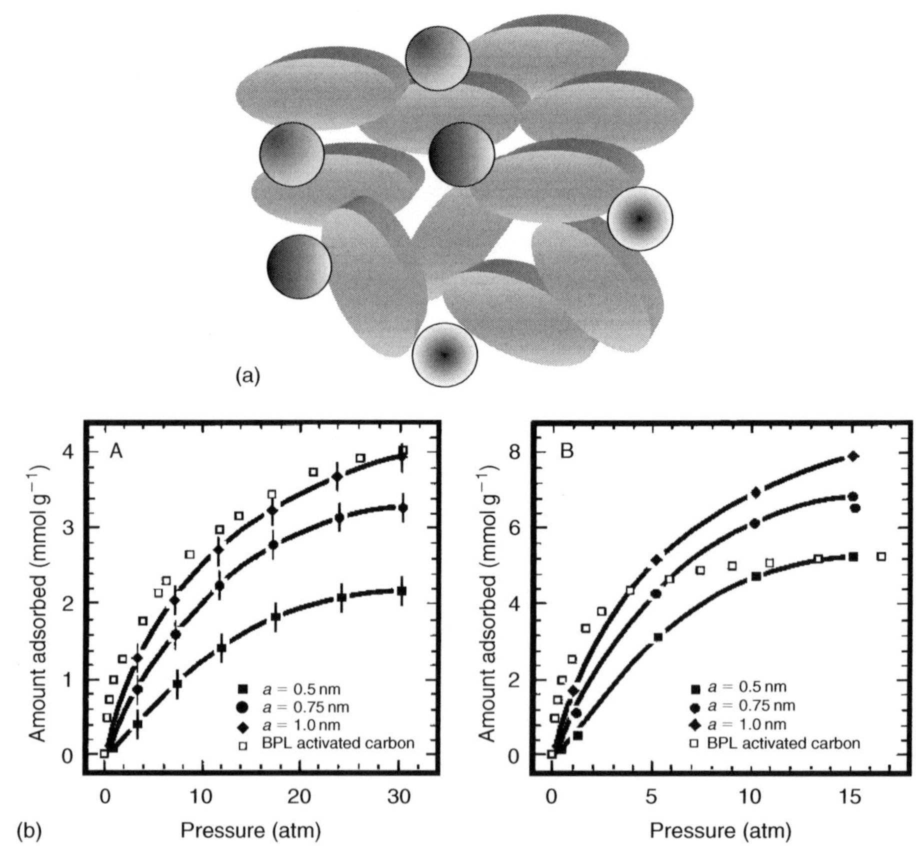

Figure 3.12. The model cylindrical disks of microcrystallites of porous carbon (adapted from Segarra and Glandt, 1994). (b) Comparison of experimental and simulated adsorption isotherms for model carbons of different platelets sizes (*a* in nm): (A) methane at 301.4 K and (B) ethane at 300.0 K (Segarra and Glandt, 1994).

In order to obtain some experimental justification for their model, these authors measured the adsorption of methane at 301.4 K and of ethane at 300 K at pressures up to 3.0 and 1.5 Mpa, respectively, the isotherms being as shown in Figure 3.12(b). Theoretical isotherms were calculated from the model assuming platelet diameters of 0.5, 0.75 and 1.0 nm, and the experimental and theoretical isotherms were compared. The theoretical isotherms are also in Figure 3.12(b) and indicate quite close agreement for methane adsorption assuming the largest of the platelets of 1.0 nm diameter. For the ethane adsorption, no agreement was obtained.

However, there are so many assumptions and variables in the model of Segarra and Glandt (1996), including the number of stacked graphene layers to form crystallites, the uniformity of surfaces, and an arbitrary decision for platelet size and size distribution and platelet alignment that the potential of this model to predict accurately isotherm shape and capacity, meaningfully, is limited. As the assumptions of the model are some distance away from the reality of structure in microporous carbons it can be considered that these manipulations are simply curve-fitting exercises where agreement with experiment data is no validation that the model is any way correct, and cannot and should not be considered as being precise modeling of an adsorption process. Further, it is difficult to understand just what is a platelet of 1.0 nm diameter.

3.4.13 Tight-binding Model of Wang et al. *(1996), Figure 3.13*

These computational methods (tight-bonding models) are reviewed by Goringe *et al.* (1997). The approach, called "tight-binding molecular dynamics", has been used by Wang *et al.* (1996). Here, the authors simulated the quenching of liquid carbon at various densities from high temperatures (6000–10,000 K) to 0 K. The basis of the calculations is to establish what structures would be generated by the cooling process based on a knowledge

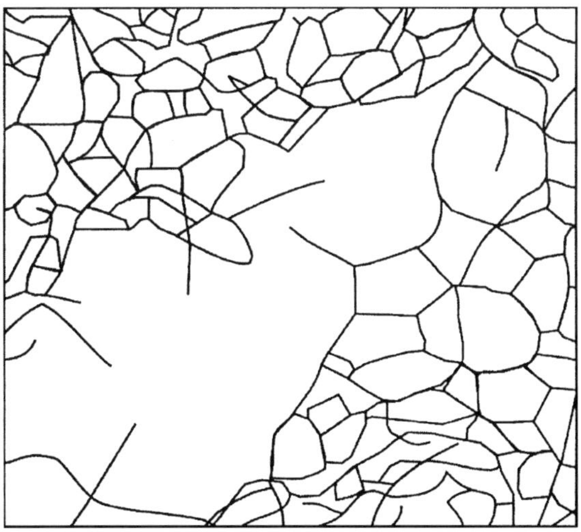

Figure 3.13. The network of amorphous carbon with an average density of 1.25 g cm^{-3} (Wang *et al.*, 1996).

of tight-bonding orbital and Coulombic repulsions of carbon nuclei. Wang *et al.* (1996) report that the type of carbon predicted from these calculations is strongly dependent on the initial assumed densities of liquid carbon. High-density carbon produced a diamond-like carbon composed with a majority of the diamond-type (sp^3) bonding. Low-density carbon produced what the authors call an amorphous carbon (it is not really amorphous) dominated by three- and twofold coordinated carbon atoms.

The structures predicted from this approach consist of graphene layers which are highly defective, and containing five-, six- and seven-membered ring systems. The presence of the five- and seven-membered rings into the graphene layer automatically introduces strain which is relieved with the curvature within the layer. The model of Wang *et al.* (1996) is reproduced here as shown in Figure 3.13. The density proposed by Wang *et al.* (1996) for their carbon is 1.25 g cm^{-3} somewhat higher than found for many of the microporous activated carbons used by industry (e.g. 0.3–0.4 g cm^{-3}).

It is interesting and important to note that such computer-generated models, based on known properties of the carbon atom, quite independent of other studies, arrive at almost the same model as those based on Raman microspectroscopy and studies of adsorption phenomena. A question raised is whether or not it may eventually be possible to predict intensities and widths of Raman spectra based on such a model.

The model, so generated, shows a heterogeneity of density of carbon atoms; there appear to be volume elements of mesoporosity, even macroporosity. Is this model indicating how mesoporosity may exist within an activated carbon? Mesoporosity cannot all be associated with cone-shaped (wedge-shaped) porosity at surfaces of particles. Although this model approximates the structures in carbon networks, it does not predict pore-size distributions, that is the molecular space networks, a matter of some importance to adsorption studies. Nevertheless, the possibility of doing this seems to be realistic.

The use of the "match-stick" model facilitates clarity in visualizing the structural arrangements within the model.

3.4.14 Computer-generated Models of Acharya et al. *(1999), Figure 3.14*

As indicated above, advances in computer power and program design enabled quite sophisticated models of carbon structures to be generated. Acharya *et al.* (1999) used "The Chemically Constrained Model" to generate structures within a microporous carbon. The approach and assumptions of these authors closely resembles the way structure is envisaged in carbons as outlined in Chapter 2. The entire family of carbons can be considered, as a continuity, as being composed of segments of a single graphene layer (not a three-dimensional segment of stacked graphene layers) of different sizes reaching nanometer dimensions, shape and degree of perfection (holes, curvature, both concave and convex, linear carbon atoms (dangling bonds) and the presence of heteroatoms) bonded together in an infinite number of ways. Segments can, in one sense, be more like polycyclic aromatic compounds of various complexities bonded into and forming part of a three-dimensional, cage-like network of carbon atoms but, and this is emphasized most strongly, not existing in isolation and so not manifestively identifiable as such, but forming part of a continuity.

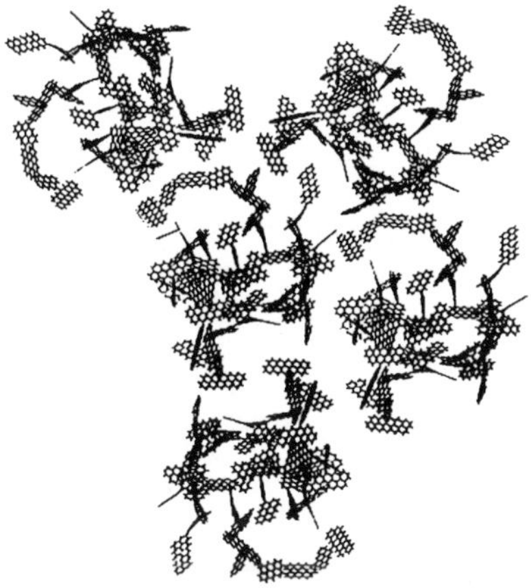

Figure 3.14. Computer-generated models showing porosity as existing between assemblies of graphene layers (microcrystallites) (adapted from Acharya *et al.*, 1999).

For their model, Acharya *et al.* (1999) assumed that the chemical composition of microporous carbons is made up of hydrogen and carbon atoms only. A further assumption is that all of the C—C bonding is sp^2 as in a perfect graphene layer (single-crystal graphite). Their computer program first generates fragments of a graphene layer of nanometer dimensions. An optimum size of fragment is calculated with hydrogen atoms bonded to the edges of the fragments, amounts based on analytical data. The authors further used a subroutine called "Signature" which joins together molecular fragments, in this case bonding the fragments at positions of nearest unsaturated carbon atoms (dangling bonds) and also at positions of adjacent unsaturated carbon atoms no matter the distance apart. The "Signature" program generated many structures but acceptable realistic structures had to fit to known densities of these carbons. The model was then refined to produce a structure of minimum internal energy. Acharya *et al.* (1999) based on their considerations on the carbonization of polyfurfuryl alcohol in the temperature range 400–800 °C. Figure 3.14 describes the type of model generated.

Figure 3.14 is a montage made up of the structure of carbon of lowest HTT to give some idea of how porosity may be generated by pasting together copies of the original model in a random way. The model shows the interconnectivity of porosity of different sizes and (with some imagination) within three dimensions. This model of Acharya *et al.* (1999) includes layers which are curved, the curvature being caused by the presence of five membered rings within the layers. This is in agreement with the structures of fullerenes where the spherical shape is only possible because of the five-membered ring systems. Similarly, for the nanotubes with closed ends, the closure is accommodated by the presence of five-membered ring systems. Additional experiments to test the model, using neutron diffraction to obtain radial distribution functions, confirmed the existence of layers which were severely non-planar (Petkov *et al.*, 1999).

However, the way that Acharya *et al.* (1999) present the model indicates that it may be more suitable for a carbon of HTT $\gg$ 1000 °C. Overall, the layers are too graphitic like and perhaps are more like the structures of carbon nanotubes with their five-membered ring systems. Raman spectroscopy of model of Acharya *et al.* (1999) may indicate a D-band of too low an intensity compared to the carbon itself as prepared from polyfurfuryl alcohol.

Another problem with the model is that the adsorption of nitrogen on such surfaces would indicate relatively low enthalpies of adsorption comparable to what is observed for adsorption by a carbon of high HTT, perhaps as high as 2000 °C. Certainly, a carbon of HTT of, for example, 600–800 °C would not possess such graphitic-like (microcrystallite) surfaces. Although this model approximates closely to structures in carbon networks, it does not predict pore-size distributions, that is the molecular space networks, a matter of some importance to adsorption studies.

However, the model is sufficiently advanced to allow it to be used constructively to compare and contrast other models. A passing note is that the final stages of modeling pore-size distributions of microporous carbons are some distance away from being realized.

3.4.15 Model of Glassy Carbon of O'Malley et al. *(1998), Figure 3.15*

Glassy carbons (Jenkins and Kawamura, 1976) are a form of carbon with high oxidation resistance, high mechanical strength, high elasticity, low helium density and zero accessible porosity. XRD studies give a very broad (002) diffraction band indicating that of all carbon forms which have been studied, it is the glassy carbons which give a minimum of structural order. Such carbons are made by carbonizing a resin or mixtures of resins in sealed vessel, such that no volatiles are released during the process. As a result, all of the carbon of the original resin forms part of the resultant solid carbon (i.e. the glassy carbon).

There are two major effects resulting from the sealing or confinement of the carbonizing system. In a closed system, the volatile decomposition products cannot escape from the

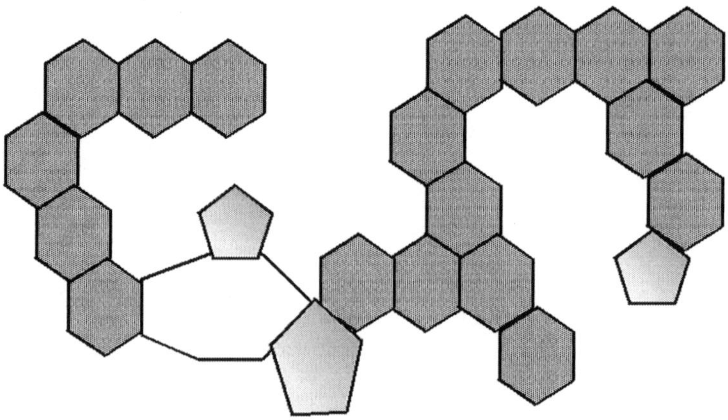

Figure 3.15. The basics of the model of a defective graphene layer (non-planar) showing vacancies, and five- and seven-membered ring systems (adapted from O'Malley *et al.*, 1998).

system and eventually decompose within the "space" from which they decomposed to form a carbon. The carbonization process is not reversible and the original macromolecular system is never reformed. A result is that the molecular space, as formed in normal carbonizations, never materializes and resultant glassy carbons have no accessible porosity.

Second, the restricted movement within the carbonizing system which appears to prevent the migration of carbon atom; as a result, they then bond into carbon networks which are more disordered than when less restricted atom movement is possible. Although the structures within glassy carbons may be highly strained they are in positions of stable quasi-equilibria, the carbon atoms of the structures not being able to move elsewhere because of the high activation energies involved (probably more than 4000 kJ mol^{-1}).

O'Malley *et al.* (1998) used a reverse Monte Carlo (RMC) approach to simulate structure of a glassy carbon. In the RMC approach (Da Silva *et al.*, 2001), movements are made of single atoms, or groups of atoms and the concept of graphite microcrystallites is also used. Rather than calculate from considerations of statistical mechanics what could be possible structures of minimum energy (but of zero reality) severe constraints were applied to the modeling. For carbons, these restraints include bond angles between carbon atoms, the relative positions of carbon atoms in a carbon network and the possibility of forming something approaching a graphene layer. O'Malley *et al.* (1998) presented a computer visualization of their conclusions in the form of a single layer (Figure 3.15). The constraints applied necessitated the formation of a buckled layer containing multiple ring systems of carbon atoms. The larger ring systems can be considered as being "holes". Any stacking of such layers would not be that of the ABABAB graphite structure but would be much less perfect with "true" molecular space between the layers (unlike in single-crystal graphite). These authors calculate the average bond angle in glassy carbon as 117° in agreement with calculations by Galli *et al.* (1989) who simulated a carbon structure that possessed 85% of sp^2-bonded atoms, the remaining atoms being sp^3 bonded with an average bond angle of 117°.

The relevance of the above discussion to understanding structure in activated carbons lies with the agreement between computer modeling that the bonding of carbon atoms in these carbons results in buckled arrays of carbon atoms with bonding in multi-carbon ring systems some being large enough to be considered as a "hole" or as a site of adsorption (pore) in studies of surface chemistry.

> *In fact, if this layer of Figure 3.15(a) is folded over on to itself several times and tied with cross-linkages to create a three-dimensional cage-like model, this may approximate closely to the carbon networks of a microporous carbon.*

3.4.16 Model of Glassy Carbon of Pikunic et al. *(2001, 2002), Figure 3.16(a, b)*

Following on from the work of O'Malley *et al.* (1998) and Pikunic *et al.* (2001, 2002) used the RMC simulations to model structure in a 2500 °C glassy carbon, of density 1.57 g cm^{-3}, as well as saccharose-based carbons. The computer visualization of the glassy carbon is in Figure 3.16 and again shows a multi-ring system representing a structure that is "frozen" or stabilized under the conditions of formation. The individual carbon atoms released during the decomposition of the macromolecular polyfurfuryl alcohol probably

(a)

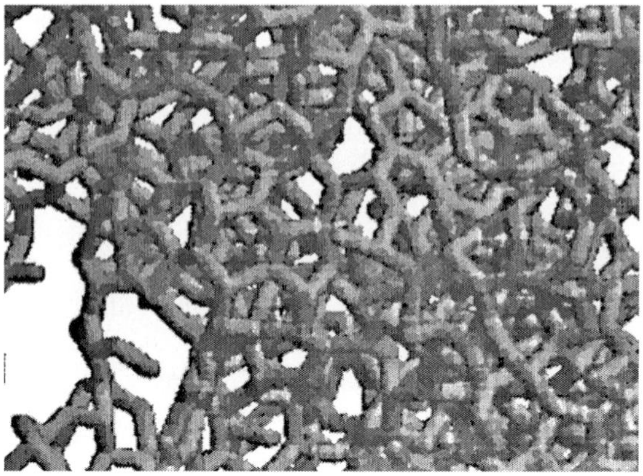

(b)

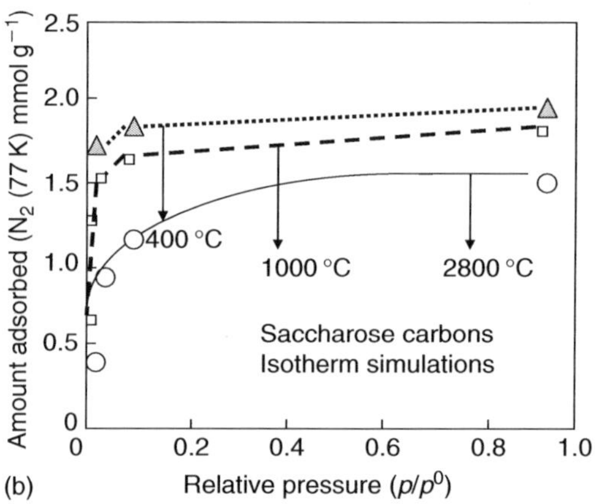

Figure 3.16. (a) The model of glassy carbon as an assembly of "threads" of carbon atoms, so creating porosity (adapted from Pikunic *et al.*, 2001). (b) Nitrogen adsorption isotherms at 77 K obtained from GCMC simulations of models for saccharose-based carbons, HTT 400, 1000 and 2800 °C (Pikunic *et al.*, 2002).

adopt these positions immediately after their release following the decomposition of the macromolecular polymer network.

These computer visualizations are in agreement with energy, density and structural restraints obtained from the carbons themselves. This model moves quite some distance from the concept of slit-shaped pores created as molecular space between graphitic microcrystallites. This model goes so far as to almost eliminate the need for any concept of surface (the word being used in its non-scientific sense). The model is suggesting that structure here is a "tangle" of threads of carbon atoms arranged in various ring structures, spaces between the spaces constituting porosity of various sizes. All of these spaces are interconnected to all others and this is of major importance when considering the modeling of porosity rather than direct modeling of carbon networks, although it is realized that one depends closely and intimately on the other. In addition, this model of Pikunic *et al.*

(2001) depicts the co-existence of porosities of all sizes from micro- to mesoporosity in the way that these "threads" approach and depart from each other in the network, and that the "threads" contain holes of eight-membered ring systems, or even larger.

This study with glassy carbon (non-porous) was extended to three carbons prepared by pyrolysis of saccharose to 400, 1000 and 2800 °C. Pikunic *et al.* (2002) used a simulation protocol based on RMC to construct models for the three above carbons. Radial distribution functions of the three models reveal that the size of the graphene layers increases with increasing HTT and are more ordered (something about carbons which has been known for some time). Also were performed GCMC simulations of nitrogen at 77 K in the resulting structural models to predict the effects of heterogeneity on the adsorption properties of carbons.

The simulated nitrogen isotherms are shown in Figure 3.16(b) and are Type-I, characteristic of microporous carbons. It is seen that the adsorption capacity decreases with increasing carbonization temperature by about 25% (400–2800 °C). The modeling indicates that the volume available for nitrogen is about 7.5% of the simulation box (Figure 3.15(a)) for the three models. (*Note*: A microporous carbon, with a nitrogen BET (Brunauer–Emmett–Teller) surface area (77 K) adsorbs 10 mmol g^{-1} (0.34 $cm^3 g^{-1}$ liquid nitrogen) of nitrogen at p/p^0 of 1.0, see Section 4.1.3.) It is further reported that carbon atoms "block" together with increasing HTT so reducing the surface area which is available for adsorption.

These conclusions appear to be in conflict with the experiences of adsorption chemists, who, when studying the effect of HTT on available surface areas of carbons, generally find that a carbon of HTT 2800 °C is unable to adsorb any nitrogen at 77 K (see Section 2.6.2.1). Figure 3.16(b) indicates that the adsorption capacity (N_2 at 77 K) of the saccharose carbon 400 °C is 2.0 mol g^{-1} of carbon, and of the 2800 °C carbon is 1.5 mol g^{-1} of carbon. These capacities are equivalent to surface areas of 2.0×10^5 and $1.5 \times 10^5 m^2 g^{-1}$, values which are significantly different for surface area values of carbons, as found in the literature (see Section 4.1.3).

Pikunic *et al.* (2002) publish "snapshots" of the locations of nitrogen atoms within these saccharose carbons indicating an almost random distribution of the nitrogen atoms. It is generally accepted that porosity in carbons is slit shaped. However, the structure of porosity in carbons is complicated and it is perhaps too much to expect, at the moment, for computer-based modeling to be able to follow closely the finesse of the detail of change which occurs within a carbon with progressive heat treatment.

3.4.17 Model of Porous Carbon of Petersen et al. *(2003), Figure 3.17*

These workers also followed the studies of O'Malley *et al.* (1998) and studied a glassy carbon with an HTT of 2500 °C using a modified RMC algorithm. In the establishment of their models, use was made of their neutron diffraction data as well as the electron diffraction data of McCulloch *et al.* (1999). These sets of data of atom distribution functions agreed with each other thereby strengthening the validity of both sets of experimental data. This agreement further suggests that the structure of this glassy carbon was essentially homogeneous because the volume element of the carbon irradiated by the neutrons

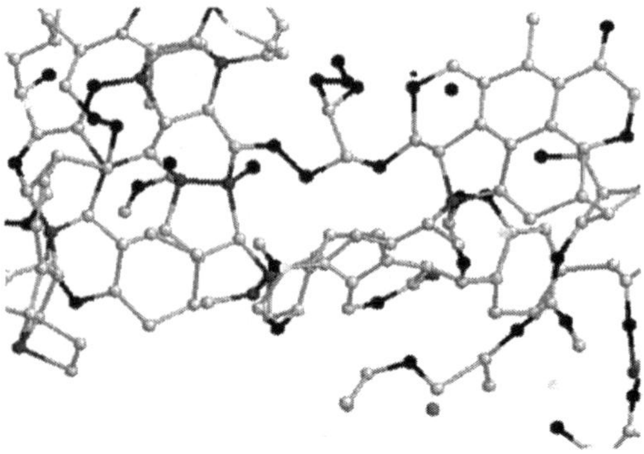

Figure 3.17. Model of assemblies of carbon atoms within porous carbon (Petersen *et al.*, 2003).

is many times larger than the volume element irradiated by the electrons. This statement is supported by transmission electron microscopy studies of glassy carbons (Jenkins and Kawamura, 1976). Petersen *et al.* (2003) used three simulation programs: the RMC, a Hybrid-RMC (HRMC) and HRMC-SC where a simple cubic lattice of 8000 atoms was used for the starting configuration. Computer visualization models of the results of this analysis are shown in Figure 3.17.

These authors report that their results are in agreement with those of O'Malley *et al.* (1998) in that the carbon network is composed of warped layers composed of sp^2-bonded carbon atoms. The sp^3-like atoms (diamond-like bonding) are to be found bonding together these warped layers. The existence of sp-bonding (linear carbon) is suggested in layers associated with ring structures made up of more than six carbon atoms. The HRMC-SC simulations indicated a number of graphite-like bonds (sp^2) orientated in the same direction. A most interesting observation from the RMC configurations of Figure 3.17 is that the structures of this figure (or very similar structures) could be found with different orientations of the simulated unit cell of the program. Figure 3.17 is not using space-filling atomic models but shows the presence of five-membered, of distorted six-membered rings as well as "graphitic" rings and linear carbon in a multi-atom ring system.

Acknowledging that the carbon under investigation is a glassy carbon, HTT 2500°C, which does not have an accessible porosity, it is reasonable to assume that structures in other microporous activated carbons will not be that significantly different. It could be that that the packing together of such units of structure as seen in Figure 3.17(a) is not so close and that there are more defects within the layers which make up the carbon network. It is highly probable that such an arrangement of atoms makes up the entire structure of macro-samples (on a larger than 1 g scale) again consistent with the evidence of homogeneity which is seen in transmission electron microscopy studies.

3.4.18 Carbon Aerogels of Gavalda et al. *(2001, 2002) and Job* et al. *(2004), Figure 3.18 (a, b)*

In recent years, a spherical form (of nanometer dimensions) of microporous carbon has been developed termed carbon aerogels as a result of their method of manufacture. They

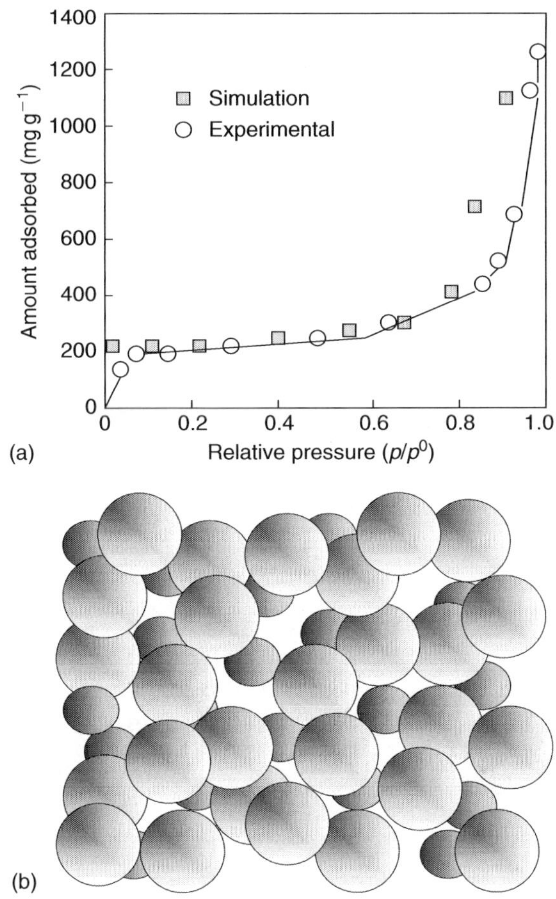

Figure 3.18. Isotherms of adsorption of nitrogen at 77 K on a carbon aerogel of diameter about 6.0 nm (Gavalda *et al.*, 2001, 2002). (b) Models of the packing of aerogel particles and of the carbon atom structure within a carbon aerogel of diameter about 6.0 nm (Gavalda *et al.*, (2001, 2002).

are prepared by polycondensation of hydroxylated benzene and an aldehyde, the most common being resorcinol and formaldehyde, in water as the solvent with Na_2CO_3 as catalyst, to produce a gel (i.e. a hydrogel). The drying of the gel is not straightforward and such methods as simple evaporation (zerogels), vacuum drying, hypercritical drying with carbon dioxide (aerogels), water replacement by a solvent and freeze-drying (cryogels) have been tried with variable success in producing microporous carbon aerogels. Job *et al.* (2004) report the successful formation of porous carbon aerogels by control of the pH of the initial aqueous resorcinol–formaldehyde gel. An optimum pH of 5.45 resulted in a carbon aerogel with a surface area (N_2 BET) of 550–600 $m^2 g^{-1}$, with diameters of the aerogel particles of about 40 nm.

Tamon (2001) summarizes the basics of carbon aerogel formation and properties. Although these carbon aerogels are dominantly mesoporous, significant microporosity can be included. For example, Tamon (2001) reports a carbon aerogel with a BET surface area of $1215\,m^2\,g^{-1}$ contained within $2.44\,cm^3\,g^{-1}$ of mesoporosity and $0.26\,cm^3\,g^{-1}$ of microporosity. Gavalda *et al.* (2001, 2002) and Job *et al.* (2004) report an isotherm shape for nitrogen adsorption at 77 K as shown in Figure 3.18(a), which is indicative of microporosity and mesoporosity.

Gavalda *et al.* (2001, 2002) used Monte Carlo type moves to simulate the packing arrangements of the carbon aerogel spheres (Figure 3.18(b)), size about 6 nm diameter. The interparticulate mesoporosity had dimensions of 3–15 nm and dimensions assigned to the microporosity of 0.7–1.5 nm.

Structure within the spheres of Figure 3.18(b) is based on the assumption of the existence of graphitic microcrystallites and indicates layers of some significant size (100–200 carbon atoms) as making up the structure. Structural analyses (Raman spectroscopy and XRD) indicate values of L_c of 0.6–0.9 nm and of L_a of 2.5–4.5 nm. Such values are difficult to reconcile with the smallness of the carbon aerogel particle of 6 nm diameter. It is possible that these aerogels not only contain some microporosity but that extents of adsorption in the low relative pressure regions contain contributions of adsorption on extremely heterogeneous external surfaces of the aerogels (which would behave like microporosity).

3.4.19 Structure of Montmorillinite as Shown by Scanning Electron Microscopy, Figure 3.19

Although montmorillinite is not a carbon, the structure of this material as shown in Figure 3.19 is helpful in appreciating the porous nature of activated carbon. Figure 3.19 shows an interconnected system of porosity separated by material. The model shows an infinite number of shapes and sizes of the porosity with a corresponding infinite variation in the adsorption potentials between adsorbate molecules which just fit into the porosity and those with room to spare. These porosities are interconnected in three dimensions.

Montmorillinite is a clay mineral belonging to the smectic group derived by the weathering of mineral ash. It is an alumino-silicate with a significant capacity for the adsorption of water. The closeness of this structure to models of Figures 3.4, 3.9, 3.13, 3.15 and 3.16 can be seen.

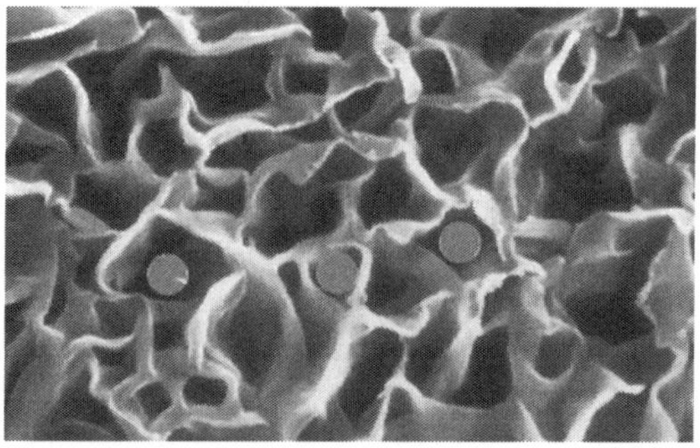

Figure 3.19. Scanning electron micrograph of the structure of montmorillinite showing a cage-like porosity with "adsorbate" molecules inserted (a close model to microporosity in carbons.

3.4.20 Images Using Scanning Electron Microscopy, Rodríguez-Reinoso (unpublished), Figure 3.20(a–d)

It is possible to obtain images from an exfoliated graphite which resemble the structure of the montmorillinite (Figure 3.19). When a graphite reacts appropriately to form an intercalation compound, e.g. graphitic oxide, the graphene layers are separated. During a heating process, in an inert atmosphere, this intercalate is forcibly distilled out of the intercalation compound and in so doing forces the layers apart to produce what can be described as a *concertina* structure. This resultant structural was imaged using scanning electron microscopy and four images are reproduced as shown in Figure 3.20(a–d).

The magnifications used are much too low to represent accurately the microporosity in an activated carbon. So, the model has immediate limitations. But, these micrographs give an indication of such microporosity, in which no porous system is duplicated, that is all pores

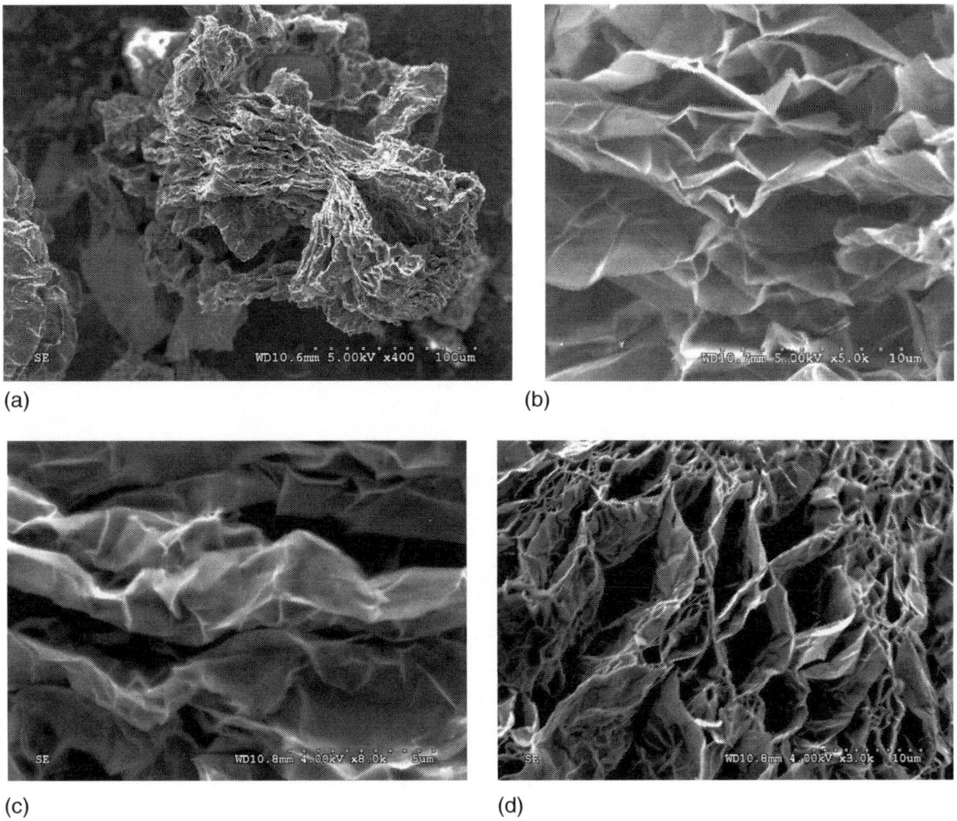

Figure 3.20. (a, b) Scanning electron micrographs of surfaces of exfoliated single-crystal graphite (to model porosity in activated carbon). (c, d) Scanning electron micrographs of surfaces of exfoliated single-crystal graphite (to model porosity in activated carbon). All four micrographs show spaces between the graphene layers, the spaces being irregular and somewhat slit shaped (Rodríguez-Reinoso, unpublished).

are different in some way. The model gives an indication of the presence of microporosity and mesoporosity in close proximity to each other. There is an indication of slit-shaped pores but the pore walls are probably too smooth and too thin, as the pore wall thickness has to approximate to the dimensions of the contained porosity.

3.4.21 Images Using High-Resolution, Fringe-imaging Transmission Electron Microscopy, Marsh et al. *(1982), Figure 3.21 (a–c)*

Transmission electron microscopy has been used successfully to elucidate structures in graphitic carbons as indicated in Chapter 2. Phase contrast imaging produces fringe images which can be related to the lamellar structures of graphitic carbon. Images can be produced from segments of the carbon specimen which are aligned parallel to the electron beam. Other parts of the specimen are not imaged. The technique has been applied to non-graphitizable carbons with some success although care has to be taken with literal interpretations of structure. The specimen in the microscope has to be several nanometers thick (otherwise, it could not be handled) and so the electron beam has to travel through several structural elements.

Figure 3.21(a) is a micrograph of a polyvinylidene chloride carbon, non-graphitizable, of surface area of $1000\,m^2\,g^{-1}$, HTT 1123 K. Figure 3.21(b) is a micrograph of a commercial activated carbon (Aqua Nuchar A) used for water treatments. Figure 3.21(c) is a fringe-image micrograph of a quite different form of carbon (PX21) as produced by the activation of cokes by KOH (Chapter 6) (Marsh *et al.*, 1982). This form of activation violently explodes and separates the structural components of the cokes. On recovery of the carbon, these components had re-assembled in a very open configuration creating cavities such as is seen at position Q. Positions P, R and S are indicative of single layers making up the surface. There is a similarity between this structure and those of the exfoliated graphite.

The fringe images of these micrographs are projections of the arrays of carbon atoms which are vertical to the plane of the paper being part of the cage-like structures, which constitute porosity. The spaces between the fringe images are of the order of 0.5 nm microporosity size. There is no indication within these micrographs of the presence of the graphitic microcrystallite as discussed in previous sections.

3.4.22 Models of Porosity in Activated Carbons as Suggested by Byrne and Marsh (1995), Figure 3.22(a–d)

Byrne and Marsh (1995) were unable to accept the concept that activated carbon possessed a structure made of graphitic microcrystallites. A major concern was the improbability of graphitic (graphene-like layers) being formed at temperatures <900 °C from organic precursors.

Figure 3.22(a–d) is what these authors consider to be reasonable models. Figure 3.22(a) is a cellulosic precursor, with Figure 3.22(b) being the structure of an intermediate in the carbonization process and still possessing an infrared spectrum. Figure 3.22(c) is a further stage when most of the functional groups have been removed and carbon atoms are beginning to

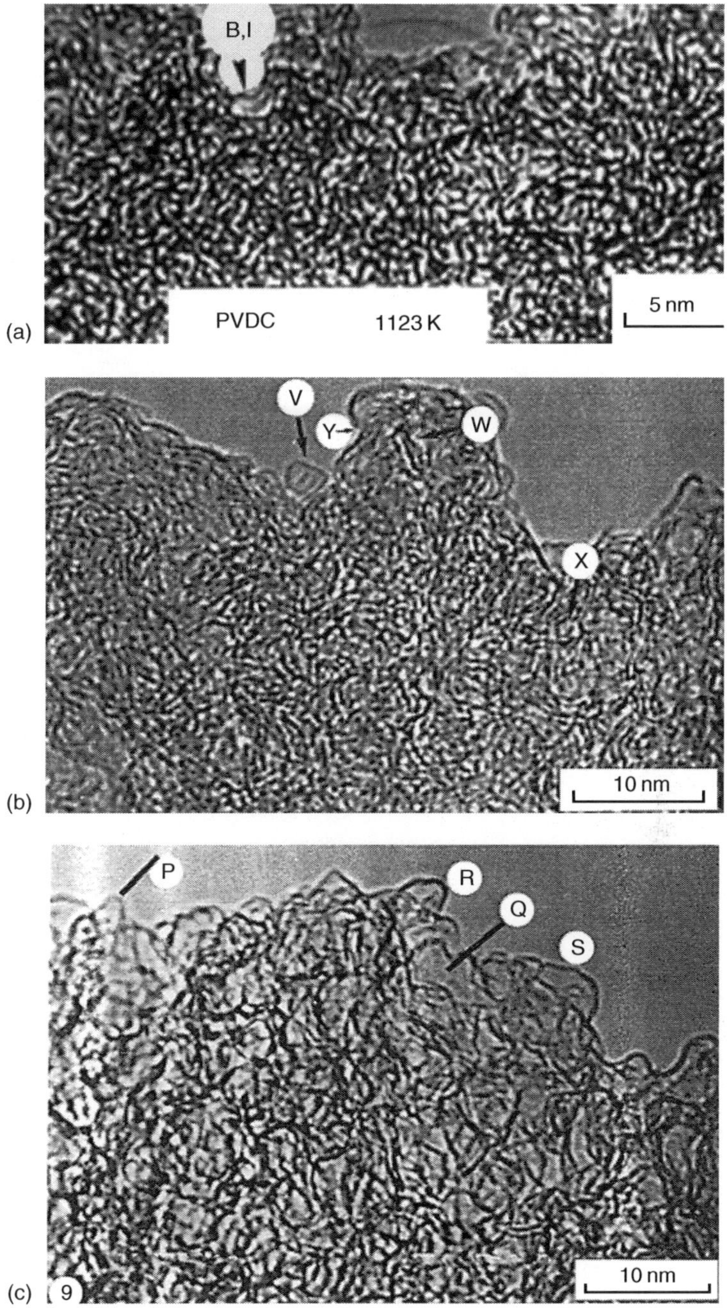

Figure 3.21. (a) Fringe imaging of a cross-section of a porous carbon from polyvinylidene chloride, HTT 1123 K, indicating the relative arrangements of defective graphene layers (Marsh *et al.*, 1982). (b) Fringe imaging of a cross-section of a commercial porous carbon, Aqua Nuchar A, indicating the relative arrangements of defective graphene layers (Marsh *et al.*, 1982). (c) Fringe imaging of a cross-section of a porous carbon from KOH activation of cokes, indicating the relative arrangements of defective graphene layers (Marsh *et al.*, 1982).

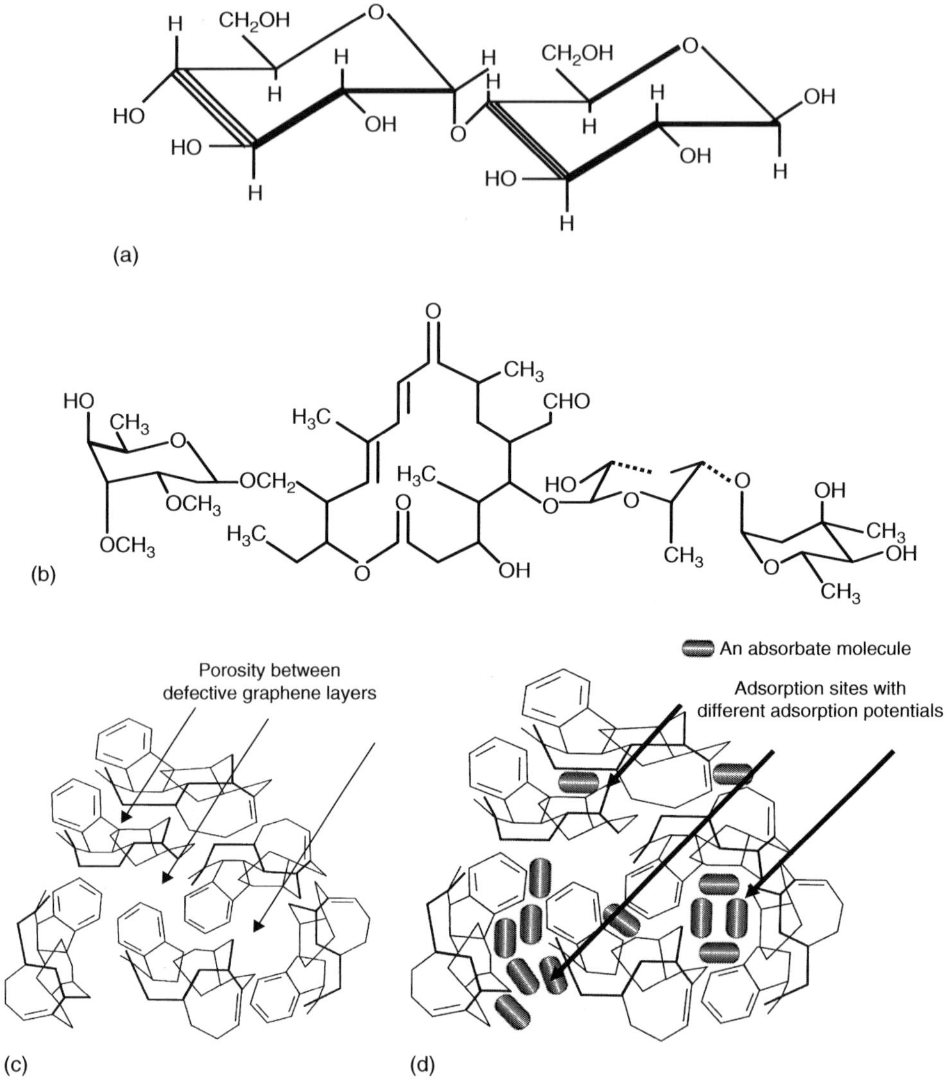

Figure 3.22. (a) Molecular structure of a cellulosic-type precursor (Byrne and Marsh, 1995). (b) A possible structure intermediate between cellulose and that of a resultant carbon (Byrne and Marsh, 1995). (c) Possible highly defective carbonaceous structures which, when interconnected, create microporosity (Byrne and Marsh, 1995). (d) Possible highly defective carbonaceous structures which, when interconnected, create microporosity, this model showing locations of adsorbate molecules (Byrne and Marsh, 1995).

stabilize themselves in five-, six- and seven-membered ring systems. These arrangements are considered not to be planar (see Figure 3.20), but to be curved to create cage-like structures capable of adsorbing an adsorbate molecule (because of the intense dispersive forces generated within these molecular-sized spaces (porosity) (Figure 3.22(d)).

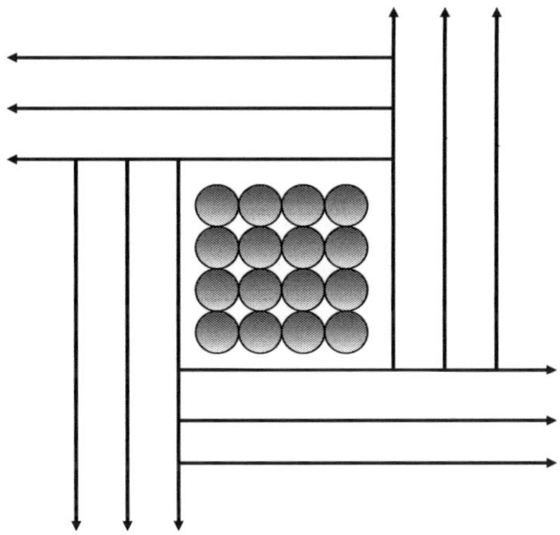

Figure 3.23. Cartoon of packing of 16 argon atoms into a square pore (Bojan and Steele, 1998).

3.4.23 Computer Simulation of Pore-filling Model of Bojan and Steele (1998), Figure 3.23

These authors assumed simple structures for porosity and simulated pore-filling mechanisms to create theoretical isotherms. Monte Carlo computer simulations were reported for the adsorption of argon at 90 K in different pores modeled as perfectly flat graphitic planes assembled to form rectangular cross-sections so allowing four layers of argon in the full pore (Figure 3.23).

A pore having dimensions 7×7 argon atoms was also studied. It was demonstrated that the presence of corners in these pores had a profound effect upon their adsorptive behavior. The dispersive forces at locations of corners will be more intense than on a plane surface and will maximize with a pore capable of adsorbing one molecule only. These are the reasons for variations of enthalpies of adsorption with coverage during the adsorption on to a microporous carbon.

3.4.24 Pore-shape Distributions by Davies and Seaton (1998), Figure 3.24

In studies similar to those of Bojan and Steele (1998) and Davies and Seaton (1998) develop ideas as to how the shape of the pore can influence calculated pore-size distributions. The concept of using slit-shaped pores as a model is based on the use of the graphitic microcrystallite to give the necessary shape. It is appreciated that carbons which are not graphitic will have pore shapes which do not possess this rigid shape and, in an attempt to diminish this problem, Davies and Seaton (1998) included "corners" in the model to indicate the lack of continuity in porosity as suggested by the slit-shaped pore. Other pore shapes were also considered including square and rectangular model pores. As a reference material, the authors used an industrial activated carbon, BPL, and a methane adsorption isotherm obtained at 308 K. This study, centrally, analyses differences in calculated pore-size distributions which result from the use different model shapes and identifies those

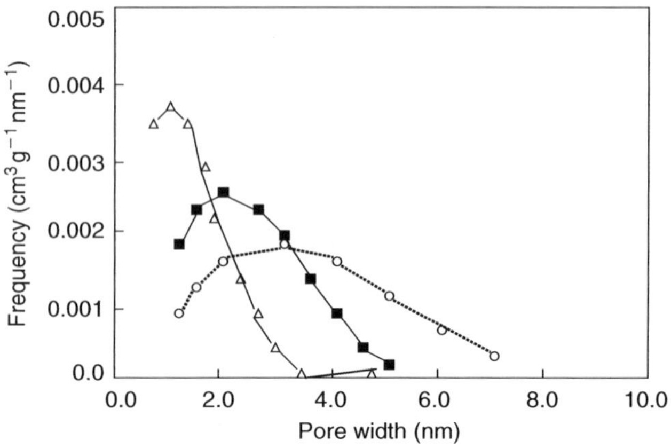

Figure 3.24. Comparison of pore-size distributions based on slit shaped (Δ, square (^{0}) and rectangular (■) model pores (Davies and Seaton, 1998).

aspects of the internal structure of the BPL carbon that can be reliably inferred from the pore-size distribution.

Davies and Seaton (1998) found that pore-size distributions depend strongly on the shape of pore assumed in the modeling. Pores, with square and rectangular shapes, have the same role in the analysis based on slit-shaped pores. This causes pore-size distributions based on these models to be flatter and to move to larger pore sizes compared with those determined using slit-shaped pores. These authors found that it was possible to resolve square and rectangular pores into an equivalent distribution of slit-shaped pores, this suggesting that slit-shaped pores represent the clearest, least complicated description of the internal structure of activated carbon.

Davies and Seaton (1998) conclude that the average pore size based on square pores is always twice as large as that determined using slit-shaped pores, and that this can be related to structure as it exists in an activated carbon. When the activated carbon is highly lamellar and if is assumed that the degree of interconnections in the pores and misaligned planes is negligible, then the average pore size is likely to be close to that determined for the slit-shaped pores. As extents of cross-linking within a carbon increase, as corners become more prevalent, so the average pore size increases.

It is relevant to note (perhaps confirm again) that an experimental (measured and calculated) pore-size distribution is a function of pore shape. However, in industrial activated carbons, and all others for that matter, at the moment, there is no way of knowing with some precision, on a statistical basis, what is the distribution of pore shapes.

Experience would suggest that, as microporosity is of restricted size and shape, governed by the structure of the carbon, there exists a limited facility for broad-based pore-shape distributions. Accordingly, pore-size distributions, as calculated from experimental isotherms, will not deviate far from reality (noting that reality is something which may never be realized). It is only when adsorption data, using a range of adsorbates of increasing molecular

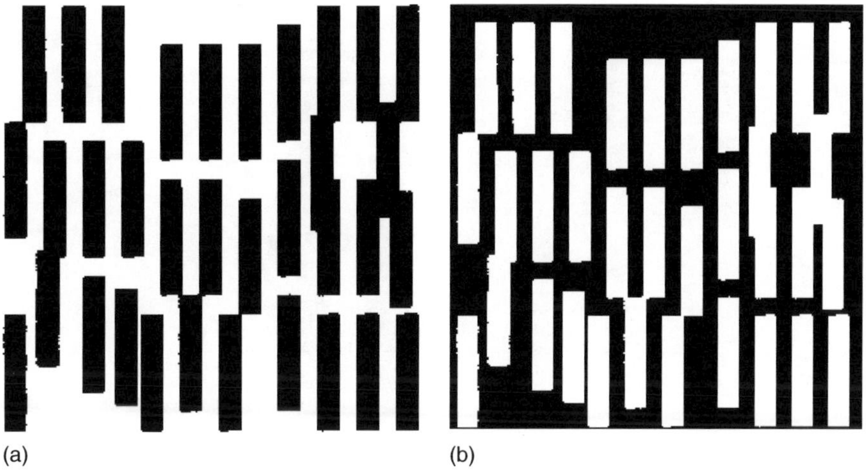

Figure 3.25. (a, b) A simple schematic to demonstrate, for microporous carbons of surface area of about $1000\,m^2g^{-1}$, that the volume elements of carbon and of porosity are essentially interchangeable. In Figure 3.25(a), the black of the carbon elements become the white of the porosity in Figure 3.25(b).

size, indicate that the pore-size distribution associated with that carbon is not correct, that attempts will be needed to look at the pore-shape distributions.

Figure 3.24 is a comparison of these pore-size distributions based on slit shaped, square and rectangular pores. Pore-size distributions modeled on square and rectangular pores are flattened and shifted to larger pore sizes than the distributions based on slit-shaped pores. Hence, the pore-size distributions modeled on the slit shaped and square shapes are the limits to possible distribution shapes. This is because the slit-shaped pore contains no corners, with the square pore having a maximum number of corners.

3.4.25 Nearly Space-filling Fractal Networks of Carbon Nanopores (Micropores) of Pfeifer et al. *(2002), Figure 3.25(a–f)*

Pfeifer *et al.* (2002) organized a "round-robin" study of a series of activated carbons prepared from olive stones. The objective of such a study was to make a comparative assessment of the results of studies using several appropriate techniques. A summary of the findings is given, initially, as below.

Small-angle X-ray scattering, nitrogen adsorption at 77 K and scanning tunneling microscopy showed that a series of activated carbons host an extended fractal network of channels with dimension $D_p = 2.8–3.0$ (pore fractal), channel width 1.5–2.0 nm (lower end of scaling), network diameter 300–340 nm (upper end of scaling) and porosity of 0.3–0.6. These data are interpreted to describe a network of quasi-planar invasion percolation clusters, formed by oxidative removal of pore walls between closed voids of diameter about 1.0 nm and held in registry by fibrils of the biological precursor, and point out unique applications.

Figure 3.25. (c, d) A simple schematic to demonstrate, for microporous carbons of surface area of $<1000\,m^2\,g^{-1}$, that the volume elements of carbon and of porosity are NOT essentially interchangeable but that the "shape format" of the porosity must be linked to the "shape format" of the carbon volume elements. In Figure 2.25(c), the black of the carbon elements becomes the white of the porosity in Figure 3.25(d).

This study of Pfeifer *et al.* (2002) is based on the unique properties of microporous carbons whereby the "volume elements" (shape and size) of the porosity are closely dependent on the volume elements (shape and size) of the constituent carbon material. No other known porous material has this type of internal relationship. Figure 3.25(a, b) are simple schematics which demonstrate this aspect for carbons with a surface area in microporosity of about $1000\,m^2\,g^{-1}$. For carbons of lower surface area ($<1000\,m^2\,g^{-1}$) this equality is reduced as pore walls are thicker relative to the volume elements of the porosity, as suggested in Figure 3.25(c, d). It was these considerations which led to the initiation of the fractal study of the microporosity of these carbons from olive stones.

Pfeifer *et al.* (2002) report that since the first experimental studies of fractal surfaces of disordered solids (Avnir *et al.*, 1984), it has been conjectured that situations may exist in which the pore space – as opposed to the solid, or the surface alone – is fractal. Such fractal networks of channels criss-crossing the solid, termed pore fractals or "negative image" of mass fractals, have attracted interest as a laboratory for unusual dynamics of confined processes, induced by the long-range correlation of the pore space. The scaling laws predicted relate the dynamic exponents to the fractal and spectral dimensions (space-filling and branching properties) of the network and include anomalous diffusion, reaction, free motion, phase transitions, electric conduction of pore fluid and hydrodynamic flow. Here, is reported the first well-documented case of a pore fractal.

The network can be considered as a new member in the family of nano-structured (terminology of authors) carbons. Its channel width is 1.5–2.0 nm, comparable to the width of single-wall carbon nanotubes, but instead of forming an assembly of freestanding tubes or bundle of tubes, the channels are embedded in a solid and are interconnected (the authors are essentially describing a microporous carbon). The network is of multiple interest: (i) Its synthesis differs vastly from that of isolated nanotubes. Created by controlled oxidation, a mainstay of mass production of porous carbons, it promises to be a low-cost competitor of isolated nanotubes for gas storage. (ii) For gas storage, it has outstanding mechanical stability, nano-fluidic properties (rapid transport through branched channels) and capacity (high porosity, condensation in high dimensions) compared to nanotube bundles (see Chapter 2). (iii) It offers a stage for "chemistry in confined spaces" and control of pathways similar to zeolites and other microscopic vessels. (iv) The extended scaling regime, created by what is believed to be invasion percolation, provides a unique platform to compare predicted dynamic exponents with experiment.

Sandstones and Vycor glass have been suggested as pore fractals, but these proposals have been controversial or withdrawn. Pore fractals are more challenging to ascertain than mass or surface fractals because they do not reveal their fractality when probed with material yardsticks: the pore-size distribution of a pore fractal is a delta function (mass and surface fractals give a power law), so an intruding non-wetting liquid, capillary condensate, or adsorbed layer will either fill the entire pore space or leave it entirely empty, depending on whether the radius of curvature of the liquid, or adsorbate size, is smaller or larger than the channel. Such an experiment shows the presence of channels, but gives no information about their spatial arrangement.

The porous carbons in this fractal study are activated carbons prepared as follows (Rodríguez-Reinoso *et al.*, 1995). Olive stones were carbonized in nitrogen at 850 °C, yielding a char denoted by HO. The char was oxidized (activated) by H_2O at 750 °C ($C + H_2O = CO + H_2$) at 1 wt% loss of carbon per hour, yielding carbons with weight loss of 8–74 wt%, denoted by H8–H74. The surface geometry of the samples (ground, dried) was investigated by small-angle X-ray scattering (SAXS). The scattered intensity at momentum transfer q, $I(q)$, was measured on a Bonse–Hart spectrometer for $q < 0.001\,\text{nm}^{-1}$ and a pinhole spectrometer for $q > 0.001\,\text{nm}^{-1}$. This gives data over an unusually wide range, $0.0003 \leqslant q \leqslant 0.07\,\text{nm}^{-1}$, with one decade of overlap of the two instruments.

To understand the scattering curves and their evolution with progressive oxidation all three types of fractals need to be considered. The intensity from a surface/mass/pore fractal obeys the power law $I(q) \propto q^{D_s 2(D_m + D_p)} + 6$ where D_s, D_m and D_p are the fractal dimension of the surface, mass and pore space, respectively. However, for a more detailed account of the fractal study, appropriate information are contained with the publication of Pfeifer *et al.* (2002) and references contained therein.

Since most channels of a pore fractal are deep inside the solid, invisible from the outside and fill the solid non-uniformly, sparse (isolated) entrances at the external surface are to be expected. Scanning tunneling microscopy (STM; Figure 3.25(f)) confirms this. The channel width of 1.3 ± 0.7 nm for H37, sampled over 42 images, agrees with $L_{min} = 1.6$ nm from SAXS sampled over the whole network. The width at the entrance and in the solid is the same. Selected as a region suitable for STM imaging, Figure 3.25(f) shows little external roughness. Roughness consistent with $D_s = 2.19$ at large scales is seen in scanning electron micrographs by Iley *et al.* (1973).

What drives the formation of pore fractals in the samples of the study? Sample HO is considered and has an easily accessible external surface ($D_s = 2.23$), no open pores in which nitrogen would condense, and closed pores with mean diameter $L_o = 1.0$ nm. From V_{pore} (HO) = V_{N_2} (H8) − $0.08_{msolid/nsolid} = 0.21\,cm^3$, it is found that the volume fraction of closed pores in HO is 0.30 which gives a mean distance of 1.2 nm between centers of neighboring voids. Oxidation by H_2O removes the weakest wall between the solid's exterior and a nearby closed pore, creating a channel through which new H_2O removes the next weak wall, and so on. This carves out an invasion percolation cluster. A planar invasion percolation cluster has fractal dimension 1.90 (=1 + 43/48), and a stack of such clusters in registry has the dimension 2.90.

This model of activated carbons proposed for H8–H74, is consistent with $D_p = 2.9 \pm 0.1$, that I_p and $N(0.05)$ grow and I_o drops with increasing burn-off, that pore mouths lie on parallel fault lines (running vertically in Figure 3.25(f), left half), and that non-invaded regions, presumably pseudomorphs of fibrils of the biological precursor, are in registry transverse to the fault planes (Figure 3.25(f), lower right quadrant). Without registry, the stack would be uniform above a few planes and scatter only at $q = 2\partial$ (spacing of planes). The model predicts that L_{min} remains constant and the width of the invaded zone, L_{max}, grows with increasing burn-off, as observed. The growth stops when the accumulation of hydrogen inside the network slows the removal rate of weak walls down to the removal rate of strong walls at the external surface. The invasion front then advances in unison with the receding external surface, and D_p remains constant because strong walls stay inert until they become part of the external surface (H37–H74).

Pore fractals similar to the ones established here could be formed under a variety of conditions. However, for oxidation of HO with CO_2, or CO_2–H_2O mixtures and at higher temperatures, broad pore-size distributions and surface fractals are found (Ehrburger-Dolle *et al.*, 1997, 2005). Most likely, the conditions for "pore aggregation" to yield pore fractals will be as complex as the conditions are for particle aggregation to yield mass fractals.

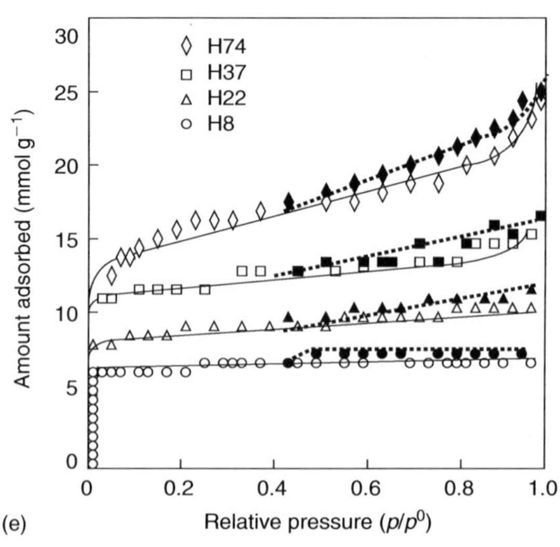

(e)

Figure 3.25. (e) Adsorption and desorption isotherms of nitrogen at 77 K, on activated olive stones of increasing burn-offs in water vapor. The non-activated sample adsorbed little nitrogen at 77 K (Rodríguez-Reinoso *et al.*, 1995).

Figure 3.25. (f) Scanning tunneling micrograph (STM) of sample of activated olive stone, H37 (32×32 nm). Solid black areas are channel entrances. The width of the entrance marked within an arrow is 1.4 nm, at that location (Pfeifer *et al.*, 2002).

3.4.26 Evaluation of Slit-like Porosity by Gun'ko and Mikhalovsky (2004) and Yang et al. (2002), Figure 3.26(a–c)

Assessments of the various models of this chapter and the discussions of carbon formation of Chapter 2 lead to the conclusion that the volume elements which make up microporosity are essentially slit shaped. This conclusion was put to the test by Gun'ko and Mikhalovsky (2004) in a semi-theoretical approach in which possible deviations (called (Δw_{slit}) of pores shapes from slit-like structures were calculated from:

$$\Delta w_{slit} = \frac{S_{BET}}{S_{slit}} - 1 \tag{3.2}$$

where S_{BET} is the specific surface area determined using the standard BET method of adsorption of nitrogen at 77 K, and S_{slit} is the surface area calculated from the differential pore-size distribution.

This study does not make a note of the problematic status of the BET equation and the way it is used to interpret isotherm shape (as explained in Chapter 4) particularly for carbons with mixed porosities of micro and mesopores. The interpretation problems are minimized for Type-I isotherms (adsorption in micropores only) but are maximized for carbons with mesopores of various shapes. Should carbons, of a microporous nature, also possess porosity of dimensions <1.0 nm, then activated diffusion effects (Chapter 4) prevent entry of nitrogen as an adsorbate at 77 K into these smallest of micropores. It is always advisable to check out the validity of the nitrogen BET surface area by comparing with the adsorption of carbon dioxide at 273 K.

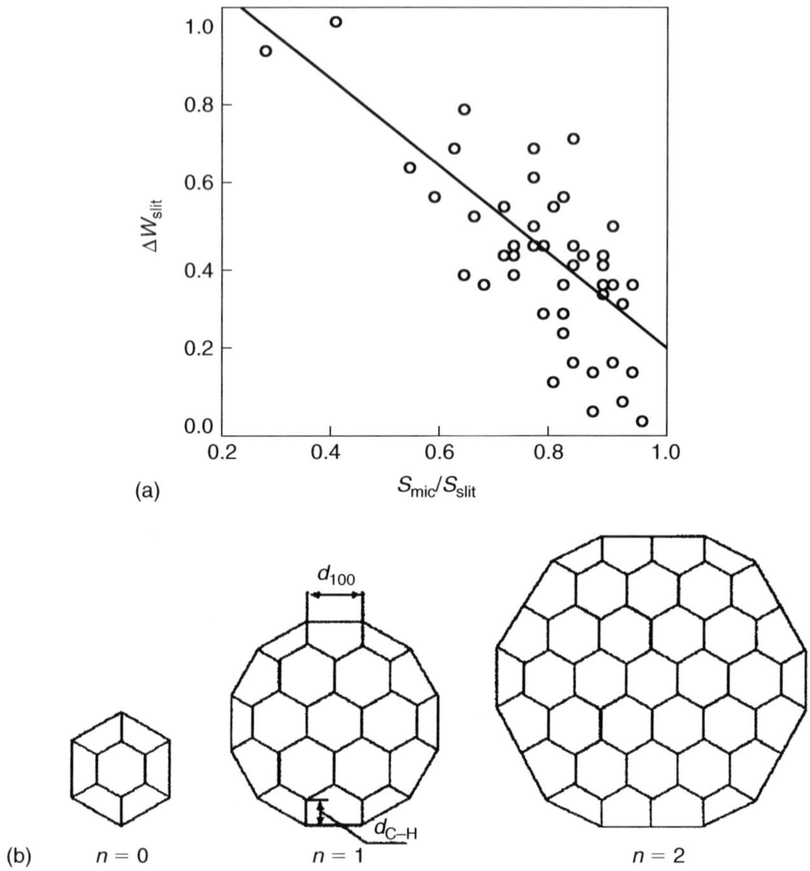

Figure 3.26. (a) The variation of Δw_{slit} with the ratio of S_{mic}/S_{slit}. The correlation coefficient $R^2 \approx 0.73$ (Gun'ko and Mikhalovsky, 2004). (b) Models of one layer of a polyhexagonal carbon atoms, including hydrogen attached to the edges. Adsorption was assumed to take place on both surfaces and at the edges (Gun'ko and Mikhalovsky, 2004). (c) The variation of idealized (calculated) surface area derived from X-ray line-broadening data (coordinate) with the BET (obtained from nitrogen adsorption at 77 K) surface areas (abscissa) (Gun'ko and Mikhalovsky, 2004).

Values of S_{slit} were calculated from:

$$S_{slit} = \int_{x_{min}}^{x_{max}} f_s(x)\mathrm{d}x = \int_{x_{min}}^{x_{max}} \frac{w}{x}\left(f_v(x) - \frac{V_p}{x}\right)\mathrm{d}x \tag{3.3}$$

where x_{min} and x_{max} are the minimum and maximum values of half-widths of pores, and given values of 0.2 nm and 100 nm, respectively. Values of the pore-size distributions were calculated as suggested by Nguyen and Do (1999) and Do and Do (2002). Values of w are assigned as being equal to 1, 2 or 3 for slit-like, cylindrical and spherical pores, respectively.

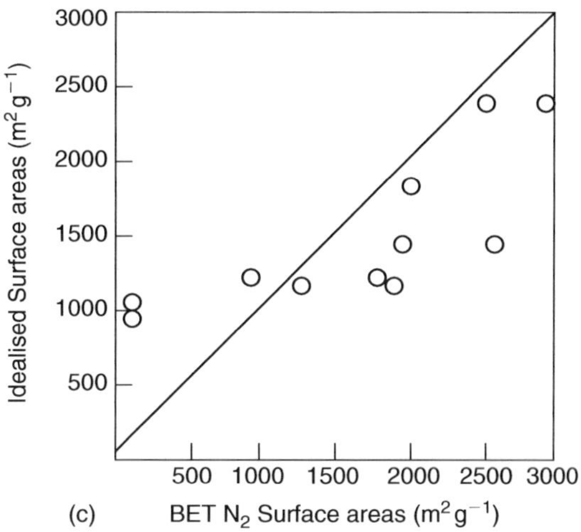

Figure 3.26. (continued)

The carbons used included activated carbons, carbon fibers, carbon blacks, industrial carbons with values of S_{BET} ranging from 47 to 2789 m^2 g^{-1}. Further details of these carbons are to be found in another 16 articles scattered within 8 different journals. (Even with some electronic access to journals, such a search is tedious and time consuming.)

Figure 2.26(a) summarizes results with reference to microporosity (of importance to activated carbons). This is a plot of Δw_{slit} against the ratio S_{mic}/S_{slit} (the determination of S_{mic} is described in several references within Gun'ko and Mikhalovsky (2004)). It is to be expected that values of Δw_{slit} will decrease with increasing values of S_{BET} because of an enhancement of contributions of micropores. The authors assume that the experimental values of S_{BET} are acceptable when values of Δw_{slit} are close to zero. This is a fair assumption as, for example for a Type-I isotherm; there is little room for error. However, for values of $\Delta w_{slit} < 0.2$ it can be assumed that the S_{BET} values have been over estimated. This is likely if an isotherm, Type-II or Type-IV, has been interpreted by computer software attached to the adsorption measuring equipment. Thus, by not visually examining the isotherms and assessing calculated monolayer capacities against their positions on the isotherm shape this flaw has been carried over into the calculations which indicate that there is a flaw. So the argument has gone full circle.

This type of analysis also indicates that values of $\Delta w_{slit} > 0.2$ result from adsorption in porosity which is not slit shaped, but may be cylindrical or spherical. Such adsorptions processes have been known for several decades with so many articles discussing pore filling mechanisms at values of $p/p^0 > 0.5$. The shapes of mesoporosity need not be as symmetrical as a cylinder or a sphere as micrographs of later chapters will show. But this is as far as such modeling can go at the moment.

Yang *et al.* (2002) manufactured a range of carbons from a bituminous coal by carbonization followed by steam activation and by carbonization and heat treatment with KOH. These authors apparently were not interested in the activation processes, but in the production of a series of carbons with a wide range of surface areas from 56 to 2540 $m^2 g^{-1}$. Crystallite sizes were determined from diffraction line-broadening information with L_a and L_c values, for example, being 2.916 and 1.533 nm (surface areas of 132 $m^2 g^{-1}$) and 3.309 and 1.007 nm (surface areas of 1784 $m^2 g^{-1}$). Clearly, there is no correlation between these numbers.

In an approach similar to that of Kaneko *et al.* (1992a) (Section 3.4.6) and Gun'ko and Mikhalovsky (2004) attempted to calculate surface areas within carbons as if these surfaces were the actual surfaces on which adsorption took place. Their models were based on layers of polyhexagonal carbon atoms as indicated in Figure 3.26(b) with their resultant correlations as shown in Figure 3.26(c). It is seen in Figure 3.26(c) that there is indeed no correlation between the idealized surface areas (as calculated) and the BET surface areas as determined form the nitrogen isotherm obtained at 77 K. It is not difficult to understand why this is, with reasons being summarized as below:

1. Experimentally, it is well established in the literature that for filling of microporosity of surface area of 1000 $m^2 g^{-1}$, the ratio of carbon atoms to adsorbate molecules is about 8:1. As extents of accessible surface area in microporosity decreases so this ratio increases to >800:1 (corresponding to a surface area of 10 $m^2 g^{-1}$). This is because much of the carbon material is not forming part of the surfaces of porosity. This basic consideration indicates the unreasonableness of attempting such a calculation. Further, any consideration of the L_a and L_c values of a non-participating carbon is equally unrealistic.
2. No serious attempts have been made to interpret adsorption data correctly. Low values of surface areas of this series of porous carbons comes from the inappropriate use of the isotherm of nitrogen adsorption at 77 K. Incidentally, all isotherms should be made available in all publications. The low values are associated with slow-activated diffusion effects in microporosities of dimensions <0.5 nm. High values of surface area ($\geqslant$1200 $m^2 g^{-1}$) result from multilayer adsorption in the wider of the micropores and in porosities of indefinite shape, approximating to cylinders, cones, spheres, etc. Hence, it seems unreasonable to relate such adsorption processes to monolayer adsorption on surfaces of graphitic microcrystallites.
3. The authors attempt to explain these deviations as being due to capillary condensation at low values of relative pressure (p/p^0). However, it is well established that the phenomena of capillary condensation does not occur at low values of relative pressure, $\geqslant$0.6 p/p^0.

The characteristics of Figure 3.26(c) can be interpreted as follows. At low values of surface area, <200 $m^2 g^{-1}$, the model predicts surface areas on a vertical line from 500 to 2500 $m^2 g^{-1}$. As surface area (BET) increases to about 1200 $m^2 g^{-1}$ so equality between the two approaches is realized, that is all the carbon atoms of the structure are participating in the adsorption process. However, beyond about 1200 $m^2 g^{-1}$ the values of surface area assigned to the carbons, 1200–2900 $m^2 g^{-1}$, are unrealistic as they include significant contributions to volume-filling mechanisms. Hence, these seven values will again lie on a vertical line placed at the 1200 $m^2 g^{-1}$ position.

All of these arguments would not have been necessary if surface areas had been correctly obtained in the first place. The assumption has to be abandoned that the BET analysis (by computer) of a nitrogen isotherm obtained at 77 K is infallible. It is considered that it may be more realistic, and less misleading, not to use the surface area concept but to use pore volumes (e.g. micropore volume) on a comparative basis.

A study which attempts to model non-existent surface areas has entered a world of mythology from which there is no escape.

3.4.27 Star-like Porosity by Py et al. *(2004), Figure 3.27(a, b)*

Py *et al.* (2004) proposed simple geometric models of porosity based on the BSU packing approach of Oberlin *et al.* (1980) and Oberlin 1989), as described in Section 3.4.10, and

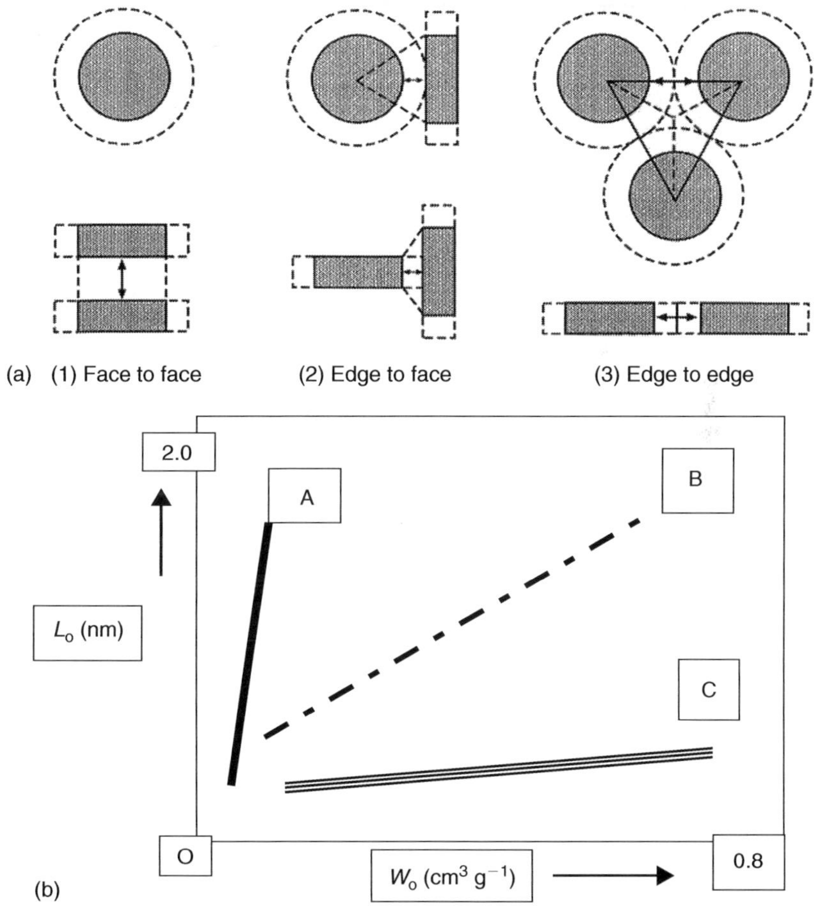

Figure 3.27. (a) Models of arrangements of graphitic microcrystallites in activated carbons used to explain the porosity. (Py *et al.*, 2004). (b) A diagram of the variation of average pore dimension (L_o in nm) obtained from pore-size distributions with specific micropore volume (cm^{3}g^{-1}) obtained from isotherm data, for a selection of porous carbons (Py *et al.*, 2004).

tested their models using activated pitches and a commercial coconut activated carbon. The models were centered around the graphitic microcrystallite theory that has been explained at length in Chapter 2 and commented upon already, unfavorably, in this chapter. The models assumed three types of assemblies for the graphitic microcrystallite as indicated in Figure 3.27(a). The model is also used to explain activation processes involving increases both in total pore volume (W_0) and mean pore size (L_0). In the first assembly (1), the disk-like BSUs use only their faces as adsorption sites (face to face). In the second assembly, the face–edge case (2), there is a BSU face and a BSU edge of relevance. Activation is assumed to involve a sole benzene ring removal. In the third assembly (3) (in Figure 3.27(a)) there are only edge–edge interactions to create a star-like arrangement. An oxidative treatment is assumed to bring about the removal of two of these BSUs.

The approach of Py *et al.* (2004) is based on two assumptions, namely that (a) carbon material is composed of graphitic microcrystallites, and that (b) it is the presentation at surfaces of these graphitic microcrystallites which controls the adsorption process.

As explained at some length in Chapter 2, and further in this chapter, the concept of the graphitic microcrystallite is unacceptable and that its use leads to explanations of properties of carbons which are totally unrealistic. This chapter uses data placed into a graph of the type of Figure 3.27(b). It is considered that porous carbons fall into three distinct groups which approximate to lines A, B and C of this graph and that these carbons are so different that they require different arrangements of the graphitic microcrystallites, as described in Figure 3.27(a). This is too simple an assumption. It is unfortunate that a much broader spectrum of porous carbons was not considered as several hundreds of carbons are characterized in the literature. In fact, if this had been done, using HTT series of carbons of different origins, as well as carbons prepared by activation processes (and many more), then the area between lines A and C of Figure 3.27(b) would be completely filled with data. That is, the availability of porous carbons is so extensive that samples of appropriate values of L_0 and W_0 could be found. For example, the carbonization of synthetic polymers such as polyfurfuryl alcohol, polyvinylidene chloride, the resorcinol–formaldehyde systems of the aerogels, low-rank coals, and woods and other organic macromolecular systems (of which there are many) provide carbons with increasing average pore dimensions and pore volumes. These values decrease on continuous heat treatment to provide yet more data to fill the area of Figure 3.27(b).

The assumptions of this chapter, leading to specific areas for carbons in Figure 3.27(b) are based on the use of too small a database; at least several hundred porous carbons should have been included.

With this mind, the need to speculate on possible arrangements of graphitic microcrystallites to explain these properties of porosity disappears.

The second assumption (B), that is, is the size and orientation of the graphitic microcrystallite which controls the adsorption process, has to be addressed very seriously. Extents and enthalpies of adsorption are massively dependent on the London Dispersion Forces (van der Waals forces) which exist within the porosity of a carbon. Such porosity is unique among solid porous materials and exhibits considerable versatility. Calculations of areas of theoretical graphitic microcrystallites, as has been reported above, with the assumption that these areas are available for adsorption, have little relevance.

3.4.28 Is a Definitive Model of Microporosity Possible? Figure 3.28(a–h)

The discussions of Chapters 2 and 3 focus attention on the following aspects of structure of porosity in porous carbons. (i) Porosity is a three-dimensional network of space of

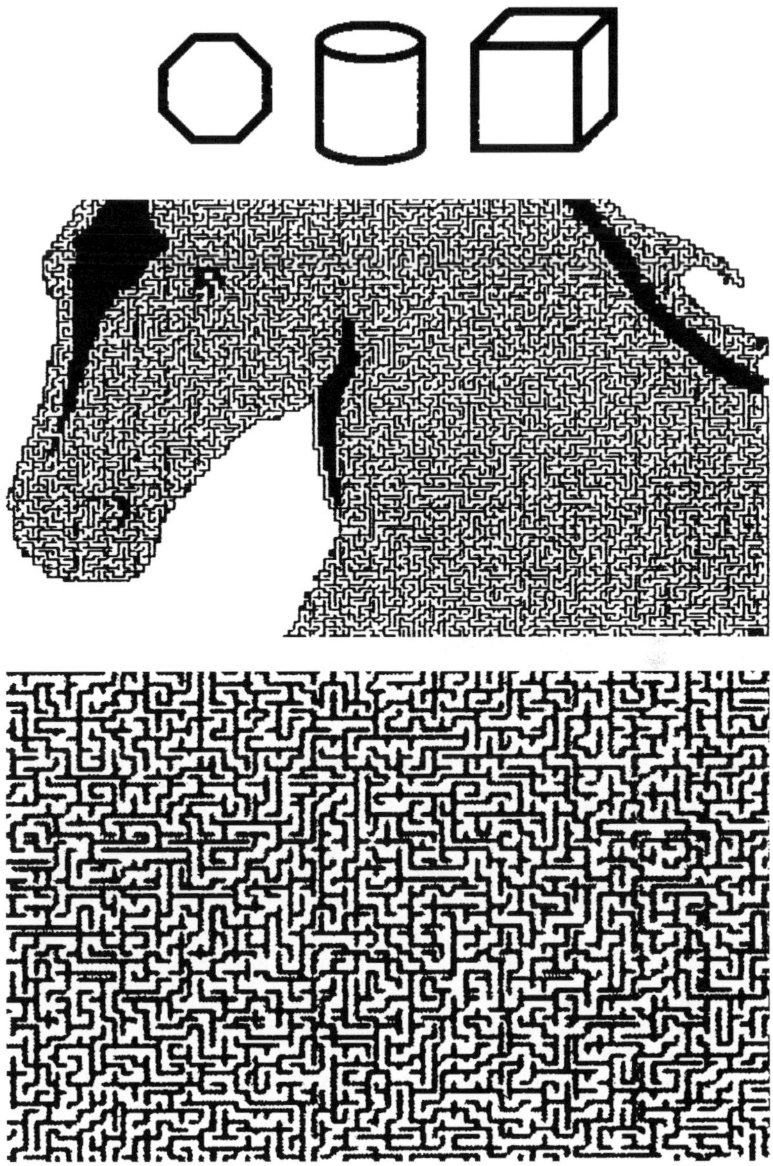

(a)

(b)

(c)

Figure 3.28. (a) Simple models to make the point that it is the entrance dimension and shape of porosity which controls the adsorption process. (b) A two-dimensional labyrinth (maze) which is proposed as a model for the network of porosity in carbons. (c) A two-dimensional labyrinth (maze) which is proposed as a model for the network of porosity in carbons. The dimensions of the quadrant correspond to 100 nm by 60 nm.

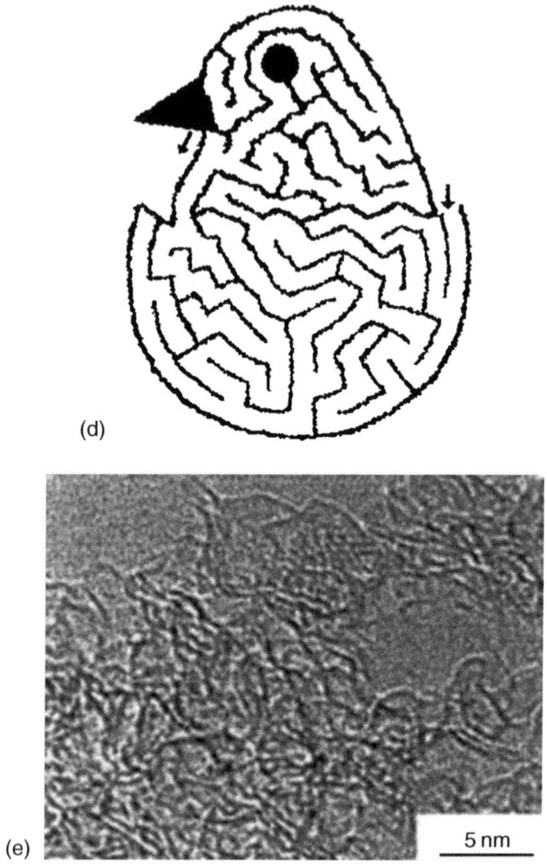

Figure 3.28. (d) A two-dimensional labyrinth (maze) which is proposed as a simple model for the network of porosity in carbons. (e) High-resolution transmission electron micrographs of a soot nanoparticle carbon indicating defective graphene layers (Rouzaud and Clinard, 2002).

intense London Dispersion Forces capable of establishing the phenomenon of physical adsorption from the gas and liquid phases. (ii) Microporosity (dominant in adsorption processes) is dominantly slit shaped with a dimension of <2.0 nm. (iii) The walls of the porosity are composed of carbon atoms which are almost totally bonded to each other in ring systems (principally six-membered, but five or seven may be present), the ring systems being conjugated into a three-dimensional defective graphene layer. Residual hydrogen, oxygen and possibly nitrogen are sulfur is to located at edges of the graphene layers. It is totally reasonable to consider that every carbon atom in a carbon particle may have a connecting path to very other carbon atoms within that particle. So, in this sense, a single but highly convoluted graphene layer constitutes structure. (iv) It is the convolutions of this defective graphene layer, the way it can turn around on itself and bond to itself at frequent intervals, which creates and stabilizes porosity in carbons as it is understood today.

A piece of information which is missing from pore specifications is the length and height of the pore. It is the entrance dimension which is always quoted. It could well be that this

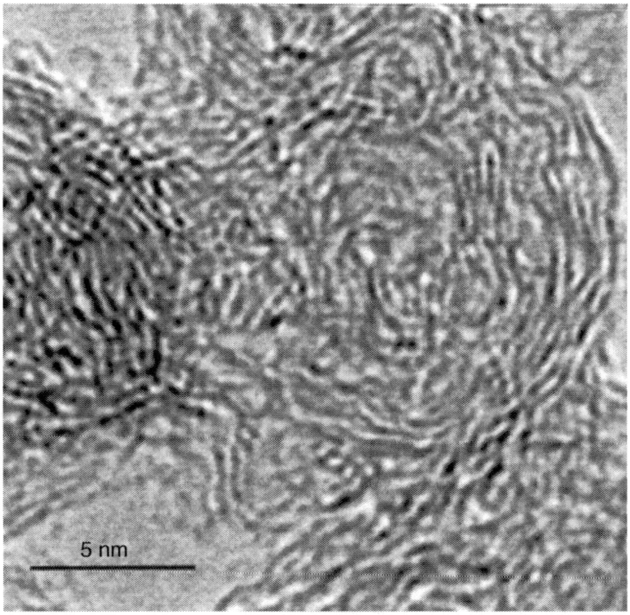

(f)

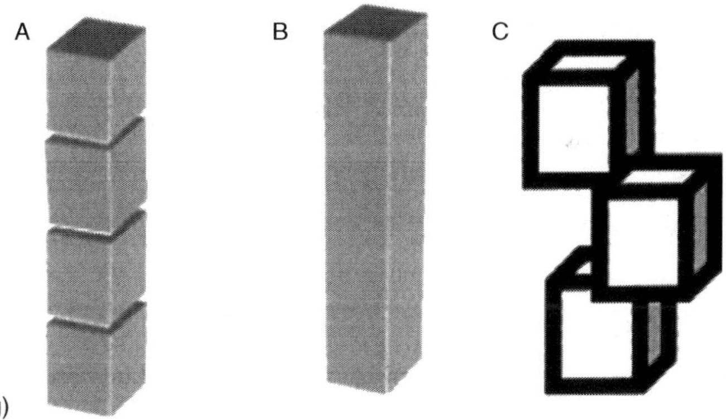

(g)

Figure 3.28. (f) High-resolution transmission electron micrographs of an activated saccharose-based carbon indicating defective graphene layers (Rouzaud and Clinard, 2002). (g) Simple models to assist with interpretations of variations in the (C_xN_y) ratio of adsorption of nitrogen at 77 K, to form a monolayer, in microporous carbons. The central model suggests a minimum in the ratio value of (C_8N), increasing to higher values as more pore walls are available (A) or as pore walls become thicker (B).

number is all important for many purposes and that length and height have no real practical value. Figure 3.28(a) illustrates this problem. In the first instance, it is the entrance dimension and shape which control the adsorption process, be this (e.g.) hexagonal (an appropriate shape), or circular or square. Once the adsorbate molecule is through the pore entrance, then the characteristics of the adsorbent take over and the isotherm is created. Is this enough? The answer is no, because the processes of physical and chemical activation have to be understood and to do this requires three-dimensional models.

The processes of physical and chemical activation of carbons, Chapters 5 and 6, extract carbon atoms from the interior of the network of porosity in carbons. It would be an advantage to have some idea of the structure of this network, in three dimensions, in order to understand the extraction (gasification) process. This three-dimensional network or porosity can be described as a labyrinth and this automatically takes us into children's maze games, some of which are complex (computer generated) as illustrated in Figure 3.28(b, c), while others are quite simple, as shown in Figure 3.28(d), designed for children under five.

The similarities between the labyrinth of a maze and of porosity in a carbon are quite remarkable although it is necessary to mention some limitations, initially. There are four limitations of importance. First, this maze, of course, is in two dimensions; second, the lines of the maze are too orientated relative to an *x*–*y* axis. Such parallelism is unlikely to exist within a porous carbon; third, this labyrinth is best suited to a microporous carbon, only and not to microporous carbon fibers. Fourth, in such a model, rates of diffusion are likely to be too slow and hence there is a need to consider the location of mesoporosity. The inclusion of mesoporosity is another matter. Mesoporosity has to promote enhanced adsorption to the interior of the fiber. As a matter of scaling, although the models of Figure 3.28(a–c) provide an impressive number of adsorption locations, it will require about 10^{19} of such models, as shown in Figure 3.28(c), to describe 1 g of carbon. The human mind cannot cope with this necessity.

But, apart from these limitations the similarities are relevant enough:

1. There is a continuous connection between the lines through out the labyrinth. All carbon atoms form part of a continuous graphene sheet.
2. There is a continuous connection of the routes (spaces) of the labyrinth. Hence, all adsorption sites are available to the adsorbate molecules.
3. The widths of the routes (spaces) of the labyrinth are not constant. Some are narrower than other. This is a very relevant point as it demonstrates, very clearly, the range of porosities, within the definition of microporosity of $<$2.0 nm, this accounting for molecular sieving effects.
4. Some of the routes of the labyrinth are barely visible (being very close to each other). This offers the suggestion that it is representative of closed porosity, that is closed to everything except helium and lithium, noting, on the way, that the term *closed porosity* is an imprecise term, meaning porosity not accessible to a specifically defined adsorbate molecule.
5. The model of Figure 3.28(c) makes the point that access to the interior of the labyrinth is available from all external surfaces. (Note to the use of a maze as a game – here there is one entrance and one exit gate.)
6. A close inspection of the edges of the lines of labyrinth (surfaces of the carbon surfaces) indicates a lack of smoothness, the edges being *rough*. This point is of importance to carbon science because the surfaces of porosity are of imperfect graphene-like layers, with surface irregularities and other defects.
7. The model also demonstrates that most of the continuous three-dimensional graphene layer is able to act as an adsorbent surface making use of both sides. Rarely, do parts of the graphene layer come together (stack) in two or three layers. This requirement

follows on from the observation that a porous carbon, of $1000\,m^2g^{-1}$, requires 10 mmol of nitrogen (adsorbate). The ratio of carbon atoms to nitrogen molecules is then $C_8(N_2)$, that is one nitrogen molecule covers one six-membered ring structure (or there abouts).

There exists one experimental technique for looking at structure within porous carbons and that is quantitative high-resolution transmission electron microscopy (Rouzaud and Clinard, 2002). Two such micrographs are shown in Figure 3.28(e, f), the former being of an activated saccharose-based carbon, the latter being of soot nanoparticles with a concentric microtexture. Note the similarity to the maze diagrams with the exceptions that the imaged graphene layers are not linear and there is no parallelism of the graphene layers.

In conclusion, a suggestion is that when it comes to defining the dimensions of a micropore, the following numbers can be given: entrance dimension, is $<2.0\,nm$, length is the dimension of the carbon particle; height will be greater than the width but is difficult to estimate.

Differences in micropore contents of carbons may be linked to the ratio of carbon atoms to nitrogen molecules (C_xN_y) of the carbon at the position of monolayer coverage. For a carbon of surface area of $1000\,m^2g^{-1}$ in microporosity this ratio is about 8.0. This ratio decreases only when multilayer adsorption occurs in mesoporosity, and this is not part of the current discussion. It is common to find that ratio has a value of >8.0 and there are two reasons (at least for this). First, the pore structure has more intervening walls, as compared in Figure 3.28(g), the (A) having more pore walls. The second reason is that the pore walls are thicker as represented in (C) of Figure 3.28(g). As values of (C_xN_y) increase so the micropore volume will decrease, relative to 1 g of carbon.

Consequently, this little-used ratio of (C_xN_y) could provide initial clues as to the structural difference between microporous carbons prepared from different parent materials. Can microporous carbons be considered, satisfactorily, in this simple way?

This discussion and analysis of modeling of structure in carbons has come a long way from the earlier discussion of drill holes in a block of wood (Section 3.4.1). This discussion should be of assistance when it comes to discussing methods of characterization, of activation of carbons and of their applications.

3.5 Model Assessments and Criteria for Acceptance

Of the above models, those of Figures 3.13–3.17 and 3.19–3.22 represent structures of maximum acceptance. These structures can be checked against the requirements of Table 3.1. Within the models, there is facility for extensive cross-linkages to create hardness, to possess heteroatoms within the carbon networks and to establish an isotropic material. Densities are incorporated into the models and the co-existence of a molecular space network within the carbon network is clear, so facilitating adsorption of gases through the interconnected networks to sites of different sizes, controlled by the various close proximities of the carbon networks. Realistic speculation is possible about the "shape" of an adsorption site and the ratio of six carbon atoms to one adsorbate molecule is quite reasonable. Physical activation requires preferential removal of carbon atoms by gasification using, e.g. carbon dioxide, and this requires that some carbon atoms should be more chemically reactive than others. The range of such "different" sites is a feature of the models. Clearly, all the requirements for molecular-sieve properties, the dynamics and enthalpies of adsorption are present even though the detail is absent. The effect of increasing HTT would be to remove defects or irregularities of the carbon atom network resulting in decreased enthalpies of adsorption on more homogeneous surfaces. Such a porous system would respond to SAXS and SANS with the range of ring structures in the carbon atom network being sufficient to explain observed Raman spectra and electron spin resonance (ESR) during the carbonization process. The use of such carbon in lithium-ion batteries has been explained in Chapter 2 and the suggestions of microporosities (<0.5 nm dimension) can be envisaged, that is the "hidden" porosity.

This chapter has reviewed modeling over the years and the progress that can be identified. It has been demonstrated that modelers, using modern computational systems, base their calculations on structure systems, usually around the graphitic microcrystallite. An objective of such modelers is the simulation of adsorption isotherms determined experimentally. Whether or not the structural models assumed for their work can be considered to be "realistic" is rarely a matter for discussion. Accordingly, the considerations of such modelers should not be adopted, uncritically, by those who have other interests in activated carbon.

There is no unique structure within an activated carbon which provides a specific isotherm, for example the adsorption of benzene at 273 K. The isotherm is a description of the distribution of adsorption potentials throughout the carbon, this distribution following a normal or Gaussian distribution. If a structure is therefore devised which permits a continuous distribution of adsorption potentials, and this model predicts an experimental adsorption isotherm, this then is no guarantee that the structure of the model is correct. The wider experience of the carbon scientist, who relates the model to preparation methods and physical and chemical properties of the carbon, has to pronounce on the "reality" or acceptance value of the model. Unfortunately, the modeler appears not to consult the carbon chemist too much, and it is left to the carbon chemist to explain the limited acceptability of the adopted structures of the modeler.

This approach of the modeler is still to be found in the latest of publications, for example that of Kowalczyk *et al.* (2004). Different models rely on different assumptions in order to obtain relationships allowing the calculation of the main characteristics of adsorbent structural heterogeneity. The assumptions of Kowalczyk *et al.* (2004) are summarized as follows:

(a) The shape of the pores is assumed to be slit like with effects of interconnectivity to be neglected.

(b) The molecule–surface interactions obey the equation of steel.

(c) The adsorbed phase can be considered as liquid-like layers between the two parallel walls of the pore, as a monolayer or as a double layer.

(d) The thickness of the adsorbed layer is determined by the distance between two parallel walls of the slit-like porosity, the density of the layers being constant between the layers and being equal to that of the bulk liquid.

(e) No gas-phase adsorbate exists within the volumes between the layers.

Earlier, Olivier (1998) used GCMC and DFT as modeling methods based on slit-like pores with graphitic surfaces, as used by Kowalczyk *et al.* (2004). Olivier (1998) concluded that the inclusion of surface heterogeneity into the model would have little effect on obtained simulations. Further, Olivier (1998) considers that the use of alternative geometries may be worth pursuing.

The approach of Pfeifer *et al.* (2002) is interesting in that several quite different approaches have been brought to the modeling of activated carbons. The concept of porosity not accessible to nitrogen at 77 K is understood. A different concept, that the surface structure (in terms of pore density) is not representative of the bulk density of porosity, can be commented upon adversely. The suggestion that the fibrils of botanical structure of the olive stones have a role to play in maintaining the mechanical strength of the stones may need to be modified in the light of modern analyses. The suggestions to describe the activation process by gasification of complete graphene layers can more or less be substantiated (see Chapter 5) as can the role of hydrogen as an inhibitor of the carbon–steam reaction.

The approach of Section 3.4.26 is limited because of confusion over what are acceptable surface area values and what are not acceptable surface area values. Section 3.4.27 makes two assumptions which can be commented upon critically, namely that carbons can be categorized into well-defined groups according to their pore volumes and mean pore dimensions, and that it is adsorption on graphitic microcrystallites which controls the creation of an isotherm. Finally, the use of two mazes, two-dimensional and three-dimensional, designed as puzzle games provides helpful models to understand the complexity of the network of microporosity within a carbon. The simple ratio of carbon atoms to nitrogen molecules (C_xN_y) adsorbed as a monolayer capacity could be useful in assessments of how the microporosity fits into the carbon layer networks.

Models designed to describe the microporous nature of carbons have been compared and contrasted leading to an assessment of the requirements of a comprehensive model to account for the properties of microporous systems. No comprehensive model has, as yet, been created. The use of a model, based on a maze, provides insights.

References

Acharya M, Strano MS, Mathews JP, Billinge SH, Petkov V, Subramoney S, Foley HC. Simulation of nonporous carbons: a chemically constrained structure. *Philos Mag B* 1999;79(10):1499–1518.

Avnir D, Farin D, Pfeifer P. Molecular fractal surfaces. *Nature (London)* 1984;308:261.

Biggs M, Agarwal P. Mass diffusion of atomic fluids in random micropore spaces using equilibrium molecular dynamics. *Phys Rev A* 1992;46:3312–3318.

Biggs M, Agarwal P. Mass diffusion of diatomic fluids in random micropore spaces using equilibrium molecular dynamics. *Phys Rev E* 1994;49:531–537.

Biggs MJ, Buts A, Williamson D. Molecular simulation evidence for solidlike adsorbate in complex carbonaceous micropore structures. *Langmuir* 2004a;20(14):5786–5800.

Biggs MJ, Buts A, Williamson D. Absolute assessment of adsorption-based porous solid characterisation methods. *Langmuir* 2004b;20(17):7123–7138.

Bojan MJ, Steele WA. Computer simulation in pores with rectangular cross-sections. *Carbon* 1998;36(10):1417–1423.

Byrne JF, Marsh H. Introductory Overview, in '*Porosity in carbons; characterization and applications*'. Ed. Patrick JW. Arnold, London, 1995, pp. 1–48.

Clar E. *Polycyclic Hydrocarbons*, Vols. 1 and 2. Academic Press, New York, 1964.

Da Silva FLB, Olivares-Rivas W, Degrève L, Akesson T. Application of a new reverse Monte Carlo algorithm to polyatomic molecular systems. 1. Liquid water. *J Chem Phys* 2001;114:907–914.

Dahn JR, Xing W, Gao T. The "falling-card model" for the structure of microporous carbons. *Carbon* 1997;35(6):825–830.

Davies GM, Seaton NA. The effect of the choice of pore model on the characterization of the internal structure of microporous carbons using pore size distributions. *Carbon* 1998;36(10):1473–1490.

Do DD, Do HD. Effects of adsorbate–adsorbate interaction in the description of adsorption isotherms of hydrocarbons in micro–mesopore carbonaceous materials. *Appl Surf Sci* 2002;196:13–29.

Ehrburger-Dolle F, Rieker T, Gonzalez MT, Molina-Sabio M, Rodríguez-Reinoso F, Pfeifer P, Schmidt PW. Oxidation processes and fractal properties of activated carbons. In: *Fractals and Chaos in Chemical Engineering*, Eds. Giona M, Biardi G, World Scientific, Singapore, 1997, 27–38.

Ehrburger-Dolle F, Fairén-Jiménez D, Berthon-Fabry S, Achard P, Bley F, Carrasco-Marín F, Djurado D, Moreno-Castilla C, Morfin I. Nanoporous carbon materials: Comparison between information obtained by SAXS and WAXS and by gas adsorption. *Carbon* 2005;43(14): 3009–3012.

Fitzer E, Kochling K-H, Boehm HP, Marsh H. Recommended terminology for the description of carbon as a solid. *Pure Appl Chem* 1995;67(3):473–506.

Galli G, Martin RM, Car R, Parrinello M. Structural and electronic properties of amorphous carbon. *Phys Rev Lett* 1989;62:555–558.

Gavalda S, Kaneko K, Thomson KT, Gubbins KE. Molecular modeling of carbon aerogels. *Colloid Surf A* 2001;187–188:531–538.

Gavalda S, Gubbins KE, Hanzawa Y, Kaneko K, Thompson KT. Nitrogen adsorption in carbon aerogels: a molecular simulation study. *Langmuir* 2002;18(6):2141–2151.

Gibson J, Holohan M, Riley HL. Amorphous carbon. *J Chem Soc* 1946;456.

Goringe CM, Bowler DR, Hernández E. Tight-bonding modelling of materials. *Rep Prog Phys* 1997;60:1447–1512.

Gun'ko VM, Mikhalovsky SV. Evaluation of slitlike porosity of carbon adsorbents. *Carbon* 2004;42(4):843–849.

Iley M, Marsh H, Rodríguez-Reinoso F. The adsorptive properties of carbonised olive stones. *Carbon* 1973;11(6):633–638.

Jenkins GM, Kawamura K. *Polymeric Carbons–Carbon Fibre Glass and Char*. Cambridge University Press, Cambridge, 1976.

Job N, Pirard R, Marien J, Pirad J-P. Porous carbon xerogels with textures tailored by pH control during sol–sol process. *Carbon* 2004;42(3):619–628.

Kaneko K, Ishii C, Ruike M, Kuwabara H. Origin of superhigh surface area and microcrystalline graphitic structures of activated carbons. *Carbon* 1992a;30(7):1075–1088.

Kaneko K, Shimixu K, Suzuki T. Intrapore field dependent microfilling of supercritical N_2 in slit-shaped micropores. *J Chem Phys* 1992b;97(11):8705.

Kowalczyk P, Ustinov EA, Terzyk AP, Gauden PA, Kaneko K, Rychlicki G. Description of benzene adsorption in slit-like pores. Theoretical foundations of the improved Horvath–Kawazoe method. *Carbon* 2004;42(4):851–864.

MacElroy JMD, Friedman SP, Seaton NA. On the origins of transport resistances within carbon molecular sieves. *Chem Eng Sci* 1999;54(8):1015–1027.

Marsh H, Crawford D, O'Grady TM, Wennerberg A. Carbons of high surface area. A study by adsorption and high-resolution electron microscopy. *Carbon* 1982;20(5):419–426.

McCulloch DG, McKenzie DR, Goringe CM, Cocaine DJH, McBride W, Green DC. Experimental and theoretical characterization of structure in thin disordered films. *Acta Crystallogr* 1999;A55:178–187.

Nguyen C, Do DD. A new method for the characterization of porous materials. *Langmuir* 1999;15:3608–3615.

Oberlin A. TEM studies of carbonization and graphitization. In: *Chemistry and Physics of Carbon*, Vol. 22, Ed. Thrower PA. Marcel Dekker Inc., New York, 1989, pp. 1–143.

Oberlin A, Villey M, Combaz A. Influence of elemental composition on carbonization: pyrolysis of kerosene shale and kuckersite. *Carbon* 1980;18(5):347–353.

Oberlin A, Bonnamy S, Rouxhet PG. Colloidal and supramolecular aspects of carbon. In: *Chemistry and Physics of Carbon*, Vol. 26, Eds. Thrower PA, Radovic LR. Marcel Dekker Inc., New York, 1999, pp. 1–148.

Olivier JP. Improving the models used for calculating the size distribution of micropore volume of activated carbons from adsorption data. *Carbon* 1998;36(10):1469–1472.

O'Malley B, Snook I, McCulloch D. Reverse Monte Carlo analysis of the structure of glassy carbon using electron microscopy data. *Phys Rev B* 1998;57:14148–14157.

Petersen T, Yarovsky I, Snook I, Dougal GM, Opletal G. Structural analyses of carbonaceous solids using an adapted reverse Monte Carlo algorithm. *Carbon* 2003;41(12):2403–2411.

Petkov V, Difrancesco RG, Billinge SJL, Acharya M, Foley HC. Local structure of nanoporous carbons. *Philos Mag B* 1999;79(10):1519–1530.

Pfeifer P, Ehrburger-Dolle F, Rieker TP, González MT, Hoffman WP, Molina-Sabio M, Rodríguez-Reinoso F, Schmidt PW, Voss DJ. Nearly space-filling fractal network of carbon nanopores. *Phys Rev Lett* 2002;88(11):115502-1–115502-4.

Pikunic J, Pellenq RJ-M, Thomson KT, Rouzaud J-N, Levitz P, Gubbins KE. Improved molecular models for porous carbons. *Stud Surf Sci Catal* 2001;132:647–652.

Pikunic J, Gubbins KE, Pellenq RJ-M, Cohaut N, Rannou I, Guet J-M, Clinard C, Rouzaud J-N. Realistic molecular models for saccharose-based carbons. *Appl Surf Sci* 2002;196:98–104.

Py X, Gullet A, Cagnon B. Nanomorphology of activated carbon porosity: geometric models confronted to experimental facts. *Carbon* 2004;42(8–9):1743–1754.

Riley HL. Amorphous carbon and graphite. *Quart Rev Chem Soc* 1947;1:59–72.

Rodríguez-Reinoso F, Molina-Sabio M, González MT. The use of steam and carbon dioxide as activating agents in the preparation of activated carbons. *Carbon* 1995;33(1):15–23.

Rouzaud JN, Clinard C. Quantitative high-resolution transmission electron microscopy: a promising tool for carbon materials characterization. *Fuel Process Technol* 2002;77–78:229–235.

Ruike M, Kasu T, Setoyama N, Suzuki T, Kaneko K. Inaccessible pore characterization of less-crystalline microporous solids. *J Phys Chem* 1994;98:9594–9600.

Ruthven DM. Principles of Adsorption and Adsorption Processes. John Wiley & Sons, New York (1984).

Segarra EI, Glandt ED. Model microporous carbons: microstructure, surface polarity and gas adsorption. *Chem Eng Sci* 1994;49(17):2953–2965.

Tamon H. *Encyclopedia of Materials. Carbon Aerogels*. Elsevier Science Ltd., Amsterdam, 2001, pp. 898–900.

Wang CZ, Qui SY, Ho KM. O(*N*) tight-bonding molecular dynamics study of amorphous carbon. *Comp Mater Sci* 1996;7:315–323.

Yang S, Hu H, Chen G. Preparation of carbon adsorbents with high surface area and a model for calculating surface area. *Carbon* 2002;40(3):277–284.

Yoshida A, Kauai Y, Hishiyama Y. Microtexture and magneto-resistance of glass-like carbons. *Carbon* 1991;29(8):1107–1111.

CHAPTER 4

Characterization of Activated Carbon

Activated carbons are uniquely complex in terms of the size, shape and variability of their porosity. Their complete characterization, accordingly, is a major challenge to the surface chemist. The techniques of adsorption from gaseous and liquid phases, kinetics and energetics, of assessments of surface polarities and X-ray and neutron scattering need to be combined to provide a complete characterization.

This chapter examines most of the methodologies and concludes that a pragmatic characterization of an activated carbon can made from three isotherms of N_2 at 77 K, CO_2 at 273 K and H_2O at 298 K or enthalpy of immersion in water. In addition, industrial performance is indicated using breakthrough behavior.

The realist requires knowledge of adsorption capacity for a specific adsorbate by an activated carbon under operating conditions, as well as knowledge of the strength of the adsorption, based on breakthrough curves. The realist says that is all we need to know. The scientist says that we need to know how to make optimized carbons and to understand all aspects of adsorption processes. The realist says that the scientist has overdone things and produced informations far in excess of what can be assimilated. Do you have a viewpoint?

4.1 Basic Concepts

Although the subject areas of porosity in solids and associated experimental and theoretical approaches (designed to further their understanding) are extremely comprehensive, it is possible to identify several basic concepts which are fundamental to these subject areas (Sing, 2004).

Is there a clear concept of the exact nature of porosity in activated carbons? The discussions of Chapter 3 and the elaborations of this chapter, although they approach closely the problem of how to describe porosity in carbons, never quite answer the question. The models of Chapter 3 provide visual images of how porosity in carbons may exist. Despite these significant and erudite research efforts, these models do not provide descriptions,

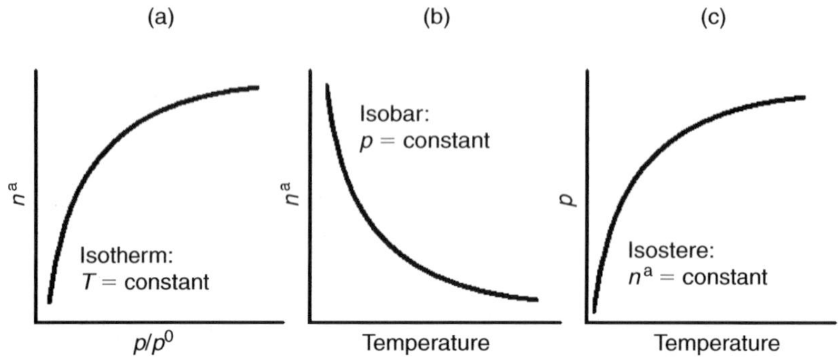

Figure 4.1. Definitions of terms used to describe adsorption phenomena.

quantitative and structural, of porosity necessary to predict applications of individual carbons and to distinguish between carbons with different origins.

Porosity, with the dimensions of nanometers or less, cannot be precisely imaged even in the most recent of transmission and scanning electron microscopes and recourse has to be made to the powerful experimental techniques of physical adsorption of gases, of immersion calorimetry and of small-angle scattering of X-rays (SAXS) and neutrons (SANS) to characterize porosity. Microporosity has the dimensions of molecules and such molecules, as adsorbates, become the experimental probes providing significant information about the adsorption site. Hence, the phenomena of porosity and adsorption are inseparable.

The basics of adsorption phenomena require definitions. In experimental adsorption systems, the gas or vapor of a gas phase, or the solute in a solution, is called the *adsorptive* (when not in the adsorbed state). It is called the *adsorbate* when existing as an adsorbed phase within a solid, and the solid is referred to as the *adsorbent*. (*Note*: The terms *absorb* and *absorption* refer to such processes as the *absorption* of light or of X-ray radiation *internally* by the entire material, where internal surfaces may not exist.) The variations of extents of adsorption (n^a in mmol g^{-1}) with relative pressure (p/p^0) of the adsorptive at constant temperature, when plotted, make up the *adsorption isotherm* (Figure 4.1(a)). The variation of extents of adsorption (n^a in mmol g^{-1}) with temperature of adsorption (T in K), at constant relative pressure, is the *adsorption isobar* (Figure 4.1(b)). The variation of the relative pressure of the adsorptive (p/p^0), with adsorption temperature (T in K), to maintain a constant amount of adsorbate adsorbed on the adsorbent (n^a in mmol g^{-1}) is the *adsorption isostere* (Figure 4.1(c)).

In adsorption studies, the experimental isotherms of nitrogen, carbon dioxide and water vapor provide significant analytical data (sufficient for most purposes) to quantitatively describe the following:

(a) Overall extents of adsorption (adsorption capacity).

(b) Pore size distributions (PSDs) of the porosity of carbons.

(c) The presence of porosity with entrance dimensions $<\sim 0.7$ nm diameter.

(d) The presence of sites of high adsorption potential, effective at low relative pressures of the adsorptive.

(e) The polarity of the surface of the carbon adsorbent (e.g. presence of surface oxygen complexes).

The adsorption of gases and vapors is a physical process, not involving formation of chemical bonds. Adsorbates differ in terms of their molecular size and polarity with the surfaces of adsorbents being energetically heterogeneous. So, selection of adsorption sites occurs during an entire adsorption process. The progressive occupancy of adsorption sites is not a random process but one of selecting the next best available site, from an energetic point of view.

Theoretical studies of the many adsorption phenomena are overwhelming in their extent. Their relevance or applicability to "real or practical" adsorption systems can be debated. Industrial users are primarily concerned with the optimization of adsorption (removal or trapping) of specific adsorbates, usually impurities, at low partial pressures, within gases and liquids, related to control of water supplies and the environment. Adsorptions, at low relative pressures/concentrations, have not received that much in the way of attention from researchers, even though this is where the action is.

Activated carbon is an industrial product and, in passing, it is noted that the techniques of adsorption of such adsorbates as nitrogen and carbon dioxide, which dominate studies to characterize activated carbons, do not feature in the American Society for Testing Materials (ASTM) book of standards covering the characterization of activated carbon (see Chapter 8). The dominant reason for this omission is that no industry is particularly interested in the commercial adsorption of nitrogen at 77 K and of carbon dioxide at 273 K.

The activated carbon industry, like any other industry, survives only on the profits it makes. Like any other industry, it has to buy lower and sell higher. Its capacity, worldwide, is in the region of about 500,000 tonnes per year. Activated carbons are not high-value products and so profit margins need to be carefully regulated. As a result, this industry needs to have stable, abundant supplies of raw (parent) materials, from a nearby source so keeping transportation costs to a minimum, and also needs to be free from the cycles of cropping. Accordingly, there is reliance on supplies of coal (lignites and sub-bituminous) as well as wood (sawdust) and coconut shells. Organic waste products, and there are many, including coffee-bean shells, banana skins, bark from tree felling, straw, fruit stones (pips) on carbonization, all produce porous carbons which can be activated to an appropriate porosity. Many such procedures are to be found in the literature. But there is much more to active carbon production than porosity. As well as physical properties and the need for regeneration, the economics of collecting, transportation and production do not allow such uses. This industry already produces some hundreds of activated carbons, from traditional sources, with quite specific applications attached to their specifications. However, there is always the need for a new carbon which will meet a demanding application.

For most purposes, a working characterization of an activated carbon, involves the determination of isotherms of nitrogen (77 K) (which fills micro- and mesoporosity), of carbon dioxide (273 K) which fills narrow microporosity and enters porosity which may not be available to nitrogen at 77 K, (because of kinetic restrictions at this low temperature) and of water vapor, or of enthalpies of immersion in water, both being indicators of the polarity of surfaces. After this, the characterization of a carbon in terms of its industrial requirement is a prerequisite, for example, as with a breakthrough curve.

A full characterization of an individual activated carbon, using the techniques listed in the standardization procedures of the ASTM (Chapter 9) would be too expensive to carry out and would take several months to complete.

Although adsorption data and calorimetric data, for example, can be measured reproducibly to within about one percent, the interpretations of such data are somewhat semi-quantitative. Such parameters as surface area and PSDs, certainly, do not possess that level of accuracy. Why this should be is explained below. It may come as a shock to readers that it is more than sufficient to quote surface areas to values $\pm 25\,m^2\,g^{-1}$.

4.1.1 The Adsorption Isotherm

In physical adsorption studies the experimental information of dominant importance is the *adsorption isotherm* which is a plot of equilibrium amounts adsorbed (n^a in $mmol\,g^{-1}$) against the relative pressure (p/p^0) of the adsorbate.

In recent times, following on from the development of automatic adsorption equipment and the computerized analysis of adsorption data, there exists a tendency to underestimate the value and importance of the adsorption isotherm. The assessment and role of what is termed "internal surface area" are assumed to be dominant. The problem with the use of computerized equipment is that it is programmed to re-assemble adsorption data into the coordinates of the Brunauer–Emmett–Teller (BET) or Dubinin–Radushkevich (DR) adsorption equations (see below) and to provide the best straight line through data, and so to obtain a value for a surface area. The adsorption isotherm is often not printed out. There is a need to do this as the isotherm, as such, provides considerable information simply from a visual inspection. No account is taken by the computer that not all of the adsorption data are appropriate. As a result, computer-based values of surface area are obtained which can be far removed from those obtained by manual calculations which use the appropriate manual points and a knowledge of adsorption processes.

A problem with interpretation of an adsorption isotherm is that, usually, it is a smooth continuous curve (Figure 4.1(a)). There is no unambiguous break in the curve to indicate a point which could be related directly and precisely to a surface area measurement. Hence, the need to use adsorption equations which contain a parameter for surface coverage (n^a_m in $mmol\,g^{-1}$ of adsorbate) leading to the parameter which is called *surface area*. A difficulty with this parameter of surface area is that it can be less informative than a qualitative inspection of the isotherm. For example, two activated carbons, of the same surface area, say $1000\,m^2\,g^{-1}$, may have quite different adsorption properties. An initial, comparative, visual inspection of the isotherms would immediately identify differences between the activated carbons.

Porosity in activated carbon offers many challenges to the researcher, always ongoing, as seen by the continuous publication of "adsorption papers" in several journals. Such challenges include the shape and dimensions of porosities, the packing arrangements of adsorbate molecules within porosity, the energetics of adsorption, etc. In association, there is the challenge of producing an activated carbon which successfully meets the requirements of an industrial application. The activated carbon has to succeed with its required task. Industrial applications of activated carbon require the removal (separation) of contaminants usually at very low levels of concentration. Under these conditions, the adsorption sites involved would be measured by nitrogen adsorption at values of relative pressure or relative concentration (p/p^0 or c/c_s) $\ll 0.01$. These are the sites of the highest adsorption potentials and it is here that activated carbons, apparently otherwise identical, may have significant differences relevant to a specific application.

However, the nitrogen isotherm dominates the literature of studies of activated carbon, and as discussed further below, the concept of "surface area" is very much in current use. So, when working with activated carbons, there is a need to maintain a critical attitude as to how surface area values are derived and used.

The adsorption isotherm results from the controlled physical adsorption of a gas or vapor or solute from a solution into an activated carbon. This process of physical adsorption is the way in which the adsorptive (gas, vapor or solute) enters into the porosity. The close proximity of carbon atoms to each other, as they constitute the boundary or surface of the porosity, results in these gas adsorbate molecules being "retained" within the porosity. The "retention" is the result of enhanced van der Waals forces (dispersion forces) within the porosity. This process of physical adsorption is exceedingly dynamic. Energetically, an adsorbed state is more stable than the gaseous state of the adsorptive at 273 K and 0.1 MPa of pressure.

The atoms of the adsorbate and adsorbent are not without internal motion; both vibrate about their bonds. The higher the temperature of the adsorption system the higher is the amplitude/frequency of these vibrations, in addition to their kinetic translation motion. Thus, an adsorbed molecule does not remain stationary at an adsorption site. It may move to another adsorption site or it may return to the adsorptive. Within the system, thermal spikes, or sites of higher than average amplitude, move throughout the surface, and these have sufficient energy (equivalent to a rise in temperature) to dislodge the molecule from a particular site. The adsorbate molecule is ejected and as the thermal spike is transitory it is replaced by another adsorbate molecule such that at all times the measured extents of adsorption remain constant.

As a result, an adsorbate molecule does not simply enter into the porosity and stay there. It will on average move in and out about 10^{15} times per second. At equilibrium, the number of molecules entering the porosity equals the number leaving. This concept of mobility and exchange is based on isotopic exchange studies. In addition, it is important to note that, for a carbon with a *surface area* (discussed in detail later) of $1000\,m^2\,g^{-1}$, there exist about 10^{20} adsorption sites every one being different from every other site by having its own potential energy well.

Figure 4.2 illustrates how the adsorption potential energy well deepens as the pore narrows, with respect to a given adsorbate. Thus, as an adsorption process is initiated, sites of highest potential (to the right of Figure 4.2) are occupied first, followed by sites of

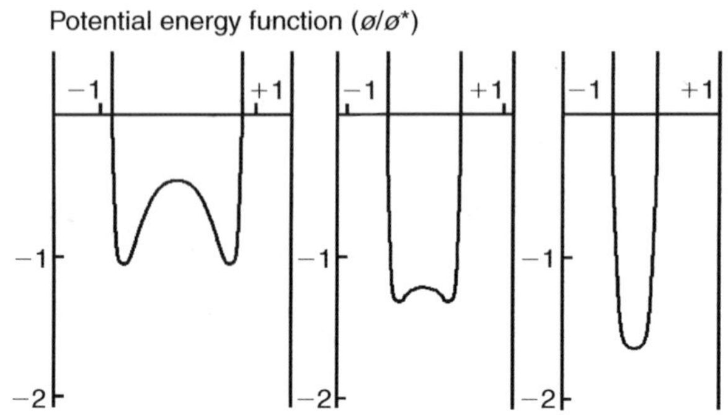

Figure 4.2. Diagrams to illustrate the change in the depth of the potential energy well, for an adsorbate molecule, as the size of a slit-shaped pore decreases (left to right) (adapted from Rouquerol *et al.*, 1999).

lower energy until a relative pressure (p/p^0) of unity is reached. The adsorption process is accordingly continuous without breaks and the adsorption isotherm is a continuous curve. However, it is a curve with special characteristics because it describes the distribution of energies of the sites (10^{20} of them) and these distributions have a Gaussian or Lorenztian shape. Such distributions occur within many naturally occurring phenomena, such as the size of particles in sedimentary mud. The adsorption isotherm contains meaningful information in considerable detail. The extraction of this information is the challenge.

4.1.2 Open and Closed Porosity

Not all of the porosity of an activated carbon has the same entrance dimensions. This is easily demonstrated from the studies relating to extents of adsorption of molecules of increasing size, for example, methane, ethane, propane, butane, etc. (molecular probe studies). Here, extents of adsorption fall off with increasing molecular size of the adsorptive. Thus, more porosity is *open* to molecules of smaller size than to molecules of larger size. Conversely, more porosity is *closed* to molecules of larger size than to molecules of smaller size. Consequently, *closed porosity* is defined as that porosity which is not accessible to a given adsorbate. Closed porosity is not an invariant parameter but approaches zero ($cm^3\,g^{-1}$), but never reaching zero for activated carbons as the adsorbate molecule decreases in size. Even with the smallest of molecules, hydrogen and helium, there is always a residual *closed porosity*.

Measurements of densities of activated carbon using helium as a displacement fluid provide values from about 1.8 to 2.1 $g\,cm^{-3}$, lower than the value for single crystal graphite of 2.26 $g\,cm^{-3}$. The existence of a PSD is the result of the disordered (not disorganized) array of imperfect, non-planar micro-graphene layers.

4.1.3 Surface Area: Fact or Fiction?

The term *surface area* has been in use since the early days of the development of the Langmuir and BET equations of adsorption (see below). Initially, the Langmuir equation described *chemisorption* phenomena involving monolayer surface coverage bonded to non-porous surfaces. The Langmuir equation confirmed the areas of these non-porous surfaces as measured independently by, for example, electron microscopy. The *physisorption* of nitrogen at 77 K, on non-porous surfaces, resulted in multilayers being formed. The BET equation described this phenomenon and predicted a value for monolayer coverage ($cm^3 g^{-1}$, of gaseous adsorptive) which could be related to the area of the non-porous solid. Further, the physisorption of nitrogen at 77 K on porous solids closely resembled that on non-porous surfaces and from the monolayer coverage, there could be calculated a value for a surface area, assuming, without any qualification, that this internal surface was similar (identical) to the external surface of the non-porous solid. Thus, the BET equation does not predict surface areas; it predicts monolayer coverage (units of $cm^3 g^{-1}$, or $mmol\ g^{-1}$ of adsorbate).

However, as is well understood subsequently, the basic assumption that an internal surface is identical to an exterior surface requires a drastic reappraisal. An inspection of the models of Figures 3.9–3.13 immediately demonstrates this need. Surface within porosity is not the surface as of a non-porous solid. Surface within porosity is "folded" surface, the close proximity of these folds creating the porosity. The understanding of surfaces, and their behavior, as they form porosity, essentially is the chemistry of porous solids.

The models of Figures 3.9–3.13 illustrate the question as to how such structures can have a *surface area*. There cannot exist a surface in the sense which has a macro-concept, as a surface of a mirror, because adsorption occurs on an atomic surface. The term *surface area* of a porous solid is (nothing more and nothing less) a convenient way to indicate adsorption capacity. The concept, at best, is a rough tool because it includes adsorptions into all porosities, whereas activated carbon needs to have all these porosities characterized in terms of the volumes of their various sizes, be they narrow and wider microporosity, <2.0 nm, etc. of entrance dimension. However, the term is here to stay, it is widely used in the literature, and it is hoped that future researchers will be aware of the limitations of its use.

The units of monolayer coverage could just as well be $mmol\ g^{-1}$ in, for example, microporosity. It is only because the units of $m^2 g^{-1}$ are easier to translate into a physical picture that they have stayed within the adsorption literature. Within the porosities of activated carbon, the "surface" is a boundary condition which limits the proximity of other atoms (adsorbed). It is a limit of electron density and as such will be curved. Figures 1.1 and 1.2 (Chapter 1) suggest this.

Figure 4.3 is of simple models of pore shapes of different sizes containing adsorbate molecules. These models demonstrate that not all of the *visualized image or* surface area of the porosity can be measured, and as a result, cannot be utilized. Figure 4.3 shows that amounts of adsorbate contained within the porosity need not relate, proportionally, to a surface area.

As with *closed porosity* discussed above there is no single unique surface area value. It is appreciated that if a relatively large molecule is used as an adsorbate, for example butane,

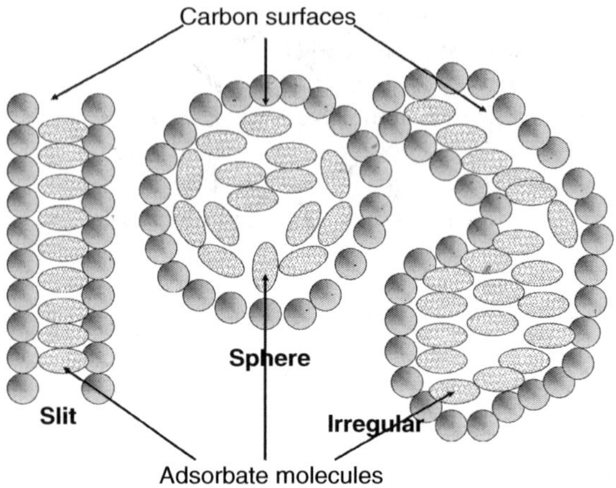

Figure 4.3. The significance of surface area in terms of mechanisms of surface coverage. This figure demonstrates the unreality of attempting to give a physical significance to the internal surface area of a porous solid.

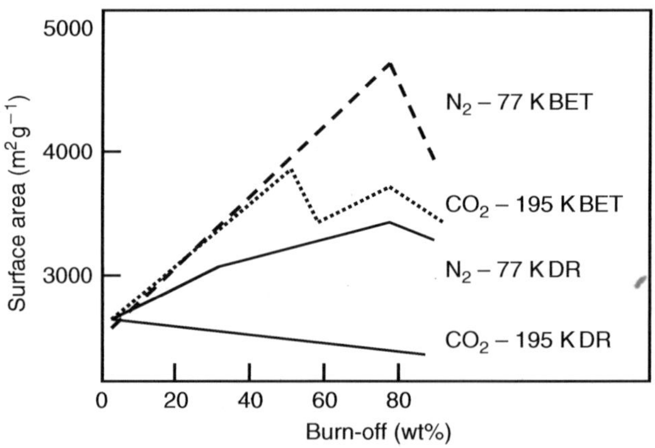

Figure 4.4. For a series of activated carbons from PVDC, 0–85 wt% burn-off, the variation of apparent surface area using the BET and DR adsorption equations based on nitrogen (77 K) and carbon dioxide isotherms (195 K) is illustrated. Is there a correct surface area? (Marsh and Wynne-Jones, 1964).

this will not be adsorbed in the same way as say nitrogen and so there will be a surface area according to butane and a surface area according to nitrogen.

Equations have been devised to simulate adsorption processes over surfaces and in porosity and which contain the term n_m^a (mmol g^{-1}) as a measure of monolayer coverage. Figure 4.4 contains plots of surface areas (apparent) of a series of activated carbons (from polyvinylidene chloride (PVDC)) against extent of activation (burn-off (wt%)) using different adsorption equations and adsorbates. The results of this study clearly indicate another problem associated with assessments of surface area. That is, the different adsorption equations (designed to describe the adsorption process (see below)) predict different values of monolayer coverage area. An explanation for this phenomenon is also discussed below. Figure 4.4 shows variations between 1000 and 5000 $m^2 g^{-1}$ for the same carbon (80 wt% burn-off) using isotherms of adsorption of nitrogen (77 K) and carbon dioxide (195 K) with interpretations using both the BET and DR equations (to be described below at length) (Marsh *et al.*, 1964).

Within some activated carbons, adsorption may occur such that several layers of adsorbate molecule(s) adsorb together. This is termed *volume filling* and occurs in the largest of the micropores. It is termed *capillary condensation* when it occurs in mesoporosity. Such *capillary condensation* must not be included in values of n^a_m. Some surface area assessments as high as 5000 $m^2 g^{-1}$ have been published. Many carbons are reported with surface areas around the 1500 $m^2 g^{-1}$ level. The literature may interpret such data as if these carbons possessed exceptionally high surface areas. If single crystal graphite is considered, taking 1 g and imagined as a single graphene layer, then the theoretical maximum surface area (both sides), using nitrogen as the adsorbate, is about 2600 $m^2 g^{-1}$. Any value beyond about 1000 $m^2 g^{-1}$ must therefore be associated with micropore volume filling and capillary condensation and should be treated with caution. The terms apparent, or equivalent or effective surface areas (ESA) could be used for surface area estimates in the range 1000 to about 5000 $m^2 g^{-1}$.

There are several ways in which adsorption capacity is quantified. They are as follows:

(a) Amount of adsorbate adsorbed in units of mmol g^{-1} or mL g^{-1} of liquid.

(b) Amount of adsorbate in units of mL g^{-1} or $m^3 t^{-1}$ of adsorbate as a gas at STP.

(c) Apparent surface area in units of $m^2 g^{-1}$.

A guide to approximate conversions of these parameters is as follows:

$$
\begin{aligned}
100\,m^2 g^{-1} &\equiv 1.0\,\text{mmol}\,g^{-1} \text{ of adsorbate } (N_2)\\
&\equiv 0.034\,\text{mL}\,g^{-1} \text{ of adsorbate } (N_2) \text{ as a liquid}\\
&\equiv 0.044\,\text{mL}\,g^{-1} \text{ of adsorbate } (CO_2) \text{ as a liquid}\\
&\equiv 22\,\text{mL}\,g^{-1} \text{ of adsorbate as a gas at STP}\\
&\equiv 22\,m^3 t^{-1} \text{ of adsorbate as a gas at STP}
\end{aligned}
$$

The units of volume adsorbed per volume of carbon (*V*/*V*) are relevant to industrial applications such as gas storage situations. Units such as ($g\,g^{-1}$), to indicate extents of adsorption, are also reported but these are of limited value.

4.1.4 Kinetics and Dynamics

The models of Chapter 3 are rigid models, giving no indication of atomic vibrations. In reality however, as with all materials, the constituent atoms and molecules are not stationary but vibrate, the amplitude of vibration being a function of temperature, decreasing with decreasing temperature. Space within the porous network of a solid is therefore not invariant but increases and decreases in size in a way which is controlled by vibrations of nearest neighbors of carbon atoms (on average about eight) as well as temperature. A similar situation exists with, for example, the nitrogen adsorptive molecule: it too will be vibrating along, and rotating about, its axis. Thus, it is most relevant, when considering an adsorption process, to envisage a vibrating molecule entering into a pore the dimensions of which are not invariant. In an adsorption process, for rapid entry, the dimensions of the pores are sufficiently large to facilitate the immediate entry of the vibrating adsorptive molecule. These phenomena point to a process which can be termed *activated*, this term being used in the reaction-kinetic context where rate increases with increasing temperature. There is thus an *activation energy of adsorption* where rapid entry means that the adsorption temperature is high enough to prevent the *activation energy of adsorption* from controlling the rate of adsorption, the rate being controlled by rates of access only. However, as the temperature of adsorption is lowered, or as the average pore dimension is decreased, a situation is reached when adsorption takes place only when the adsorptive molecule is at a minimum of bond length and the dimension of the pore is at a maximum. The rate of adsorption is then controlled by the frequency with which this concurrency happens, meaning that the activation energy of adsorption controls the rate.

One way of illustrating this effect is to consider Figure 4.5. Starting from the lowest of temperatures, 77 K, (left), using 0.1 MPa pressure of nitrogen, for a carbon which exhibits activated diffusion effects, adsorption of nitrogen cannot be detected. With increasing temperature, amounts of gas become detectable until a point is reached when true thermodynamic equilibria are established and the data points join the isobar. In Figure 4.5, data points to the left of the maximum are non-equilibrium points, where rates of adsorption are too slow to establish equilibria in a reasonable time (say 1 h per point). In Figure 4.5, the equilibrium isobar (right of maximum) describes extents of adsorption which are equilibrium points taking a time less than 1 h (normal procedures) to establish. So, for such adsorbents which adsorb very little, a decision is impossible to make as to whether or not the solid has non-accessible internal porosity (<0.5 nm) or is non-porous. Recourse has then to be made to the use of a more suitable adsorbate at a higher temperature, or to study the first adsorbate at a higher temperature.

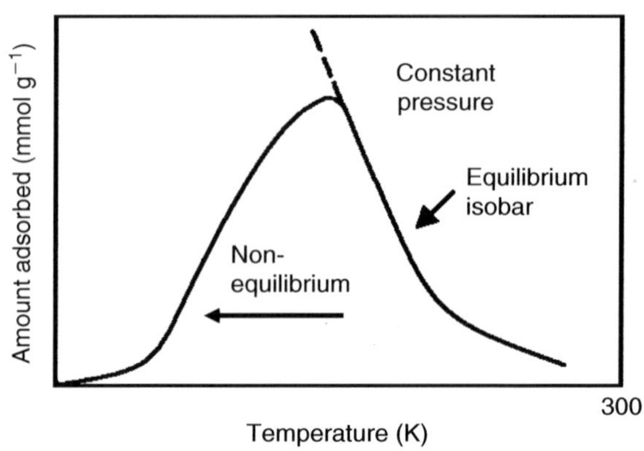

Figure 4.5. Variations of amounts adsorbed with increasing adsorption temperature when activated diffusion effects are present. At low temperatures 77 K (left) no adsorption takes place.

As indicated in Chapter 2, a microporous carbon with a surface area of 1000 $m^2 g^{-1}$ contains about 10^{20} adsorption sites. Assuming an apparent carbon density of 1.0 $cm^3 g^{-1}$ (no penetration of a displacement fluid into the carbon) means that every 1.0 cm^2 of surface of a 1 cm^3 carbon sample has in the region of 10^{14} pore entrances (a number comparable with the number of stars in the sky).

4.2 Characterization of Porosity: Gas Adsorption

4.2.1 Measurement of the Isotherm

As porosity essentially is molecular space, the most suitable methods available to elucidate the structure of porosity make use of molecules with the same dimensions as the pore entrance (photons and neutrons are in addition (see below)). Modern procedures measure extents of adsorption of gases using automatic volumetric methods which are suitable for most solids with surface areas $>1.0\,m^2 g^{-1}$. The volumetric method was used originally by Emmett and Brunauer in 1937.

Earlier, gravimetric methods make use of a balance ranging from a simple silica spring of the McBain balances to automatic computerized electromagnetic equipment using pressure transducers for pressure measurement. A diagram of McBain equipment is as shown in Figure 4.6 where the carbon sample, contained in a bucket, is attached to the lower end of a fine silica rod which in turn is attached to the lower end of a calibrated silica spring, all being contained within glass tubing which can be evacuated. It is necessary to *outgas* the carbon by controlled heating and evacuation to temperatures of about 200 °C and pressures of 1.0 Pa or less. The heater is replaced by a low-temperature thermostat (liquid nitrogen for 77 K or ice–water for 273 K).

Gas now enters the system initially through finely controlled needle valves (to prevent blow-out of the carbon sample from the bucket, and vibration of the spring) to commence the adsorption process. As the carbon adsorbs the gas (the adsorptive) its weight increases

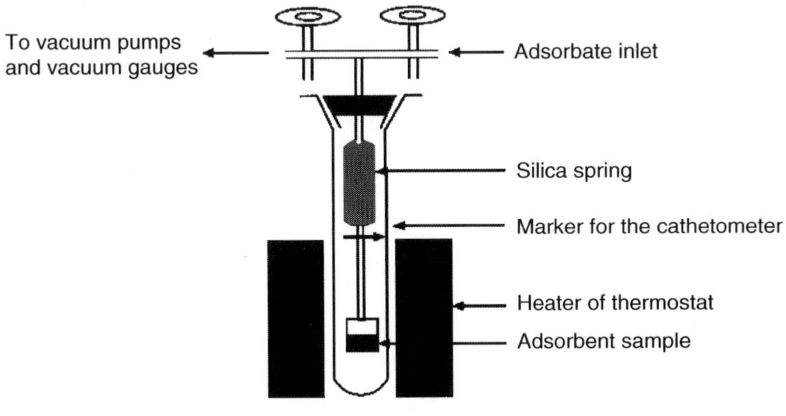

Figure 4.6. Diagram of the McBain system for the measurement of extents of adsorption, that is, the isotherm. The extensions of the calibrated spring are converted to a weight, and then to (mmol g^{-1}) of the known adsorbate.

and the end of the silica rod is lowered. New positions are monitored with a cathetometer (traveling telescope) and the distances converted into weight increases.

The length of time needed to reach an equilibrium position needs to be monitored carefully as this can vary from about 30 min to over a day to an infinite time (as explained above). Amounts adsorbed at measured pressures (converted to relative pressures) are then calculated and the plot (isotherm) of amount adsorbed (n_m^a in mmol g^{-1}) against relative pressure (p/p^0) is created.

The McBain system is useful when working with new materials of unknown properties because rates of adsorption can be monitored and the process is under the control of the operator. An important parameter is the time taken to reach equilibrium, or rates of adsorption. By so doing, knowledge is gained about contents of the smallest of microporosity, <0.5 nm entrances. Automatic adsorption equipments, commercially available, are suitable for routine measurements (for quality control) of adsorption isotherms. For research purposes, when using such equipments, an isotherm must be printed out when an experienced researcher can provide an independent interpretation. It would be an advantage if all studies of porosity published their as-obtained experimental isotherms.

The adsorption process can be reversed by lowering the pressure and establishing new positions of equilibria. This desorption process is slower than the adsorption process as the activation energies of desorption are higher. In some cases, where some types of mesoporosity are present, the desorption isotherm is not coexistent with the adsorption isotherm, this phenomenon being called hysteresis (Figure 4.7) (see below for further discussion).

Experimentally, particularly with the McBain spring balances, some difficulties may be experienced. The tube containing the sample must be well immersed in the liquid nitrogen. Thermal transpiration effects can operate when radiation from the equipment above the sample can radiate energy and raise the temperature of the sample. Similarly, as the containing tube has the characteristics of a Dewer vessel (thermos flask) sufficient time should be allowed for thermal equilibration. The carbon sample may never reach 77 K but is a little higher as experimental limitations become controlling factors.

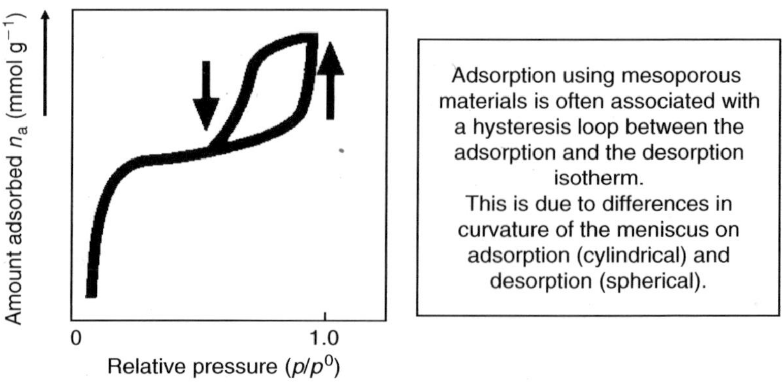

Figure 4.7. The phenomenon of hysteresis in adsorption where the adsorption and desorption isotherms are not co-linear.

4.2.2 *Qualitative Interpretation of Isotherms*

A visual inspection of isotherm shapes provides considerable useful information about extents of adsorption and the porous networks into which the adsorption process has occurred.

Such isotherms are classified into six shapes and these are as shown in Figure 4.8 (Sing *et al.*, 1985). **Type-I** isotherms reach a maximum value of adsorption without inflections and are characteristic of carbons containing microporosity only (very few do). The gradients of the initial part of the isotherm, from p/p^0 values from zero to about 0.05, are indicative of the dimensions of the microporosity, the steeper the gradient the narrower are the micropores. **Type-II** isotherms show an inflection in the region of $p/p^0 > 0.1$, and at high relative pressure, $p/p^0 > 0.9$, where extents of adsorption rise very rapidly. Such isotherms are characteristic of adsorption on open surfaces with multilayer formation (assisted condensation, and not to be considered as volume filling) occurring in the final stages of the process. Also, **Type-II** isotherms describe adsorption in mixed situations of micropores and open surfaces. **Type-III** isotherms are convex, looking upwards, and are characteristic of adsorption at sites of low adsorption potential, as is the case for adsorption on surfaces of organic polymeric systems. **Type-IV** isotherms resemble Type-II isotherms but additionally, instead of adsorption on open surfaces at high relative pressures, adsorption takes place in mesoporosity. Such isotherms may exhibit hysteresis when the mechanism of filling by capillary condensation in mesopores differs from that of mesopore emptying. It is usual for activated carbons *not* to exhibit a plateau in the high relative pressure region. **Type-V** isotherms are those of a low energy, homogeneous solid surface possessing mesoporosity. **Type-VI** isotherms are of surfaces with an extremely homogeneous structure (e.g. pyrolytic graphite) using, for example, argon and methane as adsorbates (but not nitrogen).

4.2.3 *Quantitative Interpretation of Isotherms*

The isotherm, somehow or other, has to be interpreted quantitatively to enable comparisons to be made between carbons. There is a need to know how much porosity there is in

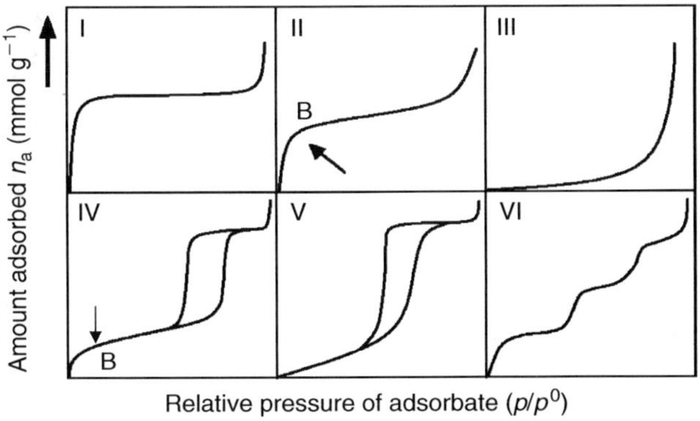

Figure 4.8. The IUPAC classification of adsorption isotherm shapes (Sing *et al.*, 1985) (see Figure 8.4).

a given carbon together with an indication of the "quality" of the porosity, namely the PSD, and surface chemistry (functionality). If possible, the data of the isotherm have to be processed to provide a single number; whether or not this number is meaningful has occupied the minds of surface scientists for decades. Although it is the surface area, based on the nitrogen isotherm, which is the single required number (perhaps not the best number) it is the structure of the porosity which is of controlling importance.

Activated carbons, in industrial applications, need to adsorb at low relative pressures (gas phase) or relative concentrations (liquid phase), and so it is the initial parts of the isotherms, at values of $p/p^0 \ll 0.001$, which are of importance to indicate the presence of sites of high adsorption energy. Extents of adsorption and the slope of the initial section of the isotherm (an indication of retentivity) are critical. Although the adsorption of nitrogen is used to characterize, in one way, a given carbon, it is the response of the carbon to the "industrial" adsorptive (impurity or contaminant, unwanted or desired) of the application which is crucial. It is the convenience of the nitrogen adsorption method which has guaranteed its widespread application. Otherwise, there is nothing so fundamental about the use of nitrogen, as an adsorbate, that it must be used.

Be that as it may, the adsorption isotherm has to be interpreted and one way to do this is to model the adsorption process, mathematically, in a way which contains an expression for monolayer coverage, that is, the amount of adsorbate required to cover the hypothetical surface with one layer of adsorbate. When the complexity of the adsorption process is compared with the over-simplified assumptions of the model equations to be described below, it is surprising that the equations do indeed appear to work. The reason for this is that the shape of an isotherm, quite a unique shape, is associated with distributions of adsorption potential (energy) within the porosity of the activated carbons. It is relevant to note that curves, shaped like isotherms and drawn manually without reference to adsorption data, are not linearized by adsorption equations. Draw a few curves and try this for yourself.

Of the three equations available to interpret adsorption isotherms, the Langmuir and BET equations are based on general models of adsorption and are not derived from concepts of porosity. They could be described as curve-fitting equations. The Langmuir and BET equations, initially, were derived to interpret adsorption isotherms of non-porous carbons. Here, the concept of a surface area has a physical interpretation. Later, the equations were applied to isotherms derived from microporous materials and appeared to work, and values for monolayer coverage could be derived. However, the meaning of monolayer coverage in a microporous solid is an issue which has not been resolved to this day.

The Langmuir equation (Equation (4.1)) is a plot of n^a against p/p^0 and provides a value for n^a_m (amount of adsorbate which forms a monolayer). For the BET equation (Equation (4.2)), a plot of the left side of the equation against p/p^0 also provided a value for n^a_m. The constants "b" and "c" describe average energies of adsorption. Values of n_m are converted to a surface area (A_s in $m^2 g^{-1}$) provided that the area effectively occupied on the surface by an adsorbed molecule (a_m in m^2 molecule^{-1}) is known (see Equation (4.4)). The BET equation does not describe the entire isotherm from p/p^0 of 0.0 to 1.0. This is asking too much for an empirical equation. The best fit is normally restricted to p/p^0 values of <0.15 for microporous carbons. It is recommended by Rouquerol *et al.* (1999) that the best linear fit should include

Point B of the isotherm, Point B being the position of n^a_m. The parameter "c" of the BET equation is indicative of the energetics of the adsorption process. Values of "c" <150 are found for adsorption on non-porous carbons with values of "c" ≫ 200 being associated with the more energetic adsorption in microporous carbons.

On the other hand, the DR equation (Equation (4.3)) has its roots in the structure of porosities and their distributions of structure (and hence of adsorption potential). The DR equation is not based on a model of the adsorption process. It is a derivative of the mathematics of Rayleigh, Gaussian or Lorenztian distributions. In fact, this DR equation is more relevant to the adsorption process than the other two equations.

Langmuir Equation

$$\frac{p/p^0}{n^a} = \frac{1}{bn^a_m} + \frac{p/p^0}{n^a_m} \tag{4.1}$$

where p = equilibrium vapor pressure (Pa), p^0 = saturation vapor pressure (Pa), n^a = amount adsorbed (mmol g^{-1}), n^a_m = monolayer capacity (mmol g^{-1}).

BET Equation

$$\frac{p}{V(p^0 - p)} = \frac{1}{V_m c} + \frac{(c-1)p}{V_m c p^0} \tag{4.2}$$

where V = amount adsorbed in volume STP (cm^3 g^{-1}), V_m = monolayer capacity in volume STP (cm^3 g^{-1}). (*Note*: Although original publications stated that the units of extents of adsorption should be in volume STP, it is more practical that amounts adsorbed should be in units of mmol g^{-1}.)

DR Equation

$$W = W_0 \exp[(-BT/\beta(\log^2(p^0/p))] \tag{4.3}$$

where W = volume of adsorbate filling micropores, W_0 = total volume of the micropore system. Values of W_0 are often assumed to equate to values of V_m and hence of n^a_m, leading to predictions of surface area. Units of W and W_0 are in cm^3 (mL) g^{-1}.

There is no doubt that the Langmuir, BET and DR equations are central to the characterization of porosity in solids. The Langmuir equation is based on the following assumptions:

1. During an entire adsorption process only monolayer adsorption occurs.
2. The adsorption process is localized with no adsorbate–adsorbate interactions.
3. The heat of adsorption is independent of surface coverage, that is, the adsorbent has an energetically homogeneous surface.

In reality, none of the above assumptions apply to activated carbon, even approximately, so it comes as a surprise that the Langmuir equation can be used to linearize adsorption isotherms. In fact, assumptions 2 and 3 probably compensate for each other, as adsorbate–adsorbate interactions increase with increasing coverage at the same time as enthalpies of adsorption decrease with increasing coverage.

Variations exist of the BET equations, designed to linearize any non-linear plots. This is done by introducing a power term n to the equations. Unfortunately, this led to assumptions that n described the number of layers in a multilayer (volume filling) situation. Similarly when the DR equation does not predict a linear plot, the $\log^2$ term is replaced by a $\log^n$ term when n usually is between 1 and 5 and need not be an integer. Such modifications are known as the Dubinin–Astakhov (DA) equation.

Such deviations occur when distributions of adsorption site energies do not fit a Gaussian-type (or related distribution function). Then, the obtained experimental isotherm will not be linearized by the conventional Langmuir, BET and DR adsorption equations. If the continuity of the distribution curve is disturbed in some way (e.g. by selective oxidation to widen some parts of the porosity during an activation process) then deviations will occur from the model equations. Elaborations of equations to obtain a better fit are mathematical devices to correct for deviations to the distribution curves but do little to explain the causes.

The adsorption equations as mentioned above predict values of monolayer capacity (n_m^a in mmol g^{-1}) which can be converted to an apparent surface area (A_s in $m^2 g^{-1}$) provided that the area effectively occupied by an adsorbed molecule (a_m in m^2 molecule^{-1}) can be estimated. Values of a_m are obtained from density data. Hence:

$$A_s = n_m^a N_a a_m \tag{4.4}$$

where N_a is the Avogadro constant, 6.013×10^{23}.

The area assigned to an adsorbed molecule is the area of projection on to a flat surface as illustrated in Figure 4.9. When adsorbate molecules occupy such sites as are modeled in Figure 4.3, the unreality of the concept of "surface area" becomes that much more apparent. The concept is essentially one of quantitatively describing comparative adsorption capacities (without providing further details of where the adsorption takes place), to satisfy the requirements of human awareness.

Areas covered by adsorbate molecules, as commonly used, are listed in Table 4.1. It has to be emphasized that such values of area do *not* reflect the actual area of projection within porosities. The problems of part filling of porosity and of volume filling have already been explained in terms as shown in Figure 4.3. Further, the packing density of adsorbate molecules is not constant under all conditions but is a function of the dimensions of the porosity into which it is adsorbed. The areas in Table 4.1 are then of an empirical nature, and are indicative.

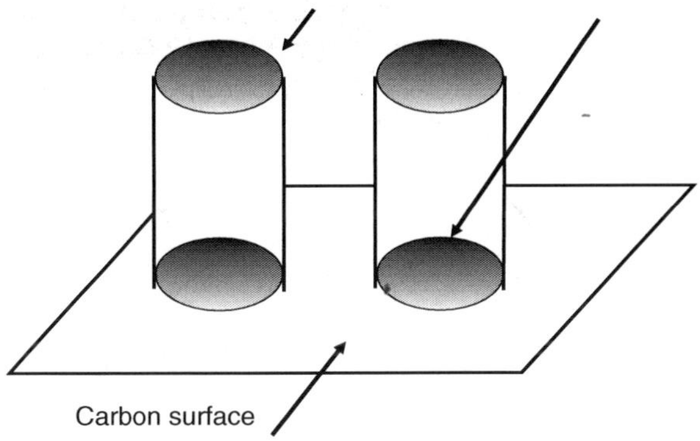

Figure 4.9. Projection of "area" of an adsorbed molecule to the surface of an adsorbent (modeled). This "area" is used to calculate specific surface of a solid (usually porous).

Table 4.1. Values of molecular areas of adsorbates. Values are taken from values of liquid density. Values with ** are obtained from isotherms of standard materials.

Adsorbate	Temperature (K)	Molecular area a_m (nm^2)
Nitrogen	77	0.162
Oxygen	77	0.143
Krypton	77	0.152
	195	0.297
Xenon	77	0.182
	90	0.25
Butane	195	0.247
	273	0.321
	273	0.444**
Pentane	273	0.466**
Benzene	273	0.43
Freon-21	273	0.264
Carbon dioxide	195	0.163
	195	0.206**

Often, from the same isotherm, the three adsorption equations, described above, predict different values of n_m^a. It is not a matter of debate as to which equation is the correct one. It is that different parts of the isotherm give different informations. A further complication, for the same adsorbent, is that different adsorbates provide different isotherms and so provide different values of n_m^a. The answer to these problems is the same, namely that the carbon is exhibiting a surface area for adsorbate **A1** and a different surface area for adsorbate **A2** because surface coverage and/or pore filling mechanisms are different. There is no right or wrong in this issue.

Figure 4.10 illustrates that the three equations of adsorption do not use identical parts of the isotherm in their predictions of monolayer coverage. The adsorption equations are

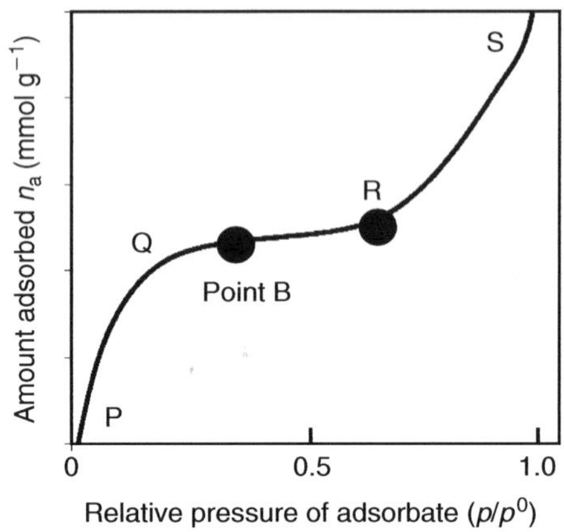

Figure 4.10. A typical adsorption isotherm and its relationship to adsorption equation (see text).

essentially equations of extrapolation. The **DR** equation extrapolates the initial segment, **P** to **Q**, of the isotherm obtained at low relative pressures. The distributions of adsorption energies predicted from the higher relative pressure regions of the isotherm may or may not exist within a given carbon. The **Langmuir** equation is based on segment **PQB** where it is assumed that monolayer coverage takes place, whereas segment **QRS**, dealing with pore filling and capillary condensation processes, forms the basis for the **BET** equation as applied to porous solids. (For non-porous solids, the segment **QRS** deals with multilayer formation on open surfaces.) Differences in monolayer capacities, from the same isotherm, for the same carbon, from different adsorption equations relate directly to the accuracy of the predictions (extrapolations) of the equations into higher levels of p/p^0. Point B of Figure 4.10 is the point of inflection in the isotherm where it is assumed that monolayer formation is complete. For most activated carbons, the limit of application of the Langmuir and DR equations is $<0.15\,p/p^0$.

Figure 4.11 which is a distribution function of potential energies of adsorption sites, is another way of describing an adsorption process. In Figure 4.11, the vertical axis is the probability function which is the probability of having adsorption sites in a defined energy range. The horizontal axis is the adsorption potential or energy release during an adsorption process. The use of effective pore entrance dimension (width?) for the horizontal axis can be considered but may be approximate as pores of the same effective dimensions may not have the same adsorption potential (the energetics of the surface having a significant relevance). The area under the curve is equivalent to the total amount adsorbed (n_m^a in mmol g^{-1}). The adsorption process, that is the filling of the curve, commences from the right of the curve and progressively fills the curve moving from right to left.

If mesoporosity is present, then the mechanism of adsorption changes with capillary condensation occurring, and hence, more adsorption occurs than is predicted from the area under the distribution curve. This is because, although the adsorption potential within a micropore, say of 2.0 nm entrance dimension, is as predicted from the distribution equation, the adsorption process is able to change this adsorption potential because adsorbed

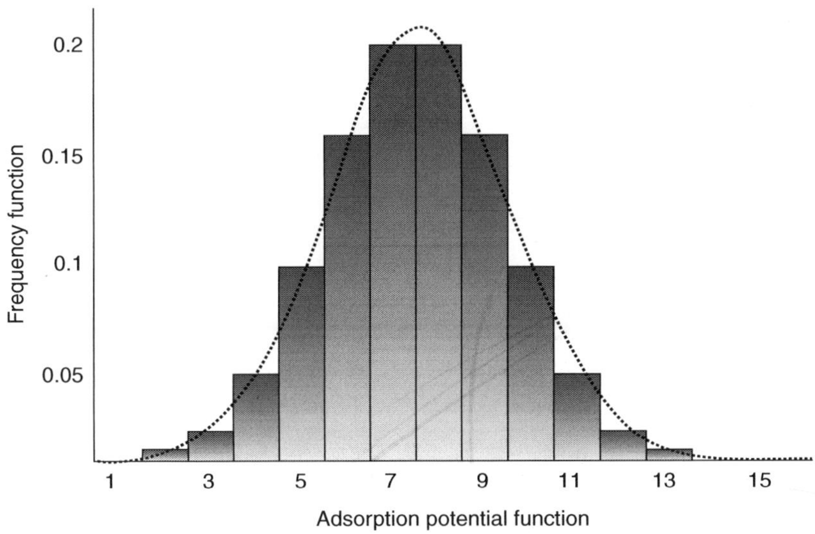

Figure 4.11. The uniqueness of the adsorption isotherm. If the distribution of adsorption site potential energies does not fit a Normal-Gaussian-type distribution function, as above in the figure, then the obtained experimental isotherm is *not* linearized by the Langmuir, BET and DR equations.

molecules effectively are decreasing (narrowing) the size of the pore and accordingly increasing its adsorption potential.

$$P(x) = \frac{1}{\sigma\sqrt{2\pi}} e^{-(x-\mu)^2/(2\sigma^2)} \qquad (4.5)$$

The above Equation (4.5) describes a Gaussian (normal) distribution of a variate (n) which in this case is the adsorption potential, with its probability P, μ is the mean, and σ^2 is the variance.

In the DR plots log V (or log W or log n^a) is plotted against $\log^2 p/p^0$, as shown in Figure 4.12 (right). To the left of Figure 4.12 are experimentally determined isotherms. To the right of Figure 4.12, these curves are redrawn in the coordinates of the DR equation and show differences between them. In interpreting these DR plots, it is the extrapolation of the linear portion of the DR plot (from right to left) to the zero value of $\log^2 p/p^0$ which gives the micropore volume (how expressed depends on the units used). It is *not*, *definitely not*, appropriate to make extrapolations of the data obtained at higher relative pressures (values of $\log^2 p/p^0$ $<\sim 2$) because this is where pore filling and capillary condensation occur in porous adsorbents, and multilayer coverage in non-porous adsorbents.

Further, Figure 4.13 illustrates how the DR plot can be used to detect adsorptions which are not fitting a statistical distribution curve, that is, the isotherms cannot be linearized over the entire experimental range of $\log^2 p/p^0$. Such a totally linearized isotherm is rare; indeed none may exist, simply because the nitrogen adsorbate (or whatever adsorbate is used)

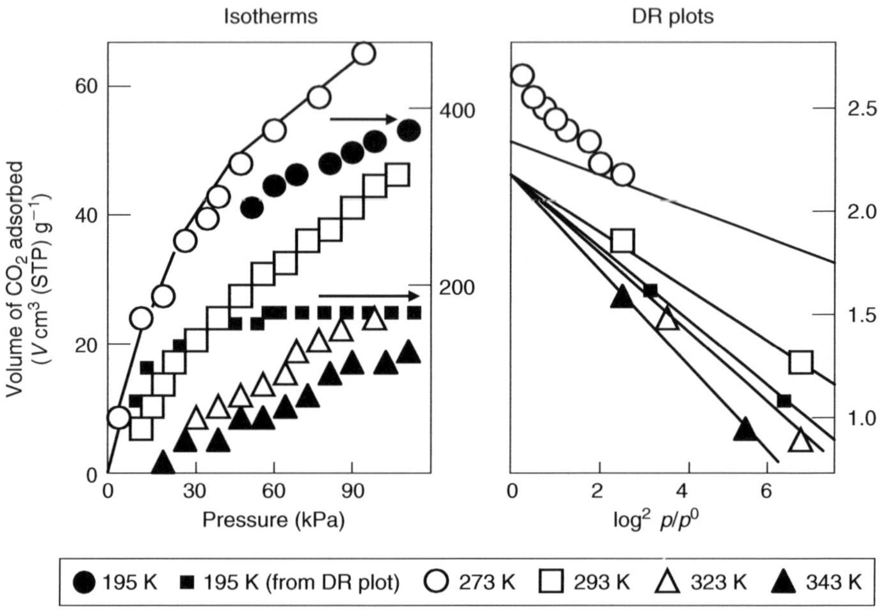

Figure 4.12. Adsorption of carbon dioxide on an activated carbon SP112, 195–343 K, interpreted by the DR equation (Marsh and Siemieniewska, 1967). Note that $\log^2 p/p^0 = \log^2 p^0/p$.

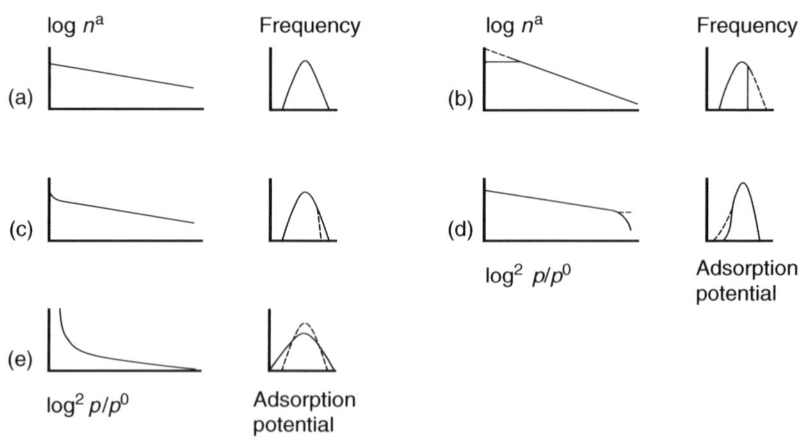

Figure 4.13. Interpretations of deviations from DR plots. A comparison of the shape of DR plots with shapes of the frequency distributions (Marsh and Siemieniewska, 1967).

cannot reach the smallest of pores at high values of $\log^2 p/p^0$. Also, most microporous carbons contain some mesoporosity. Figure 4.13(a) is a totally linear DR plot for all values of $\log^2 p/p^0$ with a complete Gaussian distribution curve (rare). Figure 4.13(b–e) show deviations from linearity; (b) indicates that the pores at the wider end of the distribution range

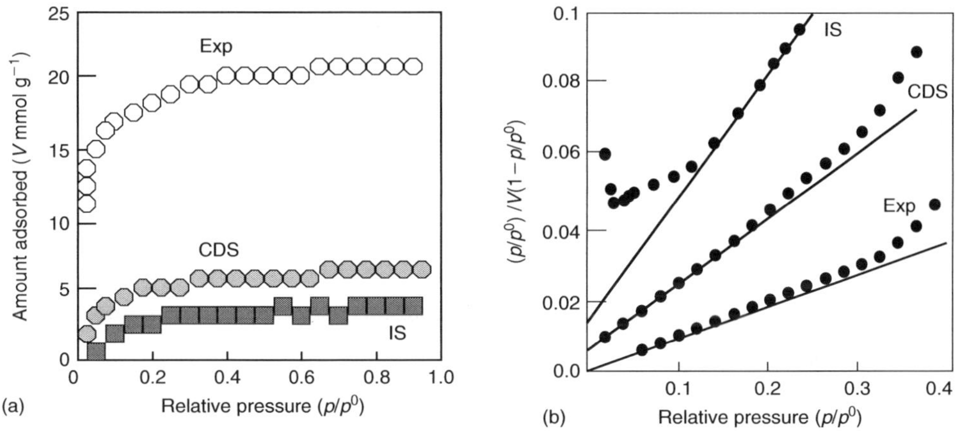

Figure 4.14. Isotherms of nitrogen (77 K) and carbon dioxide (273 K) for a porous carbon to illustrate the use of the isotherm separation method to identify non-microporous components within a porous carbon (see text) (Rodriguez-Reinoso *et al.*, 1989).

are missing; (c) indicates that the pores at the wider end of the distribution curve are wider than predicted; (d) indicates that the pores at the narrow end of the distribution range are missing, or are too narrow to allow adsorption at the experimental temperature; (e) indicates that the distribution of adsorption potentials within that particular carbon cannot be described by a classical Rayleigh, Gaussian or Lorenztian distribution.

Figure 4.14(a, b) describe how the DR equation can be used to separate micropore filling (the isotherm separation method (IS)), in an isotherm, from other adsorption processes (Rodriguez-Reinoso *et al.*, 1989). Figure 4.14(a) contains the experimental isotherm (Exp) for the adsorption of nitrogen on a porous carbon. Also, Figure 4.14(a) contains the isotherm for adsorption of carbon dioxide at 273 K (CDS), assumed to be free from pore filling effects.

When this isotherm (CDS) is subtracted from the nitrogen experimental isotherm (Exp), a residual isotherm will be obtained (IS) for the non-microporous texture, that is, the contribution from the mesoporosity. The application of this method allows the calculation of the non-microporous surface area using the BET equation. It is to be noted that this non-microporous surface is not similar to the external surface area of carbon particles (S_{ext}). The corresponding BET plots are as shown in Figure 4.15(b) from which the following surface areas values were obtained – IS 230 and CDS 500 $m^2 g^{-1}$. The Exp BET plot shows deviations from linearity at high values of relative pressure, the values of V (mmol g^{-1}) being too low according to the prediction. For the IS BET plot, deviations occur in the low relative pressure regions because values of V (mmol g^{-1}) are too low (the IS data do not contain information about the microporosity).

In order to further characterize porosity in activated carbon, to assess if pores with dimensions <0.7 nm existed in activated carbons, Setoyama *et al.* (1996) studied activated carbons from olive stones, 19–80 wt% burn-off in CO_2 by adsorption of helium at 4 K and nitrogen at 77 K. Micropore volumes, determined from the α_s analysis of the helium

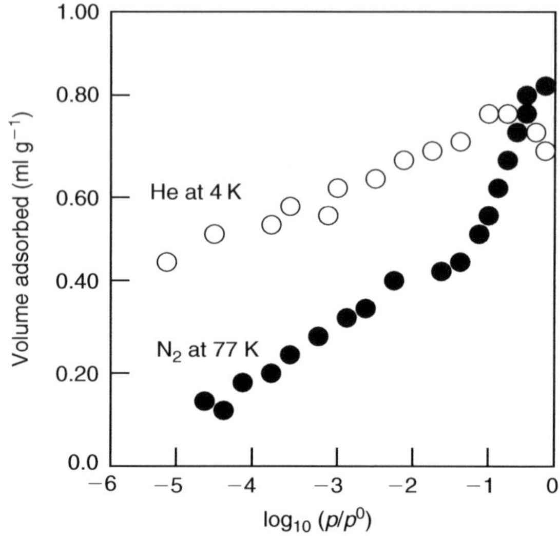

Figure 4.15. Adsorption of helium at 4 K (upper curve) and of nitrogen at 77 K (lower curve) indicating significant adsorption of helium (0.47 ml g^{-1}) at a relative pressure of 10^{-5} on P20 activated carbon fiber (Kaneko, 2000).

isotherms agreed well with those of nitrogen indicating that all of porosity of these carbons was accessible to the nitrogen. The DR equation and PSDs also supported this finding.

Not all activated carbons behave in this way as found by Kaneko (2000). Densities adapted by adsorbate molecules vary according to the dimensions and surface chemistry of the porosity. These authors state that the intermolecular structures of gases confined within microporosity (termed nanospace by Kaneko) are a function of the micropore dimensions. Using two activated carbon fibers, P5 and P20, adsorptions of helium (4 K) and nitrogen (77 K) were measured and the isotherms for P20 are as shown in Figure 4.15. Amounts adsorbed are in units of cm^3 g^{-1} of liquid, assuming a density of adsorbed helium of 0.202 g cm^{-3} (density of bulk liquid helium is 0.125 g ml^{-1}), and of nitrogen of 0.807 g ml^{-1}. The amount of helium adsorbed is much greater than that of nitrogen in the low-pressure regions (note the use of a log scale in Figure 4.15). This enhanced adsorption is explained in terms of restricted adsorption of nitrogen in microporosity, <0.5 nm dimension (see Section 4.2.5), and to accelerated bilayer adsorption of helium. The density of helium adsorbed in the cylindrical micropores of zeolites is 0.23 g cm^{-3}. As microporous carbons possess a PSD (unlike the zeolites) the densities of adsorbed helium vary from pore to pore, between 0.205 and 0.23 g cm^{-3}, higher than the bulk liquid. For nitrogen adsorption, Aukett *et al.* (1992) calculated that the density of nitrogen adsorbed in microporosity is nearer to 0.83 g cm^{-3} than 0.808 g cm^{-3} which is a value for bulk liquid nitrogen. Areas assigned to the nitrogen molecule vary in the range 0.13–0.20 nm^2 (Rouquerol *et al.*, 1999).

Additionally, Ohkubo and Kaneko (2001) obtained densities of adsorbed alcohols, water and CCl_4 on these two activated carbon fibers, P5 and P20. These authors used the technique of electron radial distribution function (ERDF), so allowing this function to elucidate

Table 4.2. Observed and calculated adsorbed densities of selected adsorbates adsorbed on activated carbon fibers P5 (surface area 790 $m^2 g^{-1}$: average pore width of 0.7 nm) and P20 (surface area 1430 $m^2 g^{-1}$: average pore width of 1.1 nm) (Ohkubo and Kaneko, 2001).

	P5			P20			Liquid (303 K)	Solid ρ_s (g cm^{-3})
	W_0 (mg g^{-1})	ρ_{obs} (g cm^{-3})	ρ_{calc} (g cm^{-3})	W_0 (mg g^{-1})	ρ_{obs} (g cm^{-3})	ρ_{calc} (g cm^{-3})	ρ_l (g cm^{-3})	
Methanol	173	0.70	0.70	762	1.00	0.95	0.79	0.98 (113 K)
Ethanol	185	0.70	0.67	867	1.10	0.95	0.79	1.03 (87 K)
1-Propanol	194	0.75	0.85	692	0.88	0.81	0.79	0.97 (146 K)
Water		0.86			0.81		0.996	0.92 (Ih) 1.17 (I_{II})
CCl_4		1.26			1.43		1.594	1.76 (fcc) 1.87 (monoc)

molecular assembly structures hidden in micropores, in particular in carbon micropores. The ERDF gives an average coordination number or molecular density at a distance r from a central molecule. The activated carbon fibers are different with P5 having an average pore dimension of 0.7 nm and P20 having an average pore dimension of 1.1 nm. Such an approach allows an assessment of densities of the same adsorbate in microporosities of different dimensions. Results are in Table 4.2 and clearly indicate the dependency of density of the adsorbate upon the dimensions of the microporosity. Further, in some samples the density is higher than that of the bulk liquid, and in other samples the opposite occurs. Such is the case for adsorption of water and CCl_4.

Discussions such as these are necessary to emphasize, most strongly, that numbers attached to a surface area parameter are only approximate. In fact, it is impossible to place an accuracy on a surface area, as distinct from a reproducibility. The number provided is an indication, only, of a maximum adsorption capacity under the conditions of the adsorption experiment. Assessments of values of n_m^a (monolayer capacity) are certainly not without controversy. To convert this number to a surface area value requires the use of a constant which includes the density of the adsorbate. As this, itself, is a variable number, being a function of the carbon under study, the consequence is to add another layer of uncertainty to the value of a surface area number.

The origins of these variable densities of adsorbates are with the adsorption potentials which exist with microporosity, as illustrated in Figure 4.2 and which will be discussed further under Section 4.6.2 (London dispersion forces).

4.2.4 Quantitative Interpretation of Isotherms Using the t- and α_s- Methods

Another method to analyze an isotherm is termed the t-plot method (t-means thickness). The basis of this approach is that adsorption isotherms for a given adsorptive on different carbon adsorbents may be superimposed by plotting them in the reduced format of n^a/n_m^a against p/p^0. This means that although the adsorptive capacity of the adsorbents may differ,

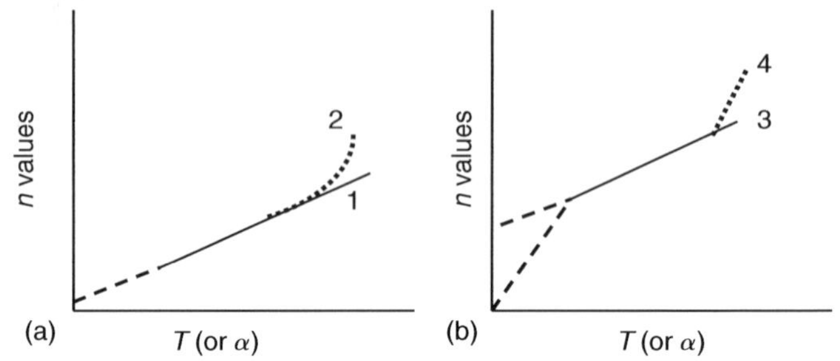

Figure 4.16. Schematic t-plots for different types of adsorbents (see text).

the shapes of the isotherms are similar, leading to a standard isotherm. As with the (IS) method, this approach enables an isotherm to be separated into the components of adsorption into microporosity and mesoporosity.

In the t-plot method, amounts adsorbed on the porous adsorbent are plotted against "t", the corresponding multilayer thickness (if the monolayer thickness is known from appropriate calculations, then n^a/n^a_m can be converted into the multilayer thickness "t") of the standard non-porous solid. Deviations in shape of the experimental isotherm from the standard multilayer thickness curve show up as a departure of this t-plot from linearity. Examples of curves of t-plots for non-porous, microporous and mesoporous solids are plotted schematically in Figure 4.16(a, b). Plots 1, and 2, Figure 4.16(a), correspond to non-porous and mesoporous solids respectively; the upward deviation of plot 2 indicates the onset of capillary condensation in mesopores. The slope, passing through the origin, provides a measure of the total surface area of the solid. Plots 3 and 4 of Figure 4.16(b) correspond to a microporous solid and to a solid containing both micropores and mesopores, respectively.

Originally, the BET equation was used to evaluate n^a_m but subsequent studies indicated that this BET n^a_m value is too high. It was considered that the amount adsorbed at p/p^0 of 0.4 would provide a more realistic approach. As a result, the values of n^a_m were replaced by n^a_s and n^a/n^a_s leading to the α_s-plot which is used in the same way as the original t-plot (Sing, 1970). It has to be recognized that this α_s-approach is based on the assumption that adsorption in micropores is complete before adsorption continues over the remainder of the surface. Often, depending on the pore structure of the carbon, there exists a wide range of relative pressures where significant amounts of adsorption occur in both micropores and mesopores. As with the use of all adsorption equations the application of α_s-plots should proceed with care.

4.2.5 *Adsorption of Nitrogen and Carbon Dioxide: A Comparison*

The partnership (interdependence) of isotherms of N_2 (77 K) and CO_2 (273 K), has developed into important new roles for both adsorptives, based on the DR and BET equations, in terms of characterization of volumes of microporosity and mesoporosity as well as availability of sites of high adsorption potential taken up at low relative pressures (concentrations) of adsorbate.

4.2.5.1 General considerations

For more than half a century the isotherms of nitrogen (77 K) and carbon dioxide (273 K) of porous carbon have been critically compared. Initially, carbon dioxide was used to overcome the problems of activated diffusion experienced by nitrogen but later comparisons of the two isotherms, when activated diffusion effects are absent, have been particularly rewarding.

These comparisons are intimately intertwined with the use of adsorption equations of which the DR equation relates directly to micropore filling, the BET equation initially being designed to describe adsorption on non-porous surfaces. The use of the DR equation to interpret isotherms, obtained at low relative pressures, of both nitrogen and carbon dioxide, provides a rapid method for the characterization of microporosity and the analysis of mechanisms of development of porosity during activation processes. The processes of adsorption of nitrogen and carbon dioxide into microporosity are different as explained below.

Nitrogen, currently, is used as the main adsorbate for the characterization of porous carbons. There is no fundamental reason as to why this should be so. It is simply that nitrogen is used because of its convenience. Liquid nitrogen is available cheaply and can be delivered to the door of the laboratory so facilitating that a carbon specimen can be maintained at 77 K, and gaseous nitrogen is also conveniently delivered, the purity of the gas being >99.999%. The nitrogen isotherm now has an established place in the modern characterizations of carbon.

The nitrogen molecule possesses a quadrupole which consists of four charges that sum to zero but have no net dipole moment (+N− −N+). The quadrupole interacts with the polarity of the carbon surface and this establishes the structure of the nitrogen monolayer. There is little influence of the carbon surface on capillary condensation by nitrogen. This facilitates the use of the t- and α_s- methods in interpretations of nitrogen isotherms. It is also for this reason that nitrogen does not provide a step-wise isotherm (Type-VI), as do argon and krypton when adsorbed on the uniform and low-energy surfaces of a graphitized carbon black, as shown in Figure 4.8.

Historically, the early use of nitrogen as an adsorbate produced conflicting results. Often, values of n^a_m (N_2 77 K) were considerably lower than values of n^a_m obtained by other methods. Conversely, values of n^a_m (N_2 77 K) were found to be much higher when compared to values obtained by other methods. The lower values are now understood to be due to the operation of activated diffusion effects (Figure 4.5). Just as in chemical reactions with a finite activation energy, rates for the process decrease exponentially with decreasing temperature. It is the same for an adsorption process. There is a small but finite activation energy (in the reaction-kinetic sense) of adsorption and at 77 K, for adsorption in microporosity of <0.5 nm diameter, rates of adsorption become impossibly slow and cannot be detected within the time span of one day. Hence, a conflict in data arises. On the other hand, to account for the higher values of n^a_m, it is that (a) the wider micropores can have several layers of nitrogen adsorbed with only two corresponding to the surfaces of the micropores, and (b) at higher relative pressures, $>0.6\,p/p^0$, the adsorbed nitrogen volume fills mesoporosity leading to these excessively high values of estimated surface area.

An alternative procedure was therefore needed to avoid these two problems. It was found that the use of carbon dioxide as an adsorbate has considerable advantages. Gaseous carbon

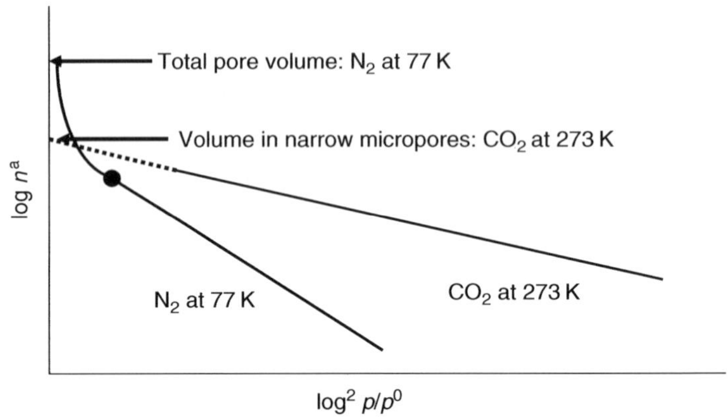

Figure 4.17. Comparisons of isotherms of nitrogen at 77 K and of carbon dioxide at 273 K, on the same porous carbon, in DR coordinates to illustrate differences in adsorption behavior (Marsh and Siemieniewska, 1967).

dioxide is supplied to the catering industries with a high purity, >99.999%, and can be delivered to the door. Adsorptions at 273 K provide isotherms, at low relative pressures, which can be *meaningfully interpreted by the DR equation*. Here, the problems with activated diffusion are almost non-existent. It would appear that the larger quadrupole of the carbon dioxide molecules facilitates monolayer adsorption and avoids volume-filling effects.

These points are illustrated by consideration of Figure 4.17 which compares the adsorption behavior of nitrogen at 77 K with that of carbon dioxide at 273 K, for the same microporous adsorbent (Marsh and Siemieniewska, 1967). The adsorption isotherms are plotted in DR coordinates. The mechanisms of filling of porosity by the two adsorbates are different. At the lowest of relative pressures carbon dioxide may be adsorbed to a larger extent than is nitrogen, within the microporosity. There are two factors to account for this (a) the larger quadrupole moment with the carbon dioxide molecule, compared with nitrogen, accounts for this behavior, and (b) the nitrogen adsorption has diffusional restrictions into narrow microporosity.

At the highest of relative pressures, nitrogen fills the mesoporosity such that the extent of adsorption now exceeds that of carbon dioxide. Experimental systems obviously influence significantly how the adsorbent responds to the adsorbate. In fact, two different "messages" are given by the carbon. The linear portions of the DR plots, at the intercept of $\log^2 p/p^0 = 0$, provide estimates of micropore volume (convertible into a surface area value, if need be) which correspond closely to each other. Differences in gradients between the two isotherms are due to differences in β, which is a shifting factor allowing superimposition of characteristic curves of different adsorbates on the same adsorbent.

The curved section of the nitrogen isotherm, in the higher relative pressure region, is usually interpreted by the BET equation. (*Note*: Never use the DR equation on this section of the isotherm.) The resultant, calculated pore volumes (surface areas) will be higher than that derived from the DR plot (using data obtained at lower values of p/p^0) due, of course, to the multilayers of volume filling of mesoporosity. The nitrogen isotherm, according to

BET, is thus providing a measure of the ESA. It is inappropriate to say that one value is correct and that the other is incorrect. Rather, both approaches are truly valid and values are chosen according to what is required to be measured.

Whereas the same adsorption isotherm provides different values of n_m^a using the different adsorption equations, different adsorbates at different adsorption temperatures also provide different values of n_m^a. The belief that the nitrogen isotherm is the only "correct" approach to measure a surface area has to be abandoned. As an example of these phenomena, microporous materials such as coals have average pore dimensions <0.5 nm, small enough to be affected by the activated diffusion effects discussed above. The adsorption of nitrogen, by such materials, predicts surface areas of $<50\,m^2\,g^{-1}$.

A good example of the problems of adsorption of N_2 at 77 K with activated carbons from almond shells was shown by Rodríguez-Reinoso *et al.* (1982). The precursor was carbonized under nitrogen, 750–900 °C, using several heating rates to prepare six chars. Conventional N_2 (77 K) adsorption isotherms were determined in manual gravimetric adsorption systems taking the usual 24 h for the complete isotherm. The resulting isotherms were quite different from each other in terms of uptake and shape, this being unexpected because the differences in experimental conditions were not large; surface areas varied from 240 to $470\,m^2\,g^{-1}$ (N_2 (BET)). Subsequently, isotherms were determined taking a total time of 100 h, 10 h for each experimental point, to reach an apparent equilibrium. The adsorption isotherms were again very different. Then, the adsorption isotherms for two samples were again repeated, this time taking 6 weeks to complete the isotherm. When the two new isotherms were plotted they come to a common curve and the uptake was larger than in the previous occasions. As it was considered that the adsorption was slow due to kinetics restrictions at the entrance of the micropores, the adsorptions were further repeated at the higher temperature of 90 K for the six chars, using the conventional equilibrium conditions (24 h per isotherm), the result being that the six isotherms were almost exactly coincidental, as expected. The higher temperature of 90 K facilitated the access of nitrogen to the interior of the micropores. In a later paper, Rodriguez-Reinoso *et al.* (1982) published results of the adsorption of CO_2 at 195 and 271 K and these confirmed the lack of equilibrium of the isotherms of nitrogen at 77 K. This idea was the basis for the work paper published in 1983 in which the isotherms of adsorption of N_2 and CO_2 were recommended for a more complete characterization of activated carbons.

4.2.5.2 Comparison of nitrogen and carbon dioxide isotherms using same samples

The combined use of isotherms of nitrogen and carbon dioxide provides for a satisfactory characterization of activated carbons. These isotherms provide informations which are complementary to each other in terms of microporosity and mesoporosity in carbons, their PSDs and the energetics of their surfaces. A comprehensive study of the comparative values of nitrogen and carbon dioxide adsorption isotherms is that of Garrido *et al.* (1987) where many references are provided. Garrido *et al.* (1987) debate and review the uncertainties of assessments of surface areas of carbons and of values which need to be assigned to the densities of the adsorbed state, be it liquid or even solid. Some of these uncertainties have been resolved in recent times, others have not. Garrido *et al.* (1987) summarized the finding of studies of more than 100 different activated carbons which led to the categorization of activated carbons into three groups, according to their adsorption behavior towards nitrogen and carbon dioxide. Further, Rodríguez-Reinoso (1989) includes a discussion of the use of N_2 and CO_2 isotherms along with other methods of characterization of activated carbons.

Group A: Nitrogen is adsorbed to a lesser extent than carbon dioxide. This group of porous carbons adsorbs smaller amounts of N_2 at 77 K than CO_2 at 273 K and contains carbonized materials, such as those (a) of heat treatment temperature (HTT) <600 °C and >800 °C, (b) the molecular sieve carbons with their limited PSDs and (c) activated carbons with <5 wt% burn-off. The restrictive activated diffusion of the nitrogen at 77 K into narrow micropores (or pore entrances) is responsible for this phenomenon.

Group B: Nitrogen and carbon dioxide are adsorbed more or less in equal amounts. This group of carbons possesses microporosity which is relatively narrow (about 1 nm dimension) and with a narrow PSD, which can accommodate only a monolayer of either adsorbate. Activated carbons with <35 wt% burn-off and some molecular sieve carbons behave in this way. Predicted micropore volumes (DR equation) are thus identical.

Group C: Nitrogen is adsorbed to a much greater extent than is carbon dioxide. The majority of activated carbons fall into this group and possess extents of activation with >35 wt% burn-off, and these have a wider microporosity and a broader PSD with some mesoporosity, than carbons with <35 wt% burn-off of Group B. The differences between the N_2 and CO_2 adsorptions become more pronounced with increasing extents of burn-off. Values of $(V_0)_{N_2}$ describe the filling of all microporosity with values of $(V_0)_{CO_2}$ describing the filling of the narrow microporosities.

At high burn-offs, the DR-plots of nitrogen adsorption are curved and do not extrapolate. Micropore volumes are obtained from the α_s-plots. Hence, the nitrogen isotherm provides the total volume in micropores, and this, together with the volume in narrow micropores provides volumes of wider micropores. Information about mesoporosity can also be obtained (Section 4.8).

To expand their data base, to further illustrate the mutual relevance of N_2 and CO_2 adsorptions, Garrido *et al.* (1987) prepared a comprehensive range of seven activated carbons from olive stones, HTT 1123 K, (D-0, D-8, D-19, D-34, D-52. D-70, D-80) activated in carbon dioxide at 1098 K up to 80 wt% burn-off at a rate of 1 wt% h^{-1}. Nitrogen isotherms were obtained at 77 K and carbon dioxide isotherms at 273 and 298 K. All of the isotherms were interpreted by the DR equation and appropriate micropore volumes calculated. Results are as in Table 4.3. In addition, the adsorption data were plotted as characteristic curves which should be superimposable if the adsorption processes are similar, using adsorbents with a common base (Figure 4.18(a–c)). Garrido *et al.* (1987) used relevant results of previously published work of adsorption of hydrocarbons (*n*-butane, iso-butane, benzene, 2,2-dimethylbutane, and iso-octane), with results for benzene being in Table 4.3.

A necessary way to compare the adsorptions of N_2 and CO_2 is to use the characteristic curves (temperature invariant) which should be the same for each carbon if the same mechanisms of adsorption occur. Such characteristic curves are shown in Figure 4.18(a–c) for the carbons D-19, D-52 and D-80, these illustrating the phenomena involved. The affinity coefficient β is not considered in these curves because of the similarity of nitrogen (0.33) and of carbon dioxide (0.35). In order to have a wide range of A^2 (A is the adsorption potential as

Table 4.3. Micropore volumes, V_0, ($cm^3 g^{-1}$) of activated olive stones (Garrido *et al.*, 1987).

Carbon sample	N_2 77 K[a]	CO_2			Benzene 298 K[e]
		273 K[b]	298 K[c]	298 K[d]	
D-0	0.02[f]	0.23			
D-8	0.26[f]	0.26	0.35	0.25	0.21
D-19	0.31	0.30	0.39	0.30	0.25
D-34	0.39	0.36	0.44	0.38	0.36
D-52	0.50	0.41	0.51	0.41	0.49
D-70	0.57[f] (0.66)[g]	0.48	0.57	0.42	0.69
D.80	0.62[f] (0.77)[g]	0.51	0.59	0.42	0.75

[a] N_2 liquid density = 0.808 g cm^{-3}.
[b] CO_2 (273 K) liquid density = 1.023 g cm^{-3}.
[c] CO_2 (298 K) liquid density = 0.700 g cm^{-3}.
[d] CO_2 (298 K) liquid density = 0.97 g cm^{-3}.
[e] Benzene (198 K) liquid density = 0.867 g cm^{-3}.
[f] Deduced from straight-line portion of DR plot at higher values of $\log^2 p/p^0$.
[g] Deduced from straight-line portion of DR plot at lower values of $\log^2 p/p^0$.

defined as $A = RT\ln(p^0/p)$ relative pressures (10^{-6}–1.0) for the nitrogen isotherm were used, employing both volumetric and gravimetric systems.

The characteristic curves for the carbon D-19 (Figure 4.18(a)) all fall on the same line for CO_2 at 273 K and N_2 at 77 and at 90 K. The linearity covers a wide range of A^2 and the extrapolation of A^2 to zero for both adsorptives, and the three temperatures used, lead to a single value of V_0.

The characteristic curves for carbon D-52 (Figure 4.18(b)) also fit the same type of data. However, the range of linearity is reduced to high values of A^2. The CO_2 (273 K) and N_2 (77 K) data give the same value of V_0 only if low-pressure data are available. When the N_2 (77 K) data obtained at values of $p/p^0 > 5 \times 10^{-3}$ are used, their extrapolation gives a value of V_0 which is larger than the extrapolation of the CO_2 data. This phenomenon is more pronounced for the carbon D-80 (Figure 4.18(c)). For carbons with <35 wt% burn-off, the use of relative pressures >0.01 for N_2 at 77 K, leads to values of V_0 larger than those of CO_2 at 273 K (Table 4.3).

Figure 4.18(b) and Table 4.3 include the CO_2 (298 K) data for carbon D-52 – taken as a typical example – calculated by using 0.700 g cm^{-3} as the density. The data lie above, and relatively parallel to, the characteristic curve of CO_2 (273 K). Due to this, a simple change in the density of CO_2 adsorbed at 298 K could bring down the plot to the characteristic curve. This is possible because, as also shown by Garrido *et al.* (1987), CO_2 at 273 K, N_2 at 77 and 90 K, and several hydrocarbons (*n*-butane, iso-butane, benzene, 2,2-dimethylbutane and iso-octane) constitute a unique characteristic curve for carbon D-52 (taken as typical example), but CO_2 (298 K) does not fit such a curve. A corrected density (0.97 g cm^{-3}), which is similar to the value (1.02 g cm^{-3}) deduced from Dubinin (1960) seems then more reasonable than 0.700 g cm^{-3}. It is that, for CO_2 at 298 K (near to the critical temperature), the density is also a function of the dimension of the microporosity. In any case, the value for the area of a CO_2 molecule, at 298 K, cannot be as large as 0.253 nm^2 and it should range

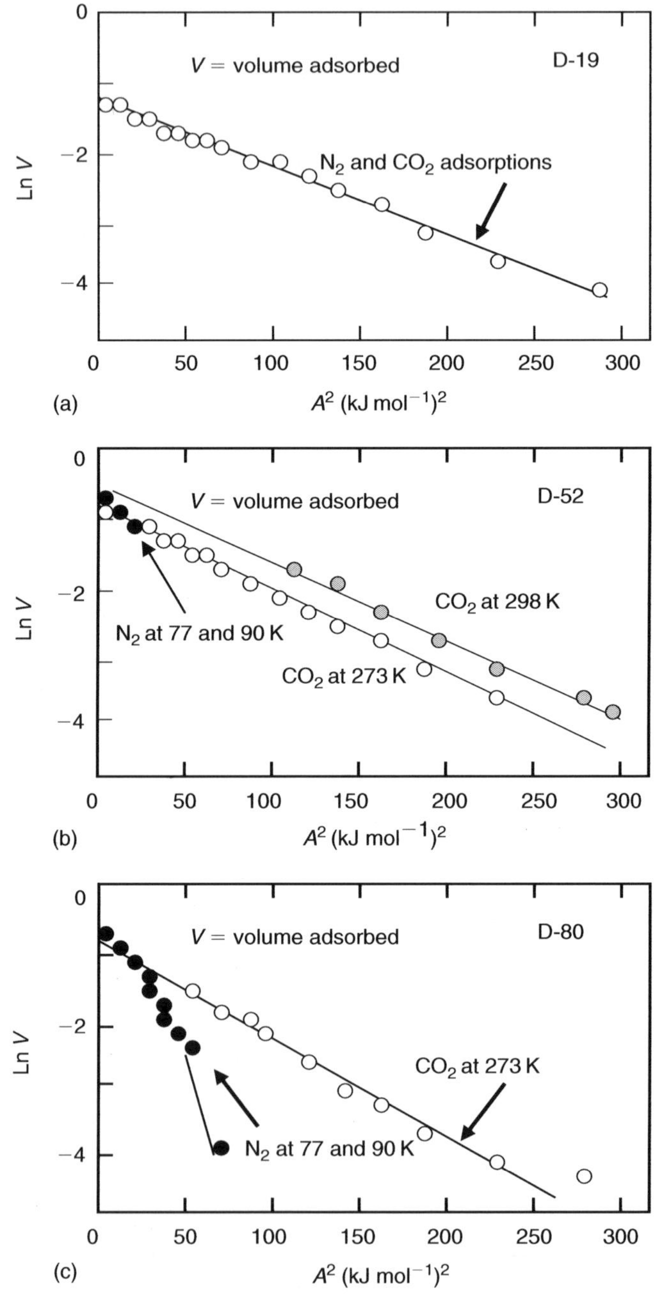

Figure 4.18. DR plots for adsorption of nitrogen and carbon dioxide by activated olive stones of increasing activation (D-19, D-52, D-80) (Garrido *et al.*, 1987).

from 0.195 to 0.220 nm^2, the former applying to low–medium activated carbons. The use of these values for surface areas measured at 298 K will therefore be closer to those determined at 273 K.

The characteristic curves of Figure 4.18(c) show an interesting additional feature. The region of A^2, <30 (kJ mol^{-1})^2 for N_2, is curved for carbon D-80. For values of A^2 >30 (kJ mol^{-1})^2 the curve lies much below that of CO_2 (273 K). There are two possible explanations for this. First, there may be restricted adsorption of nitrogen in the smallest of micropores but this is unlikely as D-80 is highly activated. The second more likely reason is that there is no remaining narrow microporosity having adsorption potentials, A, being <8 kJ mol^{-1}.

It would be unkind not to mention that certain "pitfalls" or "traps" exist for the unwary researcher into the surface chemistry of carbons. Carbon dioxide adsorption behavior can be exceptional when exceptional carbons are studied. Inagaki *et al.* (1992) prepared a carbon from a phenol resin by heating to 1000 °C in carbon dioxide. The N_2 isotherm provided a surface area of 21 m^2 g^{-1} and the CO_2 isotherm a surface area of about 230 m^2 g^{-1}. However, the CO_2 isotherm exhibited hysteresis phenomenon on desorption, extending to low relative pressures. This is not true irreversible behavior as is exhibited by N_2 adsorption at high relative pressures into mesoporosity. Rather, it is kinetic trapping when the activation energy of desorption is high enough, in this carbon of narrow microporosity, to slow down the desorption process to rates which are barely measurable. The establishment of thermodynamic equilibrium positions becomes experimentally difficult to achieve.

The use of the DR or DA equations to assess micropore volumes has to be approached with care. The problem presented by nitrogen at 77 K is its ability (willingness) to fill mesoporosity using a capillary condensation mechanism. Hence, projections of the linear part of a nitrogen DR plot, using data >0.1 p/p^0, could provide misleading information. Carrott and Carrott (1999) discuss this matter and state that the approach (using a Stoeckli derivative of the DR equation) appears to give reasonable micropore size distributions for charcoal cloth samples. They indicate that a main difficulty is in choosing the range of data to be used for calculations of the PSD and mean pore dimension. Carrott and Carrott (1999) also state that this approach is best for a series of related samples as distinct for isolated samples. Perrin *et al.* (2004), in a study of the activation of anthracites, report micropore volumes for a series of carbons in the range 1.32–1.57 cm^3 g^{-1}, yet report methane capacities at 3.5 MPa pressure as low as 75.6 V/V. An explanation for this problem is that the micropore volumes were based on extrapolations of data obtained from nitrogen isotherms at p/p^0 values where capillary condensation occurs. This size of porosity is not the most suitable for storage of methane at high pressures (see Section 6.14).

4.2.5.3 Activation processes as studied using isotherms of nitrogen and carbon dioxide
Although information has been available in the literature since about 1964 (Marsh and Wynne-Jones, 1964; Marsh and Siemieniewska, 1967) with a critique of the use of nitrogen versus carbon dioxide as adsorbates available since 1987 (Garrido *et al.*, 1987; Marsh, 1987; Rodríguez-Reinoso, 1998), the dilemma over this matter has not gone away. Hence, Lozano-Castelló *et al.* (2004) found it necessary to emphasize, once again, this matter. They conclude that, for carbons containing narrow microporosity, as for example carbon molecular sieves (CMS), nitrogen adsorption at 77 K is hardly measurable and a characterization using

nitrogen at 77 K could be useless. In contrast, the carbon dioxide adsorption at 273 K presents clear advantages including (I) a temperature high enough to avoid activated diffusion effects and (II) the ability to measure adsorption at low relative pressures (where much important information is to be found) without the use of complex equipment. Therefore, adsorption of carbon dioxide at 273 K should always be carried out for the characterization of any porous carbon as well as the adsorption of nitrogen. In other words, the characterization of a porous carbon should always consist of determination of both isotherms, not just one isotherm. This point is well illustrated by examining a series of activated carbons prepared from Saran (PVDC) and a carbon fiber (Lozano-Castelló *et al.*, 2004; Rodríguez-Reinoso, 2004).

Figure 4.19 is a plot of micropore volume ($V\,cm^3\,g^{-1}$) against burn-off (wt%) for the physical activation of Saran and a carbon fiber, using nitrogen and carbon dioxide as adsorbates when V_{N_2} increases linearly with burn-off. However, for the carbon dioxide adsorptions, values (V_{CO_2}) reach a maximum at about 60 wt% burn-off for both carbons indicating that narrow micropore volumes increase with activation to about 60 wt% burn-off and then remain constant. That is, activation initially develops narrow micropores, this process being followed by widening of the existing porosity so facilitating some wider pore filling by the nitrogen adsorbate.

Considering that in an activation process V_{N_2} and V_{CO_2} develop differently, the combination of both adsorptives helps to understand the activation process and hence to improve the applications of these materials. These considerations are relevant when comparing and contrasting activation processes by steam and by carbon dioxide at temperatures of about 750–800 °C.

The summary of this section is that the use of N_2 at 77 K (or hydrocarbons such as benzene at 298 K) should always be associated with the adsorption of CO_2 at 273 K in order to provide a complete knowledge of all porosities of an activated carbon.

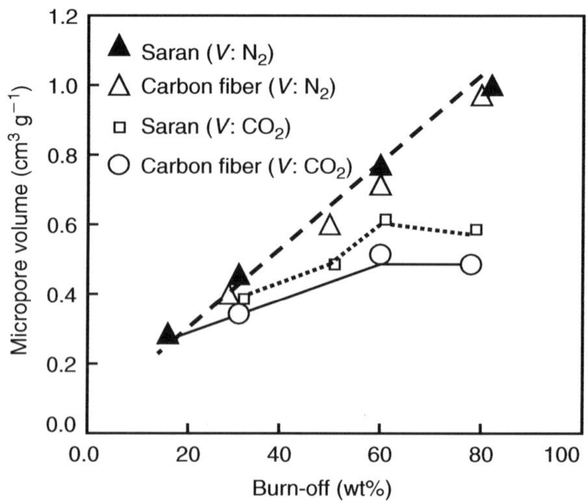

Figure 4.19. The variation of micropore volumes of adsorbed nitrogen (77 K) and carbon dioxide (273 K) on activated Saran and a carbon fiber (Lozano-Castelló *et al.*, 2004).

4.2.5.4 The complete isotherm of carbon dioxide: a standard isotherm
In terms of the characterization of activated carbons, although nitrogen (at 77 K) has been more extensively used than carbon dioxide (at 195 or 273 K) as an adsorbate, it is now appreciated that data from carbon dioxide isotherms are equally informative. In order to understand further the adsorption processes of carbon dioxide (but not specifically to characterize an activated carbon), Guillot and Stoeckli (2001) studied the adsorption of carbon dioxide at 273 K to pressures of 3.2 MPa and at 253 K to 1.9 MPa.

The CO_2 isotherm, at 273 K, describes adsorption at relative pressures $<0.03\, p/p^0$ and as such describes adsorption in the narrow micropores. However, the central assumption in the use of the DR equation is that adsorption of carbon dioxide at pressures >0.1 MPa will follow the projections of the DR plots. There was no experimental justification for this, the acceptance for this being the consistency of results relative to other data of characterization. The assumption needed conformation.

To do this, Guillot and Stoeckli (2001) adsorbed carbon dioxide at 253 and 273 K on Vulcan 3G, a carbon black, to be used to create a reference isotherm. Also included were three activated carbons, an activated carbon fiber, and a microporous carbon black XC-72. Further adsorptions included that of nitrogen at 77 K and benzene at 293 K. The high-pressure isotherms of carbon dioxide on Vulcan 3G are as shown in Figure 4.20. A BET analysis provides a monolayer capacity of 0.523 mmol g^{-1} leading to a surface area of 59 $m^2 g^{-1}$ (assuming a molecular surface area of $18.8 \times 10^{-20} m^2$). The nitrogen isotherm provides a BET surface area of 71 $m^2 g^{-1}$. Considering the assumptions that need to be made about the state of the adsorbed species, this level of agreement is acceptable. These experimental data provide confirmation of the acceptable use of the DR equation.

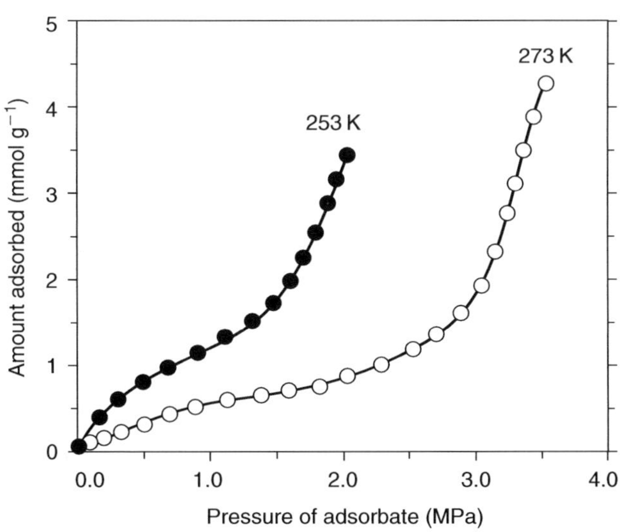

Figure 4.20. The variation of amount of carbon dioxide adsorbed at 253 and 273 K at pressure to 3.5 MPa on the carbon black Vulcan 3G (Guillot and Stoeckli, 2001).

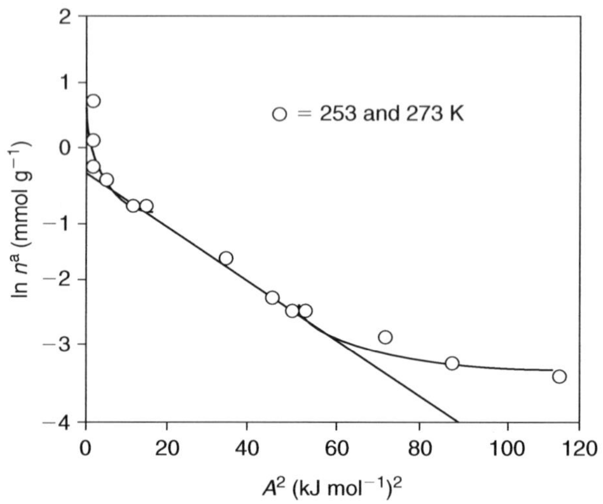

Figure 4.21. The adsorption data of Figure 4.20 are plotted using the coordinates of the DR equation (Guillot and Stoeckli, 2001).

When the data of Figure 4.20 are plotted in the coordinates of the DR equation, as in Figure 4.21 the major part of the isotherm complies with the DR equation, the exceptions being the multilayer formation at pressures >0.5 MPa, and the excessive adsorption at values of $A^2 > 50$ (kJ $mol^{-1})^2$ corresponding to $<10^{-2}$ mmol g^{-1} of adsorbate, or <1% of the surface. These adsorption sites are of exceptionally high adsorption potential, probably some form of point defect in the surface.

Figure 4.22 is a plot of characteristic curves for the adsorption of carbon dioxide at 258, 273 and 298 K on active carbon (ACB). A characteristic curve is a plot of (W in $cm^3 g^{-1}$) against ($A = RT \ln F_s/F$ in kJ mol^{-1}) where F is fugacity, and is indicative if the processes of adsorption are seriously affected by increasing adsorption temperature, such as activated effects. The superimposition of the three characteristic curves of Figure 4.22 is indicative of the acceptability of the use of carbon dioxide and interpretations by the DR equation. The authors calculated a molar volume (V_m) of adsorbed carbon dioxide of 42.90 $cm^3 mol^{-1}$, to be compared with a V_m value of 48.23 $cm^3 mol^{-1}$ (less dense) for free liquid carbon dioxide.

In summarizing the advantages of the use of the carbon dioxide isotherm, Guillot and Stoeckli (2001) consider that the isotherm has the advantage that it can be scanned accurately to the saturation pressure. Differential heats of adsorption can be obtained describing further the surface of the carbons. Of considerable importance is the realization of the non-specificity of the carbon surface. Carbon dioxide isotherms are not affected by variations in surface function oxygen contents varying from 1 to 7 mmol g^{-1}, unlike the adsorptions of methanol and of ammonia which are affected by surface oxygen functionality.

4.2.5.5 Characterization of porosity: low relative pressure (<0.0001 p/p^0)

It is stressed that a BET surface area is *not* all-important in assessments of the suitability of an activated carbon for a particular application. For example, a carbon with a BET area

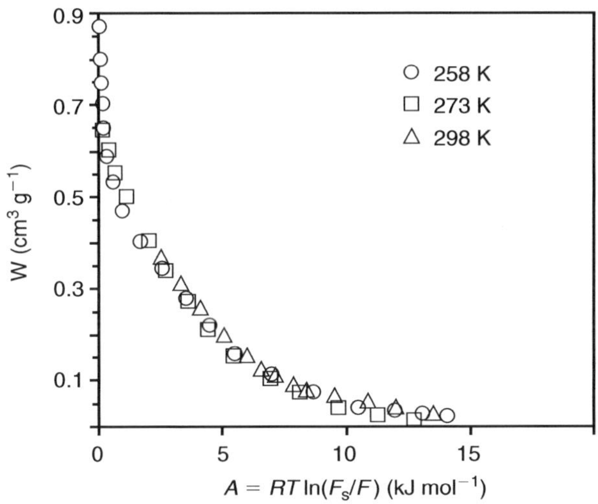

Figure 4.22. The characteristic curves for the adsorption of carbon dioxide at 258, 273, 298 K on active carbon (ACB) (Guillot and Stoeckli, 2001).

of $500\,m^2 g^{-1}$ could out-perform a carbon with a BET area of $1000\,m^2 g^{-1}$. This chapter makes the point that activated carbons are used to remove unwanted components of gaseous or liquid mixtures, these unwanted species being at low partial pressures or concentrations. As such, in terms of the isotherm (a plot of amount adsorbed versus relative pressure), it is only that part of the isotherm at relative pressures of $<0.0001\,p/p^0$ (or concentrations of $<$100 ppm) which is used. The section of the isotherm $>0.0001\,p/p^0$ is less relevant. Centeno *et al.* (2003) discuss this important point in some detail and provide methods of appropriate analyses of carbons.

The efficiency of adsorption at low relative pressures, which may not relate to BET surface areas, is discussed in terms of crossover isotherms by Mangun *et al.* (1998), Figure 4.23, and quoted by Centeno *et al.* (2003). The steepness of the isotherm at relative pressures of $<0.0001\,p/p^0$ is indicative of the high adsorption potentials for the microporosity which is active in the purification processes.

In their search for appropriate methodology to characterize the microporosity which is responsible for this crossover phenomena, Centeno *et al.* (2003) activated five Nomex-carbon fiber monoliths (ACFMs), in steam, to levels of 6, 12, 21, 30 and 40 wt% burn-off at a slow rate of 2 wt% loss per hour. In addition, they used a commercial granular activated carbon from Norit (RB3) having a BET area of $1220\,m^2 g^{-1}$ and a wide PSD. The adsorption of *n*-butane was studied at 303 K at concentrations of *n*-butane of 100, 500, 1500, 3300 ppm and 10^6 ppm. The variation of amount of *n*-butane adsorbed ($mmol\,g^{-1}$) with degree of burn-off (wt%) is as shown in Figure 4.24.

Important points to note are (1) it is only when using 100% *n*-butane do extents of adsorption continually increase with increasing burn-off, (2) for lower concentrations of *n*-butane, $<$3300 ppm, extents of adsorption show an inflexion at about 21 wt% burn-off indicating

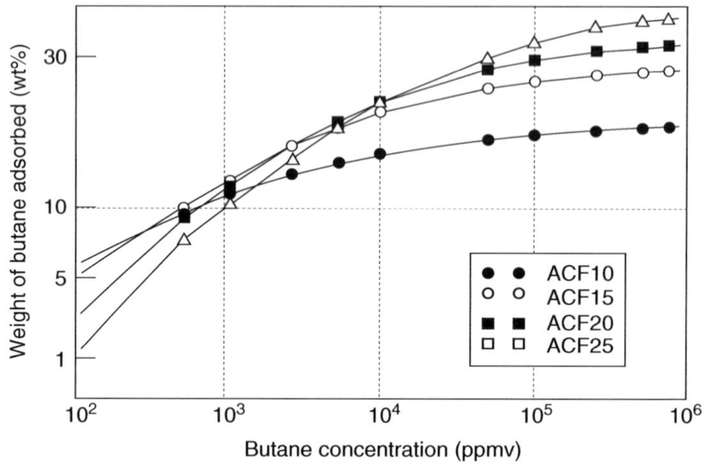

Figure 4.23. Isotherms of adsorbates at very low values of p/p^0 (~100 ppm) indicating a crossover of capacities between carbons of lower and higher BET surface area. The isotherms are of butane at 25 °C for activated carbon fibers, activated with carbon dioxide and steam, from NIPPON Kynol. ACF10 and 25 have surface areas of 900 and 1980 $m^2 g^{-1}$, respectively. Note that at 100 ppm of butane, ACF10 adsorbs twice that of ACF25 (Mangun *et al.*, 1998).

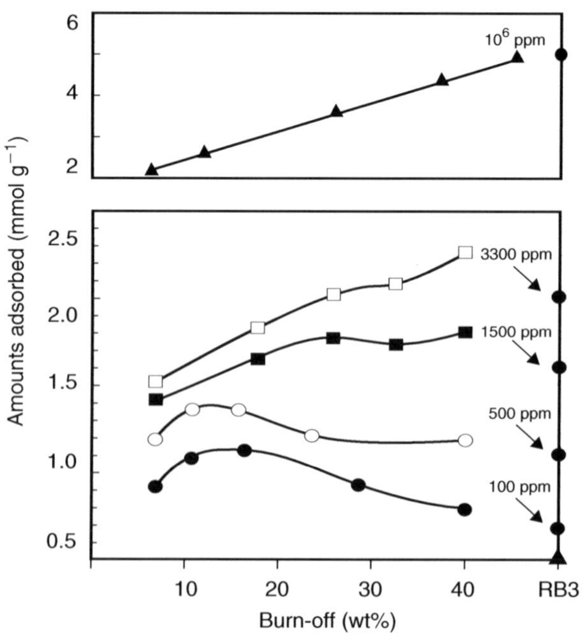

Figure 4.24. The variation of amounts of adsorption of *n*-butane with extents of burn-off for the Nomex carbon fiber monoliths, using increasing levels of *n*-butane concentration. The points on the right coordinate line are for a Norit commercial carbon (Centeno *et al.*, 2003).

that progressive burn-off need not be an advantage, (3) for the lowest of *n*-butane concentrations, <500 ppm, extents of adsorption actually decrease with increasing burn-off, indicating that an effective porosity has been removed. There is no other explanation. In fact, for the lowest of relative concentrations of adsorptive, <100 ppm, the effectiveness of activation is zero, That is, the porosity created has not increased the effective adsorption of the *n*-butane, the carbon behaving as the non-activated material.

The specific questions now required to be answered concern the dimensions of the effective porosity and how to measure them. The method of analysis of pore filling by nitrogen at 77 K, at p/p^0 values >0.6 cannot be used, obviously. Centeno *et al.* (2003) indicate that immersion calorimetry provides the means to measure micropore size distributions within an activated carbon (Section 4.7). They use a rearrangement of Equation (4.3) to calculate the micropore volume ($W_0(L_c)$) filled by a liquid of critical molecular dimensions, L_c, using:

$$W_0(L_c) = \frac{(\Delta_i H(L_c) - h_i S_e)2V_m}{\beta E_0(1 + \alpha T)\pi^{0.5}} \qquad (4.6)$$

where $\Delta_i H(L_c)$ is the experimental enthalpy of immersion of the carbon into the liquid with a dimension L_c, α and V_m are the thermal expansion coefficients and the molar volume of the liquid, and h_i is the specific enthalpy of wetting of the external (non-microporous) surface S_e.

Stoeckli and Centano (1997) showed that, for carbons of low external surface areas, the ratios between the limiting volumes filled by liquids of variable molecular dimension and the micropore volume accessible to a small molecule (when used as a reference) can be closely estimated from enthalpies of immersion. This approach describes the development of porosity during an activation process and the PSDs of the developed microporosity. As a wetting agent, Stoeckli and Centano (1997) used benzene because of the similarity of molecular size with *n*-butane which cannot be used as a liquid. The majority of the micropores were <0.8 nm in access dimension.

As a further means of characterizing microporosity, Centeno *et al.* (2003) made use of the adsorption of *n*-butane from low concentrations (from 100 to 10^6 ppm) adsorbing on ACFMs (6–40 wt% burn-off). The adsorption of *n*-butane, at low concentrations, takes place in pores of <1 nm dimension. The widening of the microporosity is seen in Figure 4.25 where for the carbon (ACFM30) which had a maximum of *n*-butane adsorption at concentrations of 100 and 500 ppm, the micropores had dimensions of 0.63–0.76 nm.

Micropores of dimension <0.63 nm could contain a monolayer of the *n*-butane, but as the micropores become wider, during activation, the ability to adsorb two layers is realized and adsorption is maximized. Further activation widens the microporosity, the adsorption potential is lowered, and extents of adsorption of *n*-butane (100 ppm level) decrease. Maximization of uptake results from a balance between the adsorption potentials of the sizes of the microporosity of the carbon, that is, the ratio of molecular size of the adsorbate and pore size. This is of extreme importance for the removal of impurities **at low levels of concentration** when the activated carbon has to be tailor-made to create a microporosity which matches the molecular size of the impurity. However, when the concentrations

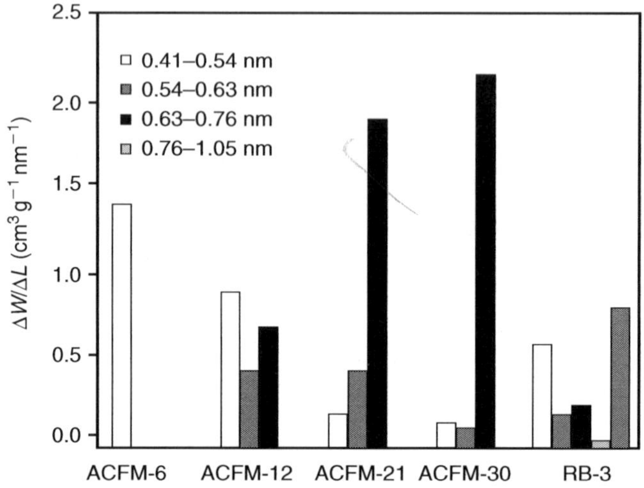

Figure 4.25. Distributions of micropore sizes in activated carbon fibers and Norit carbon (Centeno *et al.*, 2003).

of the impurity are much higher, all of the microporosity is used in the adsorption process and capacity is a function of total micropore volume.

4.2.5.6 Procedures to define microporosity in activated carbon

Adsorption of N_2 *at 77 K: and use of DR plots.* The range of linearity for DR plots is relatively large when the degree of carbon activation is low when the extrapolation to calculate the micropore volume is straightforward. Linearity decreases with increasing activation, the plots showing a clear upward deviation (indicative of the existence of wider microporosity and small mesoporosity). Any definition of the micropore volume is uncertain because of the short straight linear portion of the plot. If this short straight portion is used to calculate the micropore volume, the value may be lower than expected because it does not include wide microporosity.

The DR plots, applied to the adsorption of hydrocarbons at 298 K, provide values of micropore volume very similar to those of nitrogen when the degree of activation is low (provided there is no limitation by the small entrance of the micropore as in the case of iso-octane for D-8). However, hydrocarbons give larger values of micropore volume when the degree of activation is large (e.g. D-70 and D-80). An additional approach should then be used.

Use of the α_s *plots.* If a correct carbonaceous reference material is used, then back extrapolation of the adsorption data to calculate the micropore volume is not subject to the uncertainty of the DR plots, even for highly activated carbons. The data of Table 4.4 show that results are in agreement with DR plots only for carbons with low degree of activation (narrow microporosity), being larger in all other cases. By considering the above results for nitrogen and hydrocarbons, it is concluded that the α_s plots for the nitrogen isotherms provide the total volume of micropores whereas the DR plots for the same isotherms normally yield too low values when wide microporosity is present in the carbon.

Table 4.4. Micropore volumes, V_0 ($cm^3 g^{-1}$) of activated olive stones (supplementary to Table 4.3).

Carbon	N_2 (77 K) DR	N_2 (77 K) α	Benzene (298 K)	iso-octane (273 K)	CO_2 (273 K)	V_{NP}	V_{np}
D-8	0.26	0.25	0.21	0.03	0.26	0.26	0.16
D-19	0.31	0.31	0.25	0.21	0.30	0.31	0.24
D-34	0.39	0.39	0.36	0.35	0.36	0.38	0.31
D-52	0.50	0.53	0.49	0.50	0.41	0.51	0.41
D-70	0.57	0.64	0.69	0.68	0.48	0.52	0.47
D-80	0.62	0.76	0.75	0.76	0.51	0.52	0.50

As stated previously, the adsorption of CO_2 at 273 K, because of the low relative pressure range covered ($p/p^0 = 0.03$), defines the volume of only narrow micropores, the uncertainty in the extrapolation being much lower than for nitrogen. Results in the Table 4.4 show that it is in carbons with low degrees of activation that the DR plots for CO_2 and N_2 provide the same volume of micropores. In all other cases the CO_2 values are smaller and for carbons with a high degree of activation the difference is important.

Pre-adsorption of n-nonane. The pre-adsorption of nonane (Gregg and Sing, 1982) is a direct method to evaluate the microporosity. This is because *n*-nonane is strongly retained in narrow pores after outgassing at room temperature, leaving the rest of the surface available for nitrogen adsorption. The vertical separation of the nitrogen isotherms before and after the *n*-nonane pre-adsorption gives the micropore volume, V_{NP}. It is possible to deduce the volume of *n*-nonane retained by the carbon after outgassing at 298 K, using a density for *n*-nonane of 0.717 g cm^{-3}. Values within Table 4.4 show that for carbons with low activation the V_{NP} values are exactly coincident with the values obtained from the DR plots for N_2 and CO_2. In such carbons, the narrow micropores are completely blocked by the pre-adsorbed *n*-nonane. On the other hand, the V_{np} values for these carbons are usually lower than V_{NP} because the *n*-nonane does not fill completely the narrow micropores but effectively blocks them. As activation increases and a widening of microporosity occurs there is a change in situation, as shown Table 4.4 for carbons above D-19, because the degree of pore interconnectivity changes with activation and the values are very similar for D-80.

The volume of micropores deduced from the adsorption of CO_2 at 273 K and V_{NP} are almost coincident for carbons with low and very large activation but the CO_2 values become lower for all other carbons. The V_{np} values are lower that the CO_2 values at low and medium activation, this being explained by the different accessibility to the micropores of the two molecules on account of their molecular dimensions. When the micropores are narrow the *n*-nonane molecules block the pore entrances and do not fill them completely, whereas CO_2 molecules are able to enter and fill the narrow micropores. The widening of the micropores following increasing activation leads to an approximation of V_{np} toward V_{NP} and the values determined by CO_2 adsorption.

Consequently, from these results it is concluded that the α_S plots applied to the adsorption of N_2 at 77 K provide the total volume of micropores whereas the DR plots for the adsorption of CO_2 at 273 K provide the volume of narrow micropores. It is clear then that the volume of wide micropores can be deduced by subtracting the volume of narrow micropores from the total volume of micropores.

To summarize Section 4.2, the complementary use of N_2 (77 K) and CO_2 (273 K) isotherms is important to the characterization of activated carbon and the activation process because (1) it resolves the problems of activated diffusion effects associated with nitrogen adsorption, (2) it provides a characterization of the pore volumes (narrow and wider) of the all-important microporosity (<2.0 nm entrance dimension), (3) it characterizes mesoporosity and (4) it provides an insight into the energetics of those surfaces which adsorb at low relative pressures/ concentrations.

4.3 Characterization of Porosity: Surface Functional Groups

4.3.1 Introduction

The models of carbon structure of Chapter 3 contain only carbon atoms. Actually, activated carbons contain other elements, known as heteroatoms, such as hydrogen, oxygen, nitrogen and sulfur, in the main, as well as inorganic mineral matter should the carbon have a natural origin such as a nutshell or a coal. Hydrogen is bonded to edge atoms but the oxygen, nitrogen and sulfur can be bonded both at edges of graphene layers and also *in-ring* within the graphene layers. All of these heteroatoms influence the properties of the carbons in several ways but the greatest influence comes from the presence of oxygen, in particular edge-bonded-oxygen. An efficient industrial application of activated carbons requires an understanding, in particular, of these surface oxygen complexes.

The several types of oxygen groups to be found on carbon surfaces, because of the electronegativity of the oxygen atoms, possess dipole moments and their presence has a marked effect on the shapes of adsorption isotherms of polar adsorbates. This influence is of singular importance for systems using adsorption from solution, in particular adsorption from aqueous solutions where water molecules are competitively adsorbed at sites of the oxygen surface complexes. It is therefore of some importance that surfaces of activated carbons be analyzed in terms of amounts of oxygen present and equally in terms of the chemical composition of the several forms of oxygen which are present, collectively known as *surface oxygen complexes*. Figure 4.26 summarizes examples of oxygen functional groups to be found on carbon surfaces (Rodríguez-Reinoso, 1998).

Carbons are never free from oxygen except in extreme circumstances when they are heated to about 950 °C in vacuum or hydrogen and cooled and stored in vacuum. Otherwise, after a carbonization, exposure to air invariably produces these surface oxygen complexes, a process which can continue for months, more so if there is moisture present. Rates and extents of formation, from molecular oxygen, of these surface oxygen complexes increase significantly, rising to a maximum with increasing oxidation temperature to 300–350 °C. Above this temperature, gasification of the carbon occurs leading to ignition if an adequate supply of oxygen is available. Should the oxidation reaction be terminated at about 300–350 °C, and the carbon further heated under vacuum or an inert gas, then the surface oxygen complexes decompose to carbon monoxide and carbon dioxide, a final temperature in the region of 1000 °C being needed for removal of the complexes.

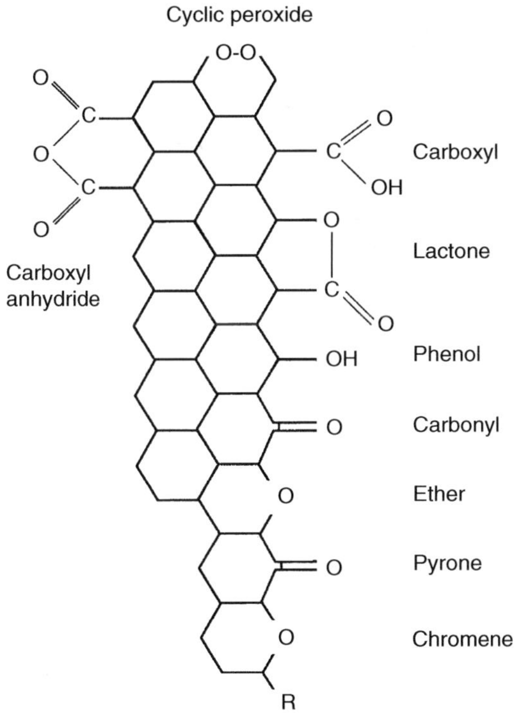

Figure 4.26. Structures of oxygen functional groups on carbon surfaces (Rodríguez-Reinoso and Molina-Sabio, 1998).

Not all oxidizing agents produce the same surface oxygen complexes. Moreno-Castilla *et al.* (2000) oxidized three activated carbon with HNO_3, H_2O_2 and $(NH_4)_2S_2O_8$. The highest total oxygen uptake was with the HNO_3, and the lowest was with the $(NH_4)_2S_2O_8$. All the oxidizing agents produced single C—O bonds, lactones, carboxyl, quinone or conjugated ketene and carboxyl-carbonate structures in all of the three carbons. However, the highest total acidity (mainly from carboxyl) was obtained using the HNO_3 and the lowest basicity (see below for a discussion of basicity) after oxidation with $(NH_4)_2S_2O_8$. The highest acid strength was obtained after the oxidation with $(NH_4)_2S_2O_8$. as shown by pH measurements and catalytic activity in the dehydration reaction of methanol. These authors consider that this effect is due to the fixation of carboxyl groups close to other groups when, due to both negative induction effects and resonance effects, the surface acidity is enhanced. Clearly, much further work is required to understand how carbon surfaces respond so differently to these different oxidizing agents.

4.3.2 Formation and Properties of Oxygen Complexes

Influence of chemisorbed hydrogen on carbon surfaces. A review of standard methods of formation of surface groups on carbon surfaces is published by Fanning and Vannice

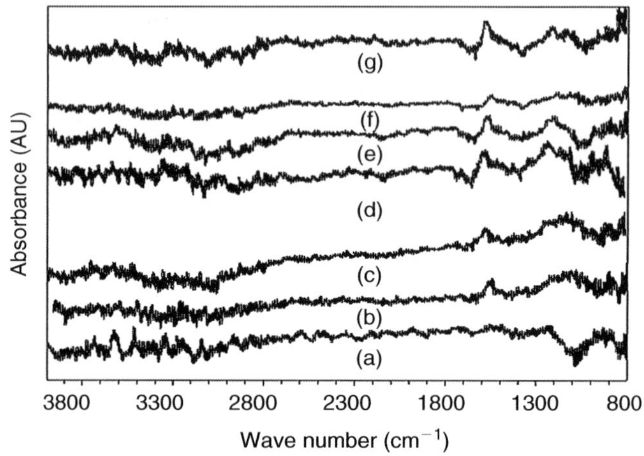

Figure 4.27. DRIFT spectra for a Saran carbon (HTT 900 °C) following increasing extents of oxidation. Carbon (a) has received no oxidation (Fanning and Vannice, 1993).

(1993). Carbon blacks and a Saran carbon (a mixture of vinyl chloride with polyvinyl chloride) were oxidized in air or by nitric acid. Figure 4.27 is a collection of DRIFT spectra (see Section 4.7.3) obtained from the Saran carbon when progressively oxidized.

However, as part of this study, Fanning and Vannice (1993) report two unexpected findings. **First**, the carbons were outgassed (cleaned) prior to oxidation studies at about 1000 °C, not under nitrogen but under hydrogen. Outgassing and cooling under hydrogen produced a surface quite unreactive towards the formation of stable surface functional groups. None were found during gasification reactions. It would appear that during the outgassing treatment and cooling process hydrogen was chemisorbed on to the carbon surfaces at sites which otherwise could have formed surface oxygen complexes. This observation is of importance when considering the physical activation of carbons by steam (see Chapter 5). **Second**, what was also found was the absence of carboxylic groups on the surfaces of the hydrogenated carbons, after being oxidized by nitric acid. Again, the hydrogenated surface may have a role to play here when cyclic hydrides are prevented from hydrolyzing to form carboxylic acids.

Formation of oxygen complexes using atomic oxygen and ozone. It has been reported that atomic oxygen is very effective in introducing oxygen groups to carbon surfaces. Activation energies are reported to be in the range of 0.0–40 kJ mol^{-1}. The use of atomic species is an effective way of introducing surface functionality to a carbon surface as well as chemisorbed hydrogen (Marsh *et al.,* 1965a, b). Although an adsorbed atom of oxygen is reported as being an intermediate in the carbon–molecular oxygen reaction, the chemical kinetics and topographical kinetics of reactions with atomic oxygen differ from those of the carbon–molecular oxygen reaction.

A second, highly reactive species of oxygen is ozone (O_3). The effects of temperature on oxidation of carbons prepared from cherry stones as well as the kinetics of the reactions with ozone are reported by Gómez-Serrano *et al.* (2002a, b). There is no doubt that if an abundance of surface oxygen complexes is needed on a carbon surface then treatment with ozone is an effective way of doing it. The usual complexes such as phenolic, hydroxyl, quinonic, carboxylic acid and ethers are generated, with concentrations of lactonic complexes being quite low, for some reason. The temperature of maximum complex formation was 100 °C with complexes being formed which would be unstable at 200–300 °C, being more unstable than when molecular oxygen is the reactant.

The HTT of the char has a critical role to play, with an HTT of 450 °C being the most reactive. A char of HTT 900 °C is unable to form ether and carbonyl groups.

Oxygen atoms are mobile over carbon surfaces (Chapter 5) and an understanding of the chemistry of formation of such a range of surface oxygen complexes is a challenge. For example, Gómez-Serrano *et al.* (2002a) suggest that the ozonolysis of aromatic structures on the carbon surface may produce phenols, quinones and aromatic acids, followed by ring rupture to give the short chain aliphatic acids and aldehydes. When using relatively high concentrations of ozone in the reacting gas stream, relatively high concentrations of carbonyl and ether structures were formed. At low ozone concentrations, ozonolysis favored the formation of hydroxyl groups. The activation energy of the ozonolysis reactions was 41.6 kJ mol^{-1}.

Gasification by hydrogen atoms proceeds with an activation energy of about 40 kJ mol^{-1}. The activation energy for reaction with nitrogen atoms is higher, at about 75 kJ mol^{-1}.

4.3.3 Analyses of Surface Functional Groups

Traditional methods and modern instrumentation has facilitated chemical analyses of surface oxygen complexes and these are summarized below.

Titration with alkaline solutions (also known as the Boehm method (selective titration)). This is a method which has been used for several decades and involves the titration of a suspension of the carbon in water with bases of increasing strengths (Gurrath *et al.*, 2000; Boehm, 2002). Experimentally, the carbon is suspended for several hours with solutions of sodium bicarbonate (carboxylic acids and anhydrides), sodium carbonate (lactone and lactol rings), sodium hydroxide (phenols) or sodium ethoxide in ethanol solution (carbonyl groups). Not all of the oxygen known to be on a carbon surface (as determined by the TPD method, see below) can be measured by these titration techniques. Some oxygen could be bonded within the graphene layers (in-ring) and hence would be non-reactive, and some carbonyl groups may not be able to react with the alkaline solutions. However, Molina-Sabio *et al.* (1991) report that access of the reactant may be limited where functionality is within narrow microporosity.

Infrared spectroscopy. This is the traditional method for structural analyses of organic compounds where infrared radiation is absorbed selectively by the various bonds within a compound.

Carbons, as such, are intrinsic absorbers (this is why carbons are black) and so advances were only possible with the development of DRIFTS (Diffuse reflectance spectroscopy) (Sellitti *et al.*, 1990). An example of a DRIFT spectrum of graphitized carbon cloth oxidized for

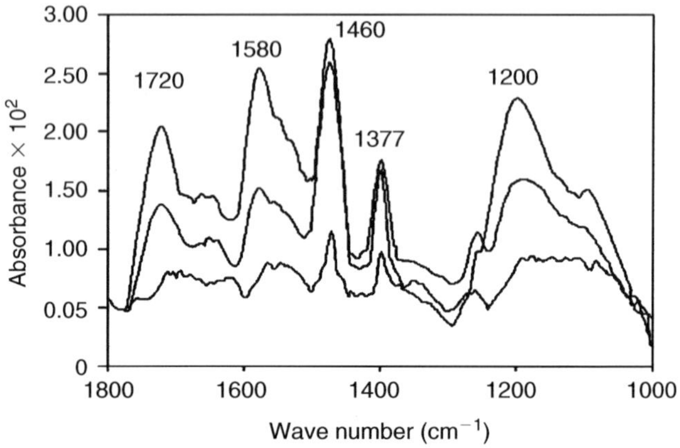

Figure 4.28. A DRIFT spectrum of graphitized carbon cloth oxidized for 25 h (top), 15 h (middle) and 5 h (bottom) (Sellitti *et al.*, 1990).

Table 4.5. Oxygen functional groups on carbon surfaces and their assignments in infrared spectra (Fanning and Vannice, 1993).

Oxygen functionality	Assignment regions (cm^{-1})		
	1000–1500	1500–2950	2950–3700
C—O stretch of ethers	1000–1300		
Ether bridge between rings	1230–1250		
Cyclic ethers containing COCOC	1025–1276		
Alcohols	1049–1276		3200–3640
Phenolic groups			
C—OH stretch	1000–1220		
O—H bend/stretch	1160–1200		2500–3620
Carbonates: carboxyl carbonates	1100–1500	1590–1600	
Aromatic C=C stretching		1585–1600	
Quinones		1550–1680	
Carboxylic acids (COOH)	1120–1200	1665–1760	2500–3300
Lactones	1160–1370	1675–1790	
Anhydrides	980–1300	1740–1880	
Ketenes (C=C—O)			2080–2200
C—H stretch			2600–3000

25 h (top), 15 h (middle) and 5 h (bottom) is as shown in Figure 4.28 with bond-assignments listed in Table 4.5.

X-ray photoelectron spectroscopy (XPS). This is a surface analysis technique whereby X-ray radiation excites the core electrons of surface atoms, and such electrons are emitted each with a characteristic kinetic energy which is analyzed in the spectrometer to provide a spectrum as seen in Figure 4.29. There is overlap of some bands which require some deconvolution before assignments are made of peaks to structures.

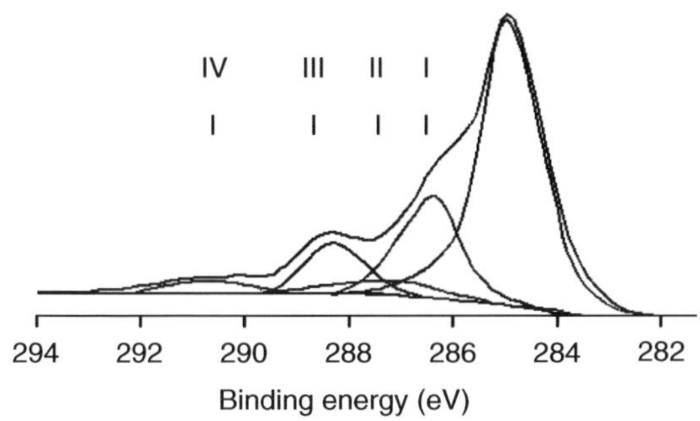

Figure 4.29. An XPS spectrum of electrochemically oxidized carbon fibers containing (I) phenols, ethers, (II) carbonyl groups, (III) carboxyl groups, (IV) plasmon peak (Yue *et al.*, 1999).

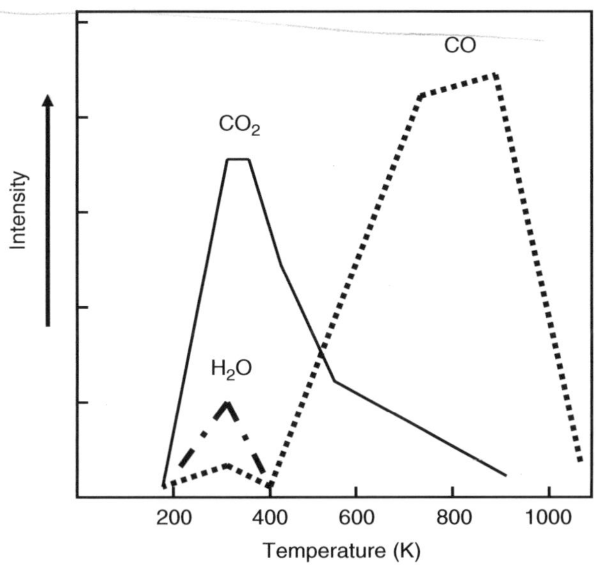

Figure 4.30. A TPD spectrum of a carbon containing surface oxygen groups (Otake and Jenkins, 1993).

Temperature programmed desorption (TPD). This is a technique requiring the programmed heating of a suitably oxidized carbon in vacuum or in a flow of helium and the facility (a mass spectrometer) to measure quantitatively the gases evolved, these usually being water, carbon dioxide and carbon monoxide, the latter desorbing at the highest of temperatures (Figure 4.30).

Carbon dioxide comes from the decomposition of carboxyl-type groups and carbon monoxide from carbonyl, hydroxyl and ether oxygen. As Figure 4.30 shows, there is overlap of the decomposition spectra. In view of the complexity of possible structures in carbons

(Chapter 3) this is not surprising. Another factor is the "instability" of a carbon surface. The removal of a carbon atom leaves a vacancy which may be filled by movement of nearby carbon atoms and so the detail of bonding energies varies during a desorption experiment.

4.3.4 Surface Acidity and Basicity

Activated carbons exhibit both surface acidity and surface basicity as monitored when carbons are placed into pure water and the system allowed to reach equilibrium. Changes in the pH of the system are then monitored. (*Note*: *pH is defined as the minus logarithm of the hydrogen ion* (H^+) *concentration. The dissociation constant,* pk_a, *of water has a value of* 10^{14} *at 25 °C, and is equal to* $[H^+][OH^-]$. *The proton is hydrated according to* $C_\pi + 2H_2O \rightarrow C_\pi \cdot H_3O^+ + OH^-$ (Rodríguez-Reinoso, 1998; Rodríguez-Reinoso and Molina-Sabio, 1998).) Thus, for water in equilibrium with a carbon, if the water has a pH value <7 (e.g. 6.9), then the water is acidic from the dissociation of H^+ from the surface oxygen groups, such as a carboxylic group, —COOH. However, for an alkaline solution, the pH has to be >7.0.

A critical statement is that it appears that this basicity is not entirely associated with surface oxygen complexes. It is not straightforward to understand surface basicity.

Carbon surfaces become basic in character following an outgassing at temperatures >700 °C. Subsequently, they must not be exposed to air until they have cooled to <200 °C. Following this treatment, carbons exhibit basic characteristics, that is, on equilibration with water, the latter becomes basic with pH values >7.0. They also exhibit positive external surface charges and require acid titrations to restore neutrality (Boehm and Voll, 1970). Leon *et al.* (1992), following a lengthy review of the nature of basicity in carbons, concluded that electron donor addition (EDA) complex formation is predominant in carbons of low oxygen contents when carbon sites are able to adsorb protons from solution in such a way as to create a positively charged surface. These carbon sites are associated with regions of the graphene layers which contain most of the π-electrons, that is, the central areas of the layer away from the edge atoms. The H_3O^+ ions in the vicinity of graphene layers exchange a secondary adsorbed water molecule for a π-electron pair which becomes partially localized, consequentially:

$$C_\pi + H_3O^+ \leftrightarrow C_\pi{-}H_3O^+ \tag{4.7}$$

Each H_3O^+ ion occupies an area associated with that of a π-electron pair (i.e. a benzenoid ring, 0.05224 nm^2). For non-graphitizable microporous carbons, the EDA interaction is restricted by the porosity and access to surfaces. As a result, for such carbons, only a small fraction of each graphene layer (defective) can hold the EDA-complexed H_3O^+ ions. For carbons with small amounts of surface oxygen complexes, this fraction increases with increasing graphene layer size and perfection of the structure within the graphene layer.

The analysis of the origins of basicity in carbons is not straightforward and the problem is ongoing. Graphitic carbons possess significant amounts of π-electrons. Molecular orbital calculations indicate that proximity to the edges of the graphene layers has a significant influence on the distribution of the mobile π-electrons. The more chemically reactive zigzag edges (Chapter 5) induce peripheral electron localization and the less reactive *armchair* edge sites lead to a more uniform electron-density distribution. Physical activation of a carbon (Chapter 5), that is the gasification of carbon atoms from a graphene layer, by reacting primarily at edge atom sites, reduces the sizes of the graphene layers and promotes the localization of the π-electrons. The presence of acidic oxygen associated with gasification further enhances the π-electron localization resulting in decreases in the basicity of the carbon.

Barton *et al.* (1997) reported that surface oxides which may be present on surfaces of carbon (a carbon black is considered) *do not* exhibit basic properties. The view is presented that the basic sites are identified as Lewis-type basic sites associated with the carbon structure itself. Barton *et al.* (1997) provide additional experimental support for the ideas put forward, as above by Leon *et al.* (1992), to account for the basicity of carbon surface. A porous carbon, supplied by Calgon BPL, after washing and drying at 110 °C, and termed AR, was oxidized with 6.0 M HNO_3 and also by anodic oxidization with 0.5 M K_2SO_4. Oxygen was stripped from the surface by heating AR to 900 °C under nitrogen and cooling to room temperature under flowing nitrogen. Four experimental procedures were adopted to study the properties of these surface oxygen complexes so formed:

1. Known weights of the carbon were added sequentially to an aqueous solution 0.1 N NaCl until the pH did not change. Here, it can be considered that dissociation of the acidic sites, producing hydrogen ions, will undergo neutralization reactions with the basic surface sites. It is assumed that the acid groups, be they carboxylic, phenolic or lactonic, have different pK_a values. Increasing amounts of chemisorbed oxygen [O], by reason of inductive effects, contributes to an increase in acid strength. The basic sites are assumed to be the Lewis base-type also with a range of pK_b values.
2. About 20 mg of carbon was equilibrated for 48 h at 25 °C in known weights of NaOH/0.1 N NaCl or HCl/0.1 N NaCl solutions. Changes in concentrations following the take up of NaOH or HCl were determined by back titration. Values of acid oxide (N_I^A) and base equivalent (N_I^B) in units of (meq g^{-1}) were calculated.
3. Using the thermal decomposition technique and analyzing the evolved gases, CO, CO_2 and H_2O with a quadrupole mass spectrometer, concentrations of total oxygen [O] (mmol g^{-1}) on the carbon surface could be assessed.
4. Heats of neutralization were obtained using a flow calorimeter by flowing solutions of 0.1 N NaOH/0.1 N NaCl and 0.1 N HCl/0.1 N NaCl over previously wetted carbons.

Table 4.5 contains the essential data of this study. It is noted that except for the lowest values of [O] the concentration of acidic sites is much greater than the concentration of basic sites. It is at a value of [O] of 1.5 mmol g^{-1} that the concentrations of both sites are equal. For carbon (OCO 13) the value of N_I^A/N_I^B is equal to 0.95 (near to unity), this carbon having a limiting pH very near to neutral, of 7.5. A value of -38 kJ eq^{-1} was calculated for the enthalpy of neutralization of the basic sites. Figure 4.31 is a plot of values of ΔH_n (J g^{-1}) against number of acidic sites (meq g^{-1}) providing a value of the enthalpy of neutralization of acidic sites of -42 kJ eq^{-1}. The two enthalpy values obtained in this way are about equal. At a value of [O] of about 1.5, where the ratio of the heats approximated to unity, the limiting pH is about 7 (Table 4.6).

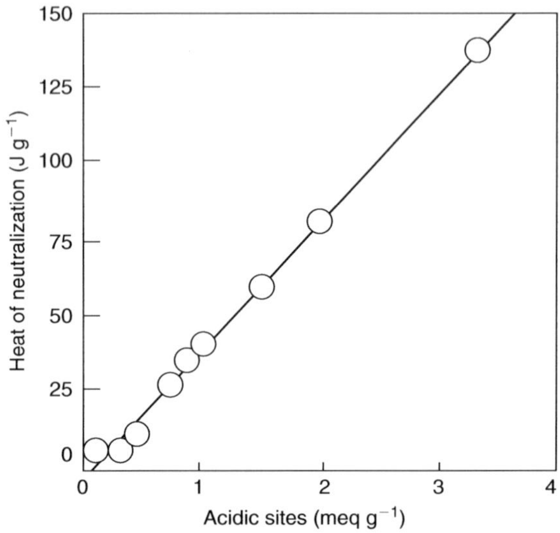

Figure 4.31. Heat of neutralization of acidic sites as a function of acidic sites for BPL carbons, providing $\Delta H_n = -42\,\text{kJ eq}^{-1}$.

Table 4.6. Acid/base characterization of carbons derived from Calgon BPL (Barton *et al.*, 1997).

Carbon	[O] (mmol g^{-1})	N_1^A (meq g^{-1})	N_1^B (meq g^{-1})	pH (pzc)	Heat of neutralization (J g^{-1})	
					Acidic	Basic
AR	1.4	0.232	0.62	9.16	7	15
BPL 900N$_2$	0.611	0.057	0.49	10.4	6	21
ECO 13	1.56	0.424	0.45	7.5	11	11
ECO 37	2.47	0.698	0.58	5.49	26	10
OX 25	3.91	0.847	0.40	4.52	35	6.3
OX 45	4.17	0.913	0.35	4.24	39	4.7
OX 65	5.99	1.33	0.29	3.91	60	3.6
OX 85	7.56	1.95	0.19	3.02	82	3.0
OX 95	10.0	3.36	0.095	2.50	138	2.0

Figure 4.32 shows the variation of concentrations of acidic and basic sites, indicating that a large decrease in acid sites is accompanied by an increase in the number of basic sites. About six acid sites are removed when one basic site is formed. These results are best interpreted by accepting that the acidic sites are associated with the surface oxygen complexes and the basic sites are best described as being concentrations of delocalized π-electrons within the graphene layers of the carbon structures.

Limiting values of pK_a and pK_b can be calculated from the data of these experiments. Figure 4.33 shows how the system changes with progressive additions of carbon to reach an equilibrium state. For the BPL carbon (900N$_2$), with the smallest of surface oxygen

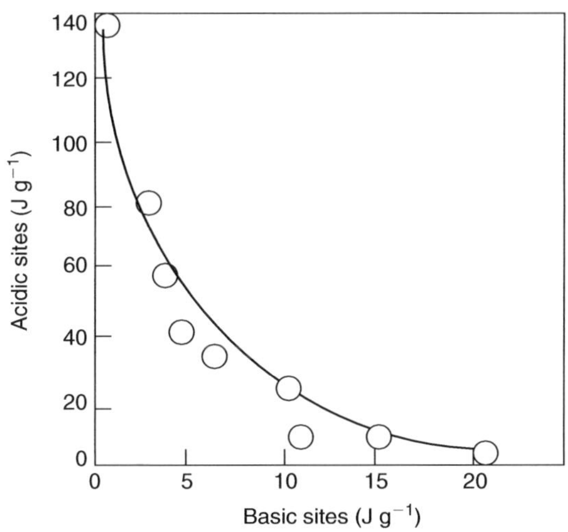

Figure 4.32. Acidic heat versus heats measured with the flow calorimeter for BLP carbons (Barton *et al.*, 1997).

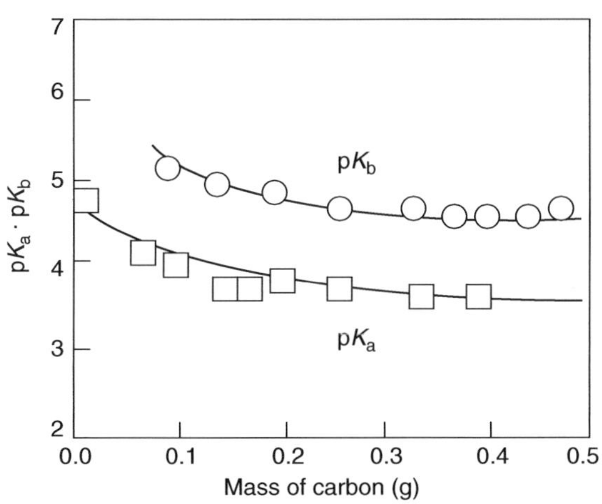

Figure 4.33. Limiting values of pK_a and pK_b calculated for the extreme cases of BPL (900N_2) (upper curve) pK_b = 4.68, and OX 95 (lower curve), pK_a = 3.75 (Barton *et al.*, 1997).

content, pK_b has a steady value of 4.68. For (OX 95), with a maximum of oxygen content, the steady value of pK_b is 3.75. For purposes of comparison, for acetic acid (CH_3COOH) pK_a is 4.76, for glycolic acid (CH_2OH—COOH) pK_a is 3.87, for glyoxylic acid (O═CH—COOH) pK_a is 3.34. Thus, the pK_a for (OX 95) is in the range known for keto and hydroxylic acids. The value of pK_b found for BPL carbon (900N_2) is close to the pK_b value of 4.75 for the Lewis base ammonia.

To summarize, the understanding of basicity in carbons is not complete. Two lines of thought dominate the issue, one stating that oxygen-containing groups such as quinine, diketone, chromene and γ-pyrones are important, the other stating that it is the delocalized π-electrons of the graphene layers which have a basic character. Again, there is no consensus about the relative strength of these basic sites. There is no doubt that increasing the nitrogen content of the graphene layers enhances considerably their basicity.

It is also a little disturbing to learn that inorganic material within a carbon, which is probably in the form of oxides, can contribute to the basicity of a suspension of the carbon in water. Currently, there is no known method which can be used to "titrate" basicity in the graphene layer. There is, however, an opinion that the strength of the basic sites in the graphene layers is small. It is considered that the intrinsic basicity of pyrones, with a wide range of basicities (0–12), their relative thermodynamic stability, and potential redox activity as well as a relative abundance in basic carbons, are suitable candidates for considerations of origins of basicity (see Montes-Morán, 2004).

4.3.5 Surface Oxygen Complexes: Effects on Adsorption Isotherms

An assessment of the effects of surface oxygen groups within microporosity, using adsorptives of different polarities was made by Rodríguez-Reinoso *et al.* (1992), this comprehensive study establishing some of the main principals. Peach (M-series) and plum stones (R-series) were carbonized and activated in CO_2 or H_2O followed by hydrogen treatment to remove any surface oxygen, followed by controlled oxidation using air, nitric acid and hydrogen peroxide. The adsorptives used were N_2, SO_2, CH_3OH and H_2O (polarities of 0, 1.6, 1.7, 1.8 D, respectively) using DR plots for analyses. For the non-polar molecules (e.g. N_2), the adsorption is mainly influenced by pore structure, but the nature and amount of oxygen surface groups are extremely important in the adsorption of polar molecules, the more important the higher the polarity.

The influence of oxygen surface functional groups on water adsorption was studied comprehensively. Water adsorption isotherms are as in Figure 4.34(a, b) where the shape changes from Type-V (M0 and R0) to a hybrid between Type-1 and Type-V for the oxidized carbons indicating little interaction for carbons without oxygen surface groups of the H_2O adsorbate molecule with the carbon surface at low relative pressures. Sample R0 has a larger micropore volume and wider microporosity with little in the way of a plateau for the R-series, compared with the M-series. With increasing oxygen content, extents of water adsorption increase at lower relative pressures with the commencement of the plateau also moving to lower relative pressures. Differences are also observed in the "steepness" of the isotherms. These authors consider that the level of the plateau is determined by the amounts of oxygen surface functional groups on the surface and that the steepness of the isotherm is determined by the chemical nature of the complexes, be they formed from air or nitric acid or hydrogen peroxide.

The results of the adsorption of N_2, SO_2, CH_3OH and H_2O (polarities of 0, 1.6, 1.7, 1.8 D, respectively) on activated carbon (from peach stones) using DR plots for analyses are in Figure 4.35. The polarity of the adsorptive prevents the filling of the micropores at low relative pressures (high values of $\log^2 p^0/p$) as noted by the increase in the ratio of slopes from

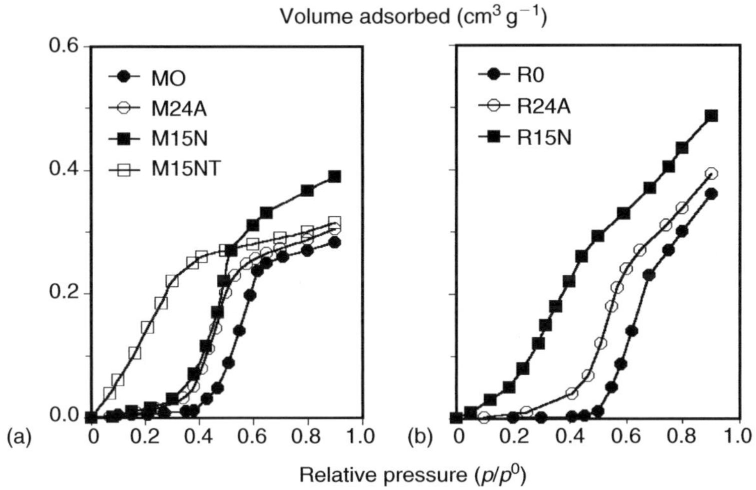

Figure 4.34. Isotherms of adsorption of H_2O at 298 K on activated carbons from peach stones (M-series) and plum stones (R-series) containing increasing amounts of oxygen surface functionality (Rodríguez-Reinoso *et al.*, 1992).

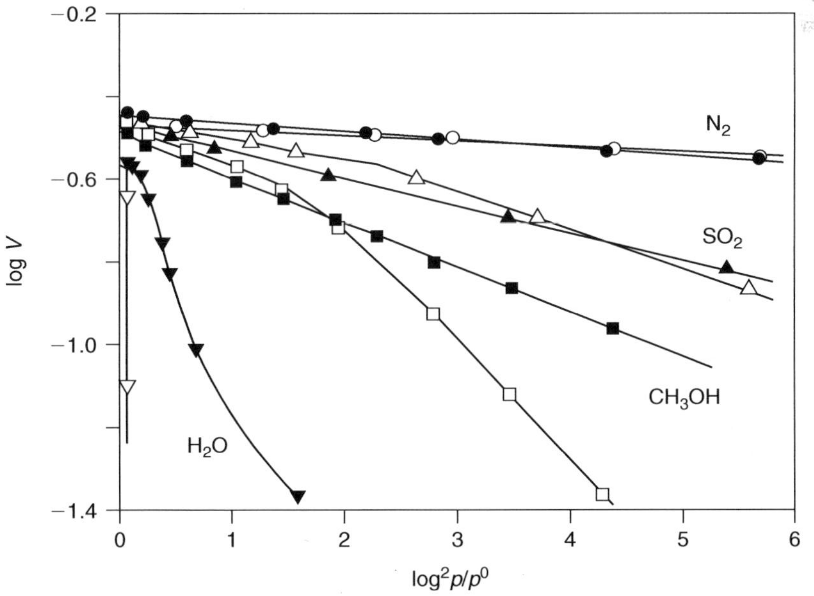

Figure 4.35. The adsorption of N_2, SO_2, CH_3OH and H_2O on a non-oxidized carbon (M0) (open symbols) and a carbon oxidized with nitric acid (M15N) (closed symbols), expressed in DR coordinates, showing major differences in extents of adsorption between polar molecules (Rodríguez-Reinoso *et al.*, 1992).

N_2 to H_2O (DR plots). It would therefore appear that polar moment is not the only factor to explain the deviations (from N_2) found for SO_2, CH_3OH and H_2O, neither is the hydrogen bond, by itself, because it is similar for water and methanol.

The most important factor seems to be a low non-specific adsorbate–carbon interaction and a high adsorptive–adsorptive interaction. Differences between CH_3OH and SO_2 suggest that intra-molecular hydrogen bonding causes a larger deviation than the simple polar moment interaction. Increasing degrees of oxidation increase the carbon–adsorbate interaction, the adsorption mechanism approaching that of N_2 micropore filling with the adsorbate as a liquid.

A comprehensive overview of the adsorption of water by activated carbons is available by Mowla *et al.* (2003).

The presence of surface oxygen complexes within porous carbons can be a mixed blessing. They can promote the adsorption of polar molecules from both the gas and liquid phases. Adsorptions from aqueous solutions, however, need special considerations because the strong competitive adsorption of the polar water molecule to the polar oxygen sites can effectively block entrances to the porosity. This effect is important with applications of activated carbons in life-supporting breathing equipment under conditions when humid air contains the contaminant. Beck *et al.* (2002) studied the influence of surface oxygen complexes on the breakthrough characteristics of hexane entrained in humid air (Section 4.5). Two carbon blacks of low nitrogen BET areas were used, 10 and 74 $m^2 g^{-1}$. The results of the experiments indicated that the presence of surface oxygen complexes resulted in a significant decrease in the adsorption performance of the carbons against such a non-polar adsorptive as hexane. The preferential uptake of water would exclude the adsorption and retention of the hexane in the breakthrough experiments.

An assessment of the problems associated with surface oxygen complexes in the use of activated carbons is available from Li *et al.* (2002). The objective of their work was to develop criteria for the effective use of activated carbons in the purification of drinking (potable) water. A matrix of ten activated carbons was created with different levels of activation and surface chemistry, using seven activated carbon fibers (phenolic resin) and three granular activated carbons, coal-, coconut shell- and wood-based. Two common drinking water contaminants were used as adsorbates, namely the relatively polar methyl tertiary-butyl ether (MTBE), kinetic diameter 0.62 nm, and the relatively non-polar trichloroethene (TCE), kinetic diameter 0.56 nm. There was an indication of molecular sieving with MTBE being adsorbed only in the adsorbents with larger micropore ranges, >0.9 nm dimension. It is suggested that the pore volume of micropores with widths about 1.3–1.8 times the kinetic diameter of the adsorbate, controls the adsorption capacity. The polarity of the adsorbent surface, for which the sum of the [O] and [N] contents was used, is an indicator for how the adsorbents may be used. As the polarity increased so the adsorption capacity decreased, for both the polar MTBE and the non-polar TCE adsorbates. This result is further explained in terms of enhanced water adsorption on the polar surface functional groups, leading to cluster of water molecules which block access to the entrances to the microporosity. To ensure that activated carbons are effective as purification agents in aqueous systems the total [O] + [N] contents of the adsorbent should not exceed about 2–3 $mmol\,g^{-1}$.

In summary, the presence of surface functionality (oxygen groups) does not appear to offer advantages to adsorption processes both from the gas phase and the liquid phase. For adsorptions from the gas phase, any advantages of the polarity for the adsorption of polar molecules may be minimized because of pore blockage effects. These may be overcome by selecting an activated carbon with a sufficiently large PSD. The purification of drinking water is not enhanced by the presence of polarity on the carbon surface.

4.4 Characterization of Porosity: SAXS and SANS

There is no doubt that methods of adsorption of gases and vapors, overall, are used for the characterization of porosity in carbons. This is because the molecular-sized porosity responds to the presence of the similar-sized molecules of an adsorptive. Another effective, but much less used (because of the need for specialty equipment), approach is that of SAXS and SANS. These two approaches provide informations which, generally, are in agreement with adsorption methods described above, as well as additional informations. They are non-destructive approaches where the probes are photons of X-ray radiation and neutrons. SAXS involves scattering of photons at interfaces with pores and tends to be more diffuse (because electron density is not zero within pores) than the scattering of neutrons. The technique is quite different to wide-angle X-ray diffraction (XRD), a technique most suitable for crystalline materials. An in-depth review of this subject area of scattering phenomena is provided by Hoinkis (1997).

The technique of SAXS has been used successfully for well over a half of a century. It formed part of the earlier investigations of the methodologies for assessment of surface areas, mainly of coals and cokes. It was part of the debate of the use of heat of wetting methods and of the use of nitrogen at 77 K, both of which presented problems of interpretation, at that time. Always, surface areas as calculated from the scattering at interfaces, are the highest reported of the several methods used. This is now understood to be due to the effects of scattering at surfaces of what was then termed "closed porosity" (not totally closed usually, but inaccessible to the probe molecules of an adsorption process). A significant section of the review of Hoinkis (1997) is devoted to studies of coals. The techniques of SAXS and SANS are particularly useful to monitor changes in carbon materials induced, for example, by activation and by heat treatment which (a) closes accessible porosity in porous carbons and (b) changes crystallographic structure and porosity during graphitization processes.

Otherwise, where access is possible, the established methodologies of gas-phase adsorption, heat of wetting (immersion) and traditional density measurements may suffice.

These scattering techniques measure heterogeneity of composition in a sample, the heterogeneous scattering centers being a second solid phase, a colloidal material, or porosity where the interface of the porosity with "carbon substance" is responsible for the scattering. Scattering dimensions are in the range 1–200 nm. The experimental data is in the form of variations of intensity with scattering angle (Figure 4.36). These scattering curves are from activated carbons made from anthracite and tar. The increases in scattering at low values of log φ are attributed to increases in mesoporosity, and the shoulder at values of log φ 1.0 onwards is attributed to the development of microporosity.

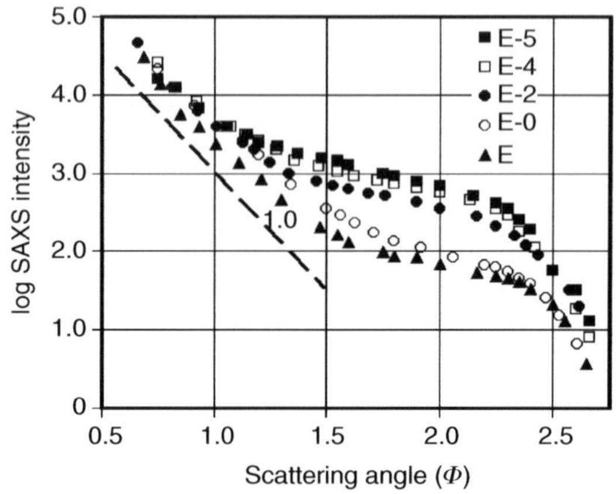

Figure 4.36. The variation of the logarithm of the SAXS intensity with scattering angle "Φ" from activated carbons prepared from anthracite and tar. Burn-offs increased from 0, 32, 48 and 56 wt% (E-0, E-2, E-4 and E-5, respectively). The intensity of scattering increases as the porosity increases both in extent and in size (Dubinin and Plavnik, 1968).

Scattering from open and closed porosity is contained within the scattering curves. To facilitate interpretations of these scattering curves, scattering from the open porosity can be removed from the carbon by prior filling with deuterated benzene (C_6D_6) leaving only scattering from closed porosity. Similarly, open porosity can be removed progressively from the scattering phenomena by progressive filling with deuterated benzene. Thus, the porosity in carbons can be studied in segments as well as in total. The mathematics of interpretations of scattering curves are available; for these, reference should be made to the specialized reviews. These scattering techniques have been used to study, for example, the microporosities of carbon fibers from polyacrylonitrile and pitch-based materials. As part of an ongoing analysis of porosity in fibers, Hoinkis and Ziehl (2003) studied the small angle scattering of cellulose, rayon and phenolic resins and report mean micropore radii of 1 ± 0.2 nm, in agreement with a study of direct measurement of porosity using scanning tunneling microscopy (Daley *et al.*, 1996). They report that the microstructure of their carbons differs from those of polyacrylonitrile and pitch-based carbon fibers.

Other studies include SANS of a Wyodak (USA) coal char and a glassy phenolic resin char as reported by Calo *et al.* (2001). Activation of the coal char was carried out in air at 673 K and of the phenolic resin char in oxygen at 743 K. Nitrogen (77 K) BET surface areas increased from 12 to 268 $m^2 g^{-1}$ for the coal char (HTT 1273 K) and from 2.2 to 27.7 $m^2 g^{-1}$ for the phenolic resin char (HTT 1173 K). These data are considered to be somewhat uncertain as nitrogen will not be significantly adsorbed by such carbon materials at 77 K (adsorption of carbon dioxide at 273 K is a better option). The use of oxygen as an activating agent is somewhat unusual as oxygen (an aggressive gasifying agent) generally only produces burn-off on external surfaces of carbon particles. However, the SANS approach is not dependent on these aspects but describes the changes which were brought about by these treatments. For the coal char, the use of air is less aggressive than pure oxygen as used for the phenolic resin char. Hence, there appears to be more opening of pores for the coal char (based on the nitrogen adsorption data).

The SANS data indicated that for the coal char, total scattering generally increased with activation. The coal char contained considerable volumes of closed microporosity which became accessible leading to the development of wider pores by gasification of separating pore walls. The activation behavior of the phenolic resin was different. There was no evidence of significant widening or alteration of the underlying porosity or the development of new porosity up to 21 wt% burn-off. In fact, very little activation has taken place, the recorded burn-off being essentially external burn-off.

An advantage that the scattering techniques has over adsorption methods is the ability to "see" into the structure of a carbon which has been heated to HTTs from 600 to >2000 °C. In this temperature range, microporosity becomes closed following structural redevelopments within the carbon. Gibaud *et al.* (1996) studied changes in structure with sugar-based carbons in the HTT range 600–2100 °C, using SAXS. They confirmed that macropores contribute to 10^{-3} of the available BET surface area. For an HTT of 900 °C and beyond, the microporosity closes with about 90% disappearing as the graphene structures become less defective. The radius of gyration (a measure of pore size) increased from 0.3 nm (600 °C) to 1.6 nm (2100 °C).

The SAXS and SANS techniques demonstrate that during activation the development of mesoporosity may occur simultaneously with the loss of microporosity. The simultaneous creation of new porosity (where none existed before), the opening of closed porosity and the widening of existing porosity represent the advantages of scattering techniques. Further, whereas the traditional adsorption techniques examine a carbon sample, overall, the development of SAXS using a 2 μm X-ray beam (as at the European Synchrotron Radiation Facility, Grenoble France), facilitates the examination of an identifiable part (2 μm diameter) of a carbon specimen, for example, that of an activated carbon fiber, distinguishing the center from a skin (Lozano-Castelló *et al.*, 2002).

4.5 Characterization of Porosity: Breakthrough Curves

4.5.1 Introduction

The performance of an activated carbon, in industrial use, depends not only on the capacity of an activated carbon but also on its retentivity, that is for how long can the carbon retain the adsorbed species, under operating conditions, without it being subsequently desorbed. Activated carbons are used in beds with the gas, vapor or liquid passing through the bed. From a practical point of view it is desirable to operate by minimization of the frequency of replacement of the activated carbon bed. The characterizations, which have been elaborated upon (Section 4.2) clearly indicate the desirability to maximize the porosity of maximum adsorption potential accessible to the adsorptive at those low relative concentrations of adsorptive which are found in industrial situations. Other factors include rates of adsorption which are dependent on intra- and inter-diffusion of the adsorbate and granules and bed construction. At the same time, these adsorbed materials must remain within the activated carbon for as long as possible.

The working life of a selected activated carbon is measured by its *breakthrough curve*. Applications include the purification of air in life support systems, environmental protection, adsorption from solution and in solvent recovery.

In a breakthrough test, the gas/vapor stream, usually air containing the adsorptives, is passed through a bed of granules of activated carbon. The composition of the emergent stream is analyzed for the adsorptives. Initially, no adsorptives are within the emergent stream, but after a time (T_B) the adsorptives emerge as the carbon bed is saturated with the adsorbates (Figure 4.37). Of course, the same considerations apply to adsorption from liquid-phase processes.

4.5.2 Breakthrough Curves and Times

Breakthrough curves can be considered as the last of the essential characterizations of an activated carbon. Equilibrium isotherm data provide information of the capacity of a carbon. Next, the kinetics of the adsorption processes must be known, giving information of the rates at which adsorptives are adsorbed by the adsorbent. Finally, the performance of a carbon (so characterized) in an industrial situation can be simulated by making use of the breakthrough curves.

A breakthrough curve is a plot of the duration of the test against the concentration of the adsorbate in the effluent stream of a mixture of air (usually) or liquid with the adsorptive, as shown in Figure 4.37. Initially, as the mixture of vapor and air enters into a defined bed of the carbon, the vapor is adsorbed and the air passes out of the bed free from the vapor. This continues as long as the bed of carbon has capacity to adsorb. When saturation is approached, then vapor is detected in the effluent gas and its concentration increases until no further adsorption takes place and the composition of the effluent mixture is equal to the composition of the entry mixture.

For an idealized situation, the breakthrough curve is S-shaped, is symmetrical and is described in terms of its (a) midpoint, (b) steepness and (c) shape. The midpoint is a function

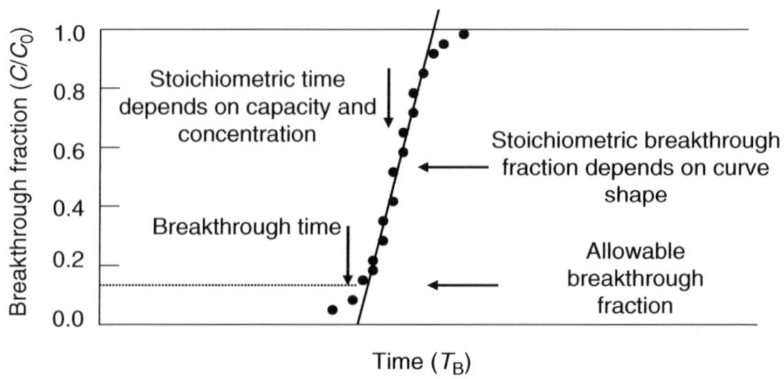

Figure 4.37. An idealized description of a breakthrough curve of a plot of time against composition of the effluent gas stream. The breakthrough time (T_B) is the intercept of the gradient line at zero C/C_0 (Wood, 2002).

of the rate of flow of the gas mixture through the bed, the concentration of the vapor in the air and the capacity of the bed (size and quantity/quality of the activated carbon in the bed) as well as the operating temperature of the bed. Higher temperatures reduce adsorption capacity and hence reduce retentivity so producing shorter breakthrough times. The steepness of the curve depends on the rate at which the mixture of air and vapor passes through the bed. The slower the rate of flow, the steeper the breakthrough curves. A symmetrical S-shaped curve is associated with microporosity without a PSD, which is a non-event for activated carbons.

More often than not, the breakthrough curves for activated carbons are not a symmetrical S-shape but are skewed, being steeper at the beginning of the curve. The reasons for this are due to heterogeneity within the bed, both on a macro-scale and on a micro-scale (nano-scale). Heterogeneity of particle size of the granular carbon resulting in differences in packing densities exists within the bed. Every known carbon has a PSD, from the narrowest of micropores to the mesoporosity. Within breakthrough experimental systems, the vapor (solute) being adsorbed is located initially in the porosity of smallest dimensions (relative to the size of the adsorptive molecule) and which has the highest of retentivity. As the narrowest of microporosity is progressively filled towards the wider pores so the retentivity (enthalpy of adsorption) decreases and this manifests itself as a decrease in the gradient of the breakthrough curve. Within the bed, the transition from the mixture to the pure component occurs within a finite volume element (known as the mass transfer zone (MTZ) volume), and it is the shape of this volume element which influences the shape of the breakthrough curve.

Deviations from a symmetrical S-shape are measured as the ratio of rates ($\partial c/\partial t$) at 1% and 10% of breakthrough. Wood (2002) estimated, from a large database, that an average value of the skew ratios was 1.22, indicating no massive deviations. One set of data is relevant to this chapter, and that is the effect of **water** entrained within the gas mixture. The mixture was entrained with water vapor to the extent of 40 wt% and results are as shown in Figure 4.38, which is a plot of the value of the skew index against the water content.

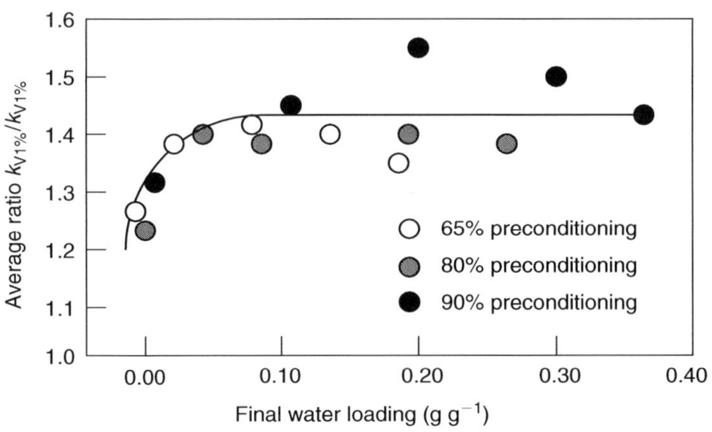

Figure 4.38. The effect of additions of water vapor to the shape of the breakthrough curve, as measured by the skew index (ordinate) (Wood, 2002).

There is an increase in the skew ratio parameter to a value of about 0.08 $g\,g^{-1}$ of water vapor, beyond which there is no further influence. An immediate explanation is that water molecules are adsorbed preferentially on any surface oxygen complexes which may be present and this could be effective in reducing the number of adsorption sites which exhibit a maximum in retentivity. Once all of these adsorption sites are occupied by water molecules, then increasing the water content of the incoming gas mixture has no further adverse effect on the skew ratio.

4.5.3 Applications of Breakthrough Curves and Times

Binary vapor mixtures. In recent times, adsorption systems have been required to remove more than one vapor component from an air stream. Vahdat (1997) made a theoretical study of the performance of an activated carbon for the purification of air mixed with a binary vapor mixture. An activated carbon can be characterized in terms of the Langmuir constants (Equation (4.1)) which are an indication of strengths of adsorption (retentivity). Vahdat (1997) was able to quantify the expected, namely that there will be competitive adsorption, the more strongly adsorbed vapor being able to displace the less strongly adsorbed vapors. Breakthrough curves are dependent on the relative concentrations of the vapors, on the relative capacities of the carbons involved, flow rate of the vapor stream and on bed geometry. The effect of concentration of the more strongly adsorbed component is to reduce the breakthrough times of the other components, and to increase the rate of replacement of the weakly adsorbed component. The effect of increasing the concentration of the weakly adsorbed compound is to decrease its rate of displacement and to reduce the breakthrough times. Dependent on the relative concentrations of the vapors of the mixture, the more strongly adsorbed component can become less strongly adsorbed. The effect of flow rate on breakthrough times of components of a mixture is similar to that of the individual components. Breakthrough times are a linear function of the reciprocal of the flow rate (increasing flow rate decreases the breakthrough time).

Lavanchy and Stoeckli (1997, 1999) published studies in the same area as Vahdat (1997) but used the equation of DA, Equation (4.8), based on Equation (4.3):

$$N_a = N_{a0} \exp[-(A/\beta E_0)^n] \tag{4.8}$$

where N_a (mmol g^{-1}) equals amount adsorbed at temperature T and relative pressure p/p^0, N_{a0} (mmol g^{-1}) is the limiting amount adsorbed, $A = RT\ln(p^0/p)$ and β and E_0 are specific parameters which depend on the adsorptive and on the activated carbon. The DA equation is derived from the DR equation (Equation (4.3)) where for typical activated carbons, the value of "n" equals 2. Accordingly, $W_0 = N_{a0}V_m$, ($cm^3\,g^{-1}$) where V_m is the molar volume of the adsorbate (but not necessarily the molar volume of the adsorbate within the porosity of the carbon). The DA equation is used on this occasion because of the limited number of parameters required to predict and describe adsorption over a wide range of temperature and relative pressure.

Two activated carbons were selected by Lavanchy and Stoeckli (1997) for the study, namely a coal-based granular carbon (UO3) with a micropore volume (W_0) of 0.495 $m^3\,kg^{-1}$, and a peat-based extruded carbon, (W_0 of 0.55 $m^3\,kg^{-1}$). A ternary mixture in air was used, and

Table 4.7. Parameters of vapors used for the simulation studies (Lavanchy and Stoeckli, 1997).

Adsorptive	Molecular mass ($kg\,mol^{-1}$)	Diffusion volume ($m^3\,mol^{-1}$)	Liquid density at 25 °C ($kg\,m^{-3}$)	Volatility at 25 °C ($kg\,m^{-3}$)	Affinity coefficient (ß)
2-Chloropropane	78.5×10^{-3}	82.9×10^{-6}	8.558×10^{2}	2.177	0.93
Carbon tetrachloride	153.8×10^{-3}	84.5×10^{-6}	1.584×10^{3}	0.944	1.05
Chlorobenzene	112.6×10^{-3}	108.2×10^{-6}	1.105×10^{3}	0.0722	1.18

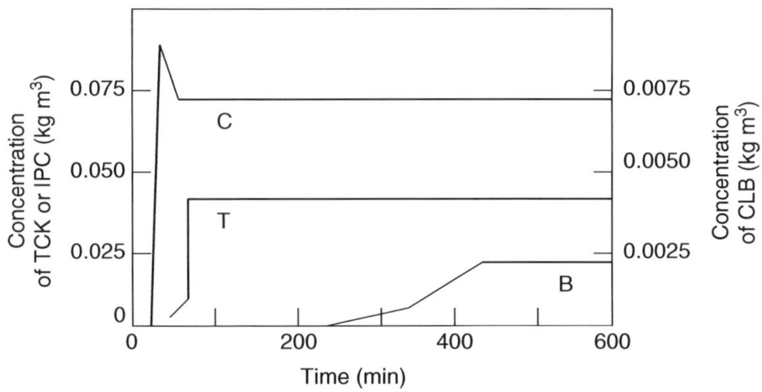

Figure 4.39. Predicted breakthrough curves for a mixture of 2-chloropropane (C), carbon tetrachloride (T) and chlorobenzene (B) in air, with relative inlet concentrations of 0.036, 0.040 and 0.032 on a bed of activated carbon UO3 (Lavanchy and Stoeckli, 1997).

the details of these adsorptives are in Table 4.7. Examples of predictions of breakthrough curves are shown in Figure 4.39, these predictions being close to experimental data. The authors indicate that little is yet known about the miscibility of the liquids of the adsorbates within porous systems or of the influence of water vapor in the carrier gas (air).

In a subsequent paper, Lavanchy and Stoeckli (1999) confirmed the agreement between experimental and predicted isotherms based on the use of the MPD equations to characterize the activated carbons. (*Note*: Myers–Prausnitz (MP) theory is an approach to characterize the carbons in association with the DA equation, hence (MPD).) However, the MPD approach is not valid for the breakthrough curve of 2-chloropropane when the carrier gas is air that contains moisture (water vapor). This is a consequence of the immiscibility of the adsorbed water and 2-chloropropane.

Adsorption of lead(II). Activated carbons, dominantly, are utilized for adsorption of organic molecules. However, they have the ability to act as ion-exchange material by reason of their acidic and basic functionality (Section 4.3). Xiu and Li (2000) studied the ability of two activated carbon fibers to remove lead ions from aqueous solution using the breakthrough curve technique. The two carbons of the study were essentially similar as far as published analytical data were concerned, having surface areas of 976 and 993 $m^2\,g^{-1}$, with basic functionalities of 0.287 and 0.419 $mmol\,g^{-1}$, respectively. However, total pore volumes were somewhat different with values of 0.58 and 0.47 $cm^3 g^{-1}$. Experimental breakthrough curves showed differences between the carbons but correlations with analytical data were not obvious. It is considered that intra-particle and external mass transfer are not the rate-controlling steps. It is possible that one carbon may have a higher mesopore content than the other. As the Pb(II) ions are likely to be hydrated, this would require a wider porosity to gain immediate access to the oxygen surface functionality of the surface. No firm conclusion is possible because the authors did not publish their nitrogen (77 K) isotherms the shapes of which would have given a clear indication of relevant differences in porosity. As suggested above, isotherms should always be provided in publications.

Pulsating flow through a bed and breakthrough times. An important application for activated carbons is their use in canisters for life-supporting gas masks. In use, the airflow is not constant but pulsates with respiration. Nir *et al.* (2002), in their review, state that previous studies had reported that the breakthrough time (T_B) was independent of flow pattern for butyl alcohol, methyl chloroform, cyclohexane and ethyl acetate. On the contrary, evidence is presented that (T_B) could be reduced by 5–30% under conditions of sinusoidal, rectangular and square profiles.

Nir *et al.* (2002) examined the effect of airflow pattern on breakthrough curves of organic vapors. The organic vapors used were dimethyl-methyl phosphonate (DMMP) and decene under conditions of fixed (stable) flow and sinusoidal flow. The ratio C_x/C_0 was used to characterize the effects of variations in airflow pattern (C_x/C_0 is the ratio of effluent to influent concentrations of the vapor in the air stream). Nir *et al.* (2002) observed a linear dependence of $\ln(C_x/C_0)$ with (T_B) in all of their experiments indicating that such concepts as *critical bed weight* (W_C) and *dynamic adsorption capacity* (W_E) (as used in the Wheeler Jones model, see Nir *et al.* (2002) for full details) were applicable to pulsating flow as well as to fixed flow. However, (W_C) was considerably larger (7–44%) when using a pulsating flow compared to a fixed flow at the same average rate. The effect of this is to shorten, significantly, values of (T_B). This confirms the need, in performance testing of activated carbon behavior, to simulate the operational patterns of flow through the carbon. For gas-mask protection at a high level, at a value of (C_0/C_x = 60,000), the protection capacity of the personal canisters, which contain the carbon, is reduced by 15% on changing from fixed to pulsating flow.

Removal of NO from a gas stream. A study of Claudino *et al.* (2004) of NO adsorption (no dilution with air) by a peat-based activated carbon (using steam at 800 °C) monitored the efficiencies of modified carbons using breakthrough curves, at temperatures between 45 and 85 °C. The NO adsorption capacity was a function of the BET surface area, PSDs, temperature and oxygen functional group content, increasing with increasing oxygen content. The adsorption was reversible and the enthalpy of adsorption was <10 kJ mol^{-1}. This study illustrates that this technique of measurement of breakthrough curves easily provides information about properties of activated carbons, such as retention, which supplements data obtained from the equilibrium adsorption isotherms.

Removal of environmental pollutants such as phenanthrene, a PAH. Polycyclic aromatic hydrocarbons are carcinogenic compounds and their release into the atmosphere has to be prevented. They are generated during combustion processes of fossil fuels, where they are associated with particles of dust and with release as vapors into the atmosphere. Particulate entrapment is not a problem but the removal of the PAH is a serious problem, with catalytic PAH catalytic decomposition and PAH adsorption on activated carbons being suitable options for their removal.

Murillo *et al.* (2004) report that extents of adsorption of pure organic vapors from hot waste gases are inversely proportional to their volatility. Also, it is reported that increasing contents of water vapor in the gas stream decrease the extents of adsorption of, for example, phenanthrene. The other major gas component in hot waste gases is carbon dioxide and preferential adsorption of this gas reduces the efficiency of phenanthrene uptake. Murillo *et al.* (2004) extended studies of phenanthrene uptake beyond the establishment of equilibrium data to the modeling of breakthrough curves and times. It is observed that the phenanthrene molecule is subject to kinetic restrictions within narrow micropores. The study incorporated three activated carbons of surface areas 273 (ICB5), 574 (ICB12) and 636 (ICB8) $m^2 g^{-1}$. The breakthrough curves are as shown in Figure 4.40. It is concluded from these studies that to improve the efficiency of phenanthrene removal (or other contaminants) using activated carbons, the latter must have a developed microporosity with a wide PSD providing high adsorption capacities and rapid kinetics. A consequence of this is that high space velocities of emission gases can be used so reducing the size of separation equipments and lowering costs.

Changes in surface chemistry and breakthrough curves. It has been established for some time that the functionality of a carbon surface can be modified by insertion of chemically bonded nitrogen, in addition to the surface oxygen complexes (Section 4.3). Such nitrogen can be inserted by reaction with ammonia or hydrogen cyanide to produce several types of nitrogen insertion, as described in Figure 4.41.

A study of Boudou (2003) was designed to assess the influence of ammonia treatment at 800 °C on the catalytic activity of a viscose-based activated carbon cloth (ACC) for the removal of H_2S and SO_2 from containment gases. At 700 °C, the micropore volume of

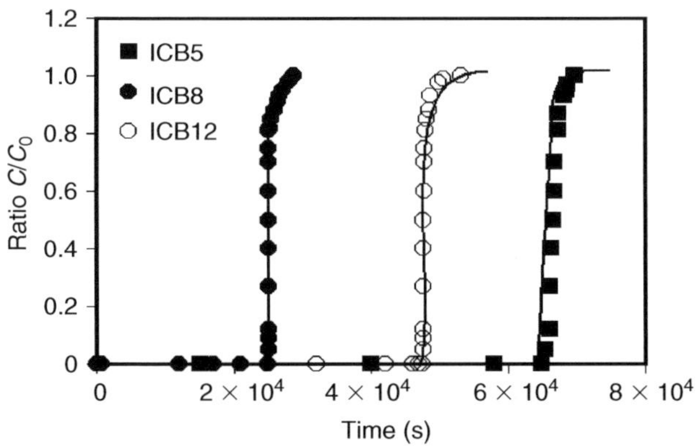

Figure 4.40. Breakthrough curves of phenanthrene in contact with three carbons (ICB5, ICB12, ICB8), of increasing surface areas of 273, 574 and 636 $m^2 g^{-1}$, and total pore volumes of 0.20, 0.33 and 0.40 $cm^3 g^{-1}$. ICB8 has the widest PSD (Murillo *et al.*, 2004).

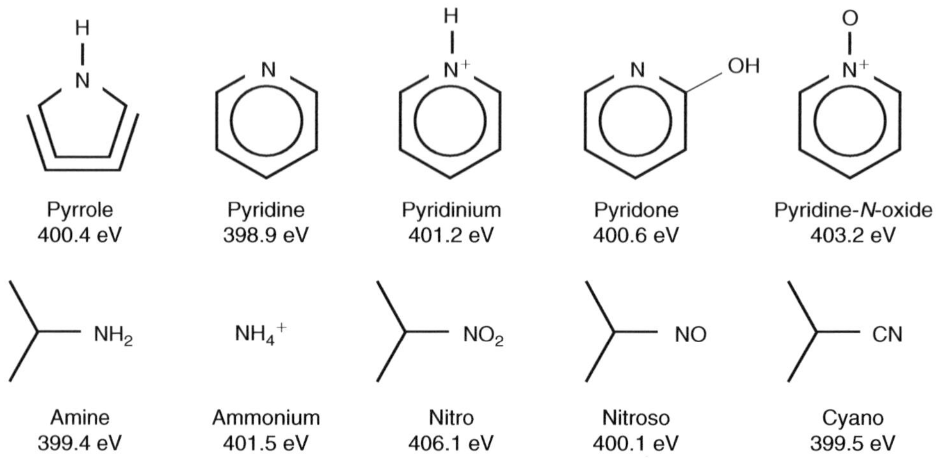

Figure 4.41. Formulae of possible nitrogen functionality existing within the structures (graphene layers) of carbons, together with their N 1 s electron binding energies (Pels *et al.*, 1995).

the carbon was unchanged. However, at >800 °C, significant increases in surface area (784–1185 $m^2 g^{-1}$) and micropore volumes occur following "gasification" of carbon atoms from the surface. Treatment with ammonia caused the removal (displacement) of acidic oxygen groups but leaving basic oxygen groups. Changes in activity of these carbonaceous surfaces, as monitored by retention of H_2S and SO_2, were measured by making use of breakthrough curves. There was an enhancement of H_2S and SO_2 retention due to (a) enhancement of surface area and pore volumes, (b) the removal of most of the acidic groups and (c) the insertion of pyridinic and pyrole-type groups (*Note*: pyrone has the formula $C_5H_4O_2$). Thermal regeneration of such carbons is obviously important. After the adsorption of SO_2, thermal regeneration caused significant loss of carbon and of the nitrogen sites. Such events reduce the operational life of an activated carbon with adverse economics. The take-up of the H_2S was less dramatic.

4.6 Characterization of Porosity: Enthalpies of Adsorption

4.6.1 Physical Adsorption and Chemisorption

The process of physical adsorption (referred to also as physisorption) of gases, normally considered to be reversible, is an exothermic process with enthalpies of adsorption in the region of <50 kJ mol^{-1} of adsorbed adsorbate. This clearly separates the process from chemisorption when an adsorbed layer is irreversibly bonded, chemically, to the surface. A simple example illustrates chemisorption. If a pristine silver object, for example, a spoon or an award cup, is exposed to an unsuitable environment it will become discolored adopting a pale brown and then a black appearance. This is because of the irreversible formation on the surface of silver sulfide. When the silver object is placed in a bell jar (for example) which is evacuated the discoloring remains unaffected. The only way to remove the discoloration is to use a polishing cloth which physically removes the chemisorbed layer of silver sulfide. Enthalpies of formation of chemisorbed layers are in the region of 400 kJ mol^{-1}.

Figure 4.42. Schematic to illustrate the formation of transitory dipoles in molecules in close proximity to each other, these being responsible for the phenomenon of physical adsorption.

4.6.2 London Dispersion Forces

An adsorption process is a spontaneous process leading to a system of lower internal energy. The entropy change from adsorptive to adsorbate is negative requiring the enthalpy change (ΔH in kJ mol^{-1}) to be negative for the negative requirement of free energy change (ΔG in kJ mol^{-1}).

The lower energy state results from the operation of *London dispersion forces*. These forces are the weakest of intermolecular forces and result when the electrons in two adjacent atoms occupy positions that make the atoms form temporary dipoles. This force is sometimes called an induced dipole-induced dipole attraction. London dispersive forces are the attractive forces that cause non-polar substances to condense to liquids and to freeze into solids when the temperature is lowered sufficiently.

Due to the constant motion of the electrons, an atom or molecule develops a temporary (instantaneous) dipole when its electrons are distributed asymmetrically about the nucleus. A second atom or molecule, in turn, is distorted by the appearance of the dipole in the first atom or molecule (because electrons repel one another) which leads to an electrostatic attraction between the two atoms or molecules (Figure 4.42). Dispersion forces are present between two molecules (even non-polar molecules) when they are in close proximity. As an adsorbate, the molecules can be closer to each other than in the liquid state; they are certainly close to the carbon atoms of the walls of the porosity.

In terms of molecular size, dispersion forces are present between all molecules, whether they are polar or non-polar. The larger and hence heavier atoms and molecules exhibit stronger dispersion forces than smaller and lighter ones. In a large atom or molecule, the valance electrons are, on average, farther from the nucleus than in a smaller atom or molecule. They are less tightly held and can more easily form temporary dipoles. The ease with which the electron distribution around an atom or molecule can be distorted is called the *polarizability*. London dispersion forces tend to be stronger between molecules that are easily polarized and to be weaker between molecules that are not easily polarized. The shapes of molecules also affect the magnitudes of dispersion forces between them. For example, at room temperature, neopentane (C_5H_{12}) is a gas whereas *n*-pentane (C_5H_{12}) is a liquid. The London dispersive forces between *n*-pentane molecules are stronger than those between neopentane molecules even though both molecules are non-polar and have the same molecular weight. The somewhat cylindrical shape of the *n*-pentane molecules allows them to come in contact with each other more effectively than the somewhat spherical neopentane molecules.

In the narrowest of the microporosity, the dispersion forces are at their most intense, so intense, in fact, that the density of adsorbed helium exceeds that of solid helium.

It is the energy resulting from this polarization which is based in calorimetry, and measured as enthalpies of adsorption.

4.6.3 Enthalpies of Adsorption

Measurement of the exothermicity of the adsorption process is utilized to measure both surface heterogeneity and total surface area. Using calorimetric methods, the heat evolved can be measured in stages as the adsorbent surface is progressively covered with the adsorbate. This provides the *differential enthalpy of adsorption*. On the other hand, the heat evolved, as measured when the entire surface is covered by adsorbate in one operation, is termed the *integral enthalpy of adsorption*. Measurements of enthalpies of adsorption can be placed into two categories: (a) the isosteric enthalpy of adsorption (from isotherm data) and (b) differential enthalpies of adsorption using flow-type micro-calorimeters and gaseous adsorbates. Immersion calorimeters measure the enthalpy of immersion when a solid is immersed in a liquid.

4.6.3.1 Isosteric enthalpies of adsorption

Isosteric (differential) enthalpies ($\Delta_{ads}H$) of adsorption can be obtained from physical adsorption data without the use of calorimetry. It is advisable to use a series of isotherms obtained at different temperatures, not too far apart. The equivalent of the Clausius–Clapeyron equation, as Equation (4.9), can be used for the one-component, gas–liquid system. The differentiated form of the equation, as Equation (4.10) allows the calculation of ($\Delta_{ads}H$) by plotting ln(p) against T^{-1}, for values at constant amount adsorbed (n^a in mmol g^{-1}). From a practical point of view, this method requires precise measurement of partial pressures and is more reliable at low values of relative pressure.

$$\Delta_{ads}H = \frac{RT_1T_2}{T_2 - T_1} \ln \frac{p_2}{p_1} \tag{4.9}$$

$$\Delta_{ads}H_{Tn} = R\left(\frac{\partial \ln\{p\}}{\partial(1/T)}\right)_n \tag{4.10}$$

A classical curve, published by Isirikyan and Kiselev (1961) showing the variation of enthalpy of adsorption of n-hexane on a graphitized carbon black at 20 °C, is as shown in Figure 4.43. Values of ΔH (cal mol^{-1}) initially decrease as the sites of highest adsorption are occupied first. As the monolayer is established, so the dispersion forces between the adsorbed molecules of n-hexane increase in magnitude reaching a maximum on completion of the monolayer.

Subsequently, values of ΔH (cal mol^{-1}) decrease sharply as continuing adsorption establishes a second layer of n-hexane on the monolayer of n-hexane until the second layer is complete when a slight increase in ΔH (cal mol^{-1}) occurs. When benzene is used as an adsorbate, instead of the n-hexane, the initial increases in values of ΔH are not found, the data fitting, more or less, on to a horizontal line. This is attributed to much lower interactions between benzene molecules, as compared between those of n-hexane.

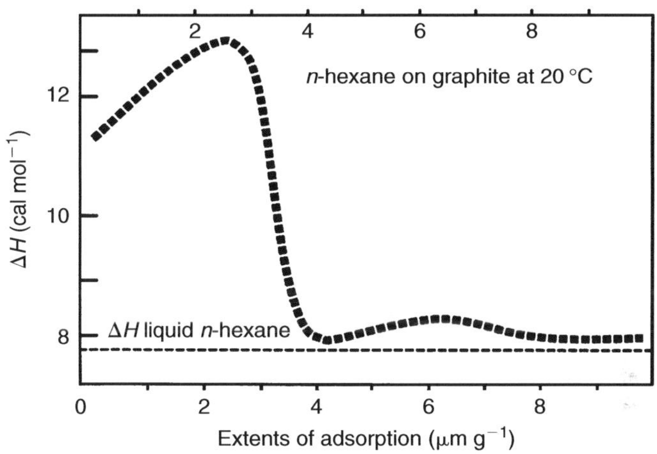

Figure 4.43. The variation of differential heats of adsorption (ΔH in cal mol^{-1}) of *n*-hexane at 20 °C on a graphitized carbon black, with coverage, (μmol m^{-2}) (Isirikyan and Kiselev, 1961) (1 cal = 4.2 J).

4.6.4 Flow-type Microcalorimetry

The development of modern calorimeters not only enables the enthalpy changes occurring in adsorption and desorption to be measured incrementally, to obtain differential enthalpies of adsorption, but also the kinetics of the processes can be followed at the same time. Such data, in association with the established characterization methods of (a) surface area, using different adsorbates, (b) PSDs and (c) TPD provide a comprehensive knowledge of the properties of an activated carbon.

An example of the facilities provided by flow microcalorimetry is described in the work of Domingo-Garcia *et al.* (2002). Here, four active carbons were studied in terms of their ability to adsorb ammonia. An ultimate objective was the recognition of acid sites on the surfaces of these activated carbons as locations of catalytic activity of the carbons.

The carbons used were (i) CAL from Calgon, (ii) BLP provided by Chemviron, (iii) GAE from CECA and (iv) GAE oxidized in an aqueous solution of $(NH_4)_2S_2O_8$. The characterization of these carbons is as in Table 4.8, containing surface areas as determined by nitrogen adsorption at 77 K (BET equation) (m^2 g^{-1}), carbon dioxide adsorption at 273 K (DA equation) (m^2 g^{-1}), mean pore widths (nm), micropore volume (cm^3 g^{-1}) and oxygen content (wt%) determined by TPD (Section 4.3.3) coupled to a mass spectrometer (MS).

Several chemical groups of different stability on the carbon surfaces were recognized as carboxylics, phenols, lactones, carbonyls, quinones and hydroquinones. The TPD profiles were distinctly different, carbon for carbon, indicating differences in the stability of the carbon surface groups. The activated carbon, CAL, has the highest contents of surface oxygen complexes.

In an experimental study, adsorption of ammonia (5 vol% in nitrogen) was studied at 150 °C, using 0.071 g of carbon. Heat was evolved for a period of 4500 s. Desorption follows

Table 4.8. Surface area, porosity and oxygen contents data of carbons of the study of Domingo-Garcia *et al.* (2002).

Carbon used	Nitrogen surface area ($m^2 g^{-1}$)	Carbon dioxide surface area ($m^2 g^{-1}$)	Mean pore width (nm)	Micropore volume ($cm^3 g^{-1}$)	Oxygen content (wt%)
CAL	870	970	1.5	0.37	11.0
BLP	1020	1150	1.7	0.44	3.4
GAE	1030	1190	1.6	0.45	3.8
GAE-oxidized	1130	1400	1.8	0.50	5.8

Table 4.9. Thermodynamic data for adsorption of ammonia by activated carbon. Domingo-Garcia *et al.* (2002).

Sample	Amount of adsorption ($\mu mol\, g^{-1}$)	Amount of desorption ($\mu mol\, g^{-1}$)	Integral heat of adsorption ($J\, g^{-1}$)	Integral heat of desorption ($J\, g^{-1}$)	Molar heat of total adsorption ($kJ\, mol^{-1}$)	Molar heat of total irreversible adsorption ($kJ\, mol^{-1}$)
CAL	18	9	−2.2	0.7	−122	−167
BLP	136	41	−16.1	2.7	−118	−141
GAE	192	76	−17.0	4.5	−86	−108
GAE-oxidized	270	72	−16.6	3.7	−62	−65

when the ammonia was removed from the carrier nitrogen gas. The amount desorbed was considerably less than the amount adsorbed indicating both reversible and irreversible components to the adsorption process. Further, the calorimetric data did not coincide with the MS data. Energy continued to be released some 1500 s after the adsorption of the ammonia was completed. This effect is explained as being due to the diffusion of ammonia within the carbon from low energy adsorption sites to sites of higher energy but of lower accessibility.

In addition, the MFL calorimeter, in this study of the kinetics of adsorption of ammonia, indicated that, after 500 s, amounts of ammonia adsorbed ($\mu mol\, g^{-1}$) were 25, 125, 145 and 160, for CAL, BPL, GAE and GAE-oxidized, respectively. The thermodynamic data from the study are summarized in Table 4.9. The carbon CAL has the highest content of functional surface oxygen and the lowest rate of ammonia uptake, indicating significant blockage by the oxygen to the microporous system. The GAE-oxidized, with its wider porosity and larger micropore volume is a more suitable adsorbent for ammonia. For the uptake of ammonia by CAL, the initial heat evolution is only 71 $kJ\, mol^{-1}$ rising to about 150 $kJ\, mol^{-1}$ in the re-arranging process. By contrast, for BLP, larger values of heat evolution (250 $kJ\, mol^{-1}$) (chemisorption) occur at low levels of uptake falling progressively to 50 $kJ\, mol^{-1}$.

Thus, the behavior of activated carbons, as described above, using the FMC technique, can be studied in depth. It is a very reliable technique which allows adsorption and thermodynamic data to be obtained under experimental conditions close to circumstances operating in industrial practice. The activated carbons of this study of Domingo-Garcia *et al.* (2002) possess a wide distribution of acid sites some of which exceed in strength those existing on acidic zeolites and some clays. The use of MFC, accordingly, provides informations of relevance to carbon utilization, which could not be obtained from other experimental methods.

Calorimetric studies of processes occurring on surfaces now have extremely high levels of sensitivity (1×10^{-15} J) (Femto $= 10^{-15}$) particularly for single-crystal adsorption calorimetry (Brown *et al.*, 1998).

4.7 Characterization of Porosity: Immersion Calorimetry

4.7.1 Introduction

Historically, immersion calorimetry has been used since about 1938. The story takes us again to the early studies of coal, a material known to be microporous, even though that description may not have been used at the time. Methane containment within coal was established, but quantitative capacities were not known. Further, coal, particularly when stored in heaps, or at the coalface when it is freshly mined and is in a reductive state, adsorbed oxygen chemically. Such a reaction is exothermic, and in the absence of suitable ventilation, this chemisorption process resulted in a rise in temperature such that spontaneous ignition could occur. Some coals were more susceptible to this effect than others creating the need to know why. Microporosity within coal varies with rank and location and so it was toward quantitative methods of gaining access to this microporosity that research was directed. Coals of different rank and origins most probably had different extents of containment of microporosity.

The calorimetric method of measuring heats of wetting, to be related to a surface area (and hence to a micropore volume) was a promising approach at that time (Griffith and Hirst, 1943). A suitable reference standard was considered to be methyl alcohol which liberated 94% of the heat of wetting within 10 min and 99% within 25 min, the reason given for this delay being a slower re-adjustment of adsorption sites for the methyl alcohol (see Section 4.6.4).

A diagram of the original ice-calorimeter used by Griffith and Hirst (1943) is as shown in Figure 4.44. The principles employed are those of today except detection methods were physical and not electronic. The sample of coal and methyl alcohol are inserted into the calorimeter "C" which is surrounded by an insulating ice–water mixture. The methyl alcohol is added to the coal (previously evacuated to remove methane) by breaking the glass-containing bulb with the plunger "T". The heat released causes movement of the mercury column "S" and, by previous calibration, movement is related to quantity of heat released. The original publication of results is reproduced as Figure 4.45 which is a plot of heat released (cal g^{-1}) (1 cal $= 4.2$ J) versus the rank of coal described as carbon content (100–75 wt%) and oxygen content (0–20 wt%). The heat of wetting shows a minimum at about 90 wt% C of 1 cal g^{-1} (4 J g^{-1}) but is as high as about 21 cal g^{-1} (90 J g^{-1}) for the lower rank coals (82 wt% C). Figure 4.45 (right), however, clearly shows that the heat of wetting is dependent on the oxygen content of the coals.

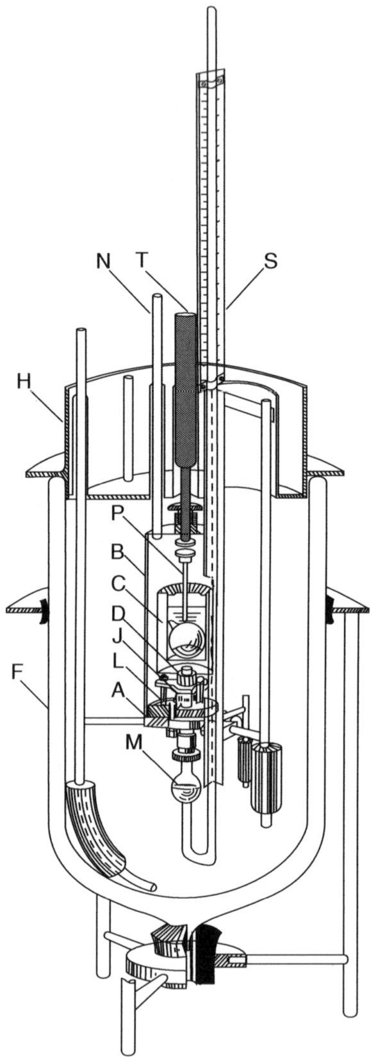

Figures 4.44. Diagram of the ice-calorimeter (left) and reproduction of early results of heats of wetting (right) by methyl alcohol of a rank range of coals (Griffiths *et al.*, 1943).

Three parameters are changing simultaneously with coal rank: (a) the micropore volume (surface area) of the coal changes with coal rank, (b) the polarity of the coal surface, caused by the oxygen content, decreases with increasing coal rank, bituminous to anthracitic coals and (c) the dimensions of microporosity of the coal change with coal rank and may maximize at about 77 wt% C. When these heats of wetting data (based on methanol) are compared with surface area data, obtained much later from isotherms using nitrogen (77 K) and carbon dioxide (298 K) as adsorbates, as shown in Figure 4.46, there is a serious mismatch in the positions of the minima and maxima (Díez Díaz-Estébanez *et al.*, 2000). The nitrogen data show a maximum at about 76 wt% C, with the carbon dioxide data showing a minimum at about 82 wt% C, the heat of wetting showing a minimum at 90 wt% carbon.

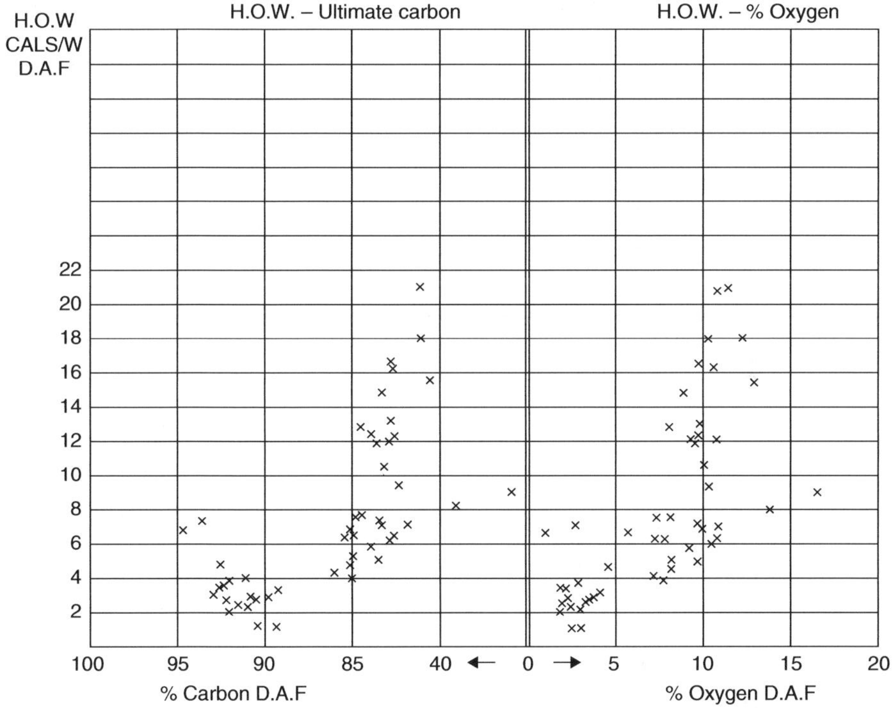

Figure 4.45. Reproduction of early results of heats of wetting by methyl alcohol of a rank range of coals (Griffith and Hirst, 1943) (H.O.W: heat of wetting).

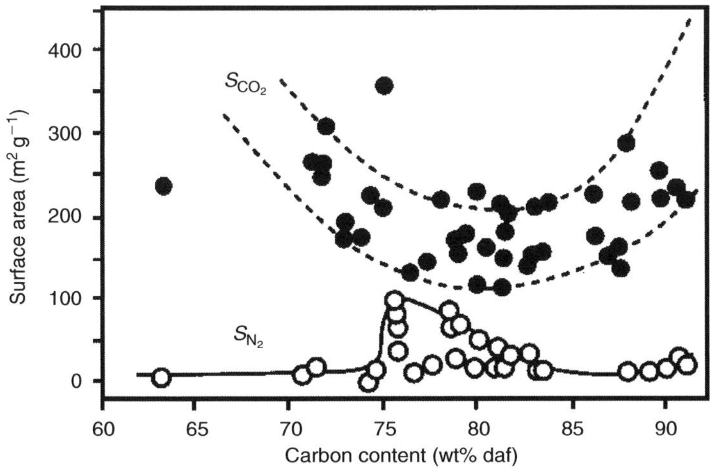

Figure 4.46. The variation of surface area of a rank range of coals. The upper set of data is surface areas determined from the adsorption of carbon dioxide at 298 K using the DR equation. The lower set of data is surface areas determined from the adsorption of nitrogen at 77 K using the BET equation (Díez Díaz-Estébanez and Marsh, 2000) (daf: dry ash-free).

Clearly, a great deal of explaining is necessary to account for such major discrepancies. Explanations are to be found in Section 4.2.5.

Initially, heat of wetting was used to measure total surface areas of solids including microporous materials. Later, this heat of wetting method was used to obtain differential heats of adsorption, as can be done from the isosteric method and by flow calorimetry.

In immersion calorimetry, when a solid is immersed into a non-reacting liquid, a given amount of heat is evolved. This "heat of immersion" or "heat of wetting" is related to the formation of an adsorbed layer of the molecules of the wetting agent on the solid surface of the solid, together with the wetting of this adsorbed layer. The heats of immersion of a given solid into different liquids are usually different (Silvestre-Albero *et al.*, 2001).

For solids with a similar surface chemistry, as for the activated carbons and CMS, without a developed surface functionality, the measurement of heats of immersion into liquids with different molecular sizes also allows for the assessment of PSD of the carbons. When polar surfaces are analyzed, both the surface accessibility of the immersion liquid, and the specific interactions between the solid surface and the molecules of the wetting agent account for the total value of the heat of immersion. The interpretation of the heat of immersion values for these latter systems is less straightforward, and care has to be taken with the selection of the liquids chosen for the measurements.

The enthalpy of immersion, ΔH_{imm} is defined as the enthalpy change, at constant temperature, that takes place when a solid is immersed into a wetting liquid in which it does not dissolve nor react. The immersion enthalpy of a given solid can be measured starting from different initial states, which must be specified. A higher value is obtained when the solid surface is essentially pristine, such that a liquid–solid interface is formed during the measurement.

In other cases, the surface may be pre-covered, totally or partially, by a liquid film prior to the heat of wetting measurement. The sample is pre-covered by the vapor of the immersion liquid at different partial pressures, and this sample is immersed in the liquid. At low partial pressures, the vapor adsorbs on high-energy sites, and the enthalpy of immersion will be lower as compared with that obtained with the outgassed solid. The enthalpy of immersion continues to decrease as the coverage becomes complete (this is achieved by increasing the partial pressure of the vapor when the sample is treated). At this point, the immersion enthalpy obtained corresponds to the formation of an adsorbed film–liquid interface.

When the solid surface is outgassed under vacuum, prior to the measurement, the enthalpy of immersion mainly depends on the following parameters:

1. Extent of the solid surface. For a given liquid–solid system, the enthalpy of immersion increases with increasing surface area of the carbon being studied.
2. Chemical nature of the liquid and of the solid surface. Specific interactions between the solid surface and the liquid increase the enthalpy of immersion. This is the case, for example, of the wetting of a polar surface with a polar liquid.
3. Porous texture of the solid.

When the liquid has a molecular size close to the dimensions of the pores, enhanced interaction yields an increase in the enthalpy of immersion. Furthermore, molecules larger than some pores will not be able to access as much of the internal surface as do smaller molecules. Thus, the use of liquids with different molecular sizes permits the estimation of the PSD of a porous solid. The shapes of the adsorbing molecule and pore shapes are also important factors. A planar molecule like benzene can wet slit-shaped pores about 0.37 nm wide, but cannot enter cylindrical pores of the same dimension.

One of the first attempts to relate the immersion enthalpy of a given solid–liquid system to the surface area of the solid (as distinct from earlier work with coals) was accomplished by Harkins *et al.* (1944). As a first approach, it can be assumed that, in the absence of complicating effects such as micropore filling, the energy of immersion is directly proportional to the surface area available to the liquid:

$$\Delta H_{\text{imm}} = S(-h_{\text{imm}}) \tag{4.11}$$

where $S(-\Delta h_{\text{imm}})$ is the area energy of immersion, characteristic of the solid–liquid system. Thus, the surface area of a solid can be estimated from its heat of immersion in a given liquid if ΔH_{imm} for the solid–liquid system has been previously evaluated. Consequently, it is necessary to use a reference material with a surface chemistry close to that of the sample under study. For carbonaceous materials without a well-developed surface functionality, non-porous carbon blacks are usually used as reference materials (Gonzalez *et al.*, 1995). However, for solids whose surface chemistry is unknown and which may affect, strongly, the magnitude of the immersion enthalpy, the choice of a reference material is not so straightforward.

The method proposed by Harkins *et al.* (1944) which they called "the absolute method", included the previous coverage of the outgassed sample with an adsorbed film (five to seven molecular layers) of the immersion liquid. During the immersion experiment, the liquid "sees" a surface with an extent equivalent to that of the solid, but with a chemical nature corresponding to that of the bulk liquid. An improvement of this method was later proposed by Partyka *et al.* (1979) who deduced that, for a number of non-porous solids, the coverage with just 1.5 molecular layers was enough to screen the solid surface without reducing the available surface area. In this "modified Harkins–Jura" technique, water was used as the immersion liquid for solids with hydrophilic surfaces and pentanol for solids with hydrophobic surfaces.

Characterization of microporous adsorbents by immersion calorimetry is not as straightforward as for non-porous adsorbents. Atkinson *et al.* (1982) measured the heats of immersion of a microporous carbon cloth and a microporous activated carbon in a series of organic liquids and, for a given solid, obtained a significant dependence of the heat of immersion with the liquid used. They concluded that the heat of immersion is a measure of the volume of pores accessible to the molecule of the immersing liquid, thus opening the possibility of using immersion calorimetry as a tool to obtain PSDs in microporous carbons.

The apparatus used for immersion experiments is shown in Figure 4.47. The sample, initially, is placed in a glass bulb with a brittle end and is outgassed at 1.3×10^{-3} Pa (10^{-5} torr) for 4 h at 523 K. While under vacuum, the tube is sealed and attached to a glass

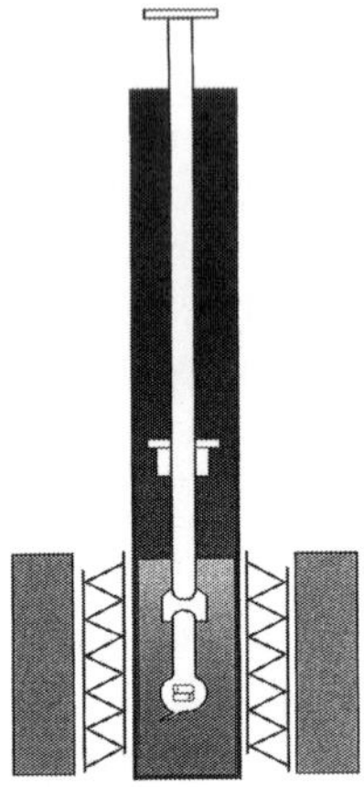

Figure 4.47. The essentials of the immersion calorimeter.

rod through a Teflon fitting and this assembly placed into the calorimeter block and closed off using an O-ring seal. The calorimeter cell contains 7 cm^3 of the immersion liquid; the system is allowed to reach a thermal equilibrium. The glass rod pushes the sample bulb to the bottom of the block where it breaks the brittle end, the solid is wetted, heat is evolved and measured as a function of time. The heat evolved is measured as an integral of the signal and corrections are applied for the breaking of the bulb and for vaporization of liquid to fill the void space of the bulb. Blank experiments with an empty bulb provide this information.

4.7.2 Immersion Calorimetry: Activated Carbons

Denoyel *et al.* (1993) proposed a method to assess the total area of microporous carbons by immersion calorimetry. It was based on the assumption of the existence of a direct relationship between the enthalpy of immersion and the total area of the solid accessible to the wetting molecules. They used a non-porous carbon black (Vulcan 3) as a reference material to obtain the area enthalpy of immersion, Δh_{imm} (J m^{-2}), of a carbonaceous surface into different liquids. In this way, and considering that the enthalpy of immersion is simply proportional to the surface area available to the immersion liquids, irrespective of its external or microporous character, and whatever the size and shape of the micropores, they obtained the surface areas available to the different liquids using Equation (4.11).

The most important feature of this approach is the assumption of simple proportionality between the surface area and the enthalpy of immersion, irrespective of the role played by micropores in the enhancement of the adsorption potential. It is established that, for slit-shaped micropores in which only one molecule of the wetting liquid can be accommodated, there is a twofold increase of the adsorption potential as compared with that in an open surface, and this would be also reflected in the enthalpy of immersion. However, such a pore has two opposite walls and, therefore, the surface area interacting with the liquid molecule is also twice that of the open surface. It follows that, in the case of slit-shaped pores, the enthalpy of immersion is proportional to the surface area accessible to the wetting liquid.

On the other hand, the calculated enhancement of the adsorption potential in a cylindrical micropore accommodating a spherical molecule of the same diameter ($2r$) is 3.679. This value is very close to 3.627, which is the ratio between the area of the pore covered by the molecule ($4\pi r^2$) and its mean cross-sectional area when it is adsorbed in a hexagonal compact assembly on an open surface ($1.1026\pi r^2$). Thus, also in this case, the enthalpy of immersion is expected to be practically proportional to the accessible surface area.

Stoeckli and Krachenbuehl (1981) proposed an alternative approach (more developed) to the interpretation of immersion enthalpy data, which is based on the Dubinin model of micropore volume filling. They obtained a mathematical relationship between the enthalpy of immersion of a microporous solid into a liquid whose vapor is adsorbed according to the DR equation and the parameters of the DR equation (Equation (4.12)).

This equation shows that the enthalpy immersion of a microporous solid into a given liquid is a function of the micropore network since E_0 is related to the average pore size, and not only depends on the micropore volume W_0.

Equation (4.12) makes it possible to calculate the external surface area of a microporous solid, S_{ext}, provided the area enthalpy of wetting (h_{imm}) is known and E_0 and W_0 are obtained from physical adsorption of the corresponding vapor. However, the contribution of microporosity to the total enthalpy of immersion is large in highly microporous carbons as compared with the contribution of the wetting of the external surface ($-\Delta h_{imm} S_{ext}$). In these cases, the estimation of the external surface from Equation (4.12) may be affected by a significant error.

$$-\Delta H_{imm}\,(\mathrm{J\,g^{-1}}) = \frac{\{\beta E_0 W_0 (1 + \alpha T)\pi^{1/2}\}}{\{2V_m\} - h_{imm} S_{ext}} \qquad (4.12)$$

where ΔH_{imm} is the measured enthalpy of immersion ($\mathrm{J\,g^{-1}}$)
W_0 is the micropore volume ($\mathrm{cm^3\,g^{-1}}$)
E_0 is the characteristic energy of the system ($\mathrm{J\,g^{-1}}$)
α is the thermal coefficient of expansion of the wetting liquid ($\mathrm{cm^3\,g^{-1}\,K^{-1}}$)
V_m is the molar volume of the liquid ($\mathrm{cm^3\,mol^{-1}}$)
Δh_{imm} is the area enthalpy of wetting of the external surface of the solid ($\mathrm{J\,cm^{-2}}$)
S_{ext} is the external surface area of the solid ($\mathrm{cm^2\,g^{-1}}$)
β is a scaling factor.

On the basis of results obtained with molecular sieves, Stoeckli and Centano (1997) showed, for active carbons, the characteristic energy E_0 ($\mathrm{kJ\,mol^{-1}}$) related to the average width (L in nm) of the slit-shaped micropores as follows:

$$L\ (\mathrm{nm}) = \frac{10.8}{(E_0 - 11.4)} \qquad (4.13)$$

For the slit-shaped micropores, their surface area is related to their volume and size through Equation (4.14) (Stoeckli and Centano,1997; Stoeckli *et al.*, 1998).

$$S_{mic}\ (m^2\ g^{-1}) = \frac{2000W_0\ (cm^3\ g^{-1})}{L\ (nm)} \tag{4.14}$$

Combination of Equations (4.13) and (4.14) gives the relationship between the microporous surface area and E_0W_0:

$$E_0W_0 = \frac{S_{mic}(10.8 + 11.4L)}{2000} \tag{4.15}$$

Following this equation, the relation between the product E_0W_0 and the microporous surface area will not be always the same, but depends on the average pore width L. Thus, for a highly microporous activated carbon, or any other adsorbent with slit-shaped pores with a low external surface area, the enthalpy of immersion is proportional to the product E_0W_0 but not to the microporous surface area S_{mic}. Anyway, E_0 can be estimated from Equation (4.15) for a given liquid, this allowing for the calculation of micropore size distributions by using liquids with different molecular sizes, Stoeckli *et al.* (1990).

The fundamentals of the two approaches described are very different; they themselves seem to be contradictory. However, both of them have been successfully applied to the characterization of the microporous structure of activated carbon and CMS. Some examples are described below.

4.7.3 Immersion Calorimetry: The Activation Process

An example of the use of immersion calorimetry is the evaluation of the development of the microporous texture of activated carbons upon activation. This is carried out by immersing the solid in a liquid able to fill the microporosity as completely as possible. Stoeckli and Ballerini (1991) used immersion into benzene to follow the evolution of the micropore volume and the pore width of a series of activated carbons, prepared from different precursors, as a function of burn-off. They deduced that for raw materials of vegetable origin (rubber tires, soft wood, barbecue charcoal, coconut shells, olive stones, banana skins and the skin of coffee beans), the main effect of activation was an increase in the micropore volume, this producing a strong increase in the enthalpy of immersion with burn-off. However, the increase of the enthalpy of immersion with burn-off was less pronounced in carbons prepared from rubber tires, and this was attributed to a change (increase) in the pore width, L. A similar study was carried out by Denoyel *et al.* (1993) with charcoals prepared from pinewood and activated to different extents with steam and phosphoric acid. They obtained the accessible surface areas by using Vulcan carbon black as the reference, as described above.

Figure 4.48 plots the enthalpies of immersion of a series of activated carbons, which were prepared from **olive stones** by carbonization in nitrogen and further activation with carbon dioxide (Rodríguez-Reinoso *et al.*, 1985). Enthalpies of immersion were measured using three liquids with different molecular sizes, namely benzene (0.37 nm);

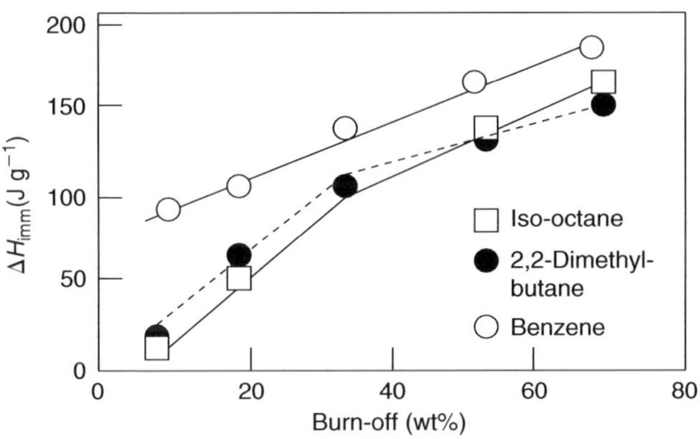

Figure 4.48. Evolution with burn-off (wt%) of experimental enthalpies of immersion (J g^{-1}) of activated carbons from olive stones into different liquids (Rodríguez-Reinoso *et al.*, 1985).

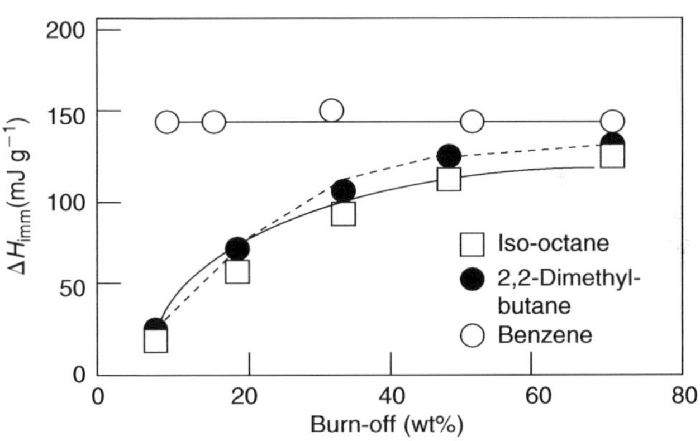

Figure 4.49. Evolution with burn-off (wt%) of the area enthalpies of immersion (mJ m^{-2}) of activated carbons from olive stones into different liquids (Rodríguez-Reinoso *et al.*, 1985).

2,2-dimethylbutane (0.56 nm) and also iso-octane (0.59 nm). Increases in the enthalpies of immersion with the activation degree (burn-off) are clearly evidenced, and these are related to increases in the micropore volume and the broadening of the micropore size distribution as activation proceeds, as deduced from nitrogen adsorption at 77 K (Rodríguez-Reinoso *et al.*, 1989). This increase of ΔH_{imm} with activation is almost parallel for the three liquids for burn-offs equal or higher than 34 wt%. For carbons with <34 wt% burn-off, enthalpies of immersion of 2,2-dimethylbutane and iso-octane increase with burn-off more quickly than that of benzene. These results point to the existence of a molecular sieving effect in these carbons, with benzene only being able to enter the porosity of the carbons.

The difficulty which 2,2-dimethylbutane and iso-octane have in accessing the whole microporosity of the less activated carbons is clearly evident from data shown in Figure 4.49 where the plotted data correspond to the enthalpies of immersion as in Figure 4.48 but divided by the BET surface area (N_2, 77 K) of the different samples. The area enthalpies of immersion (mJ m^{-2}) obtained with benzene are similar in the whole range of activation, with a small decrease as the activation degree increases. This effect has been ascribed to the interaction of the benzene molecule with both walls of very narrow micropores. Thus, the BET surface area measured with N_2, at 77 K would be underestimated in carbons with pores of a diameter close to the size of the nitrogen molecule.

However, it is obvious that the accessibility of the larger molecules to the total microporosity of the two samples, activated to a lesser extent (8 and 19% burn-off), is very limited as compared with nitrogen. This is clear evidence of the potential of immersion calorimetry to characterize, *easily*, the micropore size distribution in molecular sieves, based on the use of wetting liquids of different molecular sizes and with similar interactions with the solid surface.

In a further study, a series of CMS was prepared from coconut shells by carbonization and activation with carbon dioxide (De Salazar *et al.*, 2000). This series was characterized by carbon dioxide adsorption at 273 K and by immersion calorimetry using liquids with different molecular sizes, dichloromethane (0.33 nm), benzene (0.37 nm), cyclohexane (0.48 nm), 2,2-dimethylbutane (0.56 nm) and a-pinene (0.7 nm). Immersion data were analyzed following the two methods described above. A graphitized carbon black, V3G, with a BET surface area of 62 $m^2 g^{-1}$ (N_2, 77 K), was used as a non-porous reference to obtain the area enthalpy of immersion of a carbonaceous surface into the different liquids. With these values, and the enthalpies of immersion of the CMS into the different liquids, the surface areas accessible to the liquids were obtained. These are plotted in Figure 4.50 as a function of the molecular dimension; samples are identified by a number that indicates their activation time (De Salazar *et al.*, 2000).

For a given CMS the accessible surface area decreases as the molecular size of the wetting liquid increases, the decrease being more rapid for samples with low burn-off. Micropores in carbons CS-2, CS-4 and CS-8 are narrower than 0.37 nm. As activation proceeds, the PSD becomes wider, although Figure 4.50 clearly shows that the pore size is kept lower than 0.7 nm even for the more activated sample (CS-70).

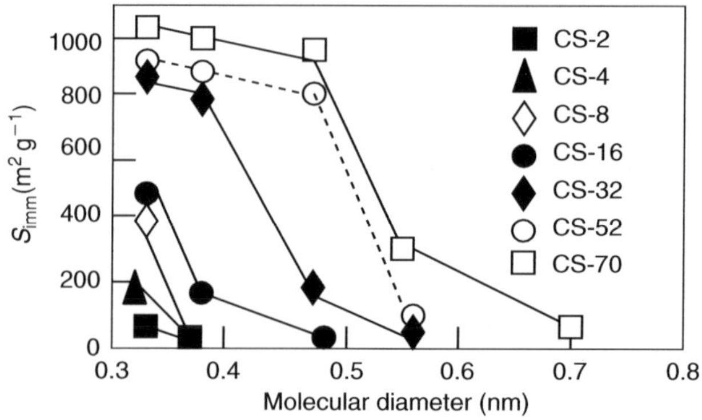

Figure 4.50. Surface areas of carbons ($m^2 g^{-1}$) from coconut shells accessible to liquids with different molecular kinetic diameters (nm), obtained by immersion calorimetry (De Salazar *et al.*, 2000).

The enthalpies of immersion of these CMS into the different liquids have also been analyzed by the approach suggested by Stoeckli and Centano (1981, 1984) who shows that the enthalpy of immersion of a solid into a given liquid is related to the micropore volume accessible to the liquid according to Equation (4.11). The last term, $-h_{imm}S_{ext}$, corresponds to the enthalpy of wetting of the external surface of the solid. For essentially microporous carbons such as those under study, this value should be much lower than the enthalpy of wetting of the microporous surface and, thus, can be ignored. Then, Equation (4.16) applies:

$$-\Delta H_{imm}\ (\mathrm{J\ g^{-1}}) = \frac{(\beta E_0 W_0 (1+\alpha T)\pi^{1/2})}{(2V_m)} \tag{4.16}$$

From this approach, the enthalpy of immersion depends not only on the micropore volume W_0 but also on the micropore size distribution through the term E_0. Thus, the micropore volume accessible to the wetting liquid cannot be directly obtained from ΔH_{imm} unless E_0 is previously evaluated from, for example, adsorption measurements (Stoeckli *et al.*, 1990; Gonzalez *et al.*, 1995, 1997). However, it is possible to estimate the molecular sieving capacity of a solid if a wetting liquid with a smaller molecular size, able to access the whole microporosity of the solid, is used as a reference. From the immersion enthalpy into this liquid, $\Delta H_{imm(ref)}$, the term $E_{0(ref)}, W_{0(ref)}$ can be obtained by application of Equation (4.17), and this term can be compared with those obtained for other wetting liquids with different molecular sizes, Stoeckli and Centano (1997):

$$\frac{E_0 W_0}{E_{0(ref)} W_{0(ref)}} = \frac{\Delta H_{imm}}{\Delta H_{imm(ref)}} \frac{\beta_{(ref)}(1+\alpha_{(ref)}T)V_m}{\beta(1+\alpha T)V_{m(ref)}} \tag{4.17}$$

Figure 4.51 plots the evolution of the $E_0W_0/E_{0(ref)}W_{0(ref)}$ ratio for some of the CMS shown in Figure 4.50, as a function of the molecular size of the wetting liquid, and using dichloromethane as the reference liquid. This Figure 4.51 indicates that the above ratio decreases as the molecular size of the wetting liquid increases, this indicating that the accessibility of the wetting liquid becomes limited as compared with dichloromethane, the reference liquid.

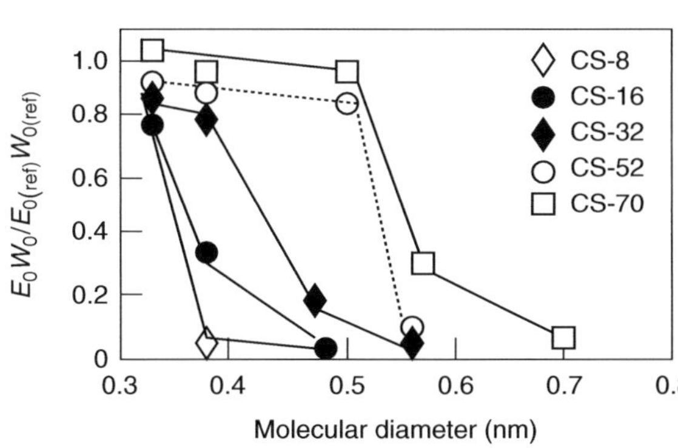

Figure 4.51. Micropore size distributions of CMS derived from immersion calorimetry (Silvestre-Albero *et al.*, 2001).

Also, the PSDs obtained by this method are comparable to those shown in Figure 4.50, which represents the surface area accessible to each wetting liquid. Sample CS-8 has a very narrow microporosity, with pores smaller than 0.37 nm. As activation proceeds, micropores become wider and, after an activation time of 70 h (sample CS-70), most of the micropores have diameters around 0.5 nm.

4.7.4 Immersion Calorimetry: Influence of Surface Chemistry

Immersion calorimetry using polar liquids gives another insight to the characterization of the solid surfaces. Here, specific interactions between the liquid molecules and active centers at the solid surface play a major role. The comparison between enthalpies of immersion of liquids with different polarity provides a refined picture of the surface properties of the solid. This point can be illustrated by the following example. Stoeckli *et al.* (1983, 1998) measured the enthalpies of immersion of two non-porous illites, with different BET surface areas (N_2 77 K) into benzene and water. Whereas the area enthalpy of immersion into benzene was similar for both samples, about 73 mJ m^{-2}, the area enthalpies of immersion into water were quite different, 371 and 782 mJ m^{-2}, respectively, Stoeckli *et al.* (1998). Clearly, the compositions of these illite surfaces differ.

Barton *et al.* (1972) used immersion calorimetry with butyl derivatives with different polarity (*n*-butanol, *n*-butyric acid, *n*-butyl chloride, *n*-butylaldehyde, *n*-butyl nitrile and *n*-butylamine) to study the polarity of the surface of graphite covered by surface oxygen complexes. For a non-polar liquid such as *n*-hexane, the enthalpies of immersion of the oxidized graphite and that of an oxygen-free sample were similar, this indicating that the surface polarity hardly affects the interaction with *n*-hexane. However, a series of different values were obtained with the butyl derivatives, which were attributed to the interaction of the electrostatic field at the surface with the dipole moment of the molecules of these wetting liquids.

The effects of oxygen-surface groups, on this graphite, on the enthalpies of immersion using benzene, water and methanol were also studied by Barton *et al.* (1972, 1975). In addition, Rodríguez-Reinoso *et al.* (1997) studied carbons from olive stones, activated to 37 wt% burn-off in steam at 730 °C, and finally oxidized to several extents with nitric acid (6N) to place surface oxygen complexes on the surfaces. This series of carbons was then heated in the range 100–900 °C (10 samples in all).

Figure 4.52 (a, b) are of the variation of enthalpies of immersion in several liquids, of the carbons of increasing HTT as prepared above, and of graphite, according to Barton *et al.* (1975). As changes in the carbon surface area affect the enthalpy of immersion in liquids, to analyze the effect of the surface groups on the heat of immersion, it is convenient to use the area enthalpy of immersion (J m^{-2}). These values are plotted as a function of the HTT in Figure 4.52(a). For comparison purposes, Figure 4.52(b) includes data obtained earlier by Barton *et al.* (1975) for a graphite outgassed at different temperatures. The similarity of the two figures is remarkable considering the large differences in the structure of the two carbon materials. Thus, considering that the enthalpy change per unit area for benzene is constant (and consequently, independent of the chemical nature of the carbon surface), and, additionally, that the two sets of values of Figure 4.52(a, b) are similar ($\approx$0.13 J m^{-2}),

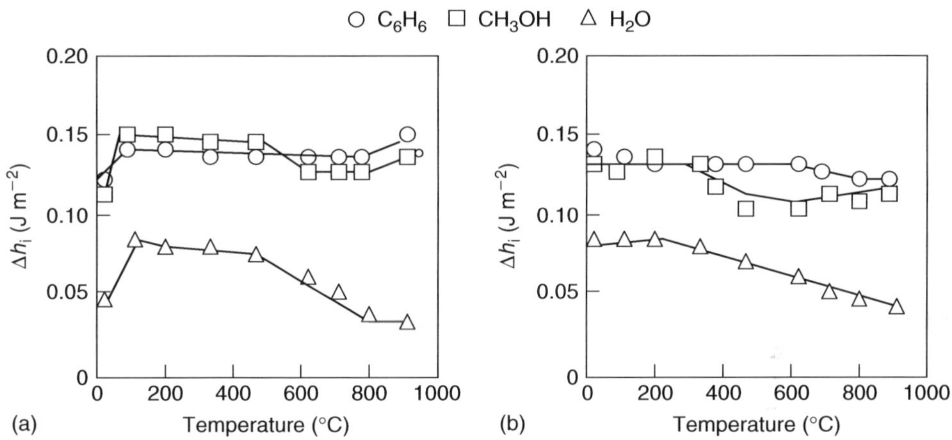

Figure 4.52. Evolution of the area enthalpies of immersion into benzene, methanol and water as functions of HTT for (a) activated carbons (Rodríguez-Reinoso *et al.*, 1997) and (b) graphite (Barton and Harrison, 1975).

then the energetic contributions due to interaction of benzene with the surface of the micropores are the same as for the non-porous surface of the carbon.

This confirms the convenience of determining the surface area of activated carbons using immersion calorimetry in benzene when a non-porous carbon is used as the reference, independently of the chemical nature of the carbon.

For methanol, the enthalpy per unit area of a carbon heat-treated to 900 °C is similar to that of graphite ($\approx 0.12\,\mathrm{J\,m^{-2}}$) and slightly increases (up to $\approx 0.14\,\mathrm{J\,m^{-2}}$) when the carbon has oxygen surface groups. The larger enthalpy of immersion registered when the carbon has groups evolving as CO_2 (at temperatures below 500 °C) indicates that the methanol molecule interacts with these groups, very probably via the formation of hydrogen bonds in addition to the non-specific carbon–methanol interactions.

As for adsorption processes, it is the enthalpy of immersion in water where the role of oxygen surface groups is more clearly shown (as commented upon above). Thus, there is a noticeable increase in ΔH_i for the oxidized carbon and there is a decrease when the heat treatment is above 500 °C. The parallelism in the evolution of Δh_i (H_2O), Figure 4.52(a), and the evolution of oxygen functional groups, shows that these groups affect the enthalpy of immersion, although the interaction of the oxygen-free carbon surface with the water molecule is also very important. In fact the latter interaction must be responsible for the adsorption of water on the activated carbon oxidized and later reduced in hydrogen at 900 °C for which adsorption starts at a relative pressure of 0.4. However, adsorption for the oxidized carbon begins at relative pressures below 0.1, as a consequence of the contribution of the interaction of the water molecules with the oxygen groups of the carbon.

The enthalpy of immersion in water for a carbon with a small amount of surface groups is rather near the value for graphite outgassed at 900 °C. If it is assumed that the structural

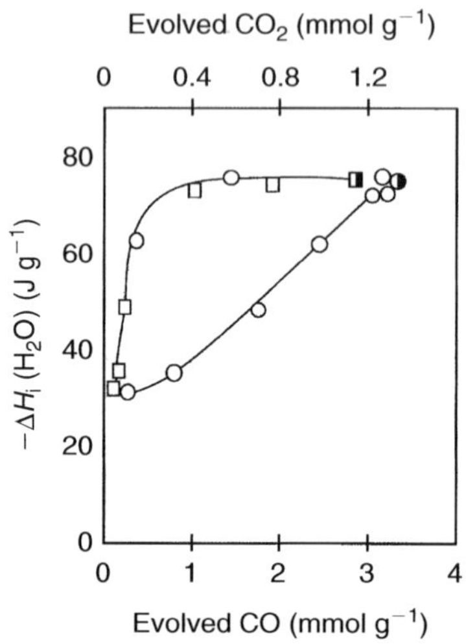

Figure 4.53. The variation of enthalpies of immersion in water of a carbon, from olive stones activated with steam to 37 wt% burn-off, and oxidized with nitric acid, with evolution of CO_2 and CO via TPD (Rodríguez-Reinoso *et al.*, 1997).

units of an activated carbon are of the order of 100 times smaller than in graphite, the proportion of carbon atoms in the edges of the structural units as compared with those on the basal plane is much larger in activated carbons. Due to the enthalpy per unit area is the same in both materials it can be assumed that once the surface groups are eliminated upon heat treatment, *there are no preferential sites (according to their position in the carbon structure) for the interaction of the water molecules with the carbon atoms*.

Whereas Figure 4.52(a, b) show the effect of HTT of carbons on enthalpies of immersion, Figure 4.53 illustrates how the chemical composition of surface oxygen complexes, on the same carbons as used for Figure 5.52(a, b) and subsequently analyzed using a TPD system (to 1100 °C) to monitor CO_2 and CO production and also measuring enthalpies of immersion in water, influences enthalpies of immersion into water, Rodríguez-Reinoso *et al.* (1997).

Figure 4.53 indicates that when the amount of CO_2 groups is large, the enthalpy of immersion in water remains constant, in contrast with oxygen groups decomposing as CO, for which there is a practically linear relationship. The extrapolated value of $\Delta H_i(H_2O)$ is near to $22\,J\,g^{-1}$ which indicates that the interaction of the carbon surface with the water molecule is not due exclusively to the oxygen groups. This value is similar to that found for non-porous carbons and graphite. This indicates that the value of ($\Delta H_i(H_2O)$ $J\,m^{-2}$) does not depend on the proportion of carbon atoms located in basal planes.

Stoeckli *et al.* (1990) related the parameters obtained from the adsorption isotherm of water on activated carbons with their enthalpies of immersion into water. In this way, they obtained a measure of the amount and type of active centers for water adsorption.

The adsorption of water by activated carbons follows Equation (4.18) (Dubinin and Serpinsky, 1981):

$$\frac{p}{p^0} = \frac{a}{(c(a_0 + a)(1 - ka))} \tag{4.18}$$

In this equation $\boldsymbol{a}$ is the amount of water adsorbed at a relative pressure $\boldsymbol{p/p^0}$, $\boldsymbol{a_0}$ is the number of primary centers of adsorption and $\boldsymbol{c}$ is the ratio between the rate constants of adsorption and desorption. The value of the constant $\boldsymbol{k}$ is fixed by the condition that the total amount of water adsorbed in the micropores at $\boldsymbol{p/p^0} = 1$ is $\boldsymbol{a_s}$. They observed that the number of primary (hydrophilic) sites in activated carbons, $\boldsymbol{a_0}$, is related to the enthalpy of immersion into water. For activated carbons outgassed at 400, 500 °C and containing a uniform type of hydrophilic centers, the enthalpy of immersion into water is given as follows:

$$\Delta H_{imm} = -25a_0 - 0.6(a_s - a_0) \tag{4.19}$$

where ΔH_{imm} is in J g^{-1} and $\boldsymbol{a_0}$ and $\boldsymbol{a_s}$ (the total amount of water adsorbed in micropores) are in mmol g^{-1}. This relationship shows that the enthalpy of immersion comes from two contributions, the primary sites and the bulk of the micropores. Furthermore, these authors deduced a relationship between the enthalpy of immersion and the parameter $\boldsymbol{c}$ of the water adsorption isotherm (Kraehenbuehl *et al.* 1986):

$$c = c_0 \exp \frac{(-\Delta H_{imm}/a_s)}{RT} \tag{4.20}$$

It is concluded from Equations (4.19) and (4.20) that the amount and strength of primary centers, $\boldsymbol{a_0}$, affect the magnitude of $\boldsymbol{c}$. The coefficient of $\boldsymbol{a_0}$ in Equation (4.19) is a measure of the strength of the interaction between the water molecule and the active centers. The value 25 in Equation (4.18) has been obtained for carbons in which the active centers are mainly carbonyl groups. In the case of carbons with only acidic groups, a better fit is obtained with Equation (4.21) which indicates higher specific interactions between the water molecules and the carbon surface:

$$+\Delta H_{imm} = -55a_0 - 0.6(a_s - a_0) \tag{4.21}$$

The analysis of the adsorption isotherm leads to $\boldsymbol{a_0}$ and $\boldsymbol{c}$. From $\boldsymbol{c}$ and the experimental enthalpy of immersion, $\boldsymbol{c_0}$ can be obtained by application of Equation (4.20). It has been observed for a series of samples that the mean values of $\boldsymbol{c_0}$ were 1.00 ± 0.11. Stoeckli *et al.* (1990) proposed to use Equation (4.20) with $\boldsymbol{c_0} = 1.00$ to estimate the value of $\boldsymbol{c}$ from the experimental enthalpy of immersion, to avoid inaccuracies in values obtained from the adsorption isotherm.

In summary, immersion calorimetry offers rapid and reproducible methodologies for the measurement of enthalpies of immersion which can be converted to PSDs, assessments of surface functionality and to units of surface area, noting the limitations which apply to this concept.

4.8 Mesoporosity

4.8.1 Definition of Mesoporosity

Mesoporosity is defined as that porosity which has an entrance dimension >2 nm and <50 nm. Such a definition is based around the way that nitrogen is adsorbed into porosity. The filling of the porosity, >2 <50 nm, is called *capillary condensation*. Adsorption occurs initially on the pore walls with the result that the pore is not only a narrower pore, but also it behaves as a narrower pore with a higher adsorption potential well in the center (Figure 4.2). This process continues with increasing relative pressure within pores of increasing diameter until the 50 nm limit (or thereabouts) is reached when the entrance dimension of the pore (gap between the pore walls) is so large that the porosity behaves as an open surface.

These transitions in pore filling mechanisms are dependent on the size of the adsorbate molecule and so the above definition (>2 nm and <50 nm) is arbitrary being a function of the adsorbate used. However, as nitrogen dominantly is the characterizing adsorptive, these definitions must suffice. From an adsorption point of view, mesoporosity is not that all-important. Its role is as a transport pore (a means of passage), particularly for adsorptions from solutions, to permit rapid access to the retaining microporosity of the adsorbate of the system.

4.8.2 Origins of Mesoporosity

A continuous distribution of pore sizes from micro to macro is not generally found in porous carbons. Rather, the distribution is tri-modal as indicated in the diagram of Figure 4.54. The question arises as how mesoporosity arises in carbons.

There are **two** ways, at least, to be considered. **First**, the carbonization process, itself, creates mesoporosity in the carbon by reason of the physical structure of the parent material, associated with shrinkage, etc., and is referred to as *indigenous mesoporosity*. The **second** method is where a non-mesoporous, but microporous carbon is "activated physically" by, for example, gasification with steam or carbon dioxide, when carbon atoms of the carbon structure (wherever they are) are gasified leaving "molecular holes" or *created*

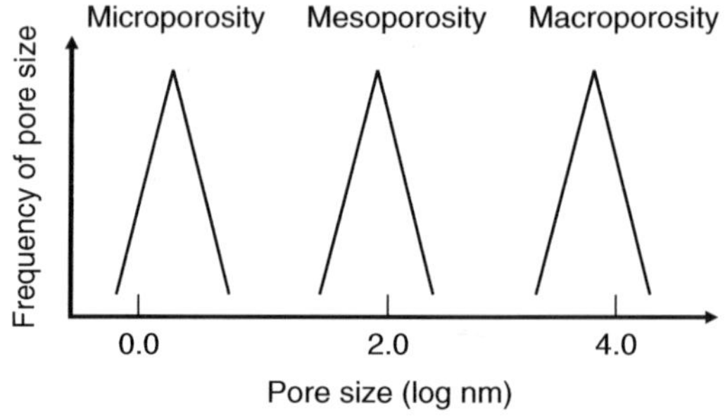

Figure 4.54. The tri-modal distribution of porosity found in carbons from natural precursors.

mesoporosity. However, both the *indigenous mesoporosity* and the *created mesoporosity* may exist in three forms, at least (simplified for modeling purposes). **One**, it may be part of a continuous distribution of pore sizes within the porous networks of the carbon, that is, the mesoporosity is intimately mixed within the microporous parts of the carbon (Figure 4.55). This mesoporosity will be totally accessible to the microporosity and may provide short cuts through the carbon so promoting accessibility. Mesoporosity is accessible via mesoporosity. **Two**, this type of mesoporosity may be a network of independent mesopores (<50 nm diameter) within the microporosity of the carbon and is accessible only via microporosity (Figure 4.56). **Third**, mesoporosity is often described as being cone-shaped and forming part of the surface of activated carbon, as can be seen in the transmission electron micrograph of Figure 4.57.

The locations of the mesoporosity, as outlined above, may influence the shape and characteristics of adsorption isotherms.

During an adsorption process, all three locations of mesoporosity within the pathways would fill with the adsorbate nitrogen (77 K) at the highest of relative pressures to give a Type-IV isotherm (Figure 4.8). For the situation of Figure 4.55, the desorption of the

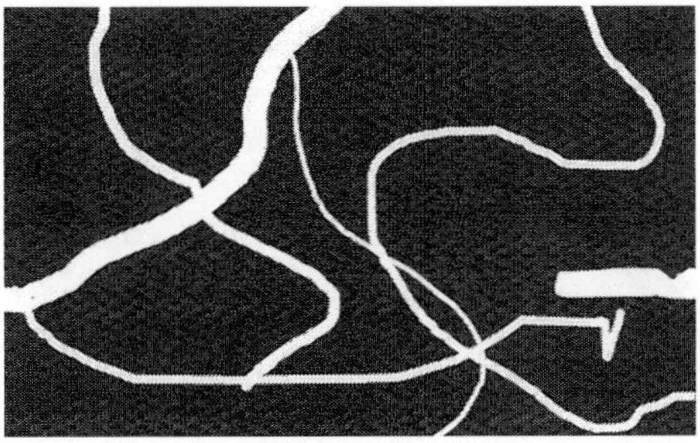

Figure 4.55. Diagram showing how mesoporosity may be intimately mixed within the microporosity of a carbon. The "surfaces" of the mesoporosity will contain "entrances" to the microporosity.

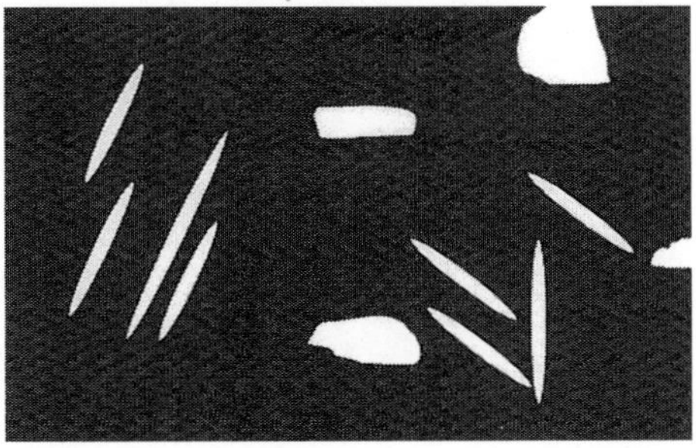

Figure 4.56. Diagram showing how mesoporosity may exist independently of the microporosity within a carbon. The "surfaces" of the mesoporosity will contain "entrances" to the microporosity.

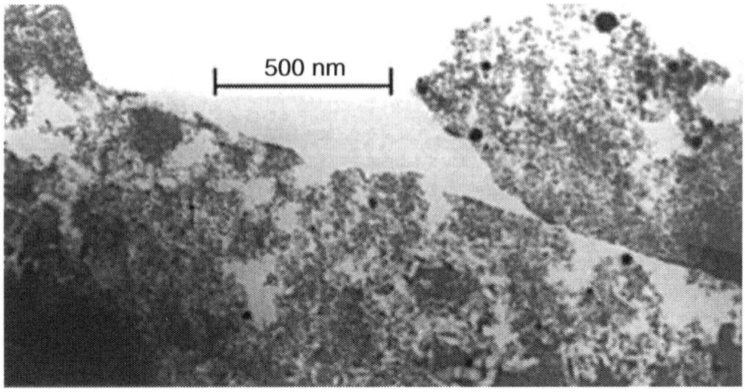

Figure 4.57. This transmission electron micrograph of a PVDC activated carbon (85 wt% burn-off) indicates the existence of mesoporosity (holes) within a matrix of microporosity and illustrates both the separation of the mesoporosity and at the same time its access to microporosity.

nitrogen would be with hysteresis. For mesoporosity within microporosity, Figure 4.56, although the adsorption isotherm would show the existence of mesoporosity, the use of molecular probing adsorbate molecules would not be able to do this because entry to the mesoporosity would be only through microporosity. On the other hand, adsorption pathways composed entirely of mesopores, Figure 4.55, would respond to the molecular probing experiments. Figure 4.55 also indicates the possibility that, throughout a sample activated carbon showing hysteresis with nitrogen adsorption, cone-shaped porosity could exist. It need not exclusively be associated with external surfaces of the sample.

Created mesoporosity is the result of controlled gasification (activation) by, for example, carbon dioxide or water vapor, of a microporous carbon at temperatures of about 700–800 °C. It is assumed that gasification occurs preferentially of carbon atoms at specific sites within the carbon particle because the activation process occurs uniformly within (throughout) the carbon sample and not just on the external surfaces or positions of initial collision. Further, it is noted that external surfaces of carbon samples may develop a mesoporosity that is cone-shaped (Figures 4.55 and 4.57). Such a mesopore would exhibit hysteresis in an adsorption experiment. However, the carbon gasification which occurs within a carbon particle, not on an external surface, may create mesoporosity which resembles a cave brought about by the *excavation* of carbon atoms (Figures 4.55 and 4.57).

In addition, preferential gasification of carbon may occur in association with an internal catalytically active impurity, so creating the mesoporosity of Figures 4.56 and 4.57.

Whether or not the adsorption of such internal mesoporosity would exhibit hysteresis effects on desorption is a matter of some uncertainty, the isotherm being of Type-IV. Gelb and Gubbins (1998) make the suggestion that hysteresis loops of nitrogen adsorption/desorption can be attributed to the presence of thermodynamically metastable clusters of adsorbate molecules within the "caves" of the porous networks, and not to kinetic effects. That is, adsorption within the structures of Figure 4.55 would contribute to hysteresis effects.

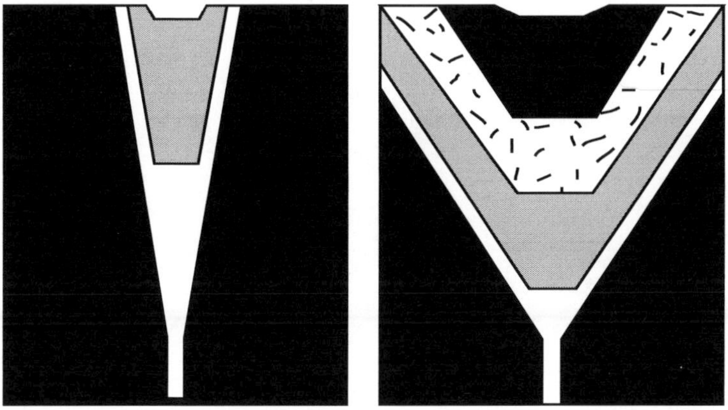

Figure 4.58. Diagram to show the progressive filling of cone-shaped porosity during an adsorption process.

It is also possible that extents of hysteresis are related to the heterogeneity of the adsorbing surfaces of the porosity.

Not too much is known about heterogeneity of structure within the graphene layers which must form all (or a major part) of the walls of porosity. Important sources of heterogeneity are five- and seven-membered ring systems and is it established that these defects are responsible for the curvature of structures as found in fullerenes, nanotubes and carbon necklaces. Porosity in carbons most probably contains these curved-wall systems, these defects making a contribution to the shape of a nitrogen isotherm (or any other isotherm). An adsorption isotherm provides informations about PSDs which are measure of carbon wall–carbon wall separation distances. Such informations must be contained with the d_{002} X-ray diffraction bands which are derived from layer–layer distances. However, not all of the layer–layer distances relate to porosity but to structure and hence it is a difficult task to separate out these components within the d_{002} bands. Pikunic *et al.* (2002) discuss these matters further.

The mechanism of capillary condensation within individual pores could therefore be too simple a model.

The matter of "hysteresis loops" requires some further discussion. Hysteresis loops have been known for quite some time and have been categorized by Sing *et al.* (1985) in an IUPAC classification. This publication covers hysteresis observed with other non-carbonaceous materials such as alumina, zeolites and porous glasses. There is a tendency in the literature to explain hysteresis in activated carbons by using the two mechanisms of irreversible filling of cone-shaped pores (Figure 4.58) and of the irreversible filling of cylindrical porosity (Figure 4.59). The cone-shaped pore, as existing on carbon surfaces and generated by easy access of a gasification gas, are considered to be of limited depth, about 200–300 nm. The total adsorptive capacity of this surface array of cone-shaped pores can account for no more than 0.01 vol% of the volume capacity of the mesoporosity. So, the overall contribution to the phenomenon of hysteresis is negligible and causes have to be sought elsewhere.

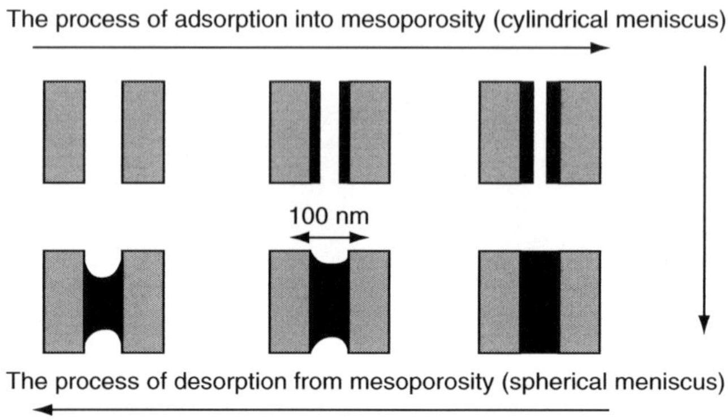

Figure 4.59. Adsorption in mesoporosity: a possible cause of hysteresis.

With the mesopore model, it is tacitly assumed that it behaves as a tube with "solid" walls. A more accurate model would be that of a convoluted tube, with of varying dimension, with other porosities leading from the "so-called" walls of the cylinder. It is therefore unlikely that processes of independent condensation and evaporation occur in the pores of such complex three-dimensional porous networks.

4.8.3 Quantitative Assessments of Mesoporosity

The very nature of porosity, as indicated in the models of Chapter 3, precludes any actual assessments of pore shape and size, even on a statistical basis and recourse therefore has to be taken to the use of models and simplifying assumptions. For the foreseeable future such approaches as these may be all that will be available. They should therefore be considered as being pragmatic and not as being precise descriptions of porosity. A severe limitation in these assessments is applicability of bulk macro-properties of materials to locations which contain only a few molecules.

4.8.3.1 The Kelvin equation

The Kelvin equation relates the vapor pressure above a meniscus of a liquid to a "diameter" of a containing pore. This section discusses the use of nitrogen isotherms to descriptions of mesoporosity <20 nm diameter.

The basic principles on which the Kelvin equation are based (and which still apply) originated from the studies of Zsigmondy (1911) who stated that an equilibrium pressure, $\boldsymbol{p}$, over a concave meniscus of a liquid (adsorptive) is less than the saturation vapor pressure, $\boldsymbol{p}^{0}$, at the same temperature. The outcome of this is that a vapor can condense as a liquid into the porosity of a solid. A further consideration is that pores of small size will fill with liquid at lower values of $\boldsymbol{p}$ than larger pores.

The Kelvin equation is an attempt to relate the variation of relative pressure of an adsorptive (the adsorbate in the vapor state) with the curvature of the meniscus of "liquid" adsorbate filling a cylindrical pore. The model picture here is of a vertical symmetrical pipe filled with a liquid, the meniscus of which is circular and is more or less perfectly formed. This is greatly over-simplified. The mesopore is more likely to resemble a badly bent "slot" of varying diameter and shape. There is no way this can be quantified and so the symmetrical pipe model has to be used. The Kelvin equation originates from two considerations, the first being that of a drop (a sphere) of liquid (A) at equilibrium within bulk liquid (B) when there is no miscibility. The pressure (p^{ex}) of the external liquid on the convex surface of the spherical bubble (radius r) is compensated by the internal pressure (p^{in}) on the concave surface plus the "pressure" associated with the surface tension at the interface:

$$p^{ex} - p^{in} = \frac{2\gamma}{r} \tag{4.22}$$

where γ *is* the surface tension.

The second consideration for the Kelvin equation comes from the relationship between the vapor pressures of a liquid with a flat meniscus and a curved meniscus. The derived expression has the following form:

$$\ln\frac{p}{p^0} = -\frac{2\gamma v^l}{r_K RT} \tag{4.23}$$

where v^l is the molar volume of the liquid and r_K is the effective radius of curvature. When using nitrogen as the adsorbate Equation (4.23) becomes Equation (4.24):

$$\frac{r_K}{\text{nm}} = -\frac{0.415}{\log_{10}(p/p^0)} \tag{4.24}$$

It is assumed that the surface tension and the molar volume are independent of the radius of curvature associated with the porosity. Corrections or adjustments can be made to the above Kelvin equation because in an adsorption process, at the stage when mesopore filling occurs, already the walls of the pore contain adsorbed material such that the effective diameter of the pore is reduced and $r_p = r_K + t$ (where t is the thickness of the adsorbed layer). Another correction is where the pore shape is not cylindrical but is perhaps slit-shaped when the meniscus is hemi-cylindrical and the effective pore width $w_p = r_K + 2t$. However, for porous carbons of mixed sizes of porosity there is always a strong case for suggesting that all of the mesopores are not of the same shape and so pore size calculations become somewhat tenuous. There is apparently no way around this problem and numbers obtained have to be accepted for their worth, checking at all times for self-consistency.

The experimental data for mesopore assessments are the isotherms at their high relative pressure ranges. With mesoporosity present, the isotherms of activated carbons adopt the Type-IV shape, usually terminating without a plateau, at position S (Figure 4.60). The p/p^0 value of 1.0 provides a limit to amounts that can be adsorbed, n_{sat} in units of mmol g^{-1}. Conversion to volumes adsorbed (V_p in cm^3 g^{-1}) is possible using an estimated molar volume, V_m, (cm^3 mmol^{-1}) of the liquid adsorbate. When different liquids are used,

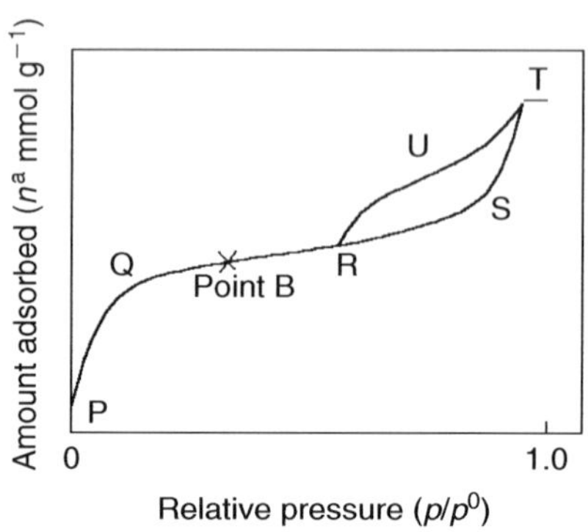

Figure 4.60. Adsorption isotherm of nitrogen, at 77 K, on a carbon containing both micro- and mesoporosity, and exhibiting a closed hysteresis loop on desorption.

with quite different chemical and physical properties, it is found for a great majority of adsorptives that, for a given carbon, values of $\boldsymbol{V_p}$ agree quite closely. This property is known as the ***Gurvitsch rule***.

In terms of procedures to be adopted for the calculation of mesopore size distributions, the methods of Barrett *et al.* (1951), Cranston and Inkley (1957), Dollimore and Heal (1970) and Roberts (1967) are mainly used. These methodologies are not that straightforward because of the many assumptions made regarding the adsorption process. This topic is summarized by Gregg and Sing (1982) in 83 pages of their book. The size and shape of possible mesoporosity can be debated; are they cylindrical, or wedge or slit shaped, do they always have an entrance opening to the outside of the sample, or is the mesoporosity without such access? These ideas can be modeled mathematically and appropriately but the reality of it all is not there. The models of Chapter 3 do not suggest the existence of such well-defined porosities for carbon materials, although they may exist in other, more ordered materials.

Results obtained and in the literature suggest that the Kelvin equation provides meaningful data describing mesoporosity. The central issues of the application of the Kelvin equation are as follows making reference to Figure 4.60 which is an adsorption isotherm of nitrogen at 77 K for a carbon containing micro- and mesoporosity and exhibiting hysteresis. Figure 4.60 is a plot of amount of nitrogen adsorbed (mmol g^{-1}) at 77 K against the equilibrium relative pressure of the nitrogen gas (vapor). The line **PQ** describes adsorption in the microporosity (and on any open surface which may be present). The smaller is the micropore size (diameter) the steeper is the line **PQ**. The line **QR** continues the adsorption process, reversibly, into the smallest of the mesopores. For a totally microporous carbon the isotherm will have leveled off from position **Q**. The line **RS** describes the progressive filling (capillary condensation) of all other mesopores in the carbon concluding with the largest of the mesopores at position **S**.

However, on lowering the pressure of the nitrogen in the equipment, the equilibrium positions, that is the desorption isotherm, do *not* follow the line **SR** but the line **SUR** to

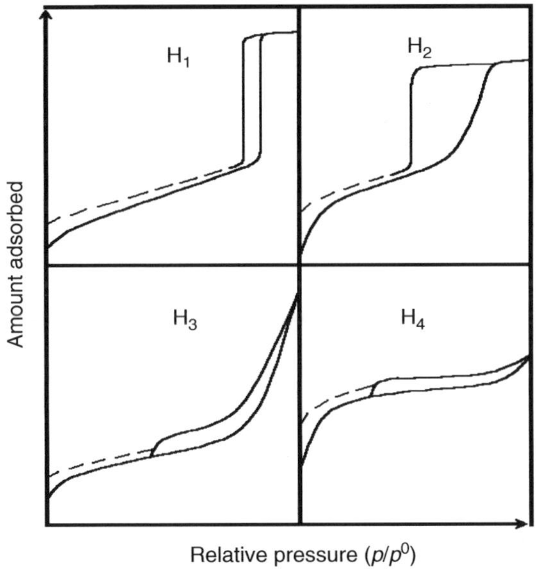

Figure 4.61. Types of hysteresis loop (irreversible adsorption–desorption processes) observed during adsorption into mesoporosity. Mesoporous carbons generally belong to the Type-H1 hysteresis loop (Sing *et al.*, 1985).

create a *hysteresis loop* closing at point **R**. The adsorption line describes equilibria of gaseous adsorbate over a curved meniscus on the internal surface of a cylinder (or so the model used says) whereas the desorption line describes equilibria of gaseous adsorbate over a curved meniscus (semi-spherical) at the end of a cylinder filled with liquid nitrogen. The lines **RS** and **SUR** are used by the Kelvin equation to relate a relative pressure to a size of mesopore. Both the adsorption and desorption parts of the isotherm are used to calculate PSDs and results are essentially in agreement. If the line **RS** is vertical then this indicates that all of the mesoporosity is of one size. The value of $\boldsymbol{p/p^0}$ of 0.90 corresponds to a pore size of 10 nm, a $\boldsymbol{p/p^0}$ value of 0.95 corresponding to a pore size of 20 nm increasing to 200 nm at value of $\boldsymbol{p/p^0}$ of 0.995. Thus, values assigned to pore size are closely dependent on the precision of measurement of relative pressure. Further, experimentally it may be difficult to maintain the sample at the boiling point of nitrogen (77.35 K). An actual temperature of 77.40 K results in an error of 40% in assignments of pore size.

In carbons, and other non-carbonaceous materials, hysteresis loops exhibit several forms which have been classified by IUPAC as shown in Figure 4.61, with carbons resembling Type-H4, and returning to zero amounts adsorbed at zero relative pressure, after desorption.

The lengthy debate about pore filling mechanisms of Gregg and Sing (1982) includes considerations of how to correct pore sizes for adsorption of nitrogen on walls prior to pore filling, etc. There would appear to be no way of knowing with any degree of precision the behavior of mesopores. The textbooks and publications all draw mesopores as being cone-shaped opening at an external surface and with closure at the narrow end. The reality must be quite different as a few *order-of-magnitude* calculations demonstrate. For granular carbon, 1 mm cube in size (volume 1.0 mm^3), the carbon having a surface area of 1,000 $m^2 g^{-1}$, the maximum volume of adsorbed liquid nitrogen is taken to about 30% of the volume of the carbon, that is 0.3 mm^3 (0.3×10^{18} nm^3). Assuming that the mesoporosity is as modeled and is all at the surface, then its volume is about 10^{14} nm^3 indicating a serious shortfall in volume assigned to the mesoporosity. An explanation is that mesoporosity must not necessarily be an external surface phenomenon but extends throughout the carbon and is part of the porous network systems within the carbon. Such a system would require new modeling. The possibility of cylindrical-, cone- or wedge-shaped mesoporosity within the bulk of a carbon seems to be remote.

Studies of mesoporous carbons serve at least two purposes. They provide informations with which to characterize the carbons in terms of their suitability for specific industrial applications. At the same time experimental data is provided with which to test the various theories and models put forward to explain adsorption phenomena. Semi-quantitative analyses of the isotherms provide valuable information on extents of possible adsorption (capacity) and of the pore-size range the carbons possess.

Confirmation of these earlier methods to calculate PSDs mainly in the mesopore range has been made by Kowalczyk *et al.* (2003). These authors prepared activated carbons from rayon and polyacrylonitrile (PAN) fibers using carbon dioxide as the activating gas. Resultant carbons were studied by adsorption of nitrogen at 77 K and isotherms are as shown in Figure 4.62 and showing hysteresis phenomena.

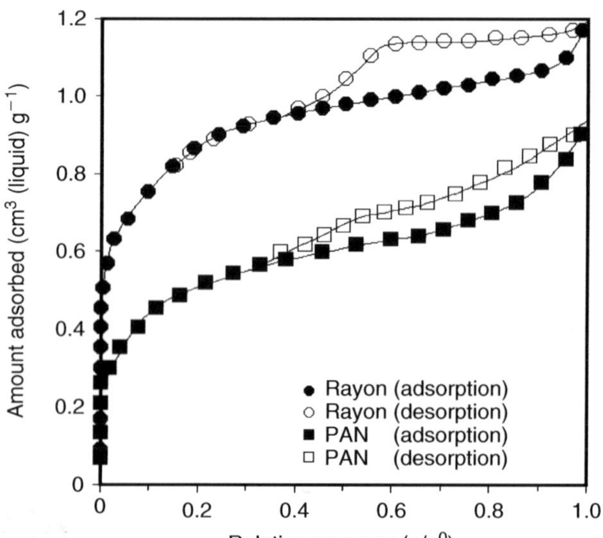

Figure 4.62. The adsorption of nitrogen at 77 K on activated Rayon and PAN fibers showing hysteresis loops on desorption (Kowalczyk *et al.*, 2003).

Conventional and recently developed methods of estimating PSDs from a single nitrogen isotherm were used. The authors further report on the use of a numerical algorithm (called adsorption stochastic algorithm (ASA)), based on an approach of Nguyen and Do (2001) and Do *et al.* (2001), (ND), and density functional theory (DFT). The latter, more sophisticated of methods lead to essentially the same results as the traditional methods mentioned above with small differences observed only in the intensity of the peak located in the upper range of micropores.

The PSDs for the Rayon and the PAN are reproduced in Figure 4.63(a, b) and show very clearly that the two parent fibers activate in quite different ways, the PAN showing a much broader distribution of pores extending to >20 nm whereas the Rayon fiber had almost no porosity beyond 3 nm. A model of structure in PAN fiber, according to Johnson (1987) is reproduced in Figure 4.64 and indicates how porosity may be located within the fiber.

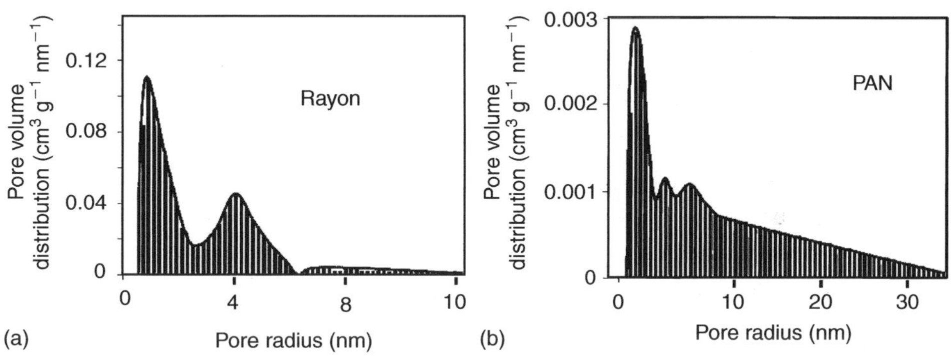

Figure 4.63. Showing PSDs for activated Rayon and PAN carbon fibers (Kowalczyk *et al.*, 2003).

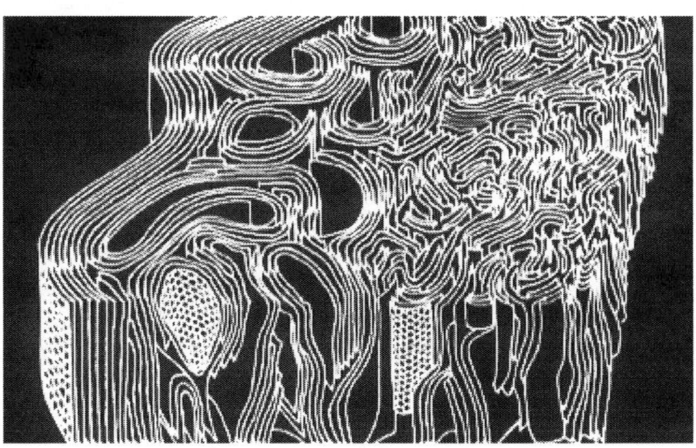

Figure 4.64. Diagram of structure in a PAN fiber showing the possible location and shape of internal porosity, based on fringe-imaging techniques, according to Johnson (1987).

4.9 Characterization of Mesoporosity: Mercury Porosimetry

4.9.1 Principles

Whereas the nitrogen isotherm provides quite reliable data for PSD with a mean size of about 20 nm or less, it is much less reliable for pores larger than this. Mercury porosimetry is the alternate technique for characterizing the sizes of the larger mesoporosity and dates back to 1921 and developed later by Ritter and Drake in 1945. The physical phenomenon of the mercury porosimeter is that mercury essentially does not wet surfaces of solids, it has a contact angle (θ) with solids of about 140° and hence an applied force (pressure ($\boldsymbol{p}$)) is needed to push mercury into tubes, or pores with size $\boldsymbol{r}_\mathbf{p}$, as shown below:

$$r_\mathrm{p} = \frac{(2\gamma \cos\theta)}{\Delta p} \tag{4.25}$$

A mercury porosimeter is a high-pressure piece of equipment fitted with the means of applying pressure to a mercury supply, noting the amount of mercury that has entered the sample and the pressure that has been applied. Such instruments are commercially available and are used for meso- and macroporous carbons, and graphitic materials having an open porosity. During an analysis, the largest of the pores fill at the lowest of the pressures. The range of pore sizes for which the porosimeter is best suited extends from about 5×10^3 nm (a macropore) at about 0.15 MPa pressure to about 7.5 nm (a mesopore) at a pressure of 200 MPa. As a safe upper limit for a pore size assessment from an isotherm is about 7.5 nm, these two methods of nitrogen adsorption and mercury porosimetry provide the facility to study the complete size range of mesoporosity in a carbon.

4.9.2 Mercury Porosimetry: Experimental Results

A set of typical data are shown in Figure 4.65 which are plots of nitrogen isotherms (77 K) of carbons from olive stones (HTT 1123 K) activated with carbon dioxide at 1098 K and with steam at 1073 K and 12.3 kPa pressure and in Figure 4.66 which are the corresponding cumulative pore volumes obtained from a mercury porosimeter. The pore filling occurs by reading Figure 4.66 from right to left. These data show that, at the higher burn-offs, carbon dioxide produces more macropores than does steam which enhances mesoporosity.

A comparison of the results of activation of the olive stones (HTT) (1123 K) with carbon dioxide (1098 K) and steam (1073 K) indicates little difference at burn-offs less than ~20 wt%. Above this value, there is considerable divergence.

4.9.3 Mercury Porosimetry: Limitations

Although the wetting angle of 140° is assumed in studies using mercury porosimetry this may not be correct for all materials, the wetting angle varying from 130° to 150°. The search for absolute values may however be in vain, and, as in Figures 4.65 and 4.66 a set of comparative data may be sufficient. However, other problems, which may occur, could be more serious and would need to be checked out. The pressures needed to force mercury into the smallest of the mesopores are high, in the order of 100–200 MPa. As a result, if the sample

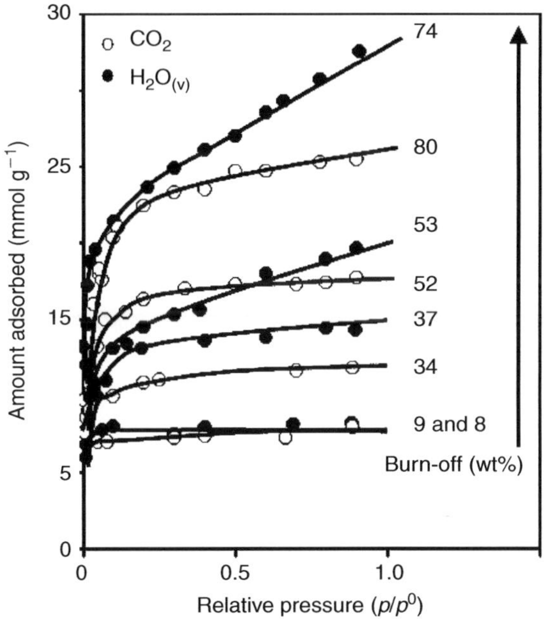

Figure 4.65. Isotherms of the adsorption of nitrogen (77 K) for carbons from olive stones prepared by activation with carbon dioxide and steam (Rodŕiguez-Reinoso, 1989).

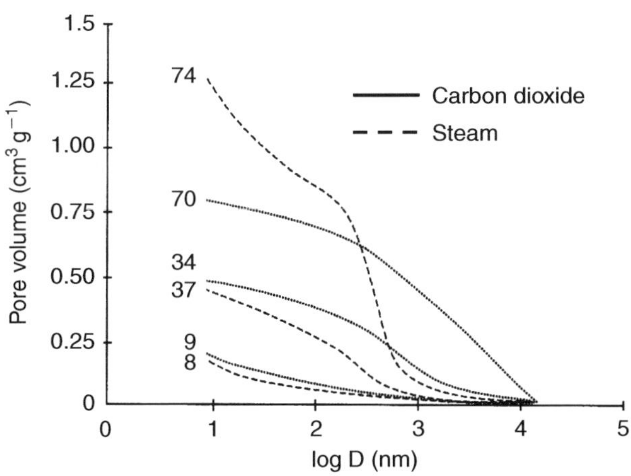

Figure 4.66. Variation of pore volume ($cm^3 g^{-1}$) with log D (pore diameter) for the activation of carbonized olive stones (HTT 1123 K) by carbon dioxide and steam to different extents (a) the isotherms for adsorption of nitrogen and (b) cumulative pore volumes as determined from mercury porosimetry (Rodŕiguez-Reinoso, 1989).

is crushed or compressed in some way then the original mesoporosity will be destroyed and results would be meaningless. Gregg and Sing (1982) report that a soft charcoal was apparently undamaged. However, for a series of graphitized cokes the volume of mercury forced into the porosity exceeded that was accessible to helium indicating that some "closed" porosity had been opened up. A similar effect was noted for some synthetic graphites. For the charcoals, the presence of mercury throughout the bulk of the sample as the pressure is increased may actually reinforce the structure and so prevent its collapse. Such a mechanism may not be possible for the graphitized cokes and graphites which have a narrower PSD of larger mesopores and macropores.

As with the adsorption/desorption of nitrogen at 77 K and the hysteresis effect, so with mercury porosimetry hysteresis is observed as the applied pressure is lowered. This may be caused, not by surface tension considerations as such, but by mercury being trapped within the network of porosity. Trapping could occur if the entry into some of the porosity is via porosity of a smaller size. A comparative study of porous carbons of this *trapping hysteresis* may assist in an understanding of interconnectivities within the porous networks. The effect apparently has been looked upon mainly as a nuisance.

4.10 The Last Word

Chapter 4 has discussed at considerable length the several approaches used to characterize microporosity. From a practical point of view, characterizations can be shortened very considerably.

Equipment. The equipments needed include facility to measure isotherms of nitrogen at 77 K, and carbon dioxide at 273 K, and microcalorimetry to measure enthalpies of immersion.

Informations provided. The nitrogen isotherm provides values of volumes of wider microporosity (0.7–2.0 nm) and volumes within mesoporosity (2–50 nm). The carbon dioxide isotherm provides values of volumes of narrow microporosity (<0.7 nm). The microcalorimetry describes the chemical composition of the carbon surface by immersion in water (polar) and organic liquids of increasing molecular size (molecular sieving).

Note: No mention has been made of the need to calculate a surface area ($m^2 g^{-1}$), the dominant reason being that such a parameter is not needed.

There are about 10^{18} adsorption sites per gram of activated carbon, with no identical sites, the adsorption potential in each being a function of the location of the adsorbate molecule within the confines of the graphene layers.

> *The contents of this chapter summarize the several methodologies used to characterize the porosity of activated carbon. The isotherms of the N_2 (77 K), CO_2 (273 K), H_2O (298 K), making use of DR and BET equations, together with a-plots, in association with enthalpies of immersion, characterize porosity in activated carbon. Equilibria data are complemented by the kinetic data from breakthrough curves.*

References

Atkinson D, McLeod AI, Sing KSW. Physical adsorption and heat of immersion studies of microporous carbons. *Carbon* 1982;20(4):339–341.

Aukett PN, Quirke N, Riddifird S, Tennison SR. Methane adsorption on microporous carbons – comparison of experiment, theory, and simulation. *Carbon* 1992;30(6):913–924.

Barrett EP, Joyner LG, Halenda PH. The determination of pore volume and area distribution in porous solids. I: Computation from nitrogen isotherms. *J Am Chem Soc* 1951;73:373–380.

Barton SS, Harrison BH. Surface studies on carbon: immersion energetics of Spheron 6 in water. *Carbon* 1975;13(1):47–50.

Barton SS, Boulton GL, Harrison BH. Surface studies of graphite: an estimation of the average polarity of the oxygen complexes. *Carbon* 1972;10(4):391–393.

Barton SS, Evans MJB, Hallop E, MacDonald JAF. Acidic and basic sites on the surface of porous carbon. *Carbon* 1997;35(9):1361–1366.

Beck NV, Meech SE, Norman PR, Pears LA. Characterisation of surface oxides on carbon and their influence on dynamic adsorption. *Carbon* 2002;40(4):531–540.

Boehm HP. Surface oxides on carbon and their analysis: a critical assessment. *Carbon* 2002; 40(2):145–149.

Boehm HP, Voll M. Basische Oberflachenoxide auf Kohkenstoffe – I. Adsorption von sauren. *Carbon* 1970;8(2):227–240.

Boudou JP. Surface chemistry of a viscose-based activated carbon cloth modified by treatment with ammonia and steam. *Carbon* 2003;41(10):1955–1963.

Brown WA, Kose R, King DK. Femtomole adsorption calorimetry on single-crystal surfaces. *Chem Rev* 1998;98:797–832.

Calo JM, Hall PJ, Antxustegi M. Carbon porosity characterization via small angle neutron scattering. *Coll Surf A: Physicochem Eng Aspects* 2001;187–188:219–232.

Carrott PJM, Carrott MML. Evaluation of the Stoeckli method for the estimation of micropore size distributions of activated charcoal cloths. *Carbon* 1999;37(4):647–656.

Centeno TA, Marbán G, Fuertes AB. Importance of micropore size distribution on adsorption at low adsorbate concentrations. *Carbon* 2003;41(4):843–846.

Claudino A, Soares JL, Moreira RFPM, José HJ. Adsorption equilibrium and breakthrough analysis for NO adsorption on activated carbons at low temperatures. *Carbon* 2004;42(8–9): 1483–1490.

Cranston RW, Inkley FA. Determination of pore structures from nitrogen adsorption isotherms. *Adv Catal* 1957;9:143–154. Academic Press, New York.

Daley MA, Tandon D, Economy J, Hippo EJ. Elucidating the porous structure of activated carbon fibers using direct and indirect methods. *Carbon* 1996;34(10):1191–1200.

De Salazar CG, Sepulveda-Escribano H, Rodríguez-Reinoso F. The use of immersion calorimetry to evaluate the separation ability of carbon molecular sieves. *Stud Surf Sci Catal* 2000;128: 303–311.

Denoyel R, Fernandez-Colinas J, Grillet Y, Rouquerol J. Assessment of the surface area and microporosity of activated charcoals from immersion calorimetry and nitrogen adsorption data. *Langmuir* 1993;9:515–518.

Díez Díaz-Estébanez M-A, Marsh H. In: *Sciences of Carbon Materials*, Eds. Marsh H and Rodríguez-Reinoso F. Universidad de Alicante, Spain, 2000, pp. 571–593.

Do DD, Nguyen C, Do HD. Characterization of micro-mesoporous carbon media. *Colloid Surf A* 2001;187–188:51–71.

Dollimore D, Heal GR. Pore-size distributions in typical adsorbent systems. *J Colloid Interf Sci* 1970;33(4):508–519.

Domingo-Garcia M, Groszek AJ, López-Garzón FJ, Pérez-Mendoza M. Dynamic adsorption of ammonia on activated carbons measured by flow microcalorimetry. *Appl Catal A: Gen* 2002;233:141–150.

Dubinin MM. The potential theory of adsorption of gases and vapours for adsorbents with energetically non-uniform surfaces. *Chem Rev* 1960;60:235–241.

Dubinin MM, Plavnik GM. Microporous structures of carbonaceous adsorbents. *Carbon* 1968;6(2):183–192.

Dubinin MM, Serpinsky VV. Isotherm equation for water vapour adsorption by microporous carbonaceous adsorbents. *Carbon* 1981;19(9):402–403.

Fanning PE, Vannice MA. A DRIFTS study of the formation of surface groups on carbon by oxidation. *Carbon* 1993;31(5):721–730.

Garrido J, Linares-Solano A, Martín-Martínez JM, Molina-Sabio M, Rodríguez-Reinoso F, Torregrosa R. Use of N_2 vs. CO_2 in the characterization of activated carbons. *Langmuir* 1987;3:76–81.

Gelb LV, Gubbins KE. Characterization of porous glasses: simulation studies, adsorption isotherms, and the Brunauer–Emmett–Teller analysis method. *Langmuir* 1998;14: 2097–2111.

Gibaud A, Xue JS, Dahn JR. A small angle X-ray scattering study of carbons made from pyrolyzed sugar. *Carbon* 1996;34(4):499–503.

Gómez-Serrano V, Álvarez PM, Jaramillo J, Beltrán FJ. Formation of oxygen complexes by ozonation of carbonaceous material prepared from cherry stones – I. Thermal effects. *Carbon* 2002a;40(4):513–522.

Gómez-Serrano V, Álvarez PM, Jaramillo J, Beltrán FJ. Formation of oxygen complexes by ozonation of carbonaceous material prepared from cherry stones – II. Kinetic study. *Carbon* 2002b;40(4):523–529.

Gonzalez JC, Gonzalez MT, Molina-Sabio M, Rodriguez-Reinoso F, Sepulveda-Escribano A. Porosity of activated carbons prepared from different lignocellulosic materials. *Carbon* 1995;33(8):1175–1177.

Gonzalez MT, Rodriguez-Reinoso F, García AN, Marcilla A. CO_2 activation of olive stones carbonised under different experimental conditions. *Carbon* 1997;35(1):159–162.

Gregg SJ, Sing KSW. The physical adsorption of gases by mesoporous solids: The Type IV isotherm (Chapter 3). In: *Adsorption, Surface Area and Porosity*. Academic Press, London, 1982, pp. 111–194.

Griffith M, Hirst W. The heat of wetting of coals in organic liquids. In: *Proceedings of a Conference on the Ultra-Fine Structure of Coals and Cokes*. The British Coal Utilization Research Association, London, 1943, pp. 80–94.

Guillot A, Stoeckli F. Reference isotherm for high pressure adsorption of CO_2 by carbons at 273 K. *Carbon* 2001;39(13):2059–2064.

Gurrath M, Kuretzky T, Boehm HP, Okhlopkova LB, Lisitsyn AS, Likholobov VA. Palladium catalysts on activated carbon supports: influence of reduction temperature, origin of the support and pre-treatments of carbon surfaces. *Carbon* 2000;38(8):1241–1255.

Harkins WD, Boyd GE. The binding energy between a crystalline solid and a liquid: the energy of adhesion and emersion. *JACS* 1942;64(11):1195–1204.

Hoinkis E. Small-angle scattering of neutrons and X-rays from carbons and graphites. *Chem Phys Carbon* 1997;25:71–241. Ed. Thrower PA. Marcel Dekker Inc., New York.

Hoinkis E, Ziehl M. A small angle neutron scattering study of activated carbon fibers. *Carbon* 2003;41(11):2047–2056.

Inagaki M, Nakashjma M. Irreversible adsorption of carbon dioxide gas on an activated carbon. *Carbon* 1992;30(7):1135–1136.

Isirikyan AA, Kiselev AV. The absolute adsorption isotherms of vapors of nitrogen, benzene and *n*-hexane, and heats of adsorption of benzene and *n*-hexane on graphitized carbon blacks. I. Graphitized thermal blacks. *J Phys Chem* 1961;65:601–607.

Johnson DF. Structural studies of PAN-based carbon fibers. *Chem Phys Carbon* 1987;20:1–58. Ed. Thrower PA. Marcel Dekker Inc., New York.

Kaneko K. Specific intermolecular structures of gases confined in carbon nanospace. *Carbon* 2000;38(2):287–303.

Kowalczyk P, Gun'ko VM, Terzyk AP, Gauiden PA, Rong H, Ryu Z, Do DD. The comparative characterization of structural heterogeneity of mesoporous activated carbon fibers (ACFs). *Appl Surf Sci* 2003;206:67–77.

Kraehenbuehl F, Stoeckli HF, Addoun A, Ehrburger P, Donnet JB. The use of immersion calorimetry in the determination of micropore distribution of carbons in the course of activation. *Carbon* 1986;24(4):483–488.

Lavanchy A, Stoeckli F. Dynamic adsorption of vapour mixtures in active carbon beds described by the Myers-Prausnitz and Dubinin theories. *Carbon* 1997;35(10–11):1573–1579.

Lavanchy A, Stoeckli F. Dynamic adsorption, in active carbon beds, of vapour mixtures corresponding to miscible and immiscible liquids. *Carbon* 1999;37(2):315–321.

Li L, Quinlivan PA, Knappe DRU. Effects of activated carbon surface chemistry and pore structure on the adsorption of organic contaminants from aqueous solution. *Carbon* 2002;40(12):2085–2100.

Lozano-Castelló D, Raymundo-Piñero E, Cozorla-Amorós D, Linares-Solano A, Müller M, Riekel C. Characterization of pore distribution in activated carbon fibers by microbeam small angle X-ray scattering. *Carbon* 2002;40(14):2727–2735.

Lozano-Castelló D, Cozorla-Amorós D, Linares-Solano A. Usefulness of CO_2 adsorption at 273 K for the characterization of porous carbons. *Carbon* 2004;42(7):1233–1243.

Leon y Leon CA, Solar JM, Calemma V, Radovic LR. Evidence for the protonation of basal plane sites on carbon. *Carbon* 1992;30(5):797–811.

Mangun CL, Daley MA, Braatz RD, Economy J. Effect of pore size on adsorption of hydrocarbons in phenolic-based activated carbon fibers. *Carbon* 1998;36(1–2):123–129.

Marsh H. Adsorption methods to study microporosity in coals and carbons – a critique. *Carbon* 1987;25(1):49–58.

Marsh H, Wynne-Jones WFK. Surface properties of carbon – I. The effect of activated diffusion in the determination of surface area. *Carbon* 1964;1(3):269–279.

Marsh H, Siemieniewska T. Adsorption of carbon dioxide on carbonised anthracite. Interpretation by Dubinin theory. *Fuel* 1967;46(6):441–457.

Marsh H, O'Hair TE, Wynne-Jones WFK. Oxidation of carbons and graphites by atomic oxygen – kinetic studies. *Trans Far Soc* 1965a;61(2):274–284.

Marsh H, O'Hair TE, Wynne-Jones WFK. Oxidation of carbons and graphites by atomic oxygen – an electron microscope study of surface changes. *Trans Far Soc* 1965b;61(2):285–293.

Molina Sabio M, Muñecas Vidal MA, Rodríguez-Reinoso F. Oxidation in porous texture and oxygen surface groups of activated carbons by oxidation. *Characterization of Porous Solids II*, 1991;62:329–339. Ed. Rodríguez-Reinoso F. Studies in Surface Science, Elsevier.

Montes-Morán MA, Suárez D, Menéndez JA, Fuente E. On the nature of basic sites on carbon surfaces: an overview. *Carbon* 2004;42(7):1219–1225.

Moreno-Castilla C, López-Ramón MV, Carrasco-Marín. Changes in surface chemistry of activated carbons by wet oxidation. *Carbon* 2000;38(14):1995–2001.

Mowla D, Do DD, Kaneko K. Adsorption of water vapour on activated carbon: a brief overview. *Chem Phys Carbon* 2003;28:220–262. Ed. Radovic LR. Marcel Dekker Inc., New York.

Murillo R, García T, Aylón E, Callén MS, Navarro MV, López JM, Mastral AM. Adsorption of phenanthrene on activated carbons: breakthrough curve modelling. *Carbon* 2004;42(10):2009–2017.

Nguyen C, Do DD. The Dubinin–Radushkevich equation and the underlying microscopic adsorption–desorption. *Carbon* 2001;39(9):1327–1336.

Nir I, Suzin Y, Kaplan D. The effect of airflow pattern on filter breakthrough in physical adsorption. *Carbon* 2002;40(13):2437–2445.

Ohkubo T, Kaneko K. Orientated structures of alcohol hidden in carbon micropores with ERDF analysis. *Coll Surf A* 2001;187–188:177–185.

Otake Y, Jenkins RG. Characterization of oxygen-containing surface complexes created on a microporous carbon by air and nitric acid treatment. *Carbon* 1993:31(1):109–121.

Partyka S, Rouquerol F, Rouquerol J. Calorimetric determination of surface areas: possibilities of a modified Harkins and Jura procedure. *J Colloid Interf Sci* 1979;68(1):21–31.

Pels JR, Kapteijn F, Moulijn JR, Zhu Q, Thomas KM. Evolution of nitrogen functionalities in carbonaceous materials during pyrolysis. *Carbon* 1995;33(11):1641–1653.

Perrin A, Celzard A, Albiniak A, Kaczmarczyk J, Marêché JM, Furdin G. NaOH activation of anthracites: effect of temperature on pore textures and methane storage ability. *Carbon* 2004; 42(14):2855–2866.

Pikunic J, Gubbins KE, Pellenq RJ-M, Cohaut N, Rannou I, Guet J-M, Clinard C, Rouzaud J-N. Realistic molecular models for saccharose-based carbons. *Appl Surf Sci* 2002;196:98–104.

Ritter HL, Drake LC. Pressure porosimetry and determination of complete macropore size distribution. *Ind Eng Chem Analyt Ed* 1945;17:782–786.

Roberts BF. A procedure for estimating pore volume and surface area distributions from sorption isotherms. *J Coll Interf Sci* 1967;23(2):266–273.

Rodríguez-Reinoso F. An overview of methods for the characterization of activated carbons. *Pure Appl Chem* 1989;61:1859–1866.

Rodríguez-Reinoso F. The role of carbon materials in heterogeneous catalysis. *Carbon* 1998; 36(3):159–175.

Rodríguez-Reinoso F, within Lozano-Castelló D, Cozorla-Amorós D, Linares-Solano A. Usefulness of CO_2 absorption at 273 K for the characterization of porous carbons. *Carbon* 2004;42(7):1241.

Rodríguez-Reinoso F, Linares-Solano A. Microporous structure of activated carbons as revealed by adsorption methods. *Chem Phys Carbon* 1989;21:1–146. Ed. Thrower PA. Marcel Dekker Inc., New York.

Rodríguez-Reinoso F, Molina-Sabio M. Textural and chemical characterization of microporous carbons. *Adv Coll Interf Sci* 1998;76–77:271–294.

Rodríguez-Reinoso F, López González JD, Berenguer C. Activated carbons from almond shells – I. Preparation and characterization by nitrogen isotherms. *Carbon* 1982;20(6):513–518.

Rodríguez-Reinoso F, López González JD, Berenguer C. Activated carbons from almond shells – II. Characterization of the pore structure. *Carbon* 1984;22(1):13–18.

Rodríguez-Reinoso F, Martin-Martinez JM, Molina Sabio M, Torregrosa R, Garrido-Segovia J. Evaluation of the microporosity in activated carbons by *n*-nonane pre-adsorption. *J Coll Interf Sci* 1985;106(2):315–323.

Rodríguez-Reinoso F, Garrido J, Martin-Martinez JM, Molina M, Torregrosa R. The combined use of different approaches in the characterization of microporous carbons. *Carbon* 1989;27(1):23–32.

Rodríguez-Reinoso F, Molina-Sabio M, Muñecas MA. Effect of microporosity and oxygen surface groups of activated carbon in the adsorption of molecules of different polarity. *J Phys Chem* 1992;96:2707–2713.

Rodriguez-Reinoso F, Molino-Sabio M, Gonzalez MT. Effect of surface groups on the immersion enthalpy of activated carbon in liquids of different polarity. *Langmuir* 1997;13:2354–2358.

Rouquerol F, Rouquerol J, Sing K. *Adsorption by Powders and Porous Solids*. Academic Press, San Diego, 1999.

Sellitti C, Koenig JC, Ishida H. Surface characterization of graphitized carbon fibers by attenuated total reflection Fourier transfer infra-red spectroscopy. *Carbon* 1990;28(1):221–228.

Setoyama N, Kaneko K, Rodríguez-Reinoso F. Ultramicropore characterization of microporous carbons by low-temperature helium adsorption. *J Phys Chem* 1996;100(24):10331–10336.

Silvestre-Albero J, Gómez de Salazar C, Sepúlveda-Escibano A, Rodríguez-Reinoso F. Characterization of microporous solids by immersion calorimetry. *Coll Surf A Physicochem Eng Aspects* 2001;187–188:151–165.

Sing KSW. In: *Surface Area Determination*, Eds. Everett DH, Ottewill RH. Butterworths, London, 1970, p. 15.

Sing KSW. Characterization of porous materials: past, present and future. *Coll Surf A* 2004;241(1):3–7.

Sing KSW, Everett DH, Haul RA, Moscou L, Pierotti RA, Rouquerol J, Siemieniewska T. Reporting physisorption data for gas/solid systems with special reference to the determination of surface area and porosity. *Pure Appl Chem* 1985;57(4):603.

Stoeckli HF. On the theoretical foundation of the Dubinin–Astakov equation. *Carbon* 1981;19(4): 325–326.

Stoeckli HF, Krachenbuehl F. The enthalpies of immersion of active carbons, in relation to the Dubinin theory of volume filling of micropores. *Carbon* 1981;19(5):353–356.

Stoeckli HF, Krachenbuehl F. The external surface of microporous carbons, derived from adsorption and immersion studies. *Carbon* 1984;22(3):297–299.

Stoeckli F, Ballerini L. Evolution of microporosity during activation of carbon. *Fuel* 1991;70(4): 557–559.

Stoeckli F, Centano TA. On the characterization of microporous carbons by immersion calorimetry alone. *Carbon* 1997;35(8):1097–1100.

Stoeckli HF, Krachenbuehl F, Morel D. The sorption of water by active carbons in relation to the enthalpy of immersion. *Carbon* 1983;21(6):589–591.

Stoeckli HF, Rebstein P, Ballerini L. An assessment of microporosity in activated carbons, a comparison of theoretical and experimental data. *Carbon* 1990;28(6):907–909.

Stoeckli F, Hugi-Cleary D, Centeno TA. The characterization of solids by adsorption and immersion techniques and by ATM/STM. *J Eur Ceram Soc* 1998;18:1177–1185.

Vahdat N. Theoretical study of the performance of activated carbon in the presence of binary vapor mixtures. *Carbon* 1997;35(10–11):1545–1557.

Wood GO. Quantification and application of skew of breakthrough curves for gases and vapours eluting from activated carbon beds. *Carbon* 2002;40(11):1883–1890.

Xiu G-h, Li P. Prediction of breakthrough curves for adsorption of lead(II) on activated carbon fibers in a fixed bed. *Carbon* 2000;38(7):975–981.

Yue ZR, Jeang W, Wang L, Gardner SD, Pittman CU. Surface characterization of electrochemically oxidized carbon fibres. *Carbon* 1999;37(11):1785–1796.

Zsigmondy R. Über die structur des gels der kieselsäure. Theorie der Entwässerung. *Z Anog Chem* 1911;71(1):356–377.

CHAPTER 5

Activation Processes (Thermal or Physical)

This Chapter, initially, provides a review of the kinetics and mechanisms of the extensively studied gasification reactions by molecular oxygen. This review provides a basis for an understanding of the mechanisms of thermal activation using carbon dioxide and water vapor. Relationships between carbon structure and reactivity are discussed and differences in activation mechanisms between carbon dioxide and steam, including inhibition reactions, are elaborated upon.

5.1 Thermal Activation: Fundamentals Considerations

Although organic macromolecular systems, as mentioned previously, carbonize to microporous carbons, they do not maximize their porous potential. That is, their adsorption capacity, measured as a micropore volume or surface area, is too low for commercial applications. In addition, there may be a need to widen existing porosity to include wider micropores and some mesoporosity. Further, porosity within the carbon, which is closed to a specific adsorbate, can be opened to allow access to larger adsorbate molecules.

There are two industrial processes used to maximize the adsorption potential of a parent organic material, natural or synthetic. **First**, modifications to a carbon, after its production, make use of the two gasifying agents carbon dioxide and water vapor, either singly or together. These agents "extract" carbon atoms from the structure of the porous carbon according to the following stoichiometric equations, Reactions 5.1, 5.2, describing these endothermic reactions: This method of activation, entirely within the gas phase, is known as *thermal or physical activation*, and is the subject of this Chapter 5.

$$C + CO_2 = 2CO \qquad \text{(Reaction 5.1)}$$

$$C + H_2O = CO + H_2 \qquad \text{(Reaction 5.2)}$$

Second, suitable modifications are made to the carbonization process by additions of such material as phosphoric acid (H_3PO_4), or zinc chloride ($ZnCl_2$) or potassium hydroxide (KOH) and potassium carbonate (K_2CO_3) (see Chapter 6). Such activation processes are said to be *chemical activations*. As some industrial adsorption processes require "fine-tuning" of the porosity of a carbon, it is possible to combine thermal and chemical activation processes to obtain a desired activated carbon.

Mechanisms of activation, to enhance porosity within a porous carbon, differ according to procedures adopted. The use of carbon dioxide or steam, to remove carbon atoms as carbon monoxide, does *not* produce identical results. This chapter elaborates on the differences exhibited by these two gasifying reagents. The chemical mechanisms of thermal activation are *not* as simple as the above stoichiometric equations suggest.

Gasification by molecular oxygen is an exothermic reaction with rates of reaction which are difficult to control. As a result, molecular oxygen, at atmospheric pressure (even air) simply consumes carbon material (causing ignition and burning, even flames). This occurs from the outside surface and does not penetrate into the interior of the carbon. As a result, no enhancement of porosity results. Lower, controlled reaction temperatures with molecular oxygen (about 600 °C) favor the formation of carbon dioxide and higher controlled reaction temperatures (about >900 °C) favor the formation of carbon monoxide, Reactions 5.3, 5.4. Clearly, changes in mechanism are involved with changes in reaction temperature.

$$2C_f + O_2 = 2CO \qquad \text{(Reaction 5.3)}$$

$$C_f + O_2 = CO_2 \qquad \text{(Reaction 5.4)}$$

C_f is the symbol used to describe a carbon atom, free from bonding with surface complexes, and which is "available" for reaction with an oxygen molecule (Bansal *et al.*, 1970).

The reactions of carbon dioxide, steam and molecular oxygen, as well as producing gaseous reaction products, also result in the formation of chemisorbed oxygen, called the *surface oxygen complexes* which possess a broad range of chemical functionality, Section 4.3. These *surface oxygen complexes* exhibit a broad range of chemical stabilities and because of this, the chemistry of these complexes is a function of temperature of formation and subsequent thermal treatments. The formation of these complexes is described by Reaction 5.5, but as no stoichiometry is involved, the equation is purely descriptive. The oxygen is chemisorbed on to the carbon surfaces, where C(O) describes surface oxygen complexes without any reference as to their chemistry.

$$C_f + CO_2 = CO + C(O) \qquad \text{(Reaction 5.5)}$$

During the course of an activation process, these surface oxygen complexes behave both as reaction intermediates in the gasification of the carbon, as well as acting as a retarder (inhibitor) to the reaction rate. During an activation, involving the use of steam (water vapor), hydrogen is similarly chemisorbed on the carbon surface, Reactions 5.2, 5.6, where C(H) describes hydrogen chemisorbed on a carbon surface:

$$C + H_2O = CO + H_2 \qquad \text{(Reaction 5.2)}$$

$$2C + H_2 = 2C(H) \qquad \text{(Reaction 5.6)}$$

The reaction rate of the carbon (also called a char) with carbon dioxide and steam is also inhibited (retarded) by the gaseous reaction products of carbon monoxide and hydrogen.

For the C—CO_2 reaction the mechanism is postulated to be:

$$C + CO_2 \leftrightarrow CO + C(O) \qquad \text{(Reaction 5.7)}$$

in which a surface oxygen complex (C(O)) is initially formed, subsequently becoming stable under the reaction conditions and acting as a retardant by blocking reaction sites:

$$C(O) \leftrightarrow C\text{—}O \qquad \text{(Reaction 5.8)}$$

It may, however, decompose leaving the surface as CO,

$$C(O) \rightarrow CO \qquad \text{(Reaction 5.9)}$$

leaving a "free" surface carbon atom available for reaction.

In this mechanism, inhibition is a consequence of the decrease in the fraction of sites available to carbon dioxide by chemisorption of oxygen as C(O) and by the reaction of the C(O) with CO, as in the reverse of Reaction 5.7.

A similar situation occurs for the C—H_2O reaction, where hydrogen inhibits the reaction rate through the formation of C(H) surface complexes. Because C(H) surface complexes are much more stable than C(O) complexes, inhibition by hydrogen is more marked. The inhibiting effect of hydrogen and carbon monoxide may not necessarily be negative because it has been shown that it prevents the much faster (more than 100 times) reaction between carbon and oxygen (present in industrial, direct-fired furnaces) thus helping to control the gasification and subsequent porosity development.

The reactions between carbons and molecular oxygen have been extensively studied from two points of view, one of chemical kinetics and the other of chemical engineering or fuel combustion engineering. It is the conversion of carbon to oxides of carbon, the carbon being part of a gas (e.g. CH_4) or a liquid (petroleum) or a solid (coal), the exothermicity of these reactions providing, almost exclusively, the energy requirements on a world scale.

5.1.1 Carbonaceous Surfaces

Today it is a prerequisite that a study of the kinetics of gasification (oxidation) of carbons by oxygen or any other gas must include a specification of the carbonaceous surface being gasified. The term "carbonaceous" covers a very wide range of materials from carbon blacks to single-crystal graphite. The kinetic parameters for these gasification reactions are a function of the physical state of the carbon as well as their chemical properties, so requiring a thorough description of the carbons being studied. Much that is already published in the literature is of limited value because so little is known of the carbon of that publication. This aspect is emphasized because the basic concept must not be overlooked that the carbonaceous material is taking part in the reaction and that carbon atoms on the surface of the carbon particle are being removed, individually, from their surface and structural environment, to form carbon monoxide and carbon dioxide in the gas phase. There are relatively few of such gas–solid phase (heterogeneous) reactions.

The structural environments of carbon atoms within carbons vary considerably. As emphasized in Chapter 2, for all carbonaceous materials (soots, carbon blacks, chars, charcoals, carbons, semi-cokes, cokes, etc.) the carbon atoms form a part of hexagonal rings (as in benzene) with these rings joined together to form a defective micro-graphene layer, including five- and seven-membered rings. These layers are *not* small, individualized, perfect sheets of graphite. Rather, the micro-graphene layers contain hydrogen and other heteroatoms such as oxygen, nitrogen, sulfur and phosphorus. Single and multiple atom vacancies exist within the layers which will, in fact, exhibit bending or curvature. Of importance is to emphasize that these micro-graphene layers (with their imperfections) are the "building-bricks" of structure within the non-graphitizable carbonaceous materials. These micro-graphene layers are bonded to each other to establish a three-dimensional network which contains the porosity. There is little parallelism between the graphene layers with spaces between layers constituting the porosity.

Surfaces of carbons, as presented to a gasifying molecule, exhibit variations in structure. As indicated in Chapter 2, the non-graphitizable isotropic carbons have to be distinguished from the graphitizable anisotropic carbons. On heating to ~3300 K the former group of carbons will *not* develop the three-dimensional symmetry of crystalline graphite as monitored by X-ray diffraction, whereas the latter group will do so. Of course, graphitizability has relevance to this chapter because of the activation requirements for pitch-based carbon fibers. The entire mechanism of thermal activation depends on how the gasification gas molecule gains access to an internal carbon atom and the facility at that site for reaction to take place leading to the removal of a carbon atom.

In the non-graphitizable carbons, to a first approximation, it is true to say that there is little symmetry to be found in the relative positions (outside of a graphene layer) of constituent carbon atoms. There is no true parallel stacking, or other forms of crystallographic order. These defective micro-graphene layers are inter-connected together chemically, in units made up of not more than five (say) layers to form the walls of the porosity, and this arrangement maintains the observed three-dimensional disorder (not amorphous) in the carbon structure as well as the marked resistance to change on heating to 3300 K. Thus, in non-graphitizable carbons, the dimension of structural order is the dimension of the defective micro-graphene layer, that is nanometers. It is within the confined space of the porosity created between these defective micro-graphene layers that the gasification gases have to extract a carbon atom from a layer.

Alternatively, in graphitizable carbons, the constituent molecules are stacked approximately parallel to each other over very much longer distances, hundreds of nanometers and micrometers. Accordingly, on heat-treatments to 3300 K it is the small movements of carbon atoms (distances of nanometers) which bring about the stacking sequences approximating to that of graphite.

Generally, non-graphitizable carbons are formed (with few exceptions) from non-melting parent substances (macromolecular), for example coconut shell and thermo-setting resins. The macro-structure within such carbons generally follows closely the original structures of the cellulose and lignitic systems or of the synthetic polymers. Such carbons often contain appreciable internal surface area (varying from 1000 to 3500 $m^2 g^{-1}$, dependent on the method of measurement) in the microporosity associated with the irregular arrangements of the defective micro-graphene layers of the carbon. The changes produced on activation are usually quite specific to that carbon.

However, the stacking order of the graphitizable carbons from pitch is established *via* the formation and stacking of lamellar, nematic liquid crystals and mesophase from the isotropic fluid phase of the pitch carbonization (Section 2.7), such graphitizable carbons having relatively low surface areas ($<\sim 5\,m^2 g^{-1}$). Hence, a gasifying gas has almost zero opportunity to penetrate the interior of a carbon and gasification will dominantly take place at external surfaces.

A significant literature exists of studies of oxidation of synthetic (poly-crystalline) graphites and natural, single-crystal graphite. Here, of course, we are dealing with a highly ordered form of carbon and it is possible to monitor oxidation processes on defined crystallographic planes of the graphite (basal planes or prismatic edges). These graphites are of very low internal surface area such that all oxidation processes occur on the geometric external surfaces of the graphite, unlike the non-graphitizing carbons (e.g. heat treatment temperature (HTT) of <1300 K) where extensive internal gasification is possible and which cannot be observed.

Thermal (physical) activation processes are unusual among chemical reactions in that the solid phase is a reactant. Unlike catalytic processes which also involve a solid phase, the catalyst surface essentially is unchanged. The gasifying reagents "pluck" a carbon atom from the surface of solid carbon. How they do this and from which part of the carbon surface the atom is taken from is an essential discussion item in activation processes. The gasifying agent does not gasify the first carbon atom it encounters on colliding with the carbon surface. There may be as many as 10^{17} encounters before a carbon atom is gasified. There is a similar number of molecules of carbon dioxide within the porosity of a carbon, during gasification, and of molecules of adsorbed nitrogen at 77 K and 0.1 MPa. Clearly, a selection process of some sensitivity is occurring. To put it another way, not all carbon atoms within a carbonaceous material are of equal reactivity.

An insight can be gained into those factors which operate within a porous carbon during an activation process, by making microscopy studies of surfaces of graphitic materials because by so doing, these factors can be imaged directly. It is not possible to do this within the interior of a porous carbon during its activation. The assumption is made here that what happens with a graphitic surface happens also, more or less, on a non-graphitic surface. Later, it is shown that this assumption can be justified.

Mechanisms of reactions with molecular oxygen are well understood because they have received considerable research investment. There exist similarities of reactions with molecular oxygen and reactions with carbon dioxide and steam (water vapor). Initially, it is irrelevant to discuss in some detail the carbon-molecular oxygen reaction, but, as is clarified later, this discussion also provides meaningful clues as to how carbon atoms are extracted in a gasification process by carbon dioxide or steam.

The following sections of this chapter describe the *chemical kinetics* of gasification processes and indicate that the reaction sequences are much more complex than given by the stoichiometric Reactions 5.1, 5.2. Subsequently, *topographical kinetics*, also called *reaction anisotropy*, is introduced and these explain how and why all carbon atoms of a carbonaceous material are not equally reactive, that is selective gasification occurs. Without such selective gasification, the process of thermal activation of carbon would not occur and there would be no removal of those carbon atoms which result in pore creation and pore widening.

Selective reactivity is a necessary prerequisite for thermal activation.

5.1.2 Thermo-chemistry of Gasification Reactions

Gasification of a char is mainly carried out at 800–1000 °C with carbon dioxide, steam, or a mixture of both. As stated above, oxygen is not normally used as an activating agent because the carbon–oxygen reaction is highly exothermic and this makes the reaction impossible to control unless extremely low partial pressures of molecular oxygen are used. The reactions of carbon with carbon dioxide and steam are endothermic and easy to control:

$$C + CO_2 \rightarrow 2CO \quad \Delta H = +159\,kJ\,mol^{-1} \qquad \text{(Reaction 5.1)}$$

$$C + H_2O \rightarrow CO + H_2 \quad \Delta H = +117\,kJ\,mol^{-1} \qquad \text{(Reaction 5.2)}$$

Both reactions are important, even for activation with steam, because at the high temperatures used for activation, the water gas shift reaction, catalysed by the surface of the carbon, is at equilibrium:

$$CO + H_2O \leftrightarrow CO_2 + H_2 \quad \Delta H = -41\,kJ\,mol^{-1} \qquad \text{(Reaction 5.10)}$$

The endothermic character of the gasification reactions with carbon dioxide and steam facilitates the accurate control of experimental conditions in the furnace, but it also makes necessary the use of direct heating in order to maintain reaction temperatures. The heat supply is generated by introducing controlled amounts of air into the furnace to burn natural gas (when added) and the gases produced during activation (CO and H_2):

$$CO + \tfrac{1}{2}O_2 \rightarrow CO_2 \quad \Delta H = -285\,kJ\,mol^{-1} \qquad \text{(Reaction 5.11)}$$

$$H_2 + \tfrac{1}{2}O_2 \rightarrow H_2O \quad \Delta H = -238\,kJ\,mol^{-1} \qquad \text{(Reaction 5.12)}$$

The combustion of the two inhibitors not only decreases their concentration but also increases the partial pressure of the two activating agents.

The analysis of the reactions involved in thermal activation indicates that, independent of the activating agent used, there is burning of carbon, according to the following reaction:

$$C + O_2 \rightarrow CO_2 \quad \Delta H = -406\,kJ\,mol^{-1} \qquad \text{(Reaction 5.4)}$$

In some of the activated carbon industries the heat supply is ensured by introducing a mixture of flue gas and steam (as activation agent), and consequently there is a combined activation by carbon dioxide and steam. In other industries the heat supply is ensured by introducing the gases evolved during carbonization into the activation furnace to substitute for natural gas. As the burning of some carbon cannot be avoided, this reduces the yield in respect to the reaction carried out in indirectly heated furnaces.

5.2 Mechanisms of the Carbon-Molecular Oxygen Reaction

5.2.1 Intermediate Stages

A discussion of this reaction provides a detailed insight as to what happens during gasification by carbon dioxide and steam. Chemical reactions, within molecular-sized space, differ from chemical reactions without space restrictions.

Molecular oxygen reacts with carbons to produce two gaseous products, namely carbon monoxide and carbon dioxide:

$$2C_f + O_2 = 2CO \qquad \text{(Reaction 5.3)}$$

$$C_f + O_2 = CO_2 \qquad \text{(Reaction 5.4)}$$

Where C_f = free site on the carbon surface where a gasification reaction is considered to be possible.

Note that in both reactions the oxygen molecule must dissociate, prior to the formation of the carbon monoxide and the carbon dioxide. It is intriguing to know, in the case of Reaction 5.4, how a carbon atom becomes centred between the two oxygen atoms of the oxygen molecule to form carbon dioxide (O—C—O).

The kinetic parameters of this reaction of molecular oxygen with carbons are extremely sensitive to the limitations imposed by energy and mass transfer processes. As the reaction is exothermic, an ignition of carbon at about 700 K quickly becomes the combustion of carbon at 1600 K. Studies primarily designed to understand the kinetics and mechanisms of the carbon–oxygen reaction must be carried out either at low pressures of oxygen or with oxygen diluted with helium or argon. That is, the extent of the reaction has to be severely limited. Reactions with carbon dioxide and steam, both endothermic reactions, do not have this problem. Temperature control for an isothermal study must be rigorous. Therefore, it is an essential characteristic of carbon-molecular oxygen reaction that it must be studied under conditions which are very far removed from industrial operations. This is not an academic nicety but a rigorous requirement to obtain meaningful data for both chemical and topographical kinetics (reaction anisotropy).

If a temperature control is not maintained then combustion (burning) occurs where rates are controlled by rates of transfer of oxygen to the carbon surface.

Without doubt, the most significant advances in our knowledge of the mechanism of reaction of molecular oxygen with carbon substance have come from the research groups working with Professor PL Walker Jr. in Pennsylvania State University, Pa, USA. To gain meaningful data, the studies were limited to only one form of carbon, and a rather unusual one too, namely *Graphon*, that is the graphitized Spheron 6 carbon black (without microporosity). Graphon was selected because of its chemical purity (free from inorganic impurities) and because it could present a graphitic surface to the molecular oxygen, the surface area being of the order of $90\,m^2\,g^{-1}$, so permitting a measurable rate of reaction and to include a study of the properties of surface oxygen complexes. It is most unlikely if comparable kinetic data could have been obtained should the Graphon have been replaced by another form of carbonaceous substance, for example a non-graphitic carbon.

Thus, a base line is established from which to study the non-graphitizable activated carbons.

Before discussing in detail the mechanistic stages of the reaction, three important aspects of the carbon-molecular oxygen reaction must be looked into initially. **One**, the reaction is inhibited by carbon monoxide. **Two**, there is a third reaction product, as well as carbon dioxide and carbon monoxide, and that is the surface oxygen complexes, C(O). These chemisorbed oxygen species may be formed without gasification of the carbon, at lower temperatures, even at room temperature on outgassed surfaces. Some of these surface oxygen complexes may be reaction intermediates. Therefore, discussions of the kinetics of oxidation must include the role of surface oxygen complexes. **Three**, there is the concept that surface oxygen complexes are *mobile* over the carbon surface. The concept of surface mobility is established in catalysis chemistry. The concept, when applied to the gasification reactions, is essential to the derivations of mechanisms from the observed kinetics.

Thus, the following reaction stages suggest mechanisms of formation of carbon monoxide and carbon dioxide:

$$C_f + O_2 \rightarrow C(O_2) \quad \text{or} \quad C(O_2)_m \qquad \text{(Reaction 5.13)}$$

$$C(O_2)_m \rightarrow C(O) + C(O)_m \quad \text{and/or} \qquad \text{(Reaction 5.14)}$$

$$C(O)_m + C(O)_m \quad \text{or/and} \qquad \text{(Reaction 5.15)}$$

$$C(O) + C(O) \qquad \text{(Reaction 5.16)}$$

$$C(O) \rightarrow CO \qquad \text{(Reaction 5.17)}$$

$$C(O)_m \rightarrow CO \qquad \text{(Reaction 5.18)}$$

$$C(O)_m + C(O)_m \rightarrow C_f + CO_2 \qquad \text{(Reaction 5.19)}$$

$$C(O)_m + C(O) \rightarrow C_f + CO_2 \qquad \text{(Reaction 5.20)}$$

$$CO + C(O) \rightarrow C_f + CO_2 \qquad \text{(Reaction 5.21)}$$

$$CO + C(O)_m \rightarrow C_f + CO_2 \qquad \text{(Reaction 5.22)}$$

$$O_2 + 2C(O) \rightarrow 2CO_2 \qquad \text{(Reaction 5.23)}$$

where
C_f = free site on the carbon surface where reaction is possible
$C(O_2)$ = chemisorbed, localized molecular oxygen
$C(O_2)_m$ = chemisorbed, mobile molecular oxygen
$C(O)$ = chemisorbed, localized atom of oxygen
$C(O)_m$ = chemisorbed, mobile atom of oxygen.

The above sequence of reactions indicates how carbon monoxide and carbon dioxide may be formed.

There is abundant evidence for the formation of the surface oxygen complexes C(O) which are stable at the temperature of reaction and under high vacuum at the reaction temperature.

5.2.2 Mobile Surface Oxygen Complexes

It is established that surface oxygen complexes become progressively unstable with increasing temperature such that at ~1300 K, under high vacuum, the surfaces of carbon are "out-gassed". This implies that the general formulae C(O) represents surface oxygen complexes with a range of thermal stabilities associated with different environments within the carbon surface. As carbon atoms attached to the surface oxygen complexes are removed by out-gassing as carbon monoxide and dioxide, so the immediate environment of surface carbon atoms, still bonded to oxygen, changes and so will their "reactivity". Such changes make it difficult to attempt correlations of thermal stability of surface oxygen complexes with surface structure of the carbon atoms. The concept of change in the energetics of carbon surfaces with gasification, probably including *mobility of carbon atoms*, has also to be accepted.

As reaction proceeds between the carbon surface and molecular oxygen there is established a standing concentration of these surface oxygen complexes. These are stable under the conditions of reaction (but see below) and take no further part in the reaction now being a reaction product rather than an intermediate. The remainder of these complexes are of a transitory nature, are probably both mobile and localized, and lead to the formation of gaseous carbon monoxide and carbon dioxide as listed in sequences, 5.13–5.23. The residence time of these complexes may be relatively short (<1 s) when they are easily removed from the surface of carbon by out-gassing at reaction temperatures. Thus, these surface oxygen complexes may be categorized as further:

$C(O)_s$ = chemisorbed, localized, stable atom of oxygen
$C(O)_r$ = chemisorbed, localized, reactive atom of oxygen
$C(O)_{ms}$ = chemisorbed, mobile, stable atom of oxygen
$C(O)_{mr}$ = chemisorbed, mobile, reactive atom of oxygen.

It is now possible to "visualize" a sequence of events, leading from the collision of an oxygen molecule with the carbon surface to the evolution of gaseous oxides of carbon. The oxygen molecule arrives on the carbon surface, it may be reflected away, or it may form a bond to the surface (nature not specified) enabling it to be mobile over the surface. At some site, where the topography for reaction is suitable (otherwise known as an activation entropy) the oxygen molecule bonds itself more firmly to the surface. The energy released by the formation of two carbon–oxygen bonds from one oxygen molecule permits the dissociation of the molecule into two atoms of oxygen both of which remain bonded to the carbon surface. The chemisorbed oxygen atom may then participate in one (of at least five) possible reaction sequences:

(i) it may remain at the site of dissociation to become a stable surface oxygen complex ($C(O)_s$);

(ii) it may migrate to another site where it also becomes a stable surface oxygen complex ($C(O)_s$);

(iii) it may remain at the site of dissociation to await further reaction ($C(O)_r$);

(iv) it may become mobile until it reacts ($C(O)_{mr}$);

(v) it may remain mobile but unreactive ($C(O)_{ms}$).

The reactive surface oxygen complexes may decompose "spontaneously" to give carbon monoxide, Reactions 5.17 and 5.18. The complexes may interact with each other to give gaseous carbon dioxide, Reactions 5.19, 5.20, or the complexes may interact with gaseous carbon monoxide or molecular oxygen also to form carbon dioxide, Reactions 5.21–5.23.

The relative or absolute values of the reaction rate constants of the above reactions are not known in detail. How structures of the carbon surfaces influence these reaction rate constants are described below.

As reaction proceeds, with the progressive removal of carbon atoms from the surface of the carbon, so changes occur in the physical and electronic structure of the carbon surface. This implies that the progressive oxidation of a carbon surface causes changes in the "strength" of bonding of surface oxygen complexes. The point to be made is that the chemistry of gasification processes, at the carbon surface, is in no sense constant during the progress of the reaction. It is thus highly probable that the relative values of reaction rate constants of the above reaction sequences change during the progress of the reaction. To complicate further this description, the level of "excitation" of the carbon surface may not be constant. This implies that the "reactivity" of a carbon atom in the carbon surface (should it not be gasified) may not be the same at the beginning of the reaction, during the reaction or when the reaction is stopped. Some relaxation phenomenon may be occurring. "Reactivity", in this sense, may be associated with the ability of the carbon atom to bring about the dissociation of the oxygen molecule, to form carbon monoxide or carbon dioxide.

5.2.3 *Use of Isotopic Oxygen: $^{16}O_2$ and $^{18}O_2$*

A significant understanding to gasification reactions came from the use of isotopic oxygen $^{18}O_2$ in conjunction with $^{16}O_2$, in particular to establish the mobility of the surface oxygen complex. To establish mobility, **Graphon** was reacted with a known mixture of $^{18}O_2$ and $^{16}O_2$. The logic behind these experiments was that if the same two oxygen atoms of the oxygen molecule became the two oxygen atoms of the carbon dioxide molecule, then what would be expected would be $^{18}O—C—^{18}O$ and $^{16}O—C—^{16}O$ in the same proportions as the original isotopic mixture. However, the results obtained did not show these proportions but rather they showed the consequences of scrambling reactions, that is a mixture of ^{16}O—C—^{18}O with ^{18}O—C—^{18}O and ^{16}O—C—^{16}O, as may be predicted statistically. In other words, an oxygen molecule dissociates on the carbon surface with the two atoms of oxygen, mobile over the surface, moving apart to independent destinies.

The behavior of these surface oxygen complexes, on a **porous carbon**, was studied using isotopic oxygen $^{18}O_2$ and $^{16}O_2$. About half of the surface of this non-graphitizable porous carbon from polyvinylidene chloride was covered with surface oxygen complexes from $^{18}O_2$, and afterwards the coverage was completed with $^{16}O_2$. Using the principle that the carbon surface is heterogeneous and that mobility is an occurrence, then the surface sites of highest adsorption energy will be occupied first with the surface sites of lowest adsorption energy being occupied last by the surface oxygen complexes. Upon desorption of these complexes, as carbon monoxide and carbon dioxide, by raising the temperature of the carbon in high vacuum and analyzing the desorbed gases with a mass spectrometer, it was found that the ratio of $^{16}O/^{18}O$ in the product gases was initially greater than one, indicating that the oxygen which was adsorbed first was desorbed last. If adsorption occurred at the site of collision of the oxygen molecule with the carbon surface, and there was no subsequent mobility, with desorption on heating coming from the same site, then the ratio of $^{16}O/^{18}O$ in the product gases would have equaled unity and would not change with progressive outgassing. Thus, the surface of a non-graphitizable carbon permits surface mobility of oxygen atoms in the same way as did the graphitic Graphon surface.

As mentioned above, the carbon-molecular oxygen reaction is inhibited by carbon monoxide, when, for example, this gas is mixed with the reacting molecular oxygen. It was at one time postulated that adsorption of carbon monoxide on the reactive sites was responsible for this inhibition, that is:

$$C_f + CO \rightarrow C(CO) \qquad \text{(Reaction 5.24)}$$

However, it is now suggested that this is not an important reaction. Rather, there is inhibition by Reactions 5.21, 5.22. Instead of blockage of reactive sites, there is the removal of surface oxygen complexes, by the carbon monoxide, as carbon dioxide with the carbon atoms, to which the oxygen was attached, remaining in the carbon surface and *not* gasified (hence the inhibition). Hydrogen is also an inhibitor of this reaction, the mechanism here probably incorporating both chemisorption of the hydrogen and the removal of surface oxygen complexes as water vapor.

5.2.4 Variation of the CO/CO_2 Partial Pressure Ratio in Product Gases

Under the rather strict conditions of the fundamental study, further experiments consolidate the above mechanisms of what happens when oxygen molecules react with a carbon surface. To do this, changes in CO/CO_2 ratio (of evolved gases) were monitored as the pressure of the reacting oxygen was increased from 3.3 Pa to 0.1 kPa. The results are as in Figure 5.1.

The logic is that increasing the oxygen pressure should also increase the standing concentration of surface oxygen complexes. As the rate of formation of carbon dioxide is

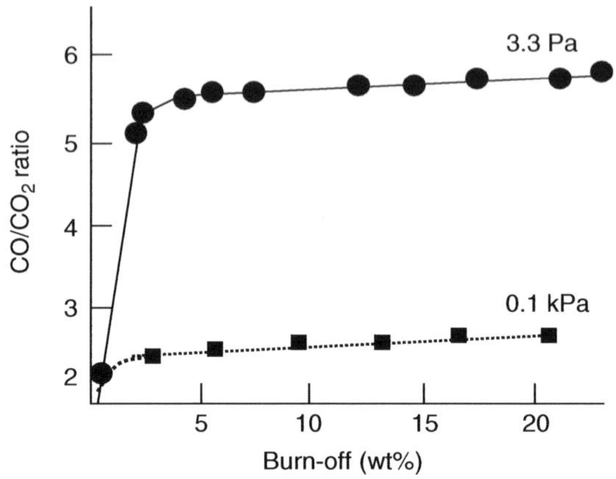

Figure 5.1. Oxidation of Graphon with molecular oxygen at 898 K. Variation of CO/CO_2 partial pressure ratio with burn-off (wt%) of Graphon, using two initial pressures (Pa) of molecular oxygen (adapted from Phillips *et al.*, 1970).

obviously a function of this standing concentration and hence the collision frequency between mobile species, any increase should decrease the observed resultant CO/CO_2 ratio. Figure 5.1 shows the decrease in the CO/CO_2 ratio from about 5.5 to 2.8 on increasing the oxygen pressure by a factor of 30 from 3.3 Pa to 100 Pa (0.1 kPa). This is confirmation of Reactions 5.19, 5.20 occurring between surface oxygen complexes.

A further experiment was designed to bring about changes in the CO/CO_2 ratio by changing the surface conditions of the Graphon. After the Graphon surface was initially "roughened," in a prior oxidation, by producing pits and other irregularities, then the facility for surface mobility is retarded. Less mobility should lead to a higher CO/CO_2 ratio.

Figure 5.2 shows the variations in the CO/CO_2 ratio by varying the concentration of surface oxygen complexes at a constant extent of carbon removal (burn-off), several burn-offs being studied. At constant burn-off it is seen that the CO/CO_2 ratio decreases with increasing concentration of surface oxygen complexes as also predicted in Figure 5.1. However, at constant concentration of surface oxygen complex it is seen that the CO/CO_2 increases with increasing extent of burn-off. This is consistent with the view that increasing extents of burn-off are associated with increasing extents of surface damage (roughening) and hence decreasing facility for mobility of the complexes and formation of carbon dioxide.

5.2.5 Gasification Reactions in Nano-sized Spaces

Gasification reactions occur in the smallest of all spaces, within volumes of $<1.0\,nm^3$. It is established that reactions under these conditions differ from those occurring in macro-volumes. Such porosity encloses enhanced dispersion forces, creating quasi-high pressure effects and a quasi-lowering of temperature. These effects are known for adsorption in porosities of dimensions $<1.0\,nm$, as for the adsorption of helium at 4 K. The properties

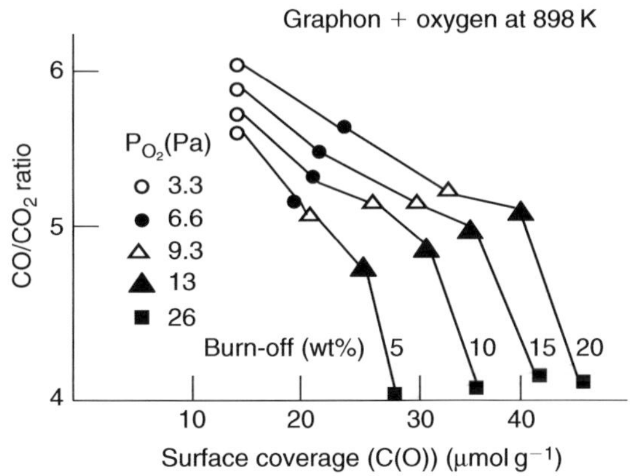

Figure 5.2. Oxidation of Graphon with molecular oxygen. Variation of CO/CO_2 partial pressure ratio with extent of coverage by surface oxygen complexes ($\mu mol\,g^{-1}$), from initial pressure of molecular oxygen (Pa) and burn-off (wt%) of the Graphon (adapted from Phillips *et al.*, 1970).

of water change when subject to these intense molecular fields. Also, levels of hydration of salt solutions change when in nm-sized porosities. It is therefore highly probable that the conditions in a micropore (volume of $<1.0\,nm^3$) will modify reaction mechanisms as carbon dioxide or steam react to remove carbon atoms from the surface of that $<1.0\,nm^3$ volume element. That this occurs is demonstrated below.

A pure porous carbon (non-activated) was prepared from polyfurfuryl alcohol (PFA), HTT 1073 K, with a surface area of $\sim 400\,m^2\,g^{-1}$ in narrow microporosity. When reacting with oxygen at 913 K and a pressure of 53 Pa, the CO/CO_2 ratio was about 20. This ratio compares with a ratio of 2.8 at 898 K under an oxygen pressure of 100 Pa, using the "open" surfaces of Graphon. These open surfaces (non-microporous) facilitate movement of mobile surface oxygen complexes, so facilitating preferential CO_2 formation.

Next, this PFA carbon was slowly gasified (about 1 wt% burn-off per hour) in a separate experiment with carbon dioxide to about 20 wt% burn-off. This treatment widened the porosity of the carbon (as seen from a CO_2 isotherm (273 K)). Upon continuation of the oxidation with molecular oxygen there occurred a marked reduction in the CO/CO_2 ratio from 20 to about 1.0. This marked effect upon CO/CO_2 ratios, as a result of a change in the porosity of a carbon is clearly seen in Figures 5.3 and 5.4. So, not only is mobility a requisite for formation of carbon dioxide, so also is the necessary "space" for the relative orientations of the O, C and O to form the O—C—O compound.

That is, there is stereo-chemical control over reaction products.

It was also observed that the CO/CO_2 ratio increased with increasing reaction temperature, all other conditions being kept constant. With increasing reaction temperature there is a decrease in the standing concentration of surface oxygen complexes. Such a decrease does not facilitate the formation of carbon dioxide and hence the CO/CO_2 ratio increases.

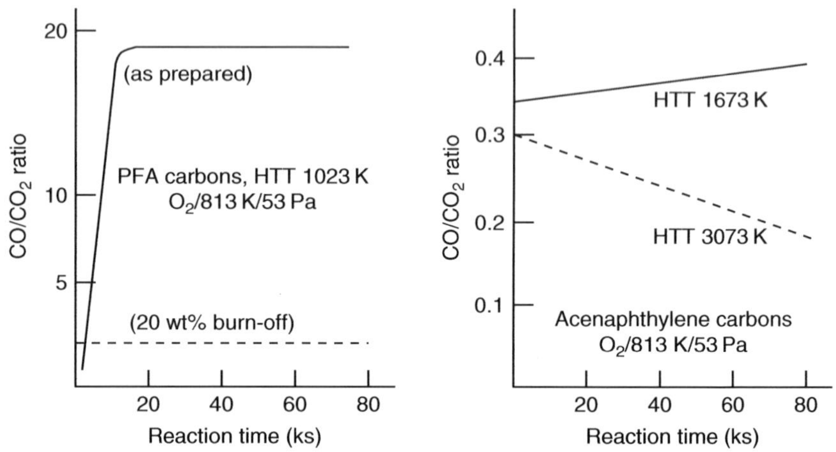

Figures 5.3 (left) and 5.4 (right). Oxidation by molecular oxygen of PFA carbon, HTT 1073 K and of acenaphthylene carbons. Variation of CO/CO_2 partial pressure ratios with reaction time (ks), with burn-off (wt%) (porosity) and with heat treatment temperature (K) (Marsh *et al.*, 1981).

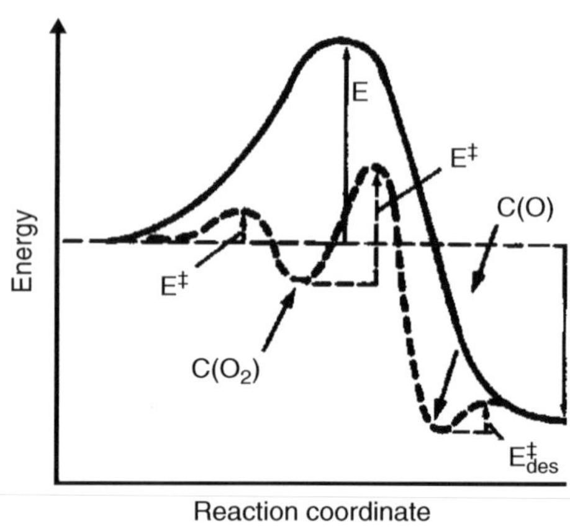

Figure 5.5. Diagram of energy profile (dotted line) as molecular oxygen reacts with a surface of carbon to form surface oxygen complexes, carbon monoxide, and carbon dioxide. E‡ describes activation energies of intermediate stage reactions.

5.2.6 The Energy Profile

Figure 5.5 outlines the energy profile for the Reactions 5.13–5.23, as shown by the dashed curve. The energy profile describes the energy content of the reacting system as it passes through the several transition states leading to the reaction products. There is a small activation energy for the chemisorption of the molecular oxygen followed by a larger activation energy for its dissociation into single atoms of chemisorbed oxygen. This stage probably

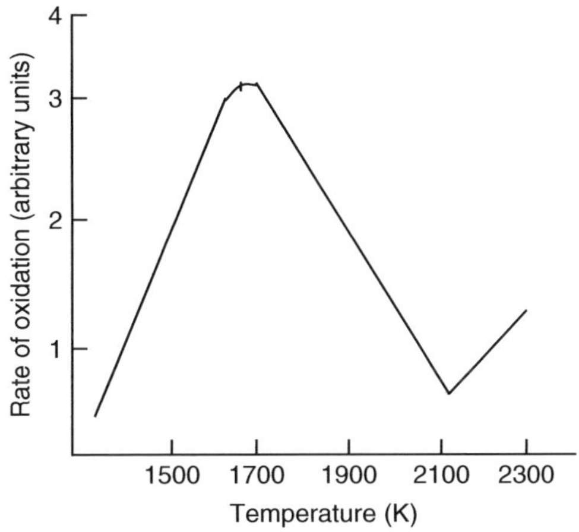

Figure 5.6. Diagram indicating how the rate of the graphite-molecular oxygen reaction changes with increasing reaction temperature, 1300–2300 K (Lewis (1970)).

does not have a single activation energy, but a spectrum of intermediate activation energies dependent on the environment, within the carbon lattice, of the reacting carbon atom. One may speculate that after dissociation of the oxygen molecule and formation of C(O) the energy made available by this process is not lost (otherwise a stable $C(O)_s$ is formed) but retained by the $C(O)_{mr}$ species to enable the carbon–carbon bonds to be broken and for the carbon of the $C(O)_{mr}$ to be extracted from the basal plane as carbon monoxide or carbon dioxide.

5.2.7 Temperature Coefficients of Reaction Rates

Figure 5.6 is a diagram of the changes of rate of the carbon-molecular oxygen reaction for a graphitic carbon, with reaction temperature (under chemical control). At temperatures <1700 K the rate progressively increases with temperature as predicted from the usual application of the Arrhenius equation. But, in the region of about 1700 K it levels off to become temperature independent, and beyond 1700 K to show a negative temperature coefficient, that is the rate of oxidation decreases with increasing temperature. Eventually, beyond about 2100 K, the rate again assumes a positive temperature coefficient. An explanation put forward to explain these effects is that up to about 1700 K any "damage" sustained by the graphite surface during oxidation remains. Beyond 1700 K, because of surface diffusion of carbon atoms, the "damage" of the graphite is "annealed" and the reactivity of the carbon exhibits a marked decrease. Eventually, with increasing temperature, the rate of diffusion of carbon atoms reaches some limiting value and the rate of oxidation begins to increase again with increasing reaction temperature.

This would suggest that changes introduced to the structure of an isotropic porous carbon during a thermal activation process will remain and will not be subsequently annealed out.

5.3 Rates of Gasification Reactions

5.3.1 Diffusion Control of Reaction Rates of Gasification

Control over the rate controlling processes for gasification, as in thermal activation, is of critical importance and an awareness of in which zone the reaction is taking place is crucial (Figures 5.7 and 5.8). These considerations are equally applicable to reactions of carbon with molecular oxygen, carbon dioxide and steam.

It has to be emphasized that if the conditions of oxidation of carbon are *not* those of a fundamental study (e.g. at pressures of about 100 Pa or less) then limitations of energy and mass transfer mean that these effects control or influence the rate of the chemical reaction

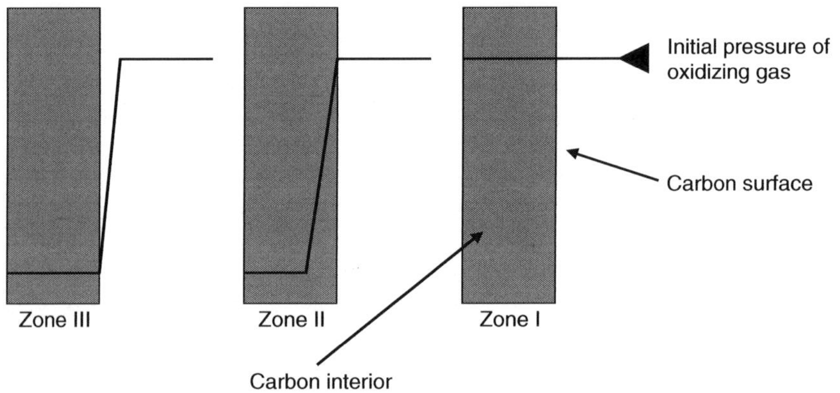

Figure 5.7. Diagram indicating how the concentration (pressure) of a reactant gas (O_2, CO_2, H_2O) varies relative to the interior (Zone I) and exterior (Zone III) of the carbon sample.

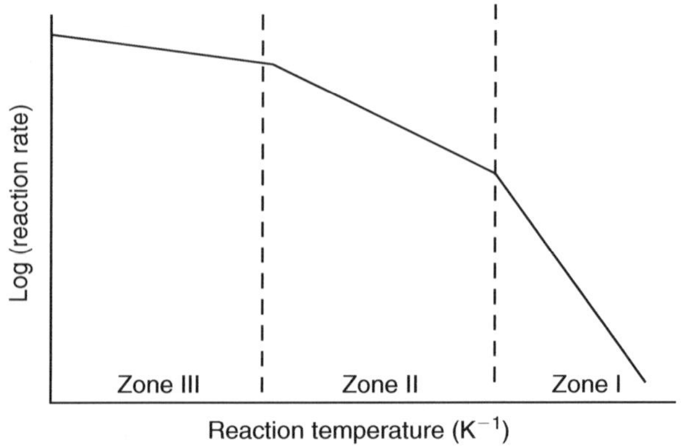

Figure 5.8. Diagram indicating how the rate (log rate) of the reaction of molecular oxygen with carbon varies with reaction temperature (T^{-1}) in Zones I–III of Figure 5.7. Zone I is of chemical control.

and kinetic interpretations become difficult. It is convenient, at this stage, to discuss the implications of these statements.

> *The implications are that meaningful chemical kinetics and mechanisms can be derived only from data obtained from a reaction operating in Zone-I.*

When a heterogeneous reaction proceeds in Zone-I, the rate controlling process is the slow stage of the sequences of the chemistry of the reaction (Figures 5.7 and 5.8). That is, there is equal probability of reaction occurring at the exterior of the surface as there is within the internal porosity. In Zone-II, where mixed conditions operate, all of the reacting gas coming to the carbon surface has reacted before it reaches the centre of the carbon specimen. Thus, an outer section of the carbon has reacted, only. In Zone-III, the reacting conditions are so harsh that all of the reaction has occurred at the outer surface. Rates of reaction (gasification) are thus controlled rates of diffusion of reactants to the surface and rates of diffusion of products away from the surface.

Figures 5.7 and 5.8 illustrate these phenomena. Figure 5.7 shows that the profile of concentration of oxygen relative to the geometry of the solid carbon may remain unchanged in the centre of the carbon (chemical control of rate) or may be reduced to zero at the surface of the carbon (diffusion, or mass transfer control of rate). Figure 5.8 illustrates how activation energies of the reaction correspondingly change as the rate controlling stage change progressively from a chemical to a physical process.

5.3.2 Reactivity of Surfaces During Gasification Reactions

A matter of fundamental importance concerns the state of the carbon surface during the course of a gasification reaction. Can the surface be considered as being unchangeable or can its reactivity (say as a rate of gasification per square meter of surface) vary during the course of the reaction? The results of an experiment to answer this question, by monitoring the reactivity of a carbon surface during the course of a gasification experiment, are as in Figure 5.9.

Carbon (Graphon) was reacted with molecular oxygen, at 900 K and 100 Pa pressure, and after formation of surface oxygen complexes the gas phase was pumped away. The temperature of the Graphon was then progressively raised from 900 K (627 °C) to 1223 K (950 °C), *without further pumping*, so allowing the gases as desorbed from the surface to remain in the reaction vessel. A mass spectrometer detected a rise in partial pressure of carbon dioxide to an out-gassing temperature of about 1000 K (727 °C) when the partial pressure reached a maximum. It then began to decrease as a result:

$$C + CO_2 \rightarrow 2CO \qquad \text{(Reaction 5.1)}$$

Because the rate of increase in out-gassing temperature of the Graphon is constant the shape of this downward curve is indicative of the rate of Reaction 5.1, that is the rate of carbon removal from a surface which is subject to continuous change due to the removal of surface oxygen complexes, that is the "excited" surface described in Section 5.2.4.

These experiments also indicate, for the carbon–carbon dioxide reaction, that the reactivity of the Graphon surface at a time when the surface is active due to removal of surface

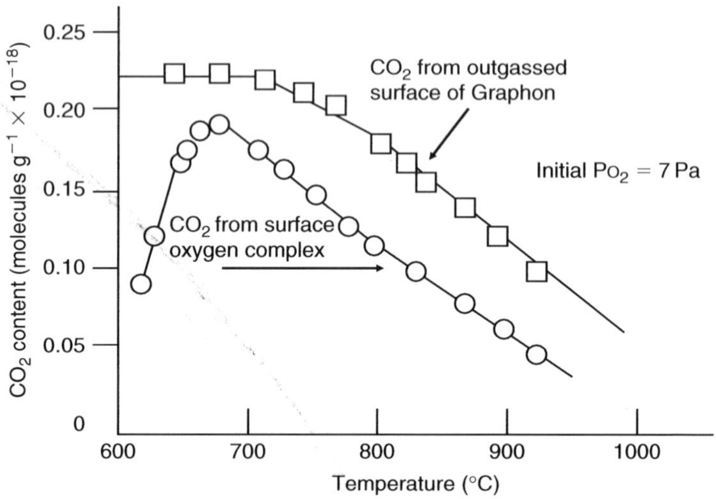

Figure 5.9. Variation of content of carbon dioxide in a system containing Graphon, with increasing reaction temperature, under conditions of removal of surface oxygen complexes (lower curve) and with outgassed Graphon (upper curve) (Phillips *et al.*, 1970).

oxygen complexes, is higher than when the surface has been stabilized by complete outgassing at 1223 K. Here then, we have direct experimental evidence of a factor influencing measured activation energies (See Section 5.2.4), albeit of the carbon–carbon dioxide reaction.

The programmed heating continued to 1223 K when all gases were pumped away; the temperature was now lowered to 900 K (627 °C) (surface stabilized and not "active") and carbon dioxide was introduced to the reaction vessel. Again, the temperature of the Graphon was raised at the same rate and the decrease in partial pressure of the carbon dioxide was monitored. In this experiment, Figure 5.9 shows the removal of carbon dioxide, 700–800 °C, for the outgassed (stabilized surface), upper curve and the removal of carbon dioxide in the initial experiment, lower curve. In this temperature range, the lower curve shows a more rapid loss of carbon dioxide. These experiments indicate that the "reactivity" of the Graphon surface at a time when the surface is "active" due to removal of surface oxygen complexes is higher than when the surface has been stabilized by complete outgassing at 1223 K.

5.3.3 Rate Equation for the Carbon-molecular Oxygen Reaction

This rate equation is rather complex because it has to take into account the above factors which control the rate of reaction:

$$-\frac{\partial c}{\partial t} = \frac{(P_{O_2})^n(\mathrm{ASA})(1-\theta)A\exp(E_A/RT)}{(1-\pi(\theta))} \qquad \text{Equation (5.1)}$$

where

$-\partial c/\partial t$	= rate of oxygen depletion with time (t).
P_{O_2}	= partial pressure of oxygen
n	= the order of reaction with respect to the oxygen.
ASA	= active surface area, which is a fraction of the (total surface area (TSA)) which is reacting.
θ	= the fraction of ASA covered by stable oxygen complexes at time t.
A	= frequency factor
E_A	= activation energy
T	= reaction temperature
R	= gas constant
$(1 - \pi(\theta))$	= participatory factor

ASA ($m^2 g^{-1}$) is defined as that part of the carbon surface which can be covered with stable surface oxygen complexes placed on the surface, from molecular oxygen, at 573 K (300 °C). These conditions are arbitrary, but the temperature of ~573 K is where maximum quantities of surface oxygen complexes are found (see Section 5.7).

5.3.4 Rate Equations for the Carbon–Carbon dioxide and Steam Reactions

The corresponding rate equation for the reaction of carbon dioxide with carbon is as Equation 5.2. This equation indicates that the reaction is inhibited by carbon monoxide which reacts with the surface oxygen complexes to form the reaction reactant, namely carbon dioxide. In addition, as well as being a reactant, leading directly to carbon monoxide, carbon dioxide forms the surface oxygen complexes. These, by blocking reactive sites on the carbon surface, reduce rates of gasification. The effective of increasing the partial pressure of the carbon dioxide is *not* one of progressively increasing the reaction rate. The effect is summarized in Figure 5.10 which is a plot of reaction order against pressure of the reactant. Initially, at low pressures of carbon dioxide (or water vapor), the reaction is first order. However, with increasing pressure of the carbon dioxide, the retarding reactions assume a major importance, causing the order to decrease to $\ll$1.0.

Rate equation: carbon dioxide–carbon reaction:

$$\text{Rate} = \frac{k_1 P_{CO_2}}{1 + k_2 P_{CO} + k_3 P_{CO_2}} \qquad \text{Equation (5.2)}$$

$$C + CO_2 = C + 2CO \qquad \text{(Reaction 5.1)}$$

$$CO + C(O)_m \rightarrow C_f + CO_2 \qquad \text{(Reaction 5.22)}$$

Rate equation: steam-carbon reaction:

$$\text{Rate} = \frac{k_1 P_{H_2O}}{1 + k_2 P_{H_2} + k_3 P_{H_2O}} \qquad \text{Equation (5.3)}$$

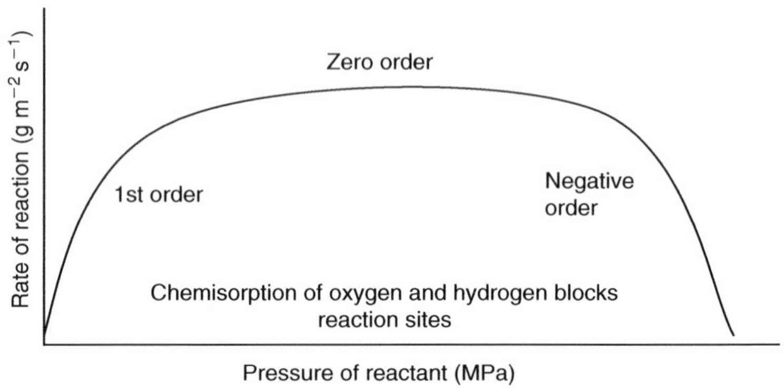

Figure 5.10. Variation of order of reaction with increasing pressure of carbon dioxide or steam to illustrate inhibition of rates of reaction (g m^{2}s^{-1}).

$$C + H_2O = CO + H_2 \qquad \text{(Reaction 5.2)}$$

$$2C + H_2 \rightarrow 2C(H) \qquad \text{(Reaction 5.6)}$$

These rate equations describe the kinetics and mechanisms of the gasification reactions of carbons by carbon dioxide and steam. As such, they describe the processes of activation by carbon dioxide and steam giving insights into mechanisms of carbon removal (the activation process), together with differences between activations by different agents and the effects of inhibition by product gases.

5.4 Carbon Structure and Gasification

5.4.1 Carbon Structure and Activation Energies of Gasification

In all gasification reactions, rates of gasification are dependent on the structure of the carbon, all other factors being normalized. As the structure becomes more ordered, for example as the micro-graphene layers lose some of their defects such as point defects, five- and seven-membered rings, etc., so it becomes more difficult to extract a carbon atom from such micro-graphene layers. The measured activation energies of gasification provide an insight into the structure of these micro-graphene layers, Table 5.1.

Marsh *et al.* (1981) prepared carbons from distilled furfuryl alcohol (FA) (isotropic, non-graphitizable) and high purity acenaphthylene and distilled vinyl acetate (both anisotropic and graphitizable). This was to ensure an absolute minimum of inorganic impurities which, if present, would invalidate kinetic data.

Table 5.1 contains activation energies for the reaction of different carbons with molecular oxygen, at a partial pressure of 1.25 kPa, in the temperature range 750–1000 K. Table 5.1 contains activation energies for the reaction of molecular oxygen with carbons from the same parent carbon, but having increasing HTT. Overall, activation energies range from

Table 5.1. The reaction of carbons of different structures with molecular oxygen, to determine activation energies of reaction (Marsh *et al.*, 1981).

Carbon sample	Activation energy/ $kJ\,mol^{-1}$	$Log_{10}k$
Spectroscopic graphite	226 ± 5	6.7
PFA 1200 K	149 ± 8	4.4
PFA 2100 K	198 ± 30	5.5
Acenaphthylene 1200 K	174 ± 25	4.4
PVA 1200 K	127 ± 12	2.2
PVA 1500 K	177 ± 6	4.5
PVA 1800 K	177 ± 5	4.0
PVA 2100 K	245 ± 13	7.0
PVA 3200 K	248 ± 8	7.0

PFA: polyfurfuryl alcohol; PVA: polyvinyl acetate.

127 ± 12 to $248 \pm 8\,kJ\,mol^{-1}$. Table 5.1 also contains rate constant data, as log k where k is derived from this equation:

$$k = \frac{(P_{O_2})^n (ASA)(1-\theta)A}{(1-\pi(\theta))} \qquad \text{Equation (5.4)}$$

The data of Table 5.1 show, beyond dispute, that the activation energies and pre-exponential terms of the carbon-molecular oxygen reaction are dependent on the structure of the carbon being oxidized. Therefore, unlike the majority of chemical reactions, for example a hydrolysis of an ester, it is not possible to publish one, single, unique value of an activation energy which is common to all studies.

The activation energy obtained from the PFA carbon (isotropic) (PFA 1200 K) is $149 \pm 8\,kJ\,mol^{-1}$ and for the polyvinyl acetate carbon (anisotropic) (PVA 1200 K), is $127 \pm 12\,kJ\,mol^{-1}$. This indicates the surface of the PVA carbon may be more defective than that of the PFA carbon. These values are below values of activation energies of graphitic carbons which are all above $226\,kJ\,mol^{-1}$. This indicates, as clearly as is possible, that no graphitic material is present in these carbons in the form of graphitic microcrystallites as discussed at some length in Chapter 2, Section 2.10.

Figures 5.11 and 5.12 contain the data of Table 5.1. Figure 5.13 is a plot of activation energy (E_A ($kJ\,mol^{-1}$)) against $\log_{10}k$. This is most intriguing relationship because it demonstrates what is termed the *compensation effect*, whereby for such heterogeneous reactions involving gasification, it is *not* possible to enhance the rate by lowering the activation energy and at the same time increasing the rate constant "k". Always, a lowering of an activation energy is compensated for by a decrease in "k" values.

These considerations apply to all gasification reactions, those of oxygen, carbon dioxide, steam, oxides of nitrogen, etc.

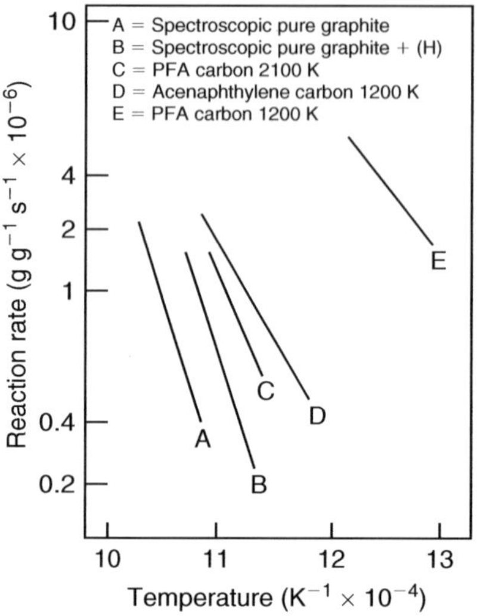

Figures 5.11 Variation of rate of oxidation by molecular oxygen (P_{O_2} = 1.25 kPa) of carbons of different structures with temperature ($T^{-1} \times 10^{-4} K^{-1}$) (Marsh *et al.*, 1981).

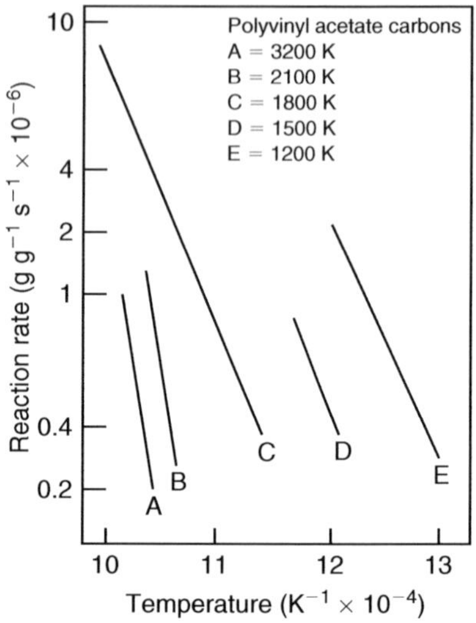

Figures 5.12 Variation of rate of oxidation by molecular oxygen (P_{O_2} = 1.25 kPa) of anisotropic carbons from PVA, of different heat treatment temperature, with reaction temperature (T^{-1} (K^{-1})) (Marsh *et al.*, 1981).

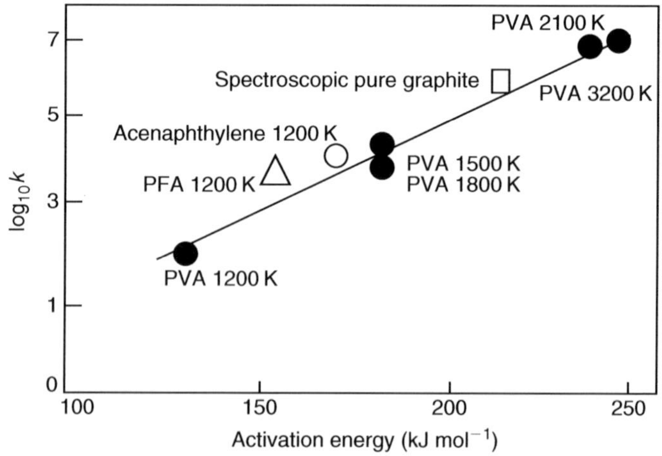

Figure 5.13. The variation of $\log_{10}k$ (k = reaction rate constant) with activation energy for the reaction of molecular oxygen (P_{O_2} = 1.25 kPa) with non-graphitizable and graphitizable carbons, to illustrate the *compensation effect* which occurs in heterogeneous reactions (Marsh *et al.*, 1981).

5.4.2 Catalysis of Gasification Reactions

Reference was made in Section 5.4.1. for the need, when studying activation energies, to use pure carbons, that is carbons with impurity levels of a few ppm. Graphites as used in nuclear reactors and the electro-graphites of the steel industry are quite pure. The regular activated carbons, from wood, from coal, from nutshells and fruit stones, all contain up to several percentage of indigenous inorganic matter (impurities). These inorganic compounds are not inert during thermal activation processes, but influence pore developments. There may be advantages to be gained from these effects.

All the carbon-oxidation reactions involving molecular oxygen, carbon dioxide, steam and oxides of nitrogen are catalyzed by inorganic material contained within the carbon. Not only are the chemical kinetics of the reactions changed by the catalytic process, but the topographical kinetics (reaction anisotropy) are also changed.

The catalytic inorganic material contained within carbons cannot be precisely defined. There exist hundreds of inorganic compounds which, if present in the original carbon, cause catalysis at reaction temperatures. To be more precise, it is the "decomposition" products (the ash) of the inorganic compounds which are catalytic. Synergetic effects caused by mixtures of catalytic material are suspected but not studied systematically.

Of these inorganic materials it is the salts, oxides and metals of the series of alkali metals, alkali earth metals and transition metals which are particularly effective as catalysts. The efficiency of a catalyst is dependent on the size of the catalyst particle and its mode of distribution within the carbon. To emphasize the importance of catalysis it can be said that about 100 ppm of for example a lead salt increases the rate of oxidation of carbon by molecular oxygen by a factor of about 10^4 when compared with a "pure" carbon. It is thus possible to appreciate an earlier statement that it is very difficult to study a carbon gasification reaction which is *not* catalyzed in some way. The preparation of pure carbons must be in conditions as strict as those of an aseptic laboratory or a radio-chemical laboratory.

It is now accepted that the catalyst acts at the position of contact with the carbon. There are *no* induction effects to produce catalysis at a position some distance from the carbon particle. Figure 5.14 models the sequences as a molecule of the reacting gas, be it oxygen,

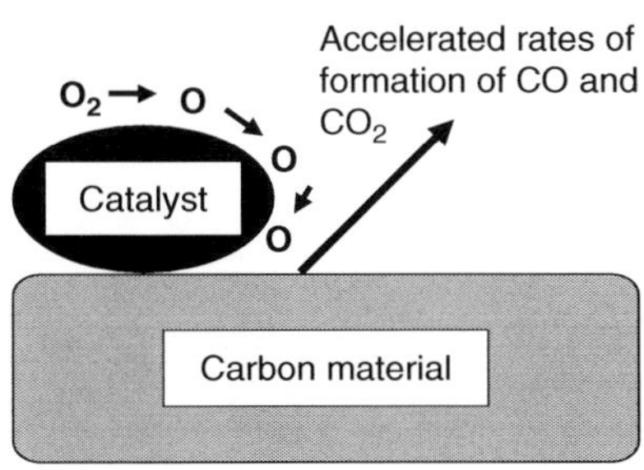

Figure 5.14. Diagram illustrating the transfer of mobile atomic oxygen over the surface of a catalyst to the carbon surface, during catalytic gasification.

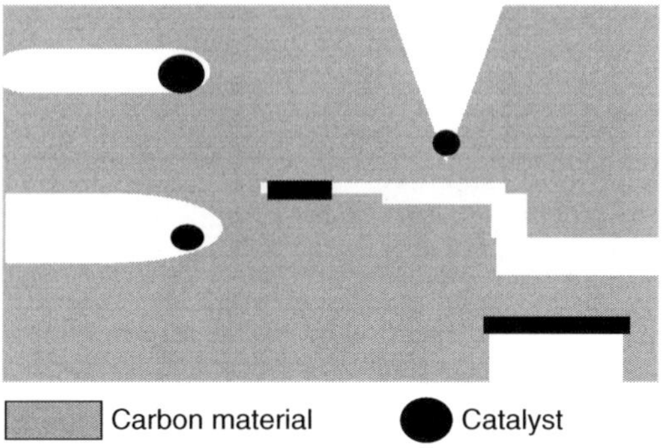

Figure 5.15. Diagram indicating that inorganic particles, acting as catalysts in gasification reactions, are mobile within the carbon material producing channels or pores the size of which approximate to the size of the particle.

carbon dioxide or water vapor, approaches the surface of the catalyst, dissociates to provide mobile species of atomic oxygen, these moving to the carbon surface to form oxides of carbon. This mechanism provides for gasification of the carbon only at the point of contact between the catalyst surface and the carbon. Figure 5.15 is a diagram indicating how these inorganic particles, acting as catalysts in gasification reactions, are mobile within the carbon material producing channels, pores or pits the size of which relate to the size of the particle.

Essentially, the catalyst acts as an oxygen carrier with the molecular oxygen, or carbon dioxide or steam dissociating preferentially on the catalyst surface, the resultant "chemisorbed" oxygen atom (on the catalyst surface as M(O)) being mobile and migrating or diffusing to the carbon-catalyst interface. In this position, the exothermicity of formation of an O—C bond breaks the M(O) bond and a surface oxygen complex C(O) is formed on the carbon surface. The enhanced rate of catalysed gasification reaction is attributed to the enhanced rate of formation of the chemisorbed oxygen, as an atom of oxygen, as compared with rates of complex formation from the gas phase, as described above as $C(O)_{mr}$ and $C(O)_r$.

This phenomenon has significant relevancy during thermal activation of carbons, using carbon dioxide or steam.

As well as influencing, so dramatically, rates of gasification, these catalytic inorganic constituents of carbon also promote the formation of CO_2 in product gases (under conditions of chemical control in Zone I, Figure 5.7). This aspect is illustrated in Figure 5.16 which shows the increase in formation of carbon dioxide when polycrystalline graphite, with potassium added, is gasified. Resultant enhanced concentrations of surface oxygen complexes (with their associated mobility) promote carbon dioxide formation, at the expense of CO formation, because of their enhanced collision frequencies on the carbon (graphite) surface, at the carbon/catalyst interface.

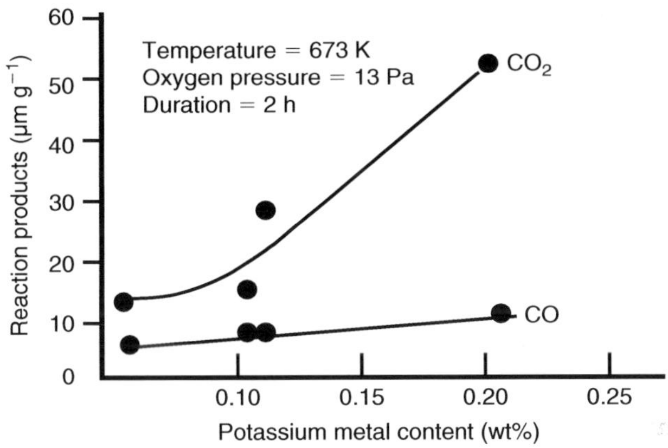

Figure 5.16. The variation of the product generation (μmol g^{-1} of CO_2) during the reaction of molecular oxygen with a graphite containing increasing amounts of potassium (wt%) (Harker, unpublished work).

5.4.3 Factors Influencing Rates of Gasification Reactions

Without elaborating the detail, the several factors which influence rates of gasification of carbon by oxidizing gases (in the region of chemical control) are as follows:

(a) Reaction temperature.

(b) Partial pressure of the reacting gas. Generally the order of reaction with respect to oxygen is less than one and values of about 0.5 are reported. Order can decrease to zero as pressure is increased.

(c) Hydrogen and carbon monoxide inhibit gasification reactions.

(d) Many inorganic compounds within the carbon act as positive catalysts for the oxidation. Some negative catalysts, for example phosphorus, are also known. The influence of some catalysts is reduced by steam in the gas phase. Synergetic effects may operate.

(e) Structure of the carbon being studied.

Research workers in this area are advised to take the greatest of care in the preparation and use of their experimental carbons. The days are over when an operator simply took a bottle labeled *charcoal* from the shelf and committed himself (herself) to 3 years of unprofitable work. The carbon used should be adequately characterized and, where possible, should be "*home-made*".

5.4.4 Topographical Changes During Gasification of Pure Carbons

In studies of gasification reactions of carbons, the elucidation of the kinetics of the reaction constitutes only half of the study. In addition, there is the study of the changes in surface topography of the carbon, induced by the gasification processes. Always, during a

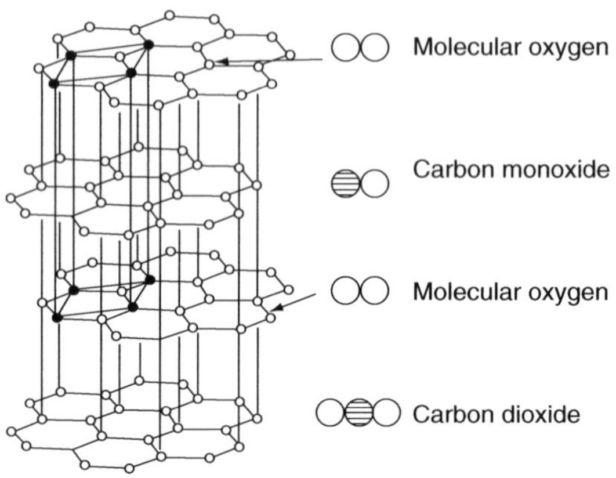

Figure 5.17. Diagram illustrating the approach of oxygen molecules to the graphene layers (basal planes) and prismatic edges of the graphite lattice to form carbon monoxide and carbon dioxide.

gasification process, there is a change in the appearance or topography of parts of the surface when selective gasification occurs, other parts apparently remaining relatively unchanged. This selective gasification of carbon atoms from a granule of carbon occurs both at the surface and within the interior of the granule. Gasification from within the interior is dependent on the presence of an accessible porosity (usually microporosity) within the granule and the experimental conditions of slow, chemical control of rate of reaction.

To explain the origins of the topographical changes of carbon gasification, certain guidelines have been established. Figure 5.17 illustrates oxygen molecules approaching the graphite lattice where there are two positions for chemical reaction involving carbon atoms of different chemical reactivity. The oxygen molecules can (i) undergo reaction at the edges of the graphene layers (prismatic edges) and (ii) on the graphene layers (basal planes), to form surface oxygen complexes at both locations, carbon monoxide and carbon dioxide. Rates of oxidation at the prismatic edges are 10^2 to 10^3 times faster than on the basal planes. Not all atoms of prismatic edges are of equal reactivity (see below). This factor helps explain many of the topographical features observed when studying graphitizable carbons.

It is possible to gain a direct insight into the structure of carbons using the technique of fringe-imaging, phase-contrast, high-resolution transmission electron microscopy. "Fringe-images" derived from the constituent structural elements of carbon structure (the defective micro-graphene layers) are obtained with resolutions of about 0.3 nm. The "fringe-images" are side-on projections of the structural elements and are not projections of the planarity of the molecules. Figure 5.18 is such an electron micrograph of a polvinylidene chloride carbon (HTT 1073 K (800 °C)) (non-graphitizing, microporous carbon) showing the rather disordered arrangement of the lamellar molecules as described in Section 5.1.1. Positions marked with arrows indicate where a gasifying molecule could enter into the microporosity of such a carbon. Reaction could occur at that site, or within the microporosity, often after the gasifying molecule has undergone many, many collisions with the internal surface. Such is the specificity of a reaction site.

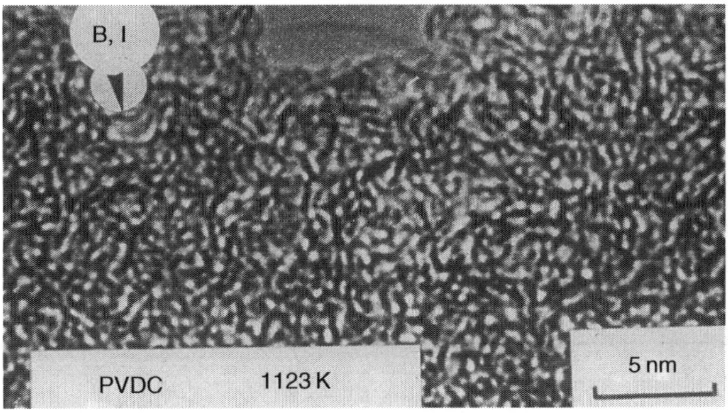

Figure 5.18. Phase-contrast, high-resolution electron micrograph of non-graphitizable polyvinylidene chloride carbon, HTT 1073 K, showing constituent structural elements and microporosity (Marsh (unpublished work) and Crawford *et al.*, 1977).

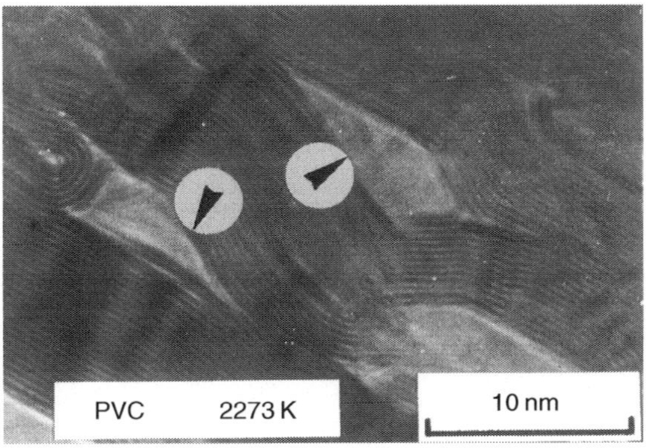

Figure 5.19. Phase-contrast, high-resolution electron micrograph of the graphitizable carbon from polyvinyl chloride (HTT 2273 K) showing the diversity of orientation of constituent graphene planes, as well as porosity, the arrow indicating possible locations for gasification. Note the folding of planes (fullerene-like) to the left of the micrograph (Marsh, unpublished work).

Figure 5.19 shows the more ordered arrangement of the defective graphene layers in this graphitizing carbon from polyvinyl chloride (HTT 2000 °C 2273 K). Note that the crystallographic order is far from perfect. Extensive twisting and folding of the layers occur. Positions marked with arrows indicate mesoporosity within the graphite. These pores will have surfaces of both graphene layers (basal planes) and prismatic edges. It is in such pores that the gasifying molecules have greater "space" for reaction (see Section 5.2.5). Note how one group of layers curves back on to itself (left-hand side of Figure 5.19). This feature will be referred to again.

Figures 5.20 and 5.21 are graphene layer (basal plane) projections of single-crystal graphite and indicate three different positions for reaction by molecular oxygen, carbon dioxide,

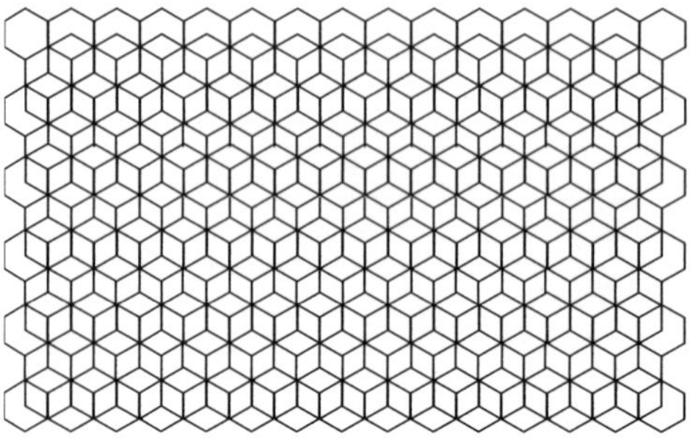

Figure 5.20. Arrangements of carbon atoms in hexagons in graphene layers (basal planes) as stacked to form hexagonal graphite, ABABA.

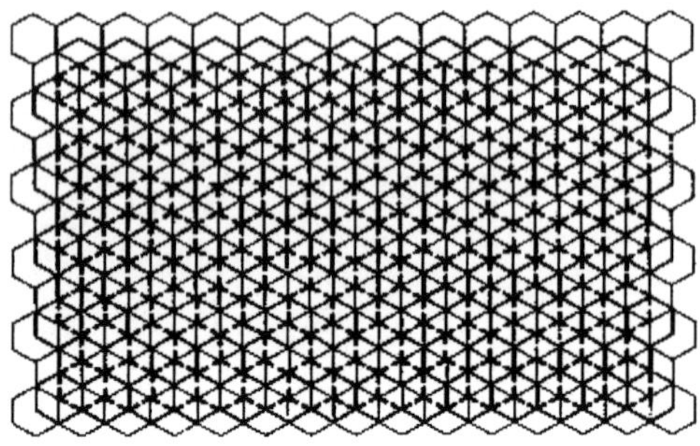

Figure 5.21. Arrangements of carbon atoms in hexagons in graphene layers (basal planes) as stacked to form rhombohedral graphite, ABCABCA.

or water vapor. These positions are the basal plane, or at *armchair* edges, or at *zigzag* edges. All positions have different "reactivities". The relevance here is that the defective micro-graphene layers which constitute the surfaces of the porosity of activated carbons may well exhibit some form of reaction anisotropy based on these considerations.

This reaction anisotropy has been illustrated by studying a graphite surface, after oxidization in molecular oxygen, using optical microscopy before, during and after reaction, and micrographs are reproduced as in Figure 5.22. At first, the gasification produced the circular pits (1 μm diameter) just visible in the optical microscope. As oxidation proceeded, these circular pits changed shape to those of hexagons, those of Figure 5.22 being about 20 μm in diameter. These hexagons all had sides parallel to twin-lines of the graphite surface (a twin-line is a line of dislocations within the original graphite surface).

These effects are best explained by reference to Figure 5.23. Not all crystallographic planes are oxidized at the same rate within the graphene layers. The zigzag or (10–10) planes oxidize at a higher rate (below about 900 °C (1173 K)) compared with the armchair or (01–20) planes. This results in a circular pit becoming polygonal with the zigzag planes receding into the corners of the polygon eventually to disappear. This type of analysis explains why, in pits within the surface of the graphite, armchair (11–20) planes are exposed running in <10–10> direction. For protuberances such as hillocks or platelets (Figure 5.23) on the graphite surface zigzag (10–10) planes are exposed running in <11–20> directions.

Hahn (2005) revisited this topic of reaction anisotropy of crystalline graphite making use of scanning tunneling microscopy, to study centers of preferential reactivity in thin graphite samples (0.2 mm), as had been done using optical microscopy by Thomas (1965). Hahn (2005) investigated the oxidative etching along the <a> and <c> directions of the graphite lattice. Results are summarized as follows: (i) Oxidation at temperatures below 875 °C produces circular pits of uniform size that form at the sites of point defects. At temperatures above 875 °C, in contrast, pit formation occurs *via* etching at both point defects and basal plane carbon atoms. (ii) As oxidation proceeded, tangential contact with next-layer pits is less frequently observed. (iii) The lateral etch rate is three orders of magnitude faster than

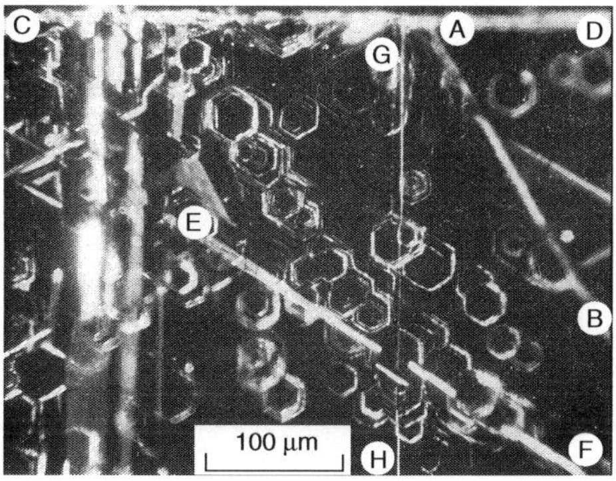

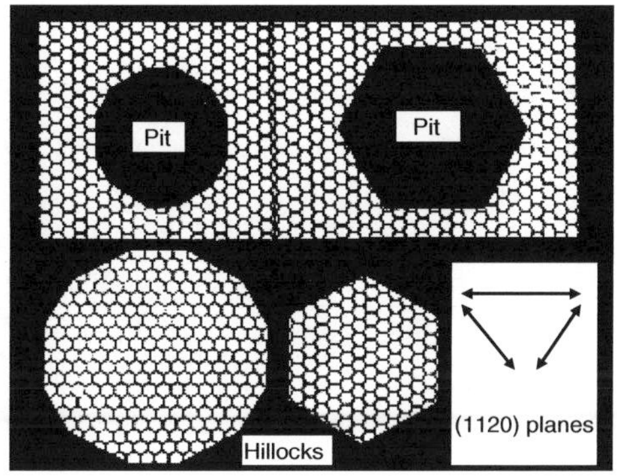

Figures 5.22 (left) and 5.23 (right). Topography of graphite crystal after oxidation at 830 °C by molecular oxygen. Locations AB and CD are surface steps and locations EF and GH are twin-lines. Note that the oxidation pits are all hexagonal-shaped and are parallel to each other (Thomas (1965)).

the vertical etch rate due to the larger frequency factor. (iv) A temperature dependence study of the etch rate showed Arrhenius behavior, the increase in etch rate with increasing temperature being also due to increases in the frequency factor.

Hahn (2005) concluded by stating that, due to the status of graphite as a model carbon material, their kinetic data for graphite oxidation are relevant to other carbons possessing six-membered rings within their structure, such as carbon fibers and carbon nanotubes.

In studies of carbons and polycrystalline graphites as used by industry (as distinct from single-crystal graphites of the more fundamental studies) the features observed in Figure 5.23 are not that clear-cut. But, that reaction anisotropy or selective gasification occurs is beyond any doubt.

Figure 5.24 is a scanning electron micrograph (SEM) of a surface of Lima graphite (nuclear-grade and used as a neutron moderator) after 16.7 wt% burn-off in air. There is pit formation (position A, right of Figure 5.24) but gasification *fissures* are the principal feature. This is because in any surface of a polycrystalline graphite the graphene layers are projected edge-on to the surface to a greater extent than are graphene layers found parallel to the surface. As rates of gasification are greater in directions parallel to the basal planes than in directions perpendicular to the basal planes, then extensive fissuring occurs and this is the essential feature of Figure 5.24 (position B). This fissuring is seen more clearly in Figure 5.25 which shows the surface of a polycrystalline graphite after gasification with carbon dioxide to 40 wt% burn-off at 1123 K (850 °C).

With a non-graphitizing carbon, (PFA, HTT 1123 K), because of the absence of systematic stacking of graphene layers, this facility for fissuring is absent. Figure 5.26 is a SEM micrograph of a non-oxidized surface, with the roughened or pitted surface of surfaces after 49 wt% burn-off, to be seen in Figure 5.27, a situation also relevant to an activation process by carbon dioxide.

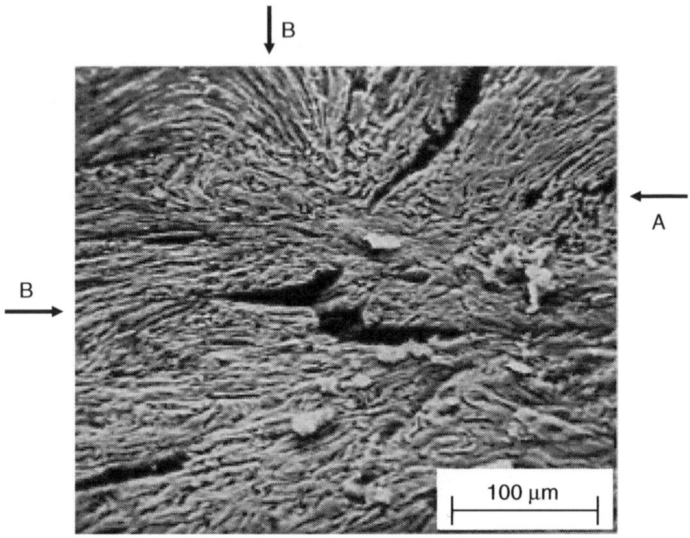

Figure 5.24. Scanning electron micrograph of surface of Lima graphite after oxidation in air to 16.7 wt% burn-off.

Figure 5.25. Scanning electron micrograph of surface of a graphitizable carbon from Ashland A200 petroleum pitch HTT 1223 K, oxidized in carbon dioxide to 40 wt% burn-off at 1123 K.

These effects, observed for the more easily studied surfaces of carbons, provide a clue as to the selective gasification observed during thermal activation of carbons. Within porosity, it can be predicted that basal planes (of the defective micro-graphene layers) will be more dominantly projected. The edge projections, more reactive, may be more difficult to find. Hence, a gasifying molecule, such as carbon dioxide, will have to undergo many collisions with the surface before finding an appropriate reactive site.

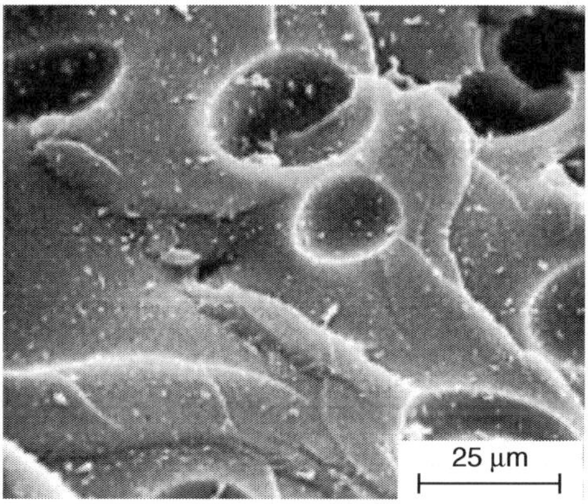

Figure 5.26. Scanning electron micrograph of surface of non-graphitizable carbon from PFA, HTT 1123 K, non-oxidized (Adair *et al.*, 1971).

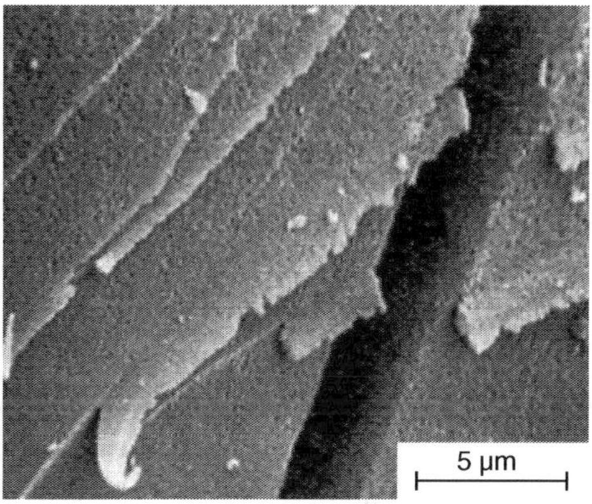

Figure 5.27. Scanning electron micrograph of surface of non-graphitizable carbon from PFA, HTT 1123 K, oxidized in carbon dioxide to 49 wt% burn-off in carbon dioxide (Adair *et al.*, 1971).

5.4.5 Topographical Changes During Gasification of Impure Carbons

The changes in topographical features, induced by catalytic gasification, are quite different from those changes induced in pure carbons. These changes are best described by examination of scanning electron micrographs of surfaces of oxidized carbons.

Use was made of non-graphitizing carbons from PFA prepared from (vacuum) distilled FA. Additions of small amounts of nitric acid to the FA caused polymerization to a brown solid which carbonized without melting. Inorganic salts (usually acetates) could be dissolved in the FA monomer which, when polymerized, contained the inorganic material (catalyst) uniformly distributed throughout the polymer (and eventually the carbon), possibly as an atomic dispersion.

Topographical features, before and after oxidation with carbon dioxide, are as seen in Figures 5.28 and 5.29. When nickel is present in the PFA carbon (Ni:C = 1:4500 atomic ratio) then oxidation with carbon dioxide at 873 K produced extensive isolated (non-uniform) pits (20 μm in diameter) as illustrated in Figure 5.28. When copper is present in the PFA carbon (Cu:C = 1:4500 atomic ratio) then oxidation with carbon dioxide at 893 K to 9 wt% burn-off produced selective (non-uniform) pitting as illustrated in Figure 5.29.

Although the nickel and copper catalysts were both added from solution, the pitting (also called etching) by carbon dioxide clearly indicated diffusion and agglomeration of the catalysts which resulted in pitting. This illustrates that such "atomic" dispersions of metal ions are mobile within the carbon matrix leading to clustering of the atoms. Within organic macromolecular parent materials, for example olive stones, then constituent inorganic material, initially finely dispersed, would tend to agglomerate during carbonization. This observation is also relevant to the process of thermal activation to produce activated carbon.

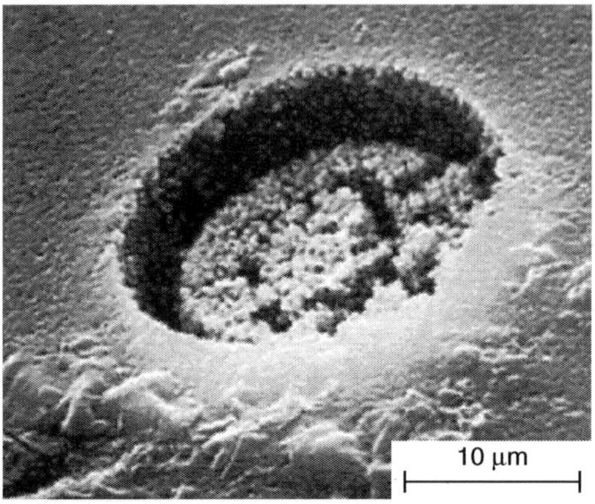

Figures 5.28. Scanning electron micrograph of surface of a non-graphitizable carbon from PFA, HTT 1123 K, with a nickel content of Ni:C of 1:4500, oxidized in carbon dioxide at 873 K to 18 wt% burn-off.

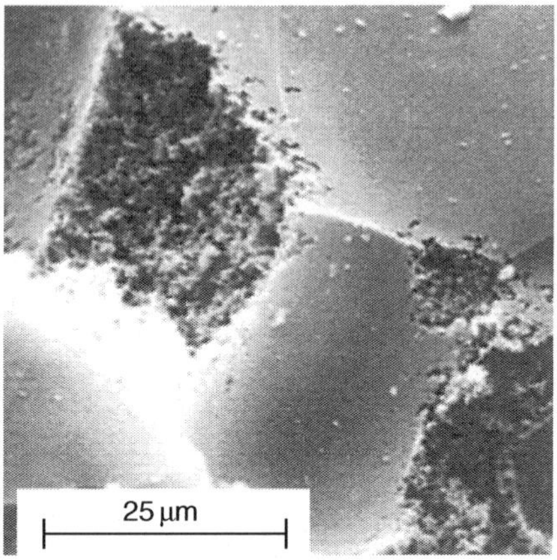

Figures 5.29. Scanning electron micrograph of surface of a non-graphitizable carbon from PFA, HTT 1123 K, with a copper content of Cu:C of 1:4500, oxidized in carbon dioxide at 873 K to 9 wt% burn-off.

5.5 Activation by CO_2 and H_2O: Inhibition by C(O) and C(H)

5.5.1 Introduction

Thermal activation of carbons by gasification using carbon dioxide or steam is the most common method of preparation of activated carbon. The stoichiometric equations for these reactions are very simple. But do not be deceived. The mechanisms are not that simple:

$$C + CO_2 = 2CO \qquad \text{(Reaction 5.1)}$$

$$C + H_2 = CO + H_2 \qquad \text{(Reaction 5.2)}$$

Early studies of these reactions soon indicated that the activated carbons from these two reactions were quite different, suggesting that the carbon atoms being removed by these two gasifying molecules are not comparable carbon atoms. The matter became further complicated when the kineticists reported multistage reaction sequences with the possibility of retardation of rate of gasification by adsorbed species and by back reactions.

Bansal *et al.* (1974) summarized earlier findings of reactions involving hydrogen, observing that the adsorption of oxygen inhibited the adsorption of hydrogen, and that a previous hydrogenation of a carbon surface would reduce the formation of surface oxygen complexes. With clean carbon surfaces (outgassed at 1273 K) it was concluded that the chemisorption of oxygen and hydrogen takes place on the same sites (called active sites) with the chemisorption of hydrogen having the larger activation energy. The carbon–hydrogen complexes are relatively stable and require high temperatures ($>$1500 K) for removal (being a gas which is evolved during graphitization processes). At low hydrogen coverage, one atom of adsorbed hydrogen prevented the adsorption of two atoms of oxygen, this effect decreasing with increasing hydrogenation of the surface until hydrogenation had little further effect on oxygen uptake. Similarly, pre-adsorption of oxygen severely restricted hydrogen uptake, at low hydrogen pressures, indicating that the same adsorption sites were involved.

5.5.2 Direct Addition of Hydrogen: Study of Walker (1996)

Inhibition, by hydrogen and carbon monoxide, of rates of activation of carbons is reviewed by Walker (1996). An activated carbon is usually produced by the partial gasification of a microporous char in CO_2 and/or H_2O. This gasification, resulting in the removal of carbon atoms, initially increases the number of open micropores and their size. At some point in the gasification process, pore walls are removed leading to a further increase in pore dimensions.

The volume of micropores and their size distributions in activated carbons, are a function of (a) the activation equipment; (b) burn-off; (c) the starting carbonaceous precursor; (d) the activating gas; (e) rate of gasification (a function of temperature); (f) rate of diffusion of activating gas(es) into and product gas(es) out of the carbon; (g) extent of product inhibition of rate; (h) amount and nature of inorganic impurities in char and (i) amounts of oxygen surface complexes present.

The University of Alicante (Spain) compared and contrasted the use of CO_2 and H_2O as activating agents (Alcañiz-Monge *et al.*, 1994; Molina-Sabio *et al.*, 1996). In the 1994 paper, petroleum pitch-based activated carbon fibers were prepared by gasification in CO_2 (0.1 MPa) and H_2O (0.05 MPa) at 1160 K to about the same burn-offs. The gasification rate in H_2O (up to 50 wt% burn-off) was about two times faster than that in CO_2. Activated carbons prepared by CO_2 gasification exhibited a larger micropore volume and a narrower micropore size distribution than those prepared from H_2O activation. Also, CO_2 activation led to a smaller decrease in fiber diameter than that using H_2O activation. Comparable experimental results were found by Molina-Sabio *et al.* (1996) for the activation of olive stones, using a common gasification rate for both CO_2 and H_2O.

Note: It is an impossibility to prepare activated carbons using different activating agents, such that the resultant carbons are strictly comparable. The criterion most commonly considered is that of equal burn-offs (yields). This can be achieved (a) by using different activation temperatures. Here, the functionalities of the surfaces are different because of different thermal stabilities of the oxygen and hydrogen complexes, and (b) by using different rates at the same activation temperature. The problem here is the knowledge that extents of activation and resultant pore size distributions are dependent on rate.

Alcañiz-Monge, *et al.* (1994) attributed the enhanced development of microporosity when using CO_2 to its more uniform gasification rate in the radial direction of the carbon fiber than was found when using H_2O. They considered two possible explanations for differences in the uniformity of reaction in the carbon fiber: (i) to differences in the extent of product inhibition in the C—CO_2 and C—H_2O reactions and (ii) to differences in diffusion coefficients of H_2O and CO_2 within the carbon fibers. The first explanation was rejected by saying that although the C—CO_2 and C—H_2O reactions are severely limited by CO and H_2, respectively, differences in inhibiting effects would not produce the differences noted. Rather, they promoted the second explanation for the greater uniformity of gasification in CO_2, that is, a higher diffusion coefficient for CO_2 than for H_2O, within the carbon.

Walker (1996) did not agree with Alcañiz-Monge *et al.* (1994) who had concluded that the higher diffusion coefficient of CO_2 was the contributory factor. Evidence was quoted from Koresh *et al.* (1980), who studied diffusion in molecular sieve carbons, that H_2O has a smaller critical molecular dimension than CO_2. Thus, activated (or configurational) diffusion of CO_2 would be lower than for H_2O. As activation (gasification) proceeds, resulting in pore enlargement, diffusion mechanisms change from activated to Knudsen flow. Again H_2O is expected to have a higher diffusion coefficient because of its lower molecular weight compared to CO_2 (Wheeler (1951)).

Walker (1996), in reviewing this topic, attributed differences in activation results between CO_2 and H_2O primarily to two causes. The first is that activation (gasification) in CO_2 proceeded at a less rapid rate than in H_2O. The second explanation discusses differences in extents of product inhibition in the C—CO_2 and C—H_2O reactions. The C—CO_2 and C—H_2O reactions are described as follows (Walker *et al.*, 1959):

The C—CO_2 reaction:

$$C_f + CO_2 = CO + C(O) \qquad \text{(Reaction 5.5)}$$

$$C(O) \rightarrow CO \qquad \text{(Reaction 5.9)}$$

The C—H_2O reaction:

$$C_f + H_2O = C(O) + H_2 \qquad \text{(Reaction 5.25)}$$

$$C(O) \rightarrow CO \qquad \text{(Reaction 5.9)}$$

$$2C_f + H_2 = 2C(H) \qquad \text{(Reaction 5.6)}$$

where C_f is a free carbon active site, C(O) is oxygen surface complex, and C(H) is chemisorbed hydrogen. Carbon monoxide inhibits the C—CO_2 reaction by removal of oxygen complex from the carbon surface (described by reverse Reaction 5.5 and not by chemisorption of carbon monoxide, C(CO)). Hydrogen inhibits the C—H_2O reaction both by removal of oxygen complexes from the carbon surface (reverse Reaction 5.25) and by dissociative chemisorption of H_2 on the carbon surface (Reaction 5.6).

Austin *et al.* (1963) had earlier shown, both experimentally and theoretically, that the build-up of small concentrations of CO (from CO_2) in the gas phase leads to non-uniformity of surfaces at gasification rates where, as the reaction is of integer order (negligible retardation by CO), uniform gasification would be expected. Using Langmuir–Hinshelwood kinetics, the steady-state rate of the C—CO_2 reaction is given by this equation:

$$\text{Rate}_{\text{C-CO}_2} = \frac{k_1 P_{CO_2}}{1 + k_2 P_{CO} + k_3 P_{CO_2}} \qquad \text{Equation (5.2a)}$$

when the reaction is described by Reactions 5.5 and 5.9 above (Strange *et al.*, 1976).

For the C—H_2O reaction, several workers had indicated that inhibition by H_2 is caused overwhelmingly by dissociative chemisorption of H_2 on to free carbon active sites and not by the reverse of Reaction 5.25. For further details see the classical foundation studies of Gadsby *et al.* (1948), Long *et al.* (1948). Giberson *et al.* (1966) studied the C—H_2O reaction under a wider range of hydrogen partial pressures than had previously been used and concluded that the rate of gasification is better given by this equation:

$$\text{Rate}_{\text{C}-\text{H}_2\text{O}} = \frac{k_1 P_{H_2O}}{1 + k_2 P_{H_2}{}^{1/2} + k_3 P_{H_2O}} \qquad \text{Equation (5.5)}$$

To compare the extent of inhibition by CO for the C—CO_2 reaction with that by H_2 for the C—H_2O reaction (k_2) values for the two reactions need to be compared. This was done by Biederman *et al.* (1976) as follows. The gasification of a high purity natural graphite was studied at temperatures between 1233 K and 1393 K at pressures of CO_2 below 14.4 Pa. For CO_2 depletion up to at least 90%, in a batch reactor, gasification rates were first order with respect to CO_2 pressure with no inhibition by the product gas, CO, being observed. Then, at 1373 K, following depletion of about 50% of the CO_2, a pressure of about 4.2 Pa of H_2 was injected into the system (Figure 5.30). At the time of injection, the CO_2 pressure had fallen from 8.7 to 4.2 Pa; and the CO pressure had built up to 9.1 Pa. At the point of injection, the reaction rate abruptly decreased. The rate constant decreased from a starting value of about 670 units to about 54 units. That is, H_2 is a much stronger inhibitor than is CO, or $k_{2H2} >> k_{2CO}$.

If inhibition by H_2 is attributed to its dissociative chemisorption on to carbon-free active sites, the rate constant, k_{2H2} will equal the equilibrium constant, *K*, for Reaction 5.6 above. Biederman *et al.* (1976) confirmed this to be the case. The adsorption isotherm of H_2 on their graphite sample was measured at 1373 K, from which *K* was determined. The equilibrium constant was also calculated from this equation:

$$\text{Rate}_{\text{C—CO}_2} = \frac{k_1 P_{CO_2}}{1 + k P_{H_2}{}^{1/2}} \qquad \text{Equation (5.6)}$$

Close agreement of equilibrium constants was found.

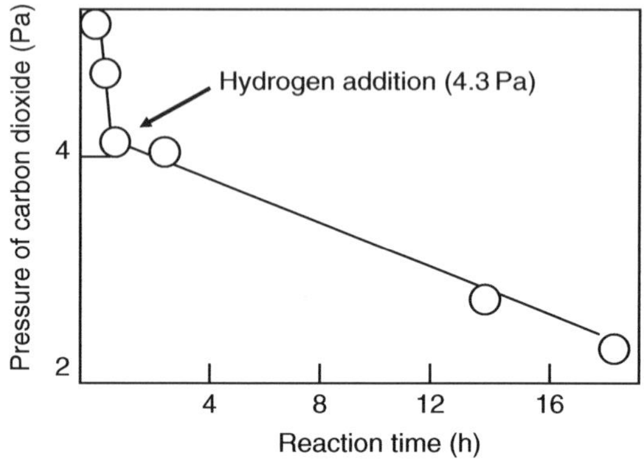

Figure 5.30. Effect of addition of 4.3 Pa of hydrogen on the rate of the C—CO_2 reaction (removal from system) at 1373 K, using high purity natural graphite (Walker (1996)).

Differences between the activation characteristics of CO_2 and H_2O are seen in the SEM micrographs of Figure 5.31 of carbons from a viscous rayon fiber. CO_2 produces a continuous development of narrow microporosity, with a slight pore widening as from 30 wt% burn-off. Steam, unlike the CO_2, leads to carbons with lower micropore volume because of the widening to large porosity as from the beginning of the activation process.

In summary, H_2 is a much stronger inhibitor of the C—H_2O reaction than is CO for the C—CO_2 reaction. Therefore, H_2 is expected to cause greater non-uniformity of gasification in the C—H_2O reaction than is CO in the C—CO_2 reaction.

5.5.3 Modern Approaches to Mechanisms of Activation by CO_2 and H_2O

What could be termed the classic exploratory experiments of the gasification kinetics of carbons established unequivocally the important role of hydrogen adsorption in the gasification (activation) of carbons by steam. More recent studies have added considerable detail to that basic understanding and these are summarized below in an attempt to create a "movie" of the dynamics of these gasification reactions at surfaces of carbons.

Lussier *et al.* (1998) studied the gasification of a char from Saran (from Dow Saran resin) and a char from No. 6 Illinois coal (coal bank number PSOC 1403. 75.3 wt% carbon (dab)) using mixtures of steam and hydrogen. Adsorbed species were studied by temperature programmed desorptions (TPD) to temperatures of 1700 K. Exposure of annealed and outgassed chars to mixtures of steam, hydrogen and argon at 1120 K and pressures to 3.1 MPa caused the immediate chemisorption of hydrogen during the initial 0.5 wt% of weight loss.

The mechanism by which hydrogen inhibits steam gasification depends on hydrogen partial pressure and extent of char conversion. At low levels of activation (burn-off) dissociative adsorption (Reaction 5.6) inhibits steam gasification by blocking reactive sites. At higher conversions where surfaces can be saturated with adsorbed hydrogen, steam gasification

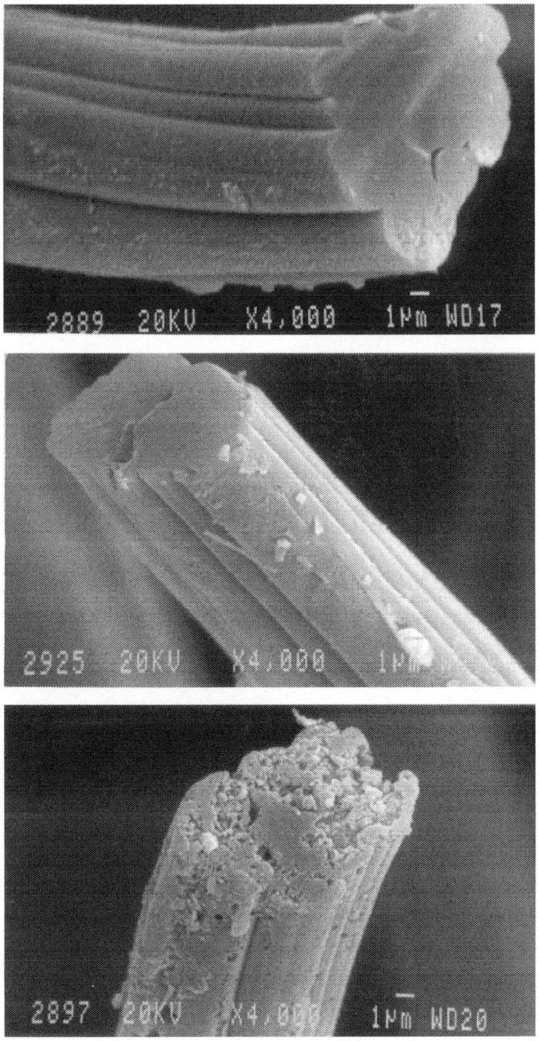

Figure 5.31. SEM micrographs of viscous rayon fibers Upper – the original char. Middle – activated in carbon dioxide. Lower – activated in steam (Rodríguez-Reinoso *et al.*, 2000a).

rate is inhibited by both Reaction 5.6 and by the reverse oxygen exchange reaction (Reaction 5.26). "Associative" hydrogen adsorption is not considered to be an inhibiting reaction (Reaction 5.27).

$$C(O) + H_2 \rightarrow C_f + H_2O \qquad \text{(Reaction 5.26)}$$

$$C + H_2 \rightarrow C(H_2) \qquad \text{(Reaction 5.27)}$$

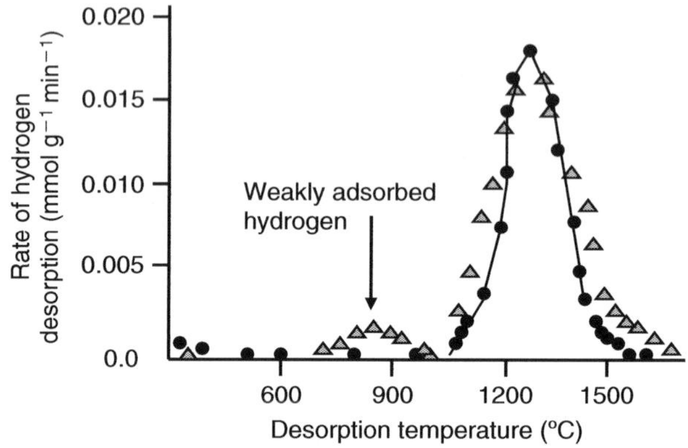

Figure 5.32. The temperature desorption profile for desorption of hydrogen from an annealed Saran char after hydrogasification at 3.1 MPa and cooling in argon ●: hydrogen ▲: (Lussier *et al.*, 1998).

This is not the end of the story however. Figure 5.32 is the hydrogen desorption profile after gasification of the Saran char with a mixture of steam and hydrogen. There appear to be two levels of stability, one, which is weak and desorbs at about 900 K and a second, much more stable complex which desorbs at about 1400 K (1130 °C). The hydrogen is also in considerable abundance on the surface compared with oxygen surface complexes, 1–4 mmol g^{-1} and <0.08 mmol g^{-1}, respectively, a ratio of about 50:1. Whether or not it is the same hydrogen, which stays on the surface during gasification, would require the use of deuterium to decide. Despite the abundance of hydrogen on the surface, there appear to be remaining sites where oxygen exchange and gasification are favored over hydrogen adsorption. A scenario is possible which sees gasification occurring around, but not involving the site of chemisorbed hydrogen. In this way, topographically, highly heterogeneous surfaces may be developed within the char particles, a process otherwise known as activation. Lussier *et al.* (1998) continue the discussion by stating that the edges of the graphene layers of structure of the chars could have somewhat zigzag and armchair configurations (Figures 5.20 and 5.21) and that these are responsible for the chemisorption of hydrogen. This leaves non-planar carbon atoms or point defects within a graphene layer as positions of gasification by the steam.

The use of deuterium (D_2) as a means of following the status of chemisorbed hydrogen during steam gasification is described by Zhang *et al.* (2000), from the same laboratory in Australia as Lussier *et al.* (1998). This reaction is further complicated (and also simplified) by considerations of the formation of methane from the use of pure hydrogen. This is not a straightforward reaction as four hydrogen atoms have to be attached to one carbon atom. It is now considered that sequential, dissociative adsorption of hydrogen occurs, to form C—H, then C—(H_2), then C—(H_3) followed by formation of CH_4. This study of Zhang *et al.* (2000) reports on hydrogen desorption energies, H—D exchange experiments and the reaction of molecular oxygen with hydrogenated surfaces, in attempts to understand further the role of adsorbed hydrogen in gasification (activation) chemistry.

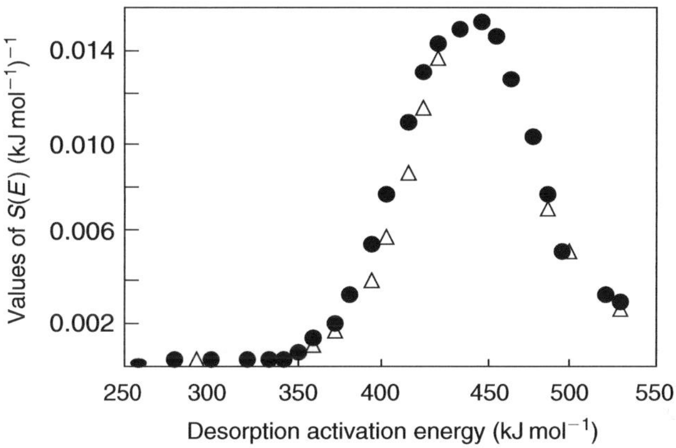

Figure 5.33. Distribution of activation energies of desorption for hydrogen on annealed Saran char. ● Gasification in hydrogen, Δ gasification in steam (Zhang *et al.*, 2000).

A desorption activation energy distribution $S(E)$ in units of $(\mathrm{kJ\,mol^{-1}})^{-1}$ was calculated using Equation 5.7 to give distributions for hydrogen gasification and steam gasification as in Figure 5.33. The similarities of the activation energy distributions for gasification by hydrogen and steam indicate that hydrogen adsorption energies are independent of reactant gas. Desorption activation energies maximize at about 440 kJ mol^{-1}, reaching as high as 550 kJ mol^{-1}, indicative of strong bond formation.

$$S(E) = \frac{\mathrm{d}\{H_2\}/\mathrm{d}t}{\{H_2\}_o R\beta(\ln(AT/\beta) - 2.1)} \qquad \text{Equation (5.7)}$$

where
$\{H_2\}_o$ the initial quantity of hydrogen adsorbed (mmol g^{-1}) prior to the TPD
R = ideal gas constant (8.31 J mol^{-1} K^{-1})
β = linear heating rate (5 K min^{-1})
A = pre-exponential factor of desorption rate constant (6×10^{14} min^{-1})
T = instantaneous char temperature (K)

The stabilities of adsorbed hydrogen- and deuterium-chemisorbed complexes were studied using H—D and D—H exchanges. At first, the char was gasified to a predetermined extent using H_2 (or D_2) to saturate the surface with H or D. The reactant gas was then switched to D_2 (or H_2) and the char continued to be gasified. The effluent gas was analyzed during this period of continued gasification. It was thus possible to know rates at which the chemisorbed hydrogen is desorbed (as H_2), exchanged (as HD) or reacted to form methane during gasification in D_2. The results indicate an extremely mobile situation on the surfaces of these chars. There is no state of permanent chemisorbed of hydrogen species, in keeping with what is known of other surface reactions over carbons.

The relevant information emerging from this isotope study includes:

1. The switch from deuterium to hydrogen, including the change in effluent methane from CH_4 to CD_4 takes place within a 5-min period.
2. HD is produced for the entire 45 min time period following the switch to hydrogen.
3. The presence of HD is clear evidence that chemisorbed hydrogen (C(H)) and deuterium (C(D)) can exchange with gas-phase hydrogen at the gasification temperature of 1120 K.
4. The TPD profiles for desorption of D_2, HD and H_2 are the same as in Figure 5.33. This indicates that H—D exchange takes place among all of the sites of chemisorbed hydrogen, irrespective of the bonding strengths in the sites.
5. These data confirm that associative chemisorption of hydrogen (as (CH_2)) does not take place.
6. The energetics of surfaces containing chemisorbed hydrogen, at 1120 K, when gasification is taking place, must be very different to surfaces containing chemisorbed hydrogen when no reaction is occurring, such as after the removal of steam/hydrogen mixtures. At 1120 K, the TPD profile shows almost no removal of hydrogen; yet, at this temperature the hydrogen is being replaced rapidly. It takes a temperature rise to 1400 °C to remove about 50% of the hydrogen. The carbon surfaces in these two situations must be in quite different energetic states. During gasification, the surface carbons atoms (at least) must be in a continuous state of excitement quite unlike the stable state during a TPD experiment. Hence, during activation in steam, the concept of a molecule of water entering into the invariant porosity of the carbon and quietly producing carbon monoxide and stable surface bonded hydrogen cannot be accepted. Rather it is more like carrying out a reaction on the surfaces of violently boiling water.

During a gasification process, the entire carbon surface must be energized. If it were ever possible to make a video recording of these movements, and to playback on the television screen, then what would be seen is something resembling the Bolshoi Ballet, atomic version. Carbon atoms would accept hydrogen atoms from the left, to pass them on to the right, and receive another partner (H) from the left, and so on, with frequencies in the order of picoseconds. When two hydrogen atoms arrive to the same carbon atom simultaneously then molecular hydrogen is formed and evolved, the site immediately accepting another hydrogen molecule, which dissociates to maintain a constant surface concentration of (C—(H)).

The possibility of mobile carbon atoms as well as mobile hydrogen atoms is another matter to be discussed below.

Another question which now arises is to ascertain if this chemisorbed hydrogen is stable enough to withstand reaction with molecular oxygen (10% in argon) at the lower temperatures of 710–750 K. Following such an oxidation, a TPD curve was immediately obtained, or gasification was continued at 1120 K with hydrogen or deuterium. The essential results are as follows:

1. In contrast to the H—D exchange, where the energy distribution of the chemisorbed hydrogen (deuterium) remains unchanged, molecular oxygen removes only the more weakly bonded hydrogen, with substantial quantities of hydrogen remaining. This is

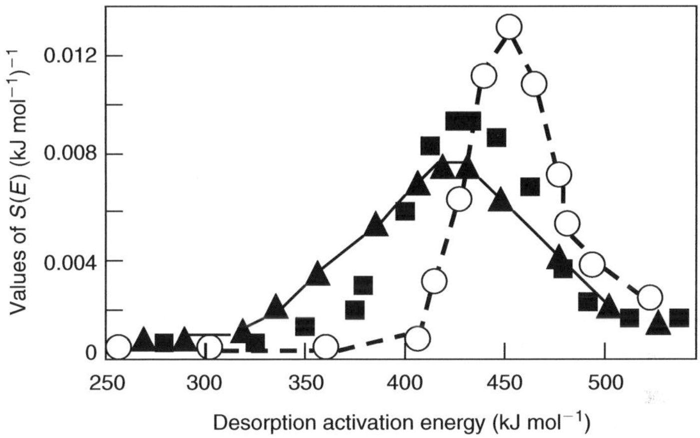

Figure 5.34. Distribution of activation energies of desorption for hydrogen on annealed Saran char following oxidation. ▲ No oxidation, ■ Oxidation in 5% O_2/Ar at 730 °C for 2.5 h, ○ Oxidation in 10% O_2/Ar at 750 °C for 1.7 h (Zhang *et al.*, 2000).

seen in the displacement to the right of the desorption profiles of the hydrogen, Figure 5.34.

2. After oxidation in molecular oxygen, the char shows an enhanced gasification rate to hydrogen. This is attributed to the generation of new reactive sites during gasification by molecular oxygen.
3. The observed shift in the energy distribution curves is attributed to the formation of more chemisorbed hydrogen on the zigzag sites of the edges of the graphene layers.

In is now quite clear, from the considerations of the results of the studies of Lussier *et al.* (1998) and of Zhang *et al.* (2000) that during steam activation of carbons, the reaction sequences take place over an extensively excited (energized) carbon surface, exhibiting considerable mobility of chemisorbed species, which although mobile, do not leave the surface.

During the gasification of the Saran char and the coal char by steam (as above) the dissociation of the water molecule provides both oxides of carbon and hydrogen atoms on the carbon surface at temperatures of 1120 K. It therefore is a useful suggestion to study the reactions, with carbons, of atoms of hydrogen (deuterium), pre-prepared. Güttler *et al.* (2004) performed such experiments. It has been known for some time that hydrogen is bonded to both the zig-zag and armchair edges of the graphene layer, and these authors found that hydrogen atoms can be bound to the graphene layer itself (0001 basal plane) and that, in order to maintain stability, this surface becomes reconstructed. This involves the puckering out of the carbon atom under the adsorbed hydrogen to the extent of about 40 pm. This is reminiscent of the *hole or claw* defects in graphite, as reported by Ubbelohde *et al.* (1960).

Güttler *et al.* (2004) generated atomic hydrogen and deuterium in ultra high vacuum (UHV) equipment and reacted the atoms with the surface of glassy carbon (much less structurally ordered than graphite). Hydrogen atoms bonded to the glassy carbon at level of about

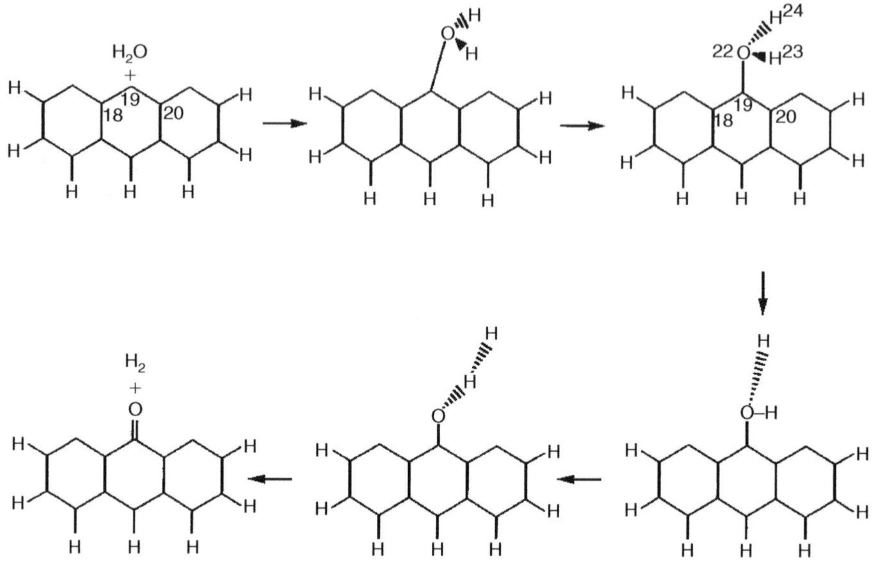

Figure 5.35. The theoretical pathway for the reaction of a molecule of water with a carbon surface to produce molecular hydrogen and carbon monoxide. The reaction sequences rotate clockwise (Zhu *et al.*, 2003).

10 wt% of the graphite. Desorption of deuterium occurred between 400 and 600 K. It was found that reactions with atomic species have little immediate relevance to the molecular reactions. A similar finding was observed for the carbon-atomic oxygen reaction (Marsh *et al.*, 1969). That is, pre-generated atomic hydrogen does not behave as chemisorbed hydrogen atoms formed "*in-situ*".

The experimental work, as described above, indicates possible reaction sequences taking place during an activation process. What such studies do not do is to offer explanations for such behavior and as a result theoretical considerations are needed. Zue *et al.* (2003) provide an excellent review of gasification reactions and offer a theoretical scheme for the reaction of a water molecule with a carbon surface, as shown in Figure 5.35.

In their review, Zhu *et al.* (2003) suggest, in Figure 5.35, a favorable energetic pathway where a molecule of water is first physically "attached" on the graphite surface without change. Chemisorption of an oxygen atom occurs at a carbon edge site with one hydrogen atom stretching away from the bonded oxygen in the transition state. This is followed by a state of local minimum energy where the stretching H is further disconnected from the oxygen with the OH group remaining on the carbon edge site. The disconnected H then rotates around the OH group to bond with the H of the OH group and to form H_2. The O atom remaining on the carbon edge site is subsequently desorbed as CO (see Section 5.6.3). The reverse pathway occurs when H_2 reacts with the surface oxygen to produce H_2O. For the total reaction $H_2O + C \rightarrow H_2 + CO$, Zhu *et al.* (2003) calculated AH to be $+146.0\,kJ\,mol^{-1}$, which is close to the experimental values $129.0\,kJ\,mol^{-1}$. Hydrogen inhibition is apparently due to the much stronger adsorption of hydrogen on edge sites compared with H_2O. The C—H bond is stronger on zigzag sites compared with armchair sites. During etching of a graphite surface with hydrogen, hexagonal etch pits were observed, in contrast to the rounded pits when using carbon dioxide (no chemisorption of hydrogen atoms), Yang *et al.* (1985).

5.5.4 Reactions with Hydrogen and Water (Liquid)

The reaction of carbon with hydrogen to form methane is endothermic with slow rates of conversion. Such a reaction is unlikely to replace "geological" methane as obtained from petroleum and coal-bed-methane. However, as reported by Cyprès *et al.* (1985) activation of a char is possible using hydrogen as the gasifying gas. Such a process will never be a replacement for conventional carbon activation methods, but raises the interesting point as to whether or not activation by hydrogen produces unusual porosities in carbons, bearing in mind the influence of chemisorbed hydrogen on gasification mechanisms.

Cyprès *et al.* (1985) prepared a char (carbon) from the Beringen Belgian coal (74.4 wt% C) at 1173 K and reacted it with hydrogen at 1173 K, 4 MPa pressure to weight loses up to 80 wt%. Surface areas increased from 230 to 720 $m^2 g^{-1}$ (55 wt% burn-off) and then decreased to 380 $m^2 g^{-1}$ (80 wt% burn-off) based on the adsorption of CO_2 at 298 K (and hence data are reliable). In the range of burn-off 40–80 wt%, the mean micropore dimension remained almost constant at about 1.5 nm, with a significant increase in the macropore surface area ($60 > D < 500$ nm) from 7 to 37 $m^2 g^{-1}$. These results are somewhat unusual and hence the subject area would benefit from being revisited.

Another experimental approach is to increase the porosity of a carbon by sequences of mild oxidation (to form surface oxygen complexes) followed by thermal treatments to remove these oxides as CO and CO_2. After about 76 such cycles a carbon with a surface area of 1670 $m^2 g^{-1}$ is reported (US Patent 5 071 920 1991). It goes without saying that such methodology is excessively energy demanding. In an attempt to reduce the energy requirements, Conesa *et al.* (2000) devised a process, using the same principles, which heated a char from Macadamia nut shell* (HTT 950 °C) in hydrogen peroxide, 100–200 °C, under a pressure of 10 MPa for about 20 min. The oxidized char was then re-heated to 950 °C to remove the chemisorbed oxygen. The original unactivated char possessed a surface area of 190 $m^2 g^{-1}$. After three of such treatments the surface area increased to about 750 $m^2 g^{-1}$ the char being highly microporous.

5.6 Surface Phenomena During Thermal Activations

5.6.1 Introduction

The carbon atoms of surfaces available to reaction by carbon dioxide, as in activation processes, are not identical but exhibit wide variations in reactivity, that is, there exists

*The Macadamia nut is native to Queensland, Australia and was imported to Hawaii in about 1881. It is also to be found in Brazil and Central America. The nut has a very hard seed coat which may make it suitable for active carbon production (resembling coconut shell); the use of hydrogen peroxide may not be the way to proceed. However, tonnage production of coconuts considerably exceeds that of Macadamia nuts, although this nut is reputed to make excellent cookies.

structural heterogeneity. Without this variation in reactivity, the process of activation would be impossible as it is dependent on carbon atom selectivity. This property is part of the overall general aspect of the uniqueness of structure of activated carbon. Without selectivity, the carbon dioxide molecule would react immediately on contact with external surfaces of the carbon without any increase in porosity. This is what happens, of course, when carbons are rapidly gasified in molecular oxygen. Accordingly, the carbon dioxide molecule penetrates the porosity of the carbon, and with extreme selectivity, attaches itself to a very specific site, ignoring the many other sites it is in contact with during its diffusion into the carbon. When considering the relative reactivities of other carbons, these are found to vary indicating that not all carbon surfaces, from different precursors, are equally heterogeneous and as a result different activated carbons are produced from different precursors.

As indicated above, it is the carbon atom which is attached to a surface oxygen complex which is removed during activation. Hence, it must be the surface oxygen complex which responds to the heterogeneity of the carbon surface, and studies of the relative stabilities of the complexes may give an insight into the surface heterogeneity. This aspect has received the experimental and theoretical attention of Kapteijn *et al.* (1994) who found, in a gasification study, by carbon dioxide, of nine different carbons, the existence of a distribution of desorption activation energies. Different gasification rates of these nine carbons, as well as being a function of CO_2 pressure and temperature, are also a function of variations in concentrations of desorbable surface oxygen complexes. It appears that two types of complex contribute to the gasification rate. A major contribution comes from highly reactive intermediates which are five times more reactive than a second group which exist in 3–4 times smaller amounts. Relative rate constants are invariant for the carbons being 0.6 and $0.034\,s^{-1}$. A further complication within these gasification sequences is seen when using isotopic carbon dioxide, $^{13}CO_2$. In such a system, it is observed that ^{13}C becomes inserted into the carbon structure. This could come from a disproportionation reaction of carbon monoxide a reaction, Reaction 5.28, which is not normally considered during activation reactions. There may be other mechanisms.

$$2CO \rightarrow CO_2 + C\downarrow \qquad \text{(Reaction 5.28)}$$

5.6.2 Transient States

There is yet another factor which requires mention and this concerns the nature of the carbon surface after outgassing. Hydrogen tends to be evolved as hydrogen gas, but molecular oxygen is never evolved from surface oxygen complexes, only as carbon dioxide (at lower temperatures) and carbon monoxide (at higher temperatures). An indication that the outgassing (cleaning) of a surface, to remove oxygen complexes, leaves behind sites of exceptional reactivity is provided by Rellick *et al.* (1975). These authors reacted a graphite with carbon dioxide at 1050 °C in three successive runs, with outgassing between each run, with results of variation in gasification rates being as in Figure 5.36. The rate of gasification increased during the first 5 h of the reaction and then subsequently decreased to a steady-state rate. Thus, the actual initial gasification produced sites of exceptionally high reactivity (>20 times above the steady-state rate). The second run showed the same effect but was less pronounced and the third run showed no transient phenomena. The authors

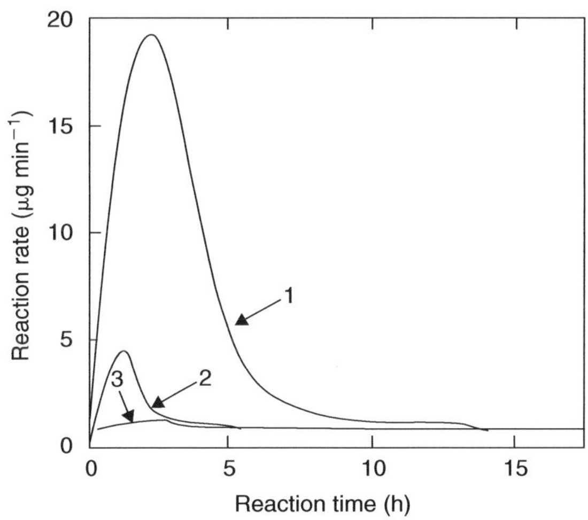

Figure 5.36. Reaction of a graphite at 1050 °C with carbon dioxide, after outgassing at 1075 °C, showing three successive runs to exhibit transient states and their stability (Rellick *et al.*, 1975).

then heated their graphite to 2500 °C at which temperature, by reason of carbon atom mobility, the sites responsible for transient rates were annealed out and no transient rates were observed on subsequent reaction with carbon dioxide.

These transient rates may be associated with active sites generated by the desorption of surface oxides from the carbon, that is the removal of carbon atoms from an otherwise stabilized structure. It would appear that such sites are able to generate additional sites of unusual reactivity so accounting for the maximum in the transient rate. It is established that as well as mobility of hydrogen and oxygen over carbon surfaces, so carbon atoms can do likewise as they move to more stable locations. This is, of course, what happens during a graphitization process.

5.6.3 Mechanisms of Desorption of Surface Oxygen Complexes as CO

Thermal activation by carbon dioxide or water vapor, have one thing in common, and that is the removal of atoms of carbon, from the solid phase, as gaseous carbon monoxide. However, thermal activations by carbon dioxide or steam produce quite different activated carbons in terms of their pore dimensions and pore size distributions. Accordingly, the mechanisms of removal of carbon atoms, as CO, must differ in some way. The most significant difference so far reported is that for steam activation, the CO being removed from a surface which, in addition, contains chemisorbed hydrogen, as distinct from a surface containing only chemisorbed oxygen, as in the activation by carbon dioxide. The question now is to understand the nature of this influence. The answer is that there is no real understanding of what happens at the carbon surface during the stages of removal of CO. When a situation exists within a study, when further experimentation is unlikely to provide that informative, the approach to adopt is to go talk to the theoretician to see what ideas are suggested.

To explore this problem further, attention now has to move towards reaction pathways and the energetics and kinetics of these gasification reactions, which extend from carbon activation ($\sim 0.6 \times 10^6$ tonnes per year) to combustion reactions, which run at about 5×10^9 tonnes per year. It is the interest in combustion processes, as in coal-fired power stations, that has given an impetus to further investigate the energetics of CO removal from carbon surfaces. As indicated above, Zue *et al.* (2003) reviewed electronic structure methods including semi-empirical, *ab initio* and density functional theory methods leading to discussions of mechanisms of activation by hydrogen- and oxygen-containing gases.

In addition, Frankcombe *et al.* (2004), specifically, discuss several possible reaction pathways leading to the removal of oxygen which is chemisorbed to a graphene layer, Figure 5.37. Direct gasification, where the two carbon–carbon bonds holding the (O) are broken more or less simultaneously, involves the high singlet and triplet energies of 366 and 469 kJ mol^{-1}, with other reaction pathways having lower energy states. Suggestions for possible reaction pathways are included in Figure 5.37, which indicates how an oxygen atom may move over carbon surfaces before eventually leaving the surface as CO.

Figure 5.37 suggests that the removal of C(O) as CO, and the location of gasification are a function of carbon structure in the vicinity of the C(O). On surfaces which are free from other adsorbed species, the CO removal may occur **from surfaces without restrictions** (remembering that it is a surface of porosity which is involved). Hence, the CO_2 activation (relative to H_2O gasification), by being equal (similar) over all parts of the surface, ***produces a narrow pore size distribution.*** On the other hand, in the H_2O activation reaction, the edge atoms in the vicinity of the C(O) will possess chemisorbed hydrogen and a different set of possible reactions pathways will be involved. The presence of the chemisorbed hydrogen effectively blocks movement of the C(O) complex to those sites.

Hence, those parts of the surface, where the hydrogen is located, ***cannot*** take part in the gasification process as it might if the hydrogen was absent. ***As a result, the CO removal***

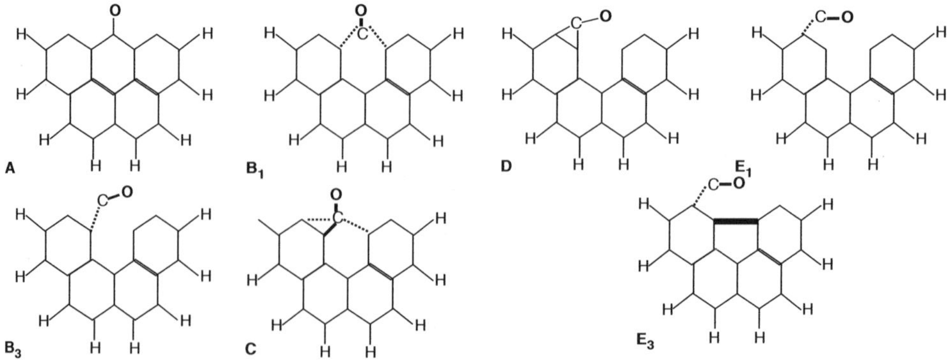

Figure 5.37. Saddle point and stable state geometries for CO extraction reactions. A is the reactant. B_1 is a single direct gasification saddle point. B_3 is a triplet direct gasification saddle point. C shows migration to a surface complex saddle point. D is a surface complex. E_1 is a singlet gasification from surface complex saddle point. E_3 is a triplet gasification (Frankcombe *et al.*, 2004).

from a hydrogenated surface will not occur uniformly over that surface resulting in a wider pore size distribution and a smaller micropore volume. In terms of the high transient reaction gasification rates reported above by Rellick *et al.* (1975), the answer to the problem must also lie with the immediate environment of the C(O) complex and where the reaction pathway and associated activation energies (and entropies) are favorable.

These theoretical approaches certainly offer viable suggestions to explain the several features of these gasification reactions. In the end, it may be difficult to prove, experimentally, that these reaction pathways do occur and self-consistency with available experimental data will be sufficient. At least, a clearer picture is available of events at the molecular level, and which have been known (if not understood) for some time.

And there remains another challenge in this area of gasification (activation) mechanism, namely to find out if the sites of high adsorption potential, which are detected at low values of p/p^0 in adsorption isotherms of nitrogen (77 K), correspond to sites of high-chemical reactivity as observed in transient gasification phenomena.

5.7 The Concept of Reactivity

Central to considerations of activation of carbons by carbon dioxide and steam are rates of gasification. Informative publications about production methods of activated carbon always indicate the rate of gasification, that is, weight loss per hour, for example $10\,wt\%\,h^{-1}$, or as used in some relevant literature, $g\,g^{-1}\,s^{-1}$. Unlike homogeneous reactions, the activation process of carbons has no economic advantages when the temperature of reaction is increased beyond critical maximum values. The rate of activation (s^{-1}) controls the nature of the porosity which is being developed, very slow rates (expensive to maintain) developing a bias towards microporosity and very high rates not producing porosity, as rates become diffusion controlled and burn-off is restricted to external surfaces.

The rate at which a carbon sample is gasified is said to depend on its "*reactivity*" and this term, like *surface area* (Section 4.1.3.) has been the centre of debate in the carbon community in recent years. The problem was clearly seen more than half a century ago when Arthur (1951), comprehensively, listed the "reactivities" of all the carbons to be found in the literature. He observed that rates (in units of s^{-1}) varied over about 5–6 orders of magnitude, with activation energies varying over two orders of magnitude for the C—O_2 reaction, and in the range 0.05–18 for C—CO_2 reaction. This wide range was narrowed when rates of gasification (reactivity) were expressed in units of $g\,s^{-1}\,m^{-2}$, that is when specific reactivities were compared. The range was narrowed further when *impure* carbons were separated from *pure* carbons and rates for pure carbons placed in an order of comparable HTT. The graphitic carbons showed the lowest reactivities and chars of low HTT (e.g. 600 °C) showed highest reactivities. No sense could be made of the reactivities of the *impure* carbons and the rational for their behavior had to await the development of an understanding of catalytic gasification mechanisms.

But still the problem persisted as *pure* carbons, of equal HTT, offered different reactivities in units of $g\,s^{-1}\,m^{-2}$, carefully checking out that the data were not made unreliable because of diffusion control effects. The conclusion was that carbons of different origins, otherwise comparable, possessed structures which had different intrinsic reactivities. The argument to explain these differences then developed around the concept of ASA which would be a measure of the percentage of reactive carbon atoms in the surfaces of these carbons, assuming that different carbon surfaces possess the same "type" of reactive carbon atoms (only their "concentrations" vary).

The next matter arising was how to measure ASA. A definitive statement was published by Causton *et al.* (1985) in which about 50 mg of a carbon sample was placed in a reaction vessel and outgassed to 1223 K at 10^{-5} Pa for 15 h. The temperature was then reduced to 423 K and molecular oxygen introduced at 0.1 MPa pressure and maintained for 12 h in order to cover the surface with chemisorbed oxygen to maximum amounts (under these defined conditions). The reaction vessel was again evacuated (at 423 K) to a pressure of 10^{-5} Pa followed by programmed heating (TPD) to desorb the surface oxides as CO and CO_2. Pressures and composition were measured in a mass spectrometer and total amounts of desorbed oxygen were converted to an equivalent surface area which, in turn, could be related to a N_2 Brunauer–Emmett–Teller (BET) surface area (noting problems which exist with this evaluation) and termed TSA. ASA values could be reproduced during several cycles of the procedure and were not sensitive to the pressure of the oxygen in the reaction vessel (0.001 to 0.1 MPa).

Results for two chars (carbons) are in Figure 5.38. It is seen that values of ASA and TSA ($m^2 g^{-1}$) for the coal char are lower than for the PVDC carbon. For both carbons, values of ASA and TSA increase with burn-off in H_2O/H_2 mixtures. For both carbons, the ratio of ASA/TSA decreases with increasing burn-off, from 0.6 to 0.18 (coal char), and from 0.8 to 0.3 (PVDC). These results (if the SA concept is accepted) indicate that the reactivity of carbons decreases with burn-off, implying that the most reactive of carbon atoms are gasified first, and *do not replace themselves*. The study of Causton *et al.* (1985) also

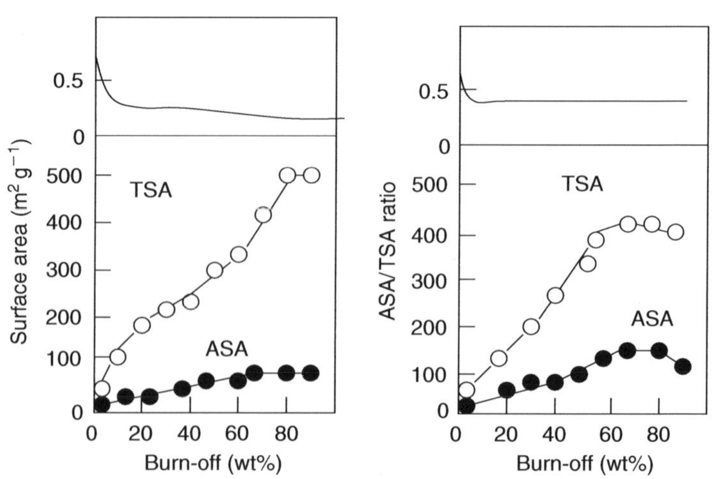

Figure 5.38. Variations of ASA and TSA with burn-off for (a) Markham Main coal char and (b) PVCD char. The variation of the ASA/TSA ratio with burn-off for these carbon materials is in the upper part of each diagram (Causton *et al.*, 1985).

indicated, from DR plots of CO_2 isotherms, that microporosity had disappeared, this indicating that the most reactive of carbon atoms were containing the microporosity. This is useful to know and can be applied to current discussions of mechanisms of activation. Some of the pioneer work in this area was also carried out by Rodríguez-Reinoso *et al.* (1974) with a subsequent appraisal of the concept of active sites by Radovic *et al.* (1983). The concept of ASA cannot be applied to impure carbons, containing inorganic gasification catalysts, where rates are controlled by catalyst concentration and dispersion, being quite independent of carbon structure.

Khan *et al.* (1987) monitored the significance of the ASAs of chars to appraise the reactivity of low- and high-temperature chars.

Attempts have been made to predict gasification rates using mathematical models. This area has been briefly reviewed by Rafsanjani *et al.* (2002) who discuss the use of (what are termed) the grain model, the random pore model, the simple particle model and the volume reaction model. They report that differential mass conservation equations are required for the oxidant gas and char particle. These authors use a simplified mathematical model (the quantise method (QM)) for activation of coal chars when both diffusion and kinetic effects have to be considered. Results are compared with other methods when it is found that QM predictions of rate are more accurate than predictions by the random pore model and the simple particle model.

The discovery and applications of **nanotubes** has enabled direct "same-area" observations of positions of gasification in the walls, etc. of multi-walled nanotubes (MWNT) with closed ends. Shimada *et al.* (2004) identified the positions of onset of gasification in the walls of the nanotubes by reacting with molecular oxygen at 823 and 1023 K, observing the same areas of a nanotube surface both before and after oxidation. The major gasification positions were in the walls of the nanotubes with the closed ends (tips) remaining intact. The micrographs, taken before gasification, revealed that positions of initial gasification were cracks and structural defects within the walls. Although the tips of the nanotubes possess five-, six- and seven-membered ring systems, necessary to contain the strain within the curved layers, this strain does not automatically enhance the reactivity of the layer plane. The entire surface of the nanotubes had become mottled by numerous pits created during the gasification, and were related to multiple defects in the original structure.

These ideas are not entirely new and are discussed in Section 5.4.4. What is of additional interest is that oxidation at 823 K causes swelling of the nanotube and complete disappearance of the (002) lattice fringe images. This does not occur when the oxidation temperature is 1023 K. (Is this related to the study of Güttler *et al.* (2004), Section 5.5.3)? Shimada *et al.* (2004) attribute this to migratory surface oxygen complexes which are formed at the defect centres and can move in between the graphene layers. At 1023 K, extents of formation of stable surface oxygen complexes are more limited and this phenomenon of swelling is not observed. Yet again, from another quite separate experimental source, there is demonstrated the mobility of oxygen species over carbon surfaces.

The relevance of these studies of gasification of nanotubes to activation of carbon is that gasification may also occur at defect positions within graphene layers as well as the edges. In fact, no carbon atom appears to be immune, it being a question of relative reactivities as to which atom is removed during activation processes.

In practical terms, no test has been devised to measure what can be described as *reactivity*, that is a measure of rate of gasification at activation temperatures. The American society for testing materials ASTM has developed *the Ignition Temperature Test* which is a measure of reactivity in some way, and involves heating a carbon sample in air until it spontaneously ignites (catches fire). This test, ASTM Designation: D 3466-76 (re-approved 1988) (see Chapter 8), provides a basis for comparing the ignition characteristics of different carbons, or the change of ignition characteristics of the same carbon after a period of service.

5.8 Surface Carbon Atom Migration

The detail of the growth of graphene layers from the parent feedstock during a carbonization process has not been clarified in detail, although infrared spectroscopy indicates the growth of aromaticity. The conversion may not be entirely within the solid phase, as the released volatiles may have a role to play as these are pyrolyzed within the solid phase. Two separate studies are outlined below which discuss growth mechanisms and although they refer to the more exotic systems of nanotube growth, the ideas within these studies can only help to understand these growth processes in carbonization systems.

The above sections of this chapter have demonstrated that, during activated processes, involving gasification mechanisms, movement of atomic (probably excited) species is a dominant aspect. During reactions with molecular oxygen, carbon dioxide and steam, movement of surface oxygen species (surface bonded oxygen), as well as bonded hydrogen, is a dominant feature of these reactions. It is a characteristic of carbon surfaces that they can bond to oxygen and hydrogen atoms firmly enough to prevent desorption but lightly enough to allow their movement from atom to atom (surface migration).

The movement of carbon atoms over surfaces is more difficult to study and there is as yet no clear assessment of its importance. A brief indication was given earlier about the incorporation of ^{13}C in a carbon lattice from the gas phase. The chemistry which is being developed around nanotubes both single walled (SWNT) and multi-walled (MWNT) has encouraged research into growth mechanisms from the vapor or plasma phase (during nanotube genesis). A comprehensive review of these studies is available, Krasheninnikov *et al.* (2004). First thoughts suggest that carbon atoms (or clusters of atoms) attach themselves to the growth edges of the nanotubes and then become part of the graphene layer. However, this is not the way that crystals, in general, grow during a crystallization process. Rather, the constituent molecules become attached to completed crystal surfaces and migrate over the surfaces to the growth edges of the crystal. In this way, predispositions of position make the attachment possible. Otherwise, the entropy changes are too high for the process to occur.

Krasheninnikov *et al.* (2004) consider that carbon atoms coming from the gas phase may become attached to the end of SWNT, should dangling covalent bonds be present. It is much more likely that those carbon atoms (groups) attach themselves to the graphene layers of the SWNT and then migrate towards the growing end. Groups of carbon atoms may detach themselves from the surface to form disorganized carbon formed in association with SWNT. A knowledge of energies of adsorption and migration are needed to develop these theories of growth of SWNT. A value of 0.1 eV is reported for the adatom migration barrier and this compares with a value of about 0.8 eV for annealing of damage in graphite caused by irradiations. Calculations indicated that adatoms are able to migrate on the zigzag edges of nanotubes where they form strong covalent bonds, and that the migration is anisotropic and adsorption energies are a function of the nanotube diameter.

Another approach to considerations of growth mechanisms of graphene layers is that of Frenklach *et al.* (2004) who suggested similarities of growth mechanisms of the graphene layer to the growth in size of polycyclic aromatic moieties. The conclusions from their Monte Carlo simulations support the assumption of five-ring migration in the growth of graphene layers. It is considered that five-membered rings are constantly formed on the growing edge of a graphene layer, and that they do not accumulate there but convert to six-membered rings.

5.9 Thermal Activation Processes: (Contributions of Rodríguez–Reinoso *et al.*, 1984)

5.9.1 Introduction

In the above sections, gasifications (activations) have been considered at the "molecular" level. It is now convenient to discuss thermal activation processes at the "macro" level.

Previous chapters of this book have summarized carbonization procedures used to make activated carbon, to understand structure in these active carbons and to understand methods of characterization of porosity. Further to these considerations, thermal activation requires knowledge of the chemistry of what occurs in the time interval between a reacting gas entering into the carbon specimen and the removal of an atom of carbon as carbon monoxide.

In reality, what happens in the time interval between a reacting gas molecule entering into the porosity of a carbon and a molecule of carbon monoxide emerging, is a closely guarded secret (by the carbon that is). Although theoretical studies have attempted to predict rates of gasification (which are crucial in activated carbon production) the prediction (on paper) of pore size distributions and pore volumes is not yet possible.

Internationally, many laboratories developed expertise in the broad area of adsorption into porous solids, both white and black. Specialization in activated carbons was commenced, in Spain, under the *aegis* of Professor Juan de Dios López-González, University of Granada (Lopez-Gonzalez *et al.*, 1972a, b, 1980). These early studies of adsorption phenomena were continued in the University of Alicante, advancing understandings of porous materials and publishing >1400 papers. Special attention is given to the use of a few parent materials so ensuring that different procedures could be realistically compared and evaluated.

Summarizing, the activation (gasification) reactions result in (a) the opening of porosity which, originally, was not accessible to an adsorbate; (b) the enhancement of micropore volumes and (c) widening of micropores into the mesopore range. Increases of burn-off produces increased pore volumes within activated carbons. The details of these processes, as in the literature, are not easily comparable because the precursors and the experimental conditions used in different laboratories vary widely from experiment to experiment (author to author). The results obtained for different precursors under comparable experimental conditions often differ. Two reasons can be suggested. **One**, the macromolecular structures of the parent organic materials differ, leading to different carbons, **Two** extents and composition differ of their mineral matter contents which catalyse the activation process, this causing changes in the porosity as compared with a pure carbon precursor (Section 5.4.2).

The developments of understandings of the activation processes, as coming from the University of Alicante, Spain, are described in an approximate chronological order, as below.

5.9.2 The Year 1984: Use of Air and CO_2

In order to establish some of the *basics* of activation processes, three different methods for the preparation of activated carbons from olive stones and almond shells (natural materials in abundance in Spain, and used consistently), 1073–1123 K, were studied: (a) carbonization in nitrogen followed by activation in CO_2; (b) direct activation in CO_2 (no previous carbonization) and (c) treatment of the raw material in air at 573 K followed by activation in CO_2 using a flow rate of $80\,cm^3\,min^{-1}$ (Rodríguez-Reinoso *et al.*, 1984). The samples from olive stones are series (A, B, C) and from almond shells (J, K, L). Adsorptive capacities were obtained from N_2 (77 K) and CO_2 (273 K) isotherms with volumes of meso-macroporosities determined *via* a Carlo-Erba 200 mercury porosimeter (7.5–15,000 nm dimension).

For both source materials, methods (a) and (b) produced almost identical results indicating that a pre-carbonization stage was not a necessary prerequisite. Method (c), involving pre-oxidation, gave the best results both in terms of adsorption capacity and overall yield. Results are summarized in Figure 5.39(a, b) which are plots of surface area ($m^2\,g^{-1}$) against yield of gasified carbon (wt%).

5.9.3 The Year 1987: Interdependence of N_2 and CO_2 Isotherms

As part of an ongoing debate or search for the *correct* isotherm to provide the *correct* value for surface area there was disagreement in the literature about the relative importance of the isotherms of nitrogen (77 K) and carbon dioxide (273 K). One school of thought pronounced the N_2 (77 K), as interpreted by the BET equation, as being totally beyond any form of criticism. A second school of thought had more understanding of the problem and could not accept that this approach (use of N_2 77 K isotherm) was one of the fundamental laws of nature. The research group at Alicante University entered this debate and presented a rational to it all. This is analyzed in depth in Section 4.2.5.2 and will not be repeated here. The last word, of course, is that it takes both isotherms for a meaningful characterization of an activated carbon. Always, the isotherms of N_2 and CO_2 should be inseparable, as are Tom and Jerry, David and Goliath, Romeo and Juliet and Derby and Joan (Garrido *et al.*, 1987).

5.9.4 The Year 1989: Methodologies

It was about this time that the debate over the meaning of "surface area" was reaching a climax. It was also the time that several methodologies, for the characterization of porosities,

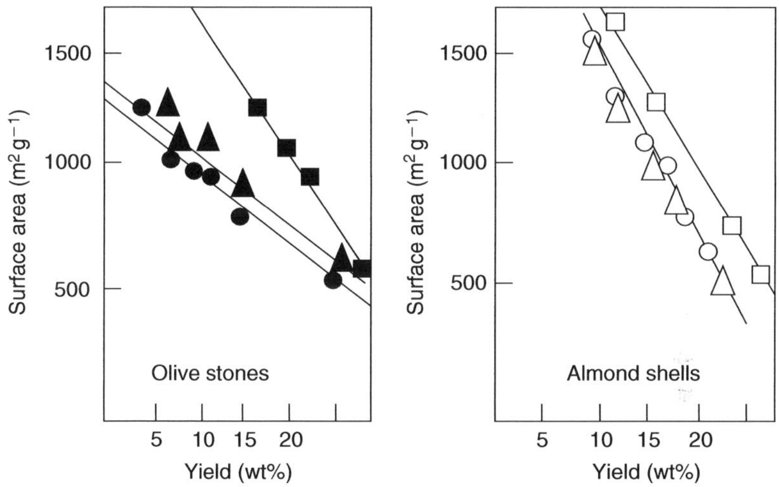

Figure 5.39. The variation of surface area ($m^2 g^{-1}$) with activation yield (wt%) for carbons from (a) olive stones are series (A, B, C) and from (b) almond shells (J, K, L) (right). The highest surface areas are for carbons prepared with pre-oxidation, Series C and L. (Rodríguez-Reinoso *et al.*, 1984). (●) Series A; (▲) Series B; (■) Series C; (○) Series J; (Δ) Series K; (□) Series L.

had reached maturity. There is no technique available today which measures surface areas of carbons (please, do not throw this book away at this point: the statement is perfectly true). What is measured, with good accuracy and reproducibility are extents of adsorption (capacities), in known pore-size distributions as well as calorimetric data (Joules released during adsorption processes, and when surfaces are wetted under controlled conditions). The adsorption isotherm measures extents of adsorption in units, for example, of mmol g^{-1}. The adsorption equations offer a "best-guess" of amounts needed to cover the available internal surfaces (again in units of mmol g^{-1}), but have nothing to say as how to convert these units into the units of $m^2 g^{-1}$ (surface area). Again, assuming various packing densities for the adsorbate molecules, the units of mmol g^{-1} can be converted to $m^2 g^{-1}$. That is to say, the amount adsorbed is equivalent to the amount needed to cover a flat surface (macro-, not atomic surface) to the extent of say 1000 $m^2 g^{-1}$. Whether or not the carbon being studied has such a surface is not a question that can be answered. A surface area value is not a measurement, it is derived parameter based on less than precise conversion factors.

Hence, arguments were promoted at this time as to the real, actual, true surface area of a porous solid and the best ways to measure this. Unfortunately, when data were presented in papers or lectures, it was somehow forgotten that *surface area* is a derived parameter and does not have the physical basis of say mass or volume. Surface area, as such, is *not* important for characterizations which need capacities ($cm^3 g^{-1}$) in defined porosities.

In order to compare/contrast methods of surface area assessment, Rodríguez-Reinoso *et al.* (1989a, b), using two series of activated carbons, made use of several different experimental approaches for their characterization and subsequent comparison. Two series of carbons were prepared, from almond shells and from olive stones which were carbonized in nitrogen at 1123 K for 2 h and subsequently activated in CO_2 at 1098 K for different periods of time to cover an 8–82 wt% burn-off range. Characterization was carried out by adsorption of nitrogen (77 K), CO_2 (273 K), benzene, 2,2-dimethylbutane (298 K), and iso-octane (2,2,4-trimethylpentane) (298 K) in gravimetric adsorption systems. The pre-adsorption of *n*-nonane was studied in the same experimental systems. It is considered that *n*-nonane is adsorbed so firmly into microporosity that it is retained during initial evacuation of the adsorption equipment so allowing further adsorption, of say nitrogen, into porosities other than microporosity. All adsorption data were analyzed using the DR Equation, the Langmuir Equation and the α_s-method (Section 4.2.4).

The results of the study indicate that no single adsorptive or theoretical approach provides complete information on the micropore structure of activated carbons. This is a major conclusion indicating that each adsorption system is unique and that the adsorption of nitrogen cannot be equated to the adsorption of, say, iso-octane. When the microporosity is narrow and uniform – and accessible to all adsorptives – all experimental and theoretical approaches lead to micropore volumes which are in relatively good agreement. For highly activated carbons – with a wide micropore size distribution – the adsorption of hydrocarbons provides the total micropore volume and the results are comparable with those deduced from the Langmuir and α_s-plots constructed from the N_2 (77 K) isotherms. When the DR-plot (N_2) uses the relative pressure range $<0.2\,p/p^0$, the micropore volume is in fairly good agreement with the α_s-plots. The adsorptions of CO_2 and the *n*-nonane pre-adsorption are restricted to narrow pores. To summarize, a basic characterization of activated carbon can be based on the CO_2 (273 K) isotherm to determine the volume of narrow micropores (primary) filling process (using the DR equation for the extrapolations) and the N_2 (77 K) isotherm, using the α_s-plots, to evaluate total micropore volumes, so enabling the volumes of the wider micropores to be evaluated.

5.9.5 The Year 1989: Major Review – Adsorption Methodologies

A few years after the opening of the Department of Inorganic Chemistry, University of Alicante, Marcel Dekker Inc, of New York in 1989, published, in Chemistry and Physics of Carbon, an overview and perspective of the use of adsorption methods as they could be used to reveal structures in microporous carbons. The work of this department over the previous 10 years, together with its placement into the general chemistry of surfaces, is contained within this review, together with 124 relevant references for further reading (Rodríguez-Reinoso *et al.*, 1989b).

5.9.6 The Year 1991: Major Review – Thermal Activation

The review of Section 5.9.5 was published in 1989, this 1991 review being presented to an international conference held in France and published as a Conference proceedings (Rodríguez-Reinoso, 1991). This review discusses the activation of carbons by carbon dioxide and steam and the development of porosities. In addition, the review considers the relevance of the reaction with oxygen in industrial activation processes.

Oxygen or air is *not* used as an activating agent in the preparation of activated carbon, but is introduced into the industrial reactor to burn the product gases, CO and H_2, for their calorific value. It is therefore useful to know how the presence of oxygen may affect the structure and porosity of activated carbons. A relevant work was published by Tomkov *et al.* (1977) who stated that during the activation of brown coal with oxygen at 673 K there is a high development of porosity at low burn-off but becomes ineffective on continuation of oxidation to medium and high burn-off. At <25 wt% burn-off, the micropore volume developed by oxygen is larger than with steam (at 1113 K). With steam, the reduction of micropore content is accompanied by a simultaneous increase in mesoporosity.

This does not happen with oxygen where the volume of mesopores remains almost constant. Tomkov *et al.* (1977) concluded that activation by oxygen occurs only at the beginning of the process, the oxygen surface groups chemisorbed at the entrances of the micropores isolating them from further activation which proceeds only in larger pores. This study was followed up by the Alicante group who carbonized and activated almond shells and olive stones in CO_2 at 1098 K to 35 and 32 wt% burn-off respectively, followed by reaction in air over a wide range of burn-offs. The results are as in Figure 5.40 showing that the micropore volume and surface area increased to a maximum around 20 wt% burn-off, and decreasing considerably after that. The volume of mesopores increased continually up to 40 wt% burn-off, decreasing thereafter.

Consequently, the development of porosity in carbons prepared by gasification in air is very different to the equivalent prepared by gasification in CO_2. A better way of comparison is included in Figure 5.41(a, b) where pore volumes have been plotted as a function of the overall burn-off so that four of the carbons prepared by air gasification are comparable to the four carbons prepared by CO_2 gasification. These carbons exhibit a continuous increase in pore volume (much more marked in the micropore volume) with increasing burn-off, whereas there is a maximum around 40–50 wt% global burn-off for carbons gasified in

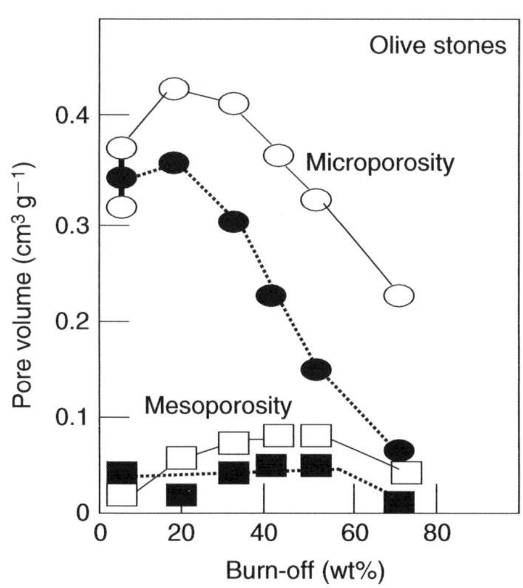

Figure 5.40. Variations in micro- (○) and meso (□) pore volumes with burn-off in air at 623 K, using olive stones: Continuous line per gram of activated carbon: broken line per gram of starting carbon (char) (Rodríguez-Reinoso (1991)).

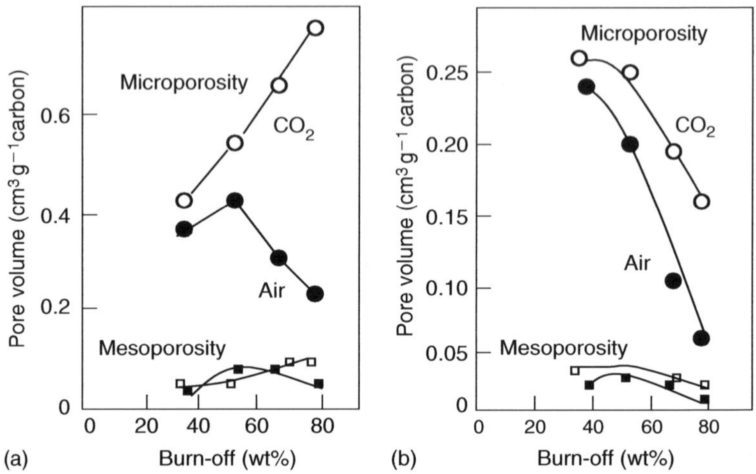

Figure 5.41. Variations in micropore and mesopore volumes with burn-off for carbons gasified in CO2 (open symbols) and air (closed symbols), using olive stones; (a) per gram of activated carbon; (b) per gram of parent char (Rodríguez-Reinoso (1991)).

air, the micropore volume decreasing rapidly. When the same data are expressed per unit weight of starting material (an active carbon with 32 wt% burn-off in CO_2 for the series gasified in air) it is observed that the mesopore volume remains almost constant for both series during gasification but that the decrease in micropore volume is much more drastic in the carbon gasified in air (Figure 5.41b). Thus, the exothermic reaction in air produces a continuous and localized ablation of the carbon particle, reducing the micropore volume to a larger extent than reaction with CO_2.

5.9.7 The Year 1995: Use of Different Precursors

As part of the scenario of preparation of activated carbons with a wide range of porosities, different source materials need to be used and various preparative procedures need to be explored. Hence, González *et al.* (1995a) investigated the behavior of six different ligno-cellulosic materials and compared and contrasted differences in porosities of the resulting activated carbons. The precursors selected were coconut shell, almond shell and peach, plum, olive and cherry stones. Carbonizations were carried out in nitrogen (80 $cm^3 min^{-1}$) at 600 °C for 1 h, followed by activation in steam at 800 °C for various times to produce various degrees of activation. The weight loss in going from 600 to 800 °C was also determined. Narrow micropore volumes were determined using the DR equation for the CO_2 (273 K) isotherms, and total micropore volumes using the α_s-plots for the N_2 (77 K) isotherms.

Elemental analyses of the original lignocellulosic materials showed a similar composition of 49 wt% C, 6.0 wt% H, about 0.2 wt% N and (by difference) 45 wt% of oxygen. The char yields were very similar, 26.5 ± 1.0 wt% at 600 °C and 23.7 ± 0.8 wt% at 800 °C. The results for the 600 °C chars are in Table 5.2 and indicate that the coconut shell, in terms of density and meso- and macropore volumes ($V_{Hg}(cm^3 g^{-1})$) stands out as being very different

Table 5.2. Characteristics of chars obtained at 600 °C from different lignocellulosic precursors (González *et al.*, 1995a).

Raw material	Mercury density ($g\,cm^{-3}$)	V_{Hg} ($cm^3\,g^{-1}$)	Activation rate ($mg\,g^{-1}\,h^{-1}$)
Coconut	1.16	0.07	11
Almond	0.99	0.19	29
Peach	0.99	0.18	12
Plum	0.89	0.29	19
Olive	0.91	0.24	11
Cherry	0.86	0.37	10

from the other five precursors. Similarly, the micropore volumes, as calculated from the adsorption data, indicate the highest micropore volumes for the coconut shell, values around 25% lower being found for plum and peach stones. Much larger differences were found for meso- and macroporosity, ranging from 0.1 to 0.6 $cm^3\,g^{-1}$ for coconut shell and olive stones respectively. Why this should be must be related to the biological anatomy of the coconut shell, and that requires a separate study. Activation rates varied from 10 to 29 $mg\,g^{-1}\,h^{-1}$ but there is no correlation between this rate and the development of microporosity.

5.9.8 The Year 1997: Use of Different Experimental Conditions of Activation

Factors which may influence the properties of an activated carbon depend on (a) the parent feedstock; (b) the heating rate; (c) the flow rate of the containing gas, usually nitrogen (d) the final HTT of carbonization; (e) the temperature of activation; (f) the activating gas; (g) the duration of activation; (h) flow rate of the activating gas and (i) the experimental equipment used. The construction of the bed of carbon, where the bed is located in the furnace, and the way the activating gas can flow around or over the carbon also are controlling influences. This implies that when activating carbons and comparing results, laboratory to laboratory, some differences are to be expected, even when the conditions (a) to (i) are fixed.

González *et al.* (1997) prepared three different chars from olive stones, 1–1.5 mm particle size, carbonized in three ways. (I) slow carbonization in a vertical furnace, heating rate 5 °C min^{-1}, nitrogen flow rate of 1275 $cm^3\,min^{-1}$, HTT of 850 °C, soak time of 1 h, giving a char yield of 26.9 wt%. (II) flash carbonization in a vertical furnace preheated to 850 °C, nitrogen flow rate of 1275 $cm^3\,min^{-1}$, HTT of 850 °C, soak time of 1 h, giving a char yield of 16.8 wt%. (III) slow carbonization in a horizontal furnace, heating rate 5 °C min^{-1}, nitrogen flow rate of 80 $cm^3\,min^{-1}$, HTT of 850 °C, soak time of 1 h, giving a char yield of 25.5 wt%.

Because the three chars were prepared under different experimental conditions, it is reasonable to expect that their reactivities towards carbon dioxide may also differ. In order to obtain comparable activated carbons, identical rates of activation were used for the three carbons and appropriate activation temperatures had to be determined. In the vertical furnace, activations were carried out at 830, 808 and 843°C, carbon dioxide flow 100 $cm^3\,min^{-1}$, for chars (I), (II) and (III) respectively, covering the range 14–68 wt% and in the horizontal furnace at 830, 800 and 850 °C, using a CO_2 flow rate of 80 $cm^3\,min^{-1}$, covering the range 12–65 wt% burn-off. The resultant activated carbons were analyzed using the N_2 (77 K) and CO_2 (273 K) isotherms. Results are summarized in Figure 5.42(a–d).

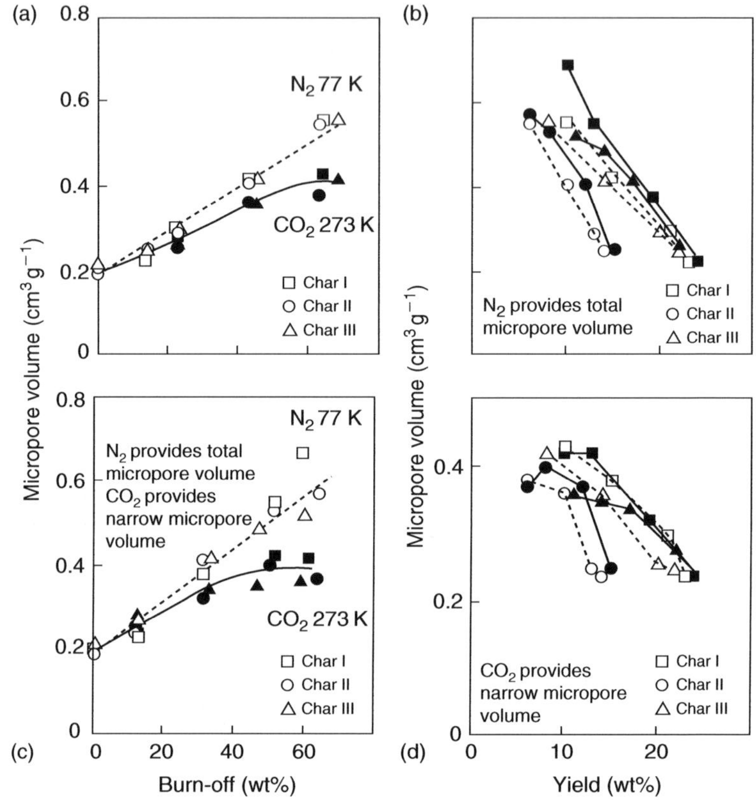

Figure 5.42. Development of micropore volume as a function of burn-off; (a) for series activated in a vertical furnace; (c) for series activated in a horizontal furnace. Development of micropore volume as a function of yield. Open symbols are for a vertical furnace; bold symbols are for a horizontal furnace. (b) N_2 adsorption at 77 K (d) CO_2 adsorption at 273 K (Gonzalez *et al.*, 1997)

Differences between the systems are best illustrated using pore volumes when, for microporosity, the DR equation, with the CO_2 isotherm (273 K), provides the narrow micropore volume (up to about 0.7 nm), and the N_2 isotherm (77 K) provides the total micropore volume (up to 2.0 nm in width). Figure 5.42a presents values of micropore volume for all samples prepared by flash activation in the vertical furnace as a function of bum-off, and Figure 5.42c includes the data for samples activated in the horizontal furnace. There is a common increase in the values of micropore volume with increasing bum-off, independent of the starting char when activation is carried out at the same gasification rate. On the other hand, Figure 5.42(a, c) also indicates that, there are no significant differences in terms of total microporosity. However, from the point of view of porosity development, there are differences between the two activation methods when the horizontal furnace seems to have advantages because both micropore volumes are larger. This is a result of external mass-transport being more limited for the horizontal furnace, and this will favor internal gasification of the char particle and the creation of micropores rather than the widening of existing micropores.

The pore volumes (narrow micropores) from the CO_2 isotherm almost equal the total volume of micropores for samples with low activations, and develop more slowly, with increasing activation than total microporosity, at medium-to-high burn-offs. Differences are small for total micropore volumes for carbons prepared in the two furnaces, increases for carbons with larger burn-off being either small or nil when prepared in the horizontal furnace. The data of Figure 5.42(a, c) are expressed per unit weight of activated carbon, independent of the yield obtained in the carbonization process because this is, commonly, the basis used by most workers in the field of activated carbon.

However, results are more clearly understood when referred to the original precursor (char), as in Figure 5.42(b, d) which are grouped by the volumes of total and narrow microporosity, respectively. Figure 5.42(b) shows that the data for the total micropore volume do not fit to a common line as well as in Figure 5.42(a, c), as the yields for carbons produced in both types of furnace are different. Chars I and III follow a similar pattern except at high bum-off (low yield), with larger micropore volumes for carbons prepared in the horizontal furnace. Char II produces similar micropore volumes but with a much lower overall yield. The development of narrow micropore volume is not straightforward. Micropore volumes are almost the same for carbons prepared in both furnaces from char I, but they differ for the other two series. Activation of chars II and III in the horizontal furnace produces a larger volume of narrow micropores in the first stages of activation (higher yield) when compared with the vertical furnace. However, Figure 5.42(d) shows that the volumes of narrow micropores for the three series decrease (char II) at the largest burn-offs (lower yield) when the horizontal furnace is used, whereas they always increase when using the vertical furnace. In any case, carbons prepared in the horizontal furnace from char I exhibit the largest values of both types of micropores.

The slower carbonization rates provided higher carbonization yields. Olive stones mainly consist of cellulose, hemicellulose, lignin and other organic macromolecular polymer systems. The slower heating rate produces a larger carbonization yield for cellulose, lignin and hemicellulose as a consequence of further dehydration and stabilization of these polymers.

5.9.9 The Year 2000: Carbon Molecular Sieves

Carbon molecular sieves (CMS) have been known and used for some time before the year 2000 when Gómez-de-Salazar *et al.* (2000) (University of Alicante) reported on the preparation of CMS using a different approach, that of controlled oxidations, as well as providing an overview of this specialized form of activated carbon. The separation ability of these CMS was studied further using immersion calorimetry by Gomez-de-Salazar *et al.* (2000) (Section 4.7) followed by a study of the use of pyrolytic carbon deposition in the preparation of CMS materials by Gómez-de-Salazar *et al.* (2005).

The separation and purification of gas mixtures by adsorption is an important unit operation in the chemical and petrochemical industries. The demand of products with high purity has increased and, therefore, attention has focused on separation processes. The two main types of adsorbents used for this purpose are zeolites and CMS. In both cases, their microporous structure allows the separation of molecules on a shape and size basis. These CMS have some advantages over molecular sieve zeolites, including shape, selectivity for planar molecules, higher hydrophobicity, higher resistance to both alkaline and acid media and thermal stability at high temperatures. CMS are prepared from a wide variety of carbonaceous raw materials, using a range of different procedures, although commercial CMS are mainly manufactured from activated carbons following deposition of pyrolytic carbon at the mouth of the pores, to "fine-tune" entrance dimensions. This procedure is not easily controllable and, without care, carbon deposits may block the entrances of narrow pores, thus decreasing the adsorption capacity of CMS. Another approach is the direct preparation of CMS from, for example, peach stone chars, by controlled oxidation using aqueous nitric acid or molecular oxygen, followed by heat treatment in an inert atmosphere.

Gómez-de-Salazar *et al.* (2000) used peach stones (as used in other studies) which were ground and sieved to a particle size between 2 and 2.8 mm, treated with dilute sulfuric acid (10 v/v) to remove mineral matter, and washed with de-ionized water to constant pH. Carbonizations were under N_2, 80 $cm^3 min^{-1}$, to 850 °C, at 2 °C min^{-1} with two intermediate isothermal stages to increase carbonization yields, soak time of 2 h. This char, so prepared, was referred to as PS. A CMS, PSAC, was prepared by activation of PS with carbon dioxide (50 vol% CO_2/N_2, 80 $cm^3 min^{-1}$, 5 °C min^{-1}, 800 °C, 3 h). PS and PSAC were then oxidized in aqueous HNO_3 (5 N) at room temperature for 3 h, washed with de-ionized water to a constant pH value and oven dried at 100 °C for 24 h. This oxidation treatment forms surface oxygen complexes in the vicinity of pore entrances and influences the effective entrance dimensions. Finally, the oxidized materials were heat treated to selected temperatures in an inert atmosphere in order to remove, sequentially, surface oxygen complexes as carbon monoxide and carbon dioxide, and so to increase effective entrance dimensions to the microporosity. These samples are PS—HNO_3(T) and PSAC—HNO_3(T), where T is the temperature of the thermal treatment (°C). Also, the char PS was oxidized with molecular oxygen by heating under helium to 850 °C and then cooled to 200 °C when pure oxygen (50 $cm^3 min^{-1}$) was introduced to the system. After 20 min, the oxygen was replaced by the helium and the temperature raised again to 850 °C in order to decompose most of the oxygen surface complexes. This sample is referred to as PS—O_2.

Amounts and type of oxygen complexes formed during oxidation, as well as their thermal stabilities, were evaluated by TPD under helium. Porosity in all carbons (previously outgassed at 250 °C) was characterized by carbon dioxide adsorption at 273 and 289 K, and CH_4, O_2 and N_2 adsorptions at 298 K (for molecular sieve studies). Heats of immersion into dichloromethane, for carbons outgassed at 250 °C under vacuum, were determined at 30 °C. Finally, the separation abilities of the different CMSs were evaluated by determining the adsorption kinetics of CO_2, CH_4, N_2 and O_2 at 298 K.

Liquid-phase oxidation, at 298 K, of PS char and PSAC CMS forms surface oxygen complexes which decompose progressively on heating in an inert atmosphere. TPD experiments carried out with the oxidized samples show typical decomposition profiles. Carbon dioxide is evolved at low temperatures, with a sharp maximum centered at 175–225 °C and a tail that extends up to about 425 °C. The CO evolution starts at about 325 °C and continuously increases to the final temperature of 850 °C. From these TPD spectra, the stability of the oxygen complexes removed by the thermal treatment was assessed. Thus, the less stable CO_2-desorbing groups are the only complexes decomposed up to 250 °C with all the CO_2 desorbing groups being removed by treatments to 400 °C. The TPD profile of sample PS—O_2 is nearly flat, with just a small amount of CO evolved at the highest temperatures (around 725 °C).

Table 5.3. Micropore volumes (V_o, DR $cm^3 g^{-1}$) and surface area accessible to dichloromethane ($S_{(DCM)}$ $m^2 g^{-1}$) for selected samples (Gómez-de-Salazar *et al.*, 2000).

Sample	PS	PS—HNO_3 (625)	PS—HNO_3 (850)	PSAC	PSAC—HNO_3 (250)	PS—O_2
V_o CO_2 273 K	0.21	0.23	0.23	0.23	0.23	0.23
$S_{(DCM)}$	–	78	316	264	35	164

Table 5.3 contains the micropore volumes of selected samples, determined by application of the DR equation to CO_2 adsorption data at 273 K, as well as the surface area accessible to dichloromethane, evaluated from the immersion enthalpy measurements and using a graphitized carbon black (V3G) as a non-porous reference. The CO_2 adsorption at 273 K, on the PS char, was lower than the other oxidized samples, and its surface area, as measured by immersion into dichloromethane, was almost zero, thus indicating the very narrow microporosity developed in this char (see also De Salazar *et al.*, 2000).

All the selected samples in Table 5.3, with the exception of PS, show the same micropore volume (CO_2 273 K, DR). However, different specific immersion enthalpies and thus accessible pore volumes (surface areas) were obtained from immersion calorimetry using dichloromethane. The values are relatively low, so indicating narrow microporosity in the samples (S), and covering a range from 35 $m^2 g^{-1}$, for sample PSAC—HNO_3 (250) to 316 $m^2 g^{-1}$ for sample PS—HNO_3 (850). Thus, although CO_2 adsorption at 273 K is not sensitive enough to detect the small textural changes developed in these samples upon oxidation and heat treatment, these textural changes can be assessed by immersion calorimetry using a liquid with a small molecular size such as dichloromethane (0.33 nm). Results reported in Table 5.3 show that the oxidation treatments with both HNO_3 and O_2, followed by the removal of the surface complexes by heat treatment, produce a relevant pore development. For the samples PS—HNO_3, microporosity is detectable only after heating at 625 °C or above, namely when all the CO_2 desorbing groups have been removed. Furthermore, the surface area accessible to dichloromethane significantly increases when the heat treatment is carried out to 850 °C. However, oxidation of PSAC with HNO_3 and thermal treatment, at only 250 °C, produces a substantial decrease in surface area when compared with the untreated PSAC. This fact can be explained as a blocking effect of the surface complexes placed at the entrance of the pores.

The adsorption kinetics of these carbons were monitored in terms of separation of mixtures of gases. Table 5.4 reports selectivity data in the form of the ratio (**s**) between the amounts adsorbed of CO_2 and CH_4 on one side, and of O_2 and N_2 on the other, measured after 2 min from the beginning of the experiment (initial pressure 0.1 MPa). Figure 5.43 compares the adsorption kinetic curves of the different gases at 25 °C for samples PS, PS—HNO_3(400), PS—HNO_3(625) and PS—HNO_3(850).

The rate of adsorption is relatively small for all four gases in the char PS, which also presents a very low adsorption capacity. As the temperature of the thermal treatment of the oxidized samples increases, and more and more surface complexes are removed, both the adsorption rate and the adsorption capacity increase, but to different extents for the various

Table 5.4. CO_2/CH_4 and O_2/N_2 selectivities (**s**) and amounts of O_2 and CO_2 (n_a (mmol g^{-1})) on selected samples in adsorption kinetic experiments at t = 2 min (Gómez-de-Salazar *et al.*, 2000).

	PS	PS—HNO_3 (400)	PS—HNO_3 (625)	PSAC	PSAC—HNO_3 (150)	PSAC—HNO_3 (250)	PS—O_2
s. CO_2/CH_4	–	–	127.9	70.6	–	–	30.1
s. O_2/N_2	–	11.3	2.8	2.5	14.2	6.1	1.91
n_a (O_2)	0.11	0.16	0.23	0.26	0.10	0.24	0.27
n_a (CO_2)	0.30	0.70	1.10	0.98	–	1.20	1.10

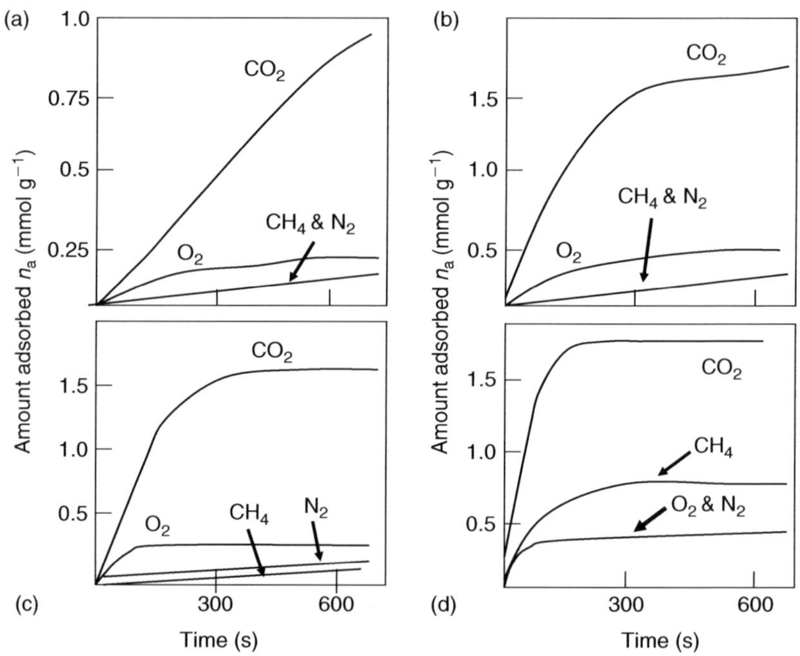

Figure 5.43. Adsorption kinetics of CO_2, CH_4, O_2, and N_2 at 25 °C on samples (a) PS (b) PS—HNO_3(400) (c) PS—HNO_3(625) and (d) PS—HNO_3(850) (Gómez-de-Salazar *et al.*, 2000).

gases. Thus, separation of N_2 and O_2 seems to the optimum in sample PS—HNO_3(400) with a selectivity of 11.3 (Table 5.4). A further increase in HTT (sample PS—HNO_3(625)) enhances the adsorption rate of both oxygen and nitrogen but decreases the selectivity. Sample PS—HNO_3(625) is the optimum for separation of carbon dioxide and methane with a selectivity as high as 128 (Table 5.4) and a relatively high adsorption capacity for CO_2. With heat treated to 850 °C, the molecular sieving properties for these gases are lost and sample PS—HNO_3(850) cannot be used for separations.

A second approach to control the entrance diameters to microporosities is to co-carbonize a parent material with a binder, as reported by Bello *et al.* (2002) using Eucalyptus globulus

Table 5.5. Characterization of micropore volumes ($V_0(cm^3 g^{-1})$) of porosity of carbons from Eucalyptus Globulus (Bello *et al.*, 2002).

Carbon sample	Precursor	Carbonization temperature (°C)	$V_0(N_2)(cm^3 g^{-1})$	$V_0(CO_2)(cm^3 g^{-1})$
C830	Eucalyptus wood	830	0.117	0.213
C870	Eucalyptus wood	870	0.024	0.237
Ci870	Ci carbon	870	0.093	0.194

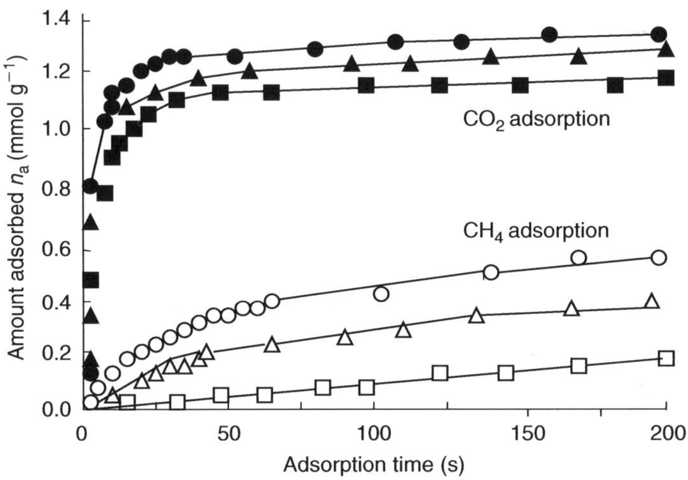

Figure 5.44. Adsorption kinetic curves for three activated carbons from Eucalyptus Globulus, upper curves for CO_2 uptake, lower curves for CH_4 uptake (Δ) = T3–10; (○) = T4–5; (□) = T6–20. Numbers 10, 5, 20 are selectivities at 30 s (Bello *et al.*, 2002).

wood (a fast growing tree) followed by steam activation. To prepare CMS, the wood (C) and a char from the wood (Ci) were carbonized at 830 and 870 °C, heating rate of 4 °C min^{-1}, soak time of 2 h, under nitrogen (300 $cm^3 min^{-1}$), thus making three carbons, C830, C870 and Ci870. These carbons were blended with a mineral tar, as a binder, stirred mechanically, dried at 110 °C, formed into pellets with a hydraulic press and activated in CO_2 (200 $cm^3 min^{-1}$) at controlled heating rates, temperature and residence time. Micropore volumes (V_0 $cm^3 g^{-1}$) were obtained from N_2 (77 K) and CO_2 (273 K) isotherms and surface oxygen contents from TPD assessments. The kinetics of adsorption of CO_2 and CH_4 were studied at room temperature. The characterizations of the C830, C870 and Ci870 are in Table 5.5 and the kinetic curves are in Figure 5.44, showing the different abilities for the CO_2/CH_4 separations, for the pellets T3 (from carbon C830), T4 (from carbon C870) and T6 (from carbon C870).

5.9.10 The Year 2000: Activation of Carbon Cloths

The studies of the Alicante group continued with the activation of carbon cloths prepared from viscous rayon, a process now patented and in commercial production in Alcoy,

Spain. Carbon cloths are also prepared in the UK, France, Japan, etc. This relatively new format for an activated carbon offers convenience for the installation of activated carbon into a wide variety of equipments. This study applied the several techniques for the assessment of variables in activation procedures which had been carefully examined in earlier publications. The objectives of this study included (i) the effect of increasing activation temperature, using carbon dioxide, on porosity and breaking load; (ii) the effect of using different activating gases, carbon dioxide and steam, on porosity and breaking load and (iii) the effect of different conditions of preparation using a two-step process, carbonization and activation and a direct, single-step activation (simultaneous carbonization and activation) on porosity and breaking load (Pastor *et al.* (1999); Rodríguez-Reinoso *et al.*, 2000a, b).

Conclusions of the study indicate that lower activation temperatures enhance adsorption capacity and micropore volumes, with slightly lower mesoporous volumes, because gasification takes place in a slower and more uniform way and thus there is a lower gradient of reactive concentration through the pore. However, a lower development of narrow micropores occurs at the expense of creating wider microporosity. There is no effect on strength.

Carbon dioxide and steam develop their porosities in different ways during activation. Carbon dioxide produces a continuous development of narrow microporosity in all of the activation stages and a slight widening from 30 wt% burn-off. Steam is different and develops smaller micropore volumes because of widening of the wider microporosity from the beginning of the process, Figure 5.45. Carbon dioxide produces a higher loss in strength during activation due to the deepening of the narrow pores.

Direct (single-step) activation of viscose rayon takes place with a higher rate than char activation, at a given temperature. When carbons obtained using similar gasification rates are compared, they are quite similar but a slightly lower surface area and micropore development takes place in the direct activation. The strength of the cloths is preserved in the initial stages of the direct process.

The breaking load of these carbon cloths is low, a major loss occurring during the carbonization stage. Improvements, therefore, must be made to this carbonization stage. To explore this, four strips of cloth were carbonized in a horizontal furnace, under nitrogen $100\,cm^3\,min^{-1}$, using heating rates of 0.5, 1, 3 and $5\,°C\,min^{-1}$ to 850 °C. A fifth strip was introduced to an isothermal stage at 230 °C (the pyrolytic decomposition temperature of the viscous rayon) and a sixth strip was heated **in air** at 230 °C for 2.7 h. The chars were activated in carbon dioxide, $100\,cm^3\,min^{-1}$, for selected periods and resultant porosities assessed as described above. Breaking loads were measured with a dynamometer (Instron 1101) following the ASTM D5034-95 standard procedure. Further, immersion enthalpies of some carbons in liquids of different molecular size, dichloromethane, benzene and 2,2-dimethyl butane were measured in a Tian-Calvet type differential micro-calorimeter (C80D, Setaram). Effects of activation on breaking load are summarized in Figure 5.46.

The slower is the heating rate of the carbonization, the more ordered is the structure within the fibers and the more ordered is the uniformity of the porosity, that is, more of the narrow microporosity. The use of the isothermal stage at 230 °C increases the total micropore volume and widens some of the narrow microporosity. The use of the pre-oxidation stage

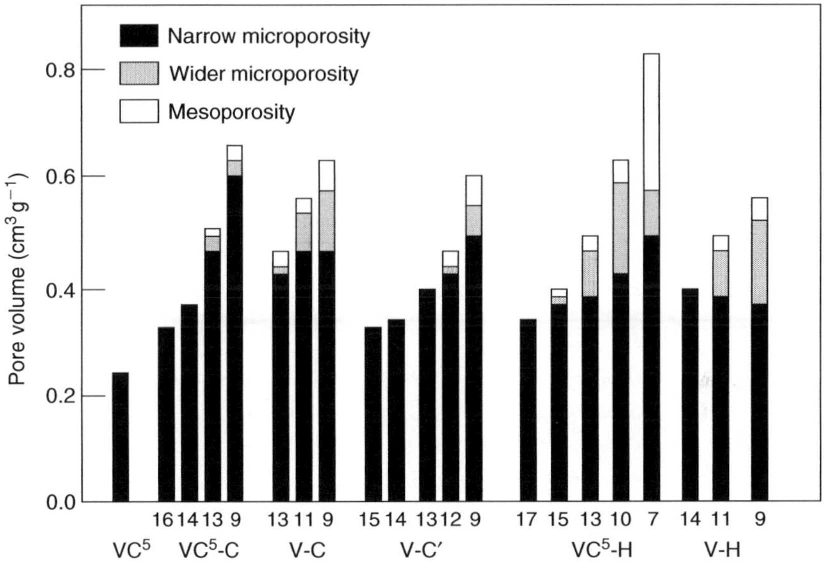

Figure 5.45. The distribution of narrow microporosity, wider microporosity and mesoporosity in viscous rayon carbons.
VC^5:char, HTT 850 °C, 5 °C min^{-1}; VC^5—C:char VC^5, activated at 825 °C in CO_2; V—C:one step activation at 825 °C in CO_2; V—C′:one step activation at 800 °C in CO_2; VC^5—H:char VC^5 activated at 750 °C in H_2O; and V—H:one step activation at 750 °C in H_2O. The horizontal numbers are yields (wt%) (Rodríguez-Reinoso *et al.*, 2000a).

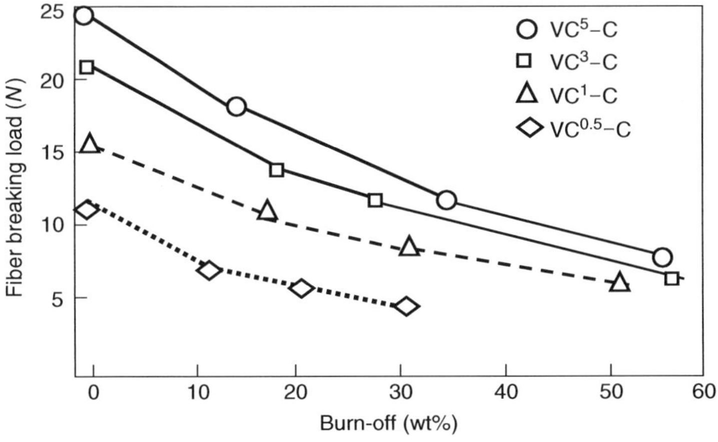

Figure 5.46. The variation in breaking load (N) of the processed viscous rayon fibers with activation yield (wt%). All samples are activated in CO_2 at 825 °C. The suffix numbers of 5, 3, 1, 0.5 refer to heating rates (°C min^{-1}) during carbonization (Rodríguez-Reinoso *et al.*, 2000b).

leads to a reduced development of narrow and total microporosity so reducing the facility for activation (gasification) by carbon dioxide. There is significant molecular sieving in all chars as shown by the heats of immersion with some of the resultant activated carbons retaining this property. The breaking loads are dependent on the initial carbonization conditions, Figure 5.46, decreasing with extent of activation and increasing rates of activation. See also Huidobro *et al.* (2001) who discuss the chemical activation of carbons from viscous rayon.

5.9.11 The Year 2001: Carbonization Under Pressure

The commercial precursors of activated carbon are brown coal, peat, wood and in general, materials containing lignin and cellulose. Carbonization of these materials involves depolymerization, cracking and dehydration of lignin and cellulose, this leading to a char with a rudimentary pore structure, and other products such as tar and volatile liquids and gases. The conditions of carbonization, such as final HTT and heating rate are set out in Section 5.9.7. However, the effects of carbonization under pressure, in terms of yield and porosity need to be explored and these effects are discussed below.

Olive stones, 2.8–3.5 mm size, were carbonized in a laboratory pilot plant with a vertical reactor to a maximum HTT of 510 °C, at 1.0 MPa pressure under nitrogen, with a heating rate of 1°C min^{-1}, Rodríguez-Valero *et al.* (2001). This operating method increases residence time of tars and volatile matter within the particle, this facilitating their conversion to solid carbon. The carbonization process was completed by heating to 850 °C in a horizontal furnace at a rate of 5 °C min^{-1} under flowing nitrogen, 80 $cm^3 min^{-1}$. Carbons were activated at 800 °C with carbon dioxide flowing at 80 $cm^3 min^{-1}$ and analyzed by adsorptions of carbon dioxide (273 K) and nitrogen (77 K).

The study concluded that the main effect of carbonizing under pressure, prior to activation, is to increase the yield and density of the final activated carbon (relative to carbonizations at 0.1 MPa), 41.8 from 32.1 wt% and 1.27 from 1.08 g cm^{-3} (Hg) respectively. The larger yield and density, due to pressure carbonization, affords the possibility of extending the activation process to a larger burn-off, thus making it possible to obtain a higher micropore volume while maintaining a reasonable yield and density. However, the extra yield and density achieved may not be enough to compensate for the additional cost of the pressure carbonizations.

5.9.12 The Year 2001: Review of Immersion Calorimetry

The use of adsorption methods to characterize porosity in carbons, and the (then considered) undisputable position of the N_2 BET isotherm appeared to be permanent. However, the advent of immersion calorimetry and its application to carbon chemistry, together with the availability of commercial calorimeters, presented a significant challenge to the supremacy of the N_2 (77 K) adsorption isotherm. The School of Adsorption, University of Alicante, made use of immersion calorimetry and reviewed their work over several years (Rodríguez-Reinoso *et al.*, 1997; Silvestre-Albero *et al.*, 2001). Immersion calorimetry, as a method of characterization, is discussed at length in Section 4.7.

Table 5.6 indicates that the sorption of nitrogen and benzene are quite comparable, with the larger two molecules, 2,2-DMB and iso-octane, indicating reduced access, especially at low degrees of burn-off.

This review concluded by stating that microporous activated carbons and CMS, can be characterized in terms of their micropore distributions and the degree of hydrophilicity of their surfaces. The value of this work is the comparison made available between data of traditional adsorption methods and calorimetric methods. Examples of this are given in Table 5.6 and Figure 5.47.

Table 5.6. Apparent surface areas ($m^2 g^{-1}$) of activated olive stones using adsorption and calorimetric methods using three molecular probes (González *et al.*, 1995b).

Carbon sample	BET N_2 (77 K)	Immersion calorimetry		
		Benzene	2,2-DMB	Iso-octane
D-8	647	754	117	66
D-19	797	817	542	463
D-34	989	1114	958	928
D-52	1271	1402	1192	1243
D-70	1426	1552	1357	1460
V3G	62			

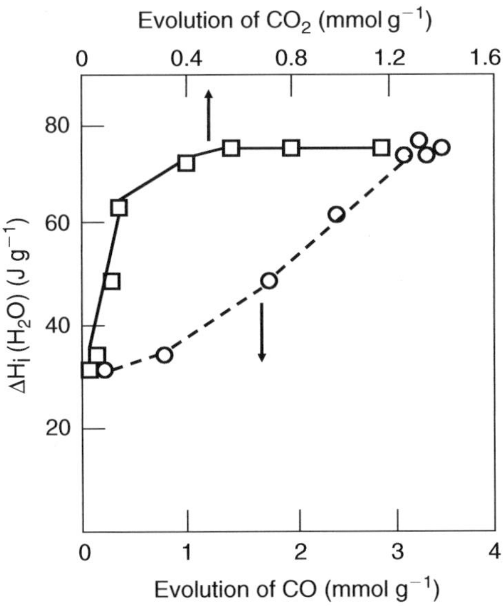

Figure 5.47. Evolution of the enthalpies of immersion (ΔH_i ($J g^{-1}$)) in water as a function of CO and CO_2 evolved from surface oxygen complexes, using an oxidized carbon from olive stones (Rodríguez-Reinoso *et al.*, 1997).

Surface polarity is of major concern with activated carbons particularly when used for adsorptions from aqueous solution. This can be assessed by water isotherms, but the process is tedious when compared to the use of an immersion calorimeter. The variation of enthalpies of immersion with coverage by surface oxygen complexes is as in Figure 5.47 which indicates the rapid increase in ΔH_i from $\sim 32\,J\,g^{-1}$, for a *clean* surface, to a maximum of $\sim 75\,J\,g^{-1}$ for carbon surfaces possessing surface oxygen complexes, for steam activated olive stones (Section 4.7.4).

5.9.13 The Year 2001: Handbook of Porous Materials

The subject area of activated carbon was by now a significant technology in several industries where the applications of carbons, prepared by thermal (as well as chemical activation) were of fundamental importance. This handbook provided a chapter (summary) of aspects which must be considered in discussions of the use of activated carbon. The chapter contains an Introduction, Production Methods, Precursors, Physical (Thermal) Activation, Chemical Activation, Combined Activations, Carbon Molecular Sieves, Activated Carbon Fibers and Cloths, Pelletized Activated Carbons, Washed, Treated and Impregnated Activated Carbon, as well as a section covering industrial production and applications. As such, this chapter is a substantial reference document and will remain so for some considerable time (Rodríguez-Reinoso, 2002).

5.9.14 The Year 2001: Carbon-sepiolite Pellets

Sepiolite is a fibrous silicate, $Si_{12}Mg_8O_{30}(OH)_4(H_2O)_4$, made up of microporous channels parallel to the fiber axis. The chemical composition and structure of sepiolite are responsible for good adsorbent behavior towards polar molecules such as water, ammonia, amines and aldehydes in both gas and liquid phases because of its hydrophilic surfaces. Activated carbon is essentially microporous and hydrophobic, making it suitable for non-polar molecules such as hydrocarbons. As these properties are complementary, a mixture of both could be useful in specific applications such as adsorption of mixtures of molecules. The preparation of discs or pellets is straightforward because in mixtures of carbon and sepiolite, the latter acts as a **binder** when adding small quantities of water.

This study of Rodríguez-Reinoso *et al.* (2001) used two series (P and C) of activated carbons prepared from olive stones using steam and KOH respectively, initially sized 2.8–3.5 mm and subsequently ground to 100 μm particles and mixed with sepiolite S (supplied by TOLSA under the name of Pangel), 30 wt% (SP30, SC30). To this mixture was added water ($1.75\,cm^3\,g^{-1}$) to obtain an extrudable mass, the pellets dried overnight at 110 °C. Higher amounts of sepiolite, 50 and 70 wt%, were used to prepare pellets SP50 and SP70. To consolidate the pellets, they were heated to 300 and 500 °C under nitrogen.

Figure 5.48 includes the adsorption isotherms of N_2 at 77 K for sepiolite S and the two original carbons (P and C). The shape of the isotherm for sepiolite is near to Type II, characteristic of this material, whereas the isotherms for the activated carbons are of Type I, as expected. Micropore volumes show differences in microporosity and micropore size distribution of the adsorbents. The similarity of $V_o(N_2\ 77\,K)$ and $V_o(CO_2\ 273\,K)$ values for sepiolite indicates the existence of both narrow and homogeneous microporosity due

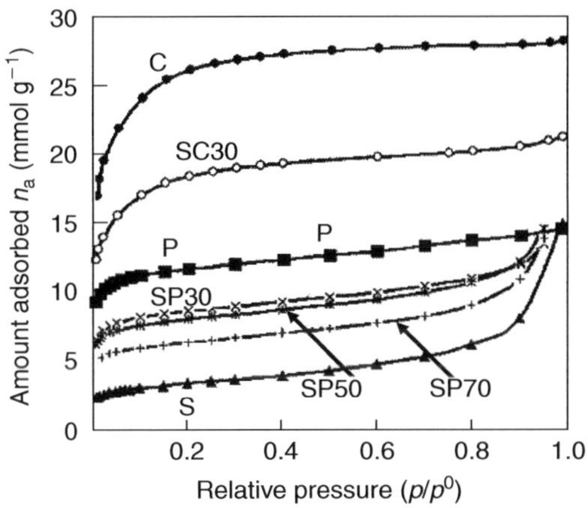

Figure 5.48. Adsorption isotherms of N_2 at 77 K for sepiolite, activated carbons, P-thermal activation – steam: C-chemical activation, SP-sepiolite-activated carbon pellets (Rodríguez-Reinoso *et al.*, 2001).

to the presence of regular 0.37 × 1.06 nm channels parallel to the fiber axis. Carbons P and C have a more developed microporosity than sepiolite, with carbon C exhibiting a wider and more heterogeneous micropore size distribution, as deduced from the larger difference in the values of $V_o(N_2)$ and $V_o(CO_2)$.

When the carbons are mixed with sepiolite there is a reduction in uptake at low relative pressures (and micropore volume), and an increase at high relative pressures, Figure 5.48. The N_2 adsorption isotherm is intermediate between those of both adsorbents, but nearer to that of the predominant component.

In order to evaluate the behavior of the pellets towards the adsorption of polar molecules, the adsorption isotherms for water vapor in air were determined at 23 °C (Figure 5.49), and show the different chemical nature of sepiolite and activated carbon. Thus, granular sepiolite S exhibits a hydrophilic behavior, whereas the type V isotherm for the activated carbon is typical of a small interaction between the carbon surface and the water molecules at low relative humidity (RH). The water adsorption isotherms, on pellets SP30 and SC30, have an intermediate shape between samples S and P or C. Furthermore, the uptake at each RH is the additive of the humidity adsorption by the individual two components, taking into account the proportion of each one in the mixture.

Adsorbents SP30, SP50 and SP70 were heat treated at 300 and 500 °C, and the micropore volumes of the pellets were calculated. In principle, these treatments are not expected to affect the porosity of the activated carbon, but the heat treatment up to 500 °C of sepiolite implies the loss of the four water molecules coordinated to magnesium two of them being lost at around 300 °C and the second two at 500 °C. The loss of water produces structural modifications in sepiolite, causing the channels to collapse to a folded structure, the microporosity now being inaccessible to nitrogen and carbon dioxide molecules. The loss of microporosity is responsible for the initial decrease observed in the volume of micropores

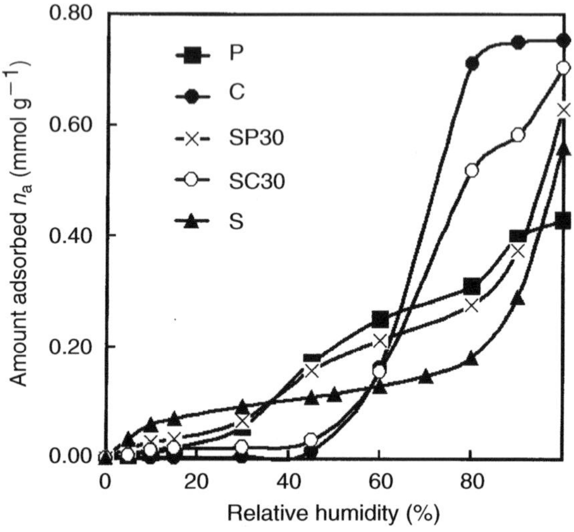

Figure 5.49. Water adsorption at 23 °C for sepiolite, activated carbons and sepiolite-activated carbon pellets (Rodríguez-Reinoso *et al.*, 2001).

of the pellet with increasing heat treatment. These experiments also show that sepiolite does not block or fill the microporosity of the carbon, the values obtained being near to those calculated from the individual components.

On the other hand, the experimental results show that the pellets resulting from the heat treatment at 500 °C are consistent when kept under water: this is not the case for pellets dried at 110 °C or heat treated at 300 °C. Thus, a multi-purpose adsorbent is easily prepared, which can he used either in gas- or liquid-phase applications in the form of pellets or monoliths.

Thus, this study illustrates, further, the versatility of activated carbons.

5.10 Activation Processes (Thermal): Summary of Effects

5.10.1 General Considerations

Typical plots of the evolution of micropore volume with increasing activation (burn-off) are depicted in Figure 5.50(a, b), for two chars, one obtained from olive stones and the second from almond shells (both with very low ash content) activated by gasification in carbon dioxide, Rodríguez-Reinoso (1991). The pore volumes, expressed per unit weight of activated carbon, increase continuously with burn-off, a similar plot being obtained for the development of surface area. This is a more conventional type of plot. From an industrial point of view it is more important to plot the parameters related to porosity expressed per unit weight of the original char, as set out in Figure 5.50(b) from the data of Figure 5.49(a). This is the general behavior found for most activated carbons prepared by thermal (physical) activation from different precursors, Rodríguez-Reinoso (1991).

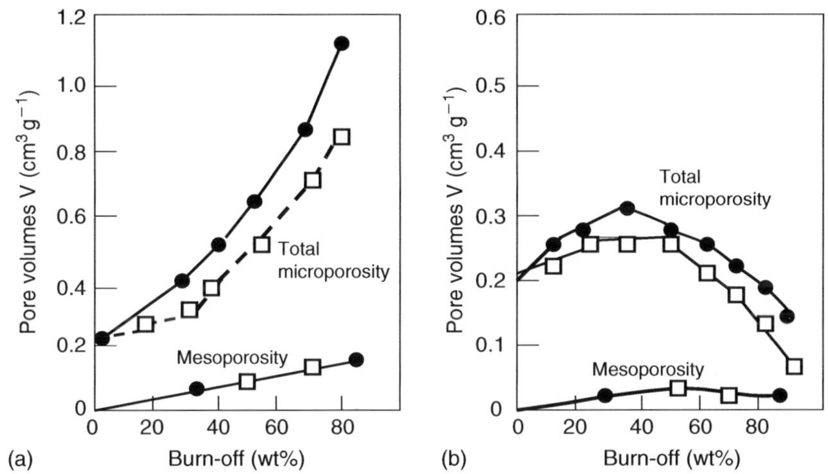

Figure 5.50. Evolution of total micropore and mesopore volume with burn-off for activated carbons obtained from almond shells (circles) and olive stones (squares): (a) per unit weight of activated carbon: (b) per unit weight of original char (Rodríguez-Reinoso (1991).

It is clear that there is an initial increase in micropore volume up to about 20 wt% burn-off, mainly produced by creation of new microporosity and the widening of existing microporosity. Thereafter, the increase in pore volume is smaller and after 40–50 wt% burn-off the micropore volume, then progressively decreases. This behavior indicates that during activation further pore enlargement, and therefore, a shift from microporosity to meso- or even macroporosity occurs. At high levels of burn-off, there is not only the widening of porosity but also the ablation of the exterior of the carbon particles. Consequently, the industrial gasification of any char very rarely goes beyond 40–50 wt%, because production is dominated by the compromise between porosity development and yield of the process.

There are three mechanisms often invoked to explain the development of porosity during gasification of a char: (I) opening of previously inaccessible pores (II) creation of new pores by selective gasification of certain structural components, and (III) widening of existing pores. The opening of inaccessible porosity is deduced from the increase in true (helium) density with burn-off, and the removal of carbon atoms from interior of the char particle is deduced from the continuous decrease of apparent (mercury) density. The fact that the increase in micropore volume (15–25%) for the char activated at around 10 wt% burn-off (Figure 5.50(a)) is larger than the fraction of carbon volume gasified is a good indication of the initial opening of previously inaccessible microporosity. The removal of more reactive carbon (e.g. from the pyrolysis of tars from the precursor) takes place at the initial stage of the activation process.

5.10.2 Comparison of Activation by Carbon dioxide and Steam

Because carbon dioxide and steam are the two main activating agents used, it is advisable to compare the porosity developed by each one. The comparison of results from the literature is difficult because the activations were carried out under different experimental conditions. However, the work of Rodríguez-Reinoso *et al.* (1995) describes the preparation of activated carbons from a common precursor (carbonized olive stones) using carbon dioxide and steam *at identical gasification rates*, in an attempt to reduce the effect of the relative differences in diffusion and accessibility of both gases to the interior of the particles. To facilitate comparisons, discussions are centred on the development of micro- and mesoporosity, with the former sub-divided into (a) narrow microporosity, as deduced form the adsorption of CO_2 at 273 K (using the DR Equation), and (b) total microporosity, as deduced from the adsorption of N_2 at 77 K (α_s-plot), Rodríguez-Reinoso *et al.* (1989a).

For olive stones, the development of total microporosity is given in Figure 5.51(a) and the volume of narrow microporosity in Figure 5.51(b), the volume being expressed per unit weight of activated carbon, Rodríguez-Reinoso *et al.* (1995). As in Figure 5.50(a), activation with carbon dioxide (series D) produces a continuous increase in the volume of micro- and mesopores. Furthermore, there is a linear relationship between the volume of narrow micropores and burn-off, and an important development of total microporosity for burn-offs above 40 wt%.

Additionally, both values of micropore volumes are coincident only up to about 20 wt% burn-off, the difference in favor of total microporosity increasing with activation, especially above 40 wt% burn-off. Consequently, the effect of activation with carbon dioxide is, simultaneously, to enhance and widen narrow micropores, the latter effect dominating above 30–40 wt% burn-off, as denoted by the increasing value of total microporosity with respect to narrow microporosity. This widening of microporosity does not correspond to a significant increase in the volume of mesopores.

Carbons activated with steam, series H in Figure 5.51(a, b) (but using the same gasification rate as for carbon dioxide activation) exhibit porosities which differ from those activated

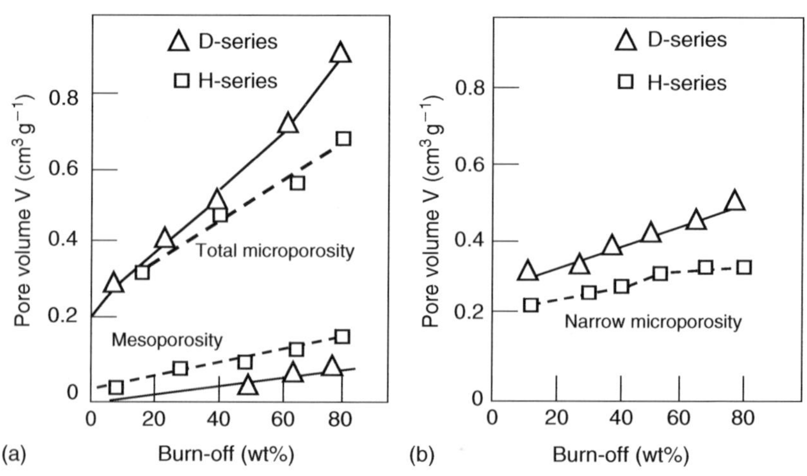

Figure 5.51. Evolution of pore volume (per unit weight of activated carbon) as a function of burn-off in carbon dioxide (series D) and steam (series H) from olive stones: (a) total microporosity, and mesoporosity: (b) narrow microporosity (Rodríguez-Reinoso *et al.*, 1995).

with carbon dioxide. The volume of narrow micropores is lower than in carbons activated with carbon dioxide, and total microporosity is similar up to 40 wt% burn-off, becoming lower thereafter. There is a slightly larger development of mesopores in carbons activated with steam.

In order to effectively analyze the creation or destruction of pores during activation, it is more convenient to use results expressed per unit weight of starting char (common for both types of activation), as in Figure 5.52(a, b). For carbons activated with carbon dioxide (series D), there is an increase in volumes of both total and narrow microporosity, up to about 20 wt% burn-off, followed by a widening when the volume of total microporosity increases and the volume of narrow microporosity decreases, up to about 40 wt% burn-off. The volume of mesopores does not change much and, in any case, it clearly decreases at higher levels of burn-off. The growth of narrow microporosity in carbons activated with steam (series H) is different, because this porosity is destroyed as from the first stages of activation. However, the volume of total microporosity increases as for carbon dioxide, up to 20 wt% burn-off, the decrease at larger degrees of burn-off taking place at a higher rate. The growth of mesoporosity is similar to that of carbons activated with carbon dioxide, although the absolute values are larger for steam (the same applies to macroporosity). Consequently, the main difference between the two activating agents is that carbon dioxide develops narrow microporosity, with a narrow pore size distribution followed, whereas steam initially widens microporosity and later produces a wider micropore size distribution.

The comparison of activation with carbon dioxide and steam is complex because both molecules have different dimensions and reactivity (the latter is not important here because

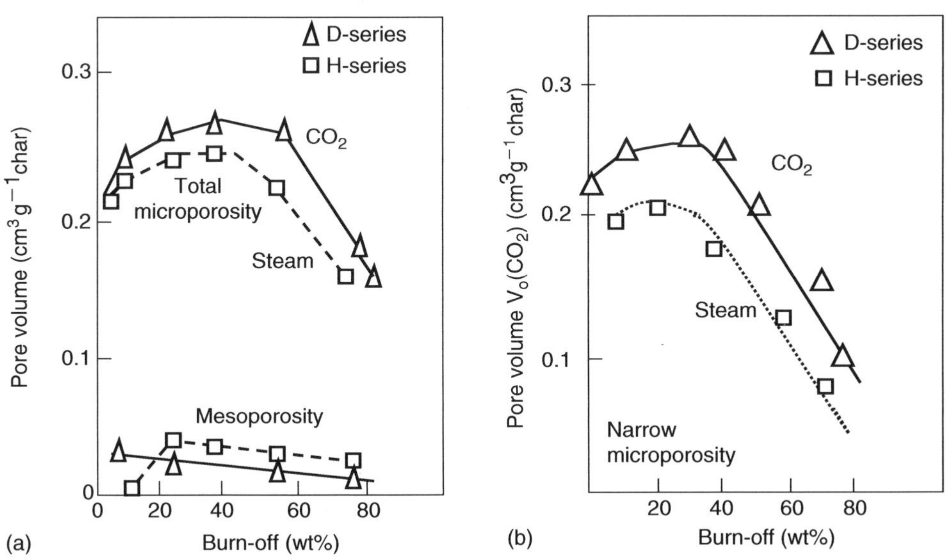

Figure 5.52. Evolution of total micropore and mesopore volumes (per unit weight of starting char) as a function of burn-off in carbon dioxide (series D) and steam (series H) from olive stones:(a) total microporosity and mesoporosity: (b) narrow microporosity (Rodríguez-Reinoso *et al.*, 1995).

the activations have been carried out at the same rate). For a given porosity, the effective diffusivity of steam is estimated to be around 50% higher than that of carbon dioxide, Koresh *et al.* (1980). Consequently, steam would be expected to develop narrow microporosity to a larger extent than carbon dioxide. As shown above and in Section 5.5.2, this is not the case and it seems as if water molecules within the porous structure react less selectively than carbon dioxide. The explanation of the differences found for carbon dioxide and steam as activating agents may be with differences in mechanisms of inhibition by chemisorbed hydrogen (C(H)) for the C—H_2O reaction compared with inhibition of the C—CO_2 reaction. Here, two inhibiting processes occur; (a) the reaction of CO with surface oxygen complexes (C(O)) and (b) the inhibition (blockage) by the (C(O)). How these three inhibition reactions for the CO_2 and H_2O reactions influence development of porosity, at the molecular reaction, is not yet understood. However, the above considerations were confirmed using activated carbon fibers and an activated carbon cloth, Alcañiz-Monge *et al.* (1994), Rodríguez-Reinoso *et al.* (2000a).

Further sets of activated carbon were prepared by activation with steam under different experimental conditions, González *et al.* (1997), indicating that the resultant pore size distributions are conditioned, principally, by the activating agent and not by an indirect influence on the final volume of pores developed.

The different porous structures produced by both activating agents have also been related to the oxygen surface groups in the carbon, Molina-Sabio *et al.* (1996). Activation with carbon dioxide creates not only a larger number of relatively stable oxygen groups (evolving as CO upon TPD), but also these groups are thermally more stable than those produced by steam activation. All these results suggest that to achieve greater uniformity of carbon gasification in the char particle, it is clear that sufficient inhibitor, CO in the case of C—CO_2 reaction or H_2 and CO in the C—H_2O reaction, needs to be added to the original reaction mixture to swamp out the effect of inhibitor concentration produced during carbon gasification, Walker (1996).

5.10.3 One-step and Two-step Activations

As mentioned in Sections 5.9.2, 5.9.8, it is possible to produce activated carbons without previous carbonization of the precursor, although the two-stage process is the one more common in industrial manufacture. In order to show the main differences between conventional and direct thermal activation, activated carbons were prepared from almond shells and olive stones using (i) the conventional carbonization (850 °C) followed by activation at 825 °C in carbon dioxide and (ii) a single (direct) activation step in carbon dioxide from room temperature to 825 or 850 °C, Rodríguez-Reinoso *et al.* (1984). Both types of activation produce a very similar yield, and this means that the direct reaction of the raw material with CO_2 during the heating from room temperature to reaction temperature is not important and does not imply a noticeable activation; in other words, its effect is similar to carbonization under nitrogen. The volume of micropores and the surface area of the resulting activated carbons (for a common yield) are relatively similar, but slightly larger for carbons prepared by direct activation. Similar results were found for other lignocellulosic materials.

5.10.4 Activation with Supercritical Water

Supercritical water (SCW) has been employed by Cai *et al.* (2004) as an efficient activating agent for the preparation of activated carbon micro-spheres (P—AC—S) with developed mesopores from phenolic resin. Several processing factors that influenced the activation reaction, including activation temperature, activation duration, supercritical pressure and water flow rate were investigated. Increasing activation temperature and duration lead to larger porosity and higher specific surface area. Supercritical pressure change has little effect on the activation process. However, there are indications that a slight increase in mesoporosity can be obtained when the pressure was raised to 36 MPa or higher. Higher water flow rate slightly enhanced the development of microporosity but had little effect on the mesoporosity. Compared with the traditional steam activation, SCW activation can produce activated carbon with more mesoporosity and higher mechanical strength. The activated carbons are dominantly microporous, with a distribution centred around 0.7 nm, with micropore volumes of $0.6\,cm^3\,g^{-1}$ ($\equiv 1200\,m^2\,g^{-1}$) with almost near vertical N_2 (77 K) isotherms at low relative pressures.

Production of activated carbon using activation of lignite or charcoal by supercritical water or carbon dioxide has been patented by Salvador-Palacios (2001) who claims that a maximum in porosity development is reached for about 45–50 wt% burn-off. When comparing supercritical activation with steam activation, it was found that the former is about 10 times faster, leading to activated carbons with larger volumes of micropores and a wider pore-size distribution, with lower ash content.

This chapter summarises the several major aspects which are necessary for an appreciation of thermal activation processes. Initially, attention is given to the kinetics and mechanisms of gasification by molecular oxygen, carbon dioxide and steam. Mobility of intermediate species over carbon surfaces is a feature of these reactions. The chemisorption of oxygen and hydrogen on carbon surfaces influences the selection of carbon atoms during gasification reactions, so controlling the nature of activation and type of porosity developed. The details of the transition states of gasification reactions are reviewed. The entire process of activation is dependent on selective gasification within a carbon particle, this being a consequence of a spectrum of reactivities of carbon atoms within their structures. The chapter concludes with a brief review of the development of understandings of thermal activation processes.

References

Adair RR, Boult EH, Freeman EM, Jasienko S, Marsh H. Catalytic gasification of doped carbon – a microscopic study. *Carbon* 1971;9(6):765–771.

Arthur J. Reactions between carbon and oxygen. *Trans Faraday Soc* 1951;47:164–178.

Bansal RC, Vastola FJ, Walker Jr PL. Kinetics of chemisorption of oxygen on Graphon. *J Coll Interf Sci* 1970;32(2):187–194.

Bansal RC, Vastola FJ, Walker Jr PL. Influence of hydrogen chemisorption on the subsequent chemisorption of oxygen on activated carbon. *Carbon* 1974;12(3):355–357.

Bello G, García R, Arriagada R, Sepúlveda-Escribano A, Rodríguez-Reinoso F. Carbon molecular sieves from Eucalyptus globulus charcoal. *Micropor Mesopor Mat* 2002;56:139–145.

Biederman DL, Miles AJ, Vastola FJ, Walker Jr PL Carbon-carbon dioxide reaction; Kinetics at low pressures and hydrogen inhibition. *Carbon* 1976;14(6):351–356.

Cai Q, Huang Z-H, Fang F, JB. Preparation of activated carbon microspheres from phenol-resin by supercritical water activation. *Carbon* 2004;42(4):775–783.

Causton P, McEnaney B. Determination of active surface area of coal chars using a temperature-programmed desorption technique. *Fuel* 1985;64(11):1447–1452.

Conesa JA, Sakurai M, Antal Jr MJ. Synthesis of a high-yield activated carbon by oxygen gasification of macadamia nut shell in hot, liquid water. *Carbon* 2000;38(6):839–848.

Crawford D, Marsh H. High resolution electron microscopy of carbon structure, *J Microsc* 1977;109(1):145–152.

Cyprès R, Planchon D, Braekman-Danheux C. Evolution of pore structure and active surface areas of coal and char during hydrogenation. *Fuel* 1985;64(10):1375–1378.

De Salazar CG, Sepúlveda-Escribano A, Rodríguez-Reinoso F. Use of immersion calorimetry to evaluate the separation of carbon molecular sieves. *Stud Surf Sci Catal* 2000;128:303–312.

Frankcombe TJ, Smith SC. On the microscopic mechanism of carbon gasification: A theoretical study. *Carbon* 2004;42(14):2921–2928.

Frenklach M, Ping J. On the role of surface migration in the growth and structure of graphene layers. *Carbon* 2004;42(7):1209–1212.

Gadsby J, Long FJ, Sleightholm P, Sykes KW. The mechanism of the carbon dioxide-carbon reaction. *Proc Roy Soc* 1948;193A:357–376.

Garrido J, Linares-Solano A, Martín-Martínez JM, Molina-Sabio M, Rodríguez-Reinoso F, Torregrosa R. Use of N_2 vs. CO_2 in the characterization of activated carbons. *Langmuir* 1987;3:76–81.

Giberson RC, Walker JP. Reaction of nuclear graphite with water vapor. Part I. Effect of hydrogen and water vapor partial pressure. *Carbon* 1966;3(4):521–525.

Gómez-de-Salazar C, Sepúlveda-Escribano A, Rodríguez-Reinoso F. Preparation of carbon molecular sieves by controlled oxidation treatments. *Carbon* 2000;38(13):1889–1892.

Gómez-de-Salazar C, Sepúlveda-Escribano A, Rodríguez-Reinoso F. Preparation of carbon molecular sieves by pyrolytic carbon deposition. *Adsorption* 2005;11:663–667.

González JC, González MT, Molina-Sabio M, Rodríguez-Reinoso F, Sepúlveda-Escribano A. Porosity of activated carbons prepared from different lignocellulosic materials. *Carbon* 1995a;33(8);1175–1177.

González MT, Sepúlveda-Escribano A, Molina-Sabio M, Rodríguez-Reinoso F. Correlation between surface areas and micropore volumes of activated carbons obtained from physical adsorption and immersion calorimetry. *Langmuir* 1995b;11:2151–2155.

González MT, Rodríguez-Reinoso F, García AN, Marcilla A, CO_2 activation of olive stones carbonized under different experimental conditions. *Carbon* 1997;35(1):159–165.

Güttler A, Zecho Th, Küppers J. Interaction of H (D) atoms with surfaces of glassy carbon: adsorption, abstraction, and etching. *Carbon* 2004;42(2):337–343.

Hahn JR. Kinetic study of graphite oxidation along two lattice directions. *Carbon* 2005;43(7): 1506–1511.

Kapteijn F, Meijer R, Moulijn JA, Cazorla-Amorós D. On why do different carbons show different gasification rates: A transient isotopic CO_2 gasification study. *Carbon* 1994;32(7):1223–1231.

Khan MR. Significance of char active surface area for appraising the reactivity of low- and high-temperature chars. *Fuel* 1987;66(12):1626–1634.

✓Koresh J, Soffer A. Study of molecular sieve carbons. Part I. Pore structure, gradual pore opening and mechanism of molecular sieving. *JCS Faraday 1* 1980;76:2457–2471. Part 2. Estimation of cross-sectional diameters of non-spherical molecules. *JCS Faraday 1* 1980;76:2472–2485. Molecular sieving range of pore diameters of adsorbents. *JCS Faraday 1* 1980;76:2567–2569.

Krasheninnikov AV, Norland K, Lehtinen PO, Foster AS, Ayuela A, Nieminen RM. Adsorption and migration of carbon atoms on zigzag carbon nanotubes. *Carbon* 2004;42(5–6):1021–1025.

Lewis JB. Thermal gas reactions of graphite. In: *Modern Aspects of Graphite Technology*, Ed. Blackburn LCF. Academic Press, London, 1970, Chapter IV, 129–199.

Long FJ, Sykes KW. The mechanism of the steam-carbon reaction. *Proc Roy Soc* 1948;193A: 377–399.

López-González JD, Rodríguez-Reinoso F. Preparation of activated carbon from olive stones. I. Carbonization and characterization of products. *An Quim* 1972a;68:135–140.

López-González JD, Rodríguez-Reinoso F. Preparation of activated carbon from olive stones. II Activation and surface area. *An Quim* 1972b;68:977–987.

López-González JD, Martínez-Vílchez F, Rodríguez-Reinoso F. Preparation and characterization of active carbons from olive stones. *Carbon* 1980;18(3): 413–418.

Lussier MG, Zhang Z. Miller DJ. Characterizing rate inhibition in steam/hydrogen gasification via analysis of adsorbed hydrogen. *Carbon* 1998;36(9):1361–1369.

Marsh H, O'Hair TE, Wynne-Jones Lord, The carbon-atomic oxygen reaction – surface-oxide formation on para-crystalline carbon and graphite. *Carbon* 1969;7(5):555–558.

Marsh H, Taylor DA, Lander JR. Kinetic study of gasification by oxygen and carbon dioxide of pure and doped graphitizable carbons of increasing heat treatment temperature. *Carbon* 1981;19(3):375–381.

✓Molina-Sabio M, Gonzalez MT, Rodríguez-Reinoso F, Sepúlveda-Escribano A. Effect of steam and carbon dioxide activation in the micropore size distribution of activated carbon. *Carbon* 1996;34(4):505–510.

Pastor AC, Rodríguez-Reinoso F, Marsh H, Martínez MA. Preparation of activated carbon cloths from viscous rayon. Part I. Carbonization procedures. *Carbon* 1999;37(8):1275–1283.

Phillips R, Vastola FJ, Walker Jr. PL. Thermal decomposition of surface oxides formed on Graphon. *Carbon* 1970;8(2):197–203.

Radovic LR, Walker Jr PL, Jenkins RG. Importance of carbon active sites in the gasification of coal chars. *Fuel* 1983;62(7):849–856.

Rafsanjani HH, Jamshidi E, Rostam-Abadi M. A new mathematical solution for predicting char activation reactions. *Carbon* 2002;40(8);1167–1171.

Rellick GS, Rodríguez-Reinoso F, Thrower PA, Walker PL Jr. Transient rates in the reaction of CO_2 with highly orientated pyrolytic graphite. *Carbon* 1975;13(1):81–82.

Rodríguez-Reinoso F, Thrower PA and Walker Jr PL. Kinetic studies of the oxidation of highly orientated pyrolytic graphites. *Carbon* 1974;12(1):63–70.

Rodríguez-Reinoso F, Linares-Solano A, Molina-Sabio M, López-González J de D. The two-stage air-CO_2 activation in the preparation of activated carbons. I. Characterization by gas adsorption. *Adsorpt Sci Technol* 1984;1:211–222.

Rodríguez-Reinoso F, Garrido, Martín-Martínez JM, Molina-Sabio M, Torregrosa R. The combined use of different approaches in the characterization of microporous carbons. *Carbon* 1989a;27(1):23–32.

Rodríguez-Reinoso F, Linares-Solano A. Microporous structures of activated carbons as revealed by adsorption methods. In: *Chemistry and Physics of Carbon*, 21, Ed. Thrower PA, Marcel Dekker Inc., New York, 1989b;1–146.

Rodríguez-Reinoso F. In: *Fundamental Issues in Control of Carbon Gasification Reactivity*. Lahaye J, Ehrburger P. (Eds), Kluwer Academic Publishers, Netherlands, Controlled gasification of carbon and pore structure development. 1991;533–571.

Rodríguez-Reinoso F, Molina-Sabio M, Gonzalez MT. The use of steam and CO_2 as activating agents in the preparation of activated carbons. *Carbon* 1995;33(1):15–23.

Rodríguez-Reinoso F, Molina-Sabio M, González MT. Effect of oxygen surface groups on the immersion enthalpy of activated carbons in liquids of different polarity. *Langmuir* 1997;13: 2354–2358.

Rodríguez-Reinoso F, Molina-Sabio M. Textural and chemical characterization of microporous carbons. *Adv Colloid Interf Sci* 1998;76–78:271–294.

Rodríguez-Reinoso F, Pastor AC, Marsh H, Martínez MA, Preparation of activated carbon cloths from viscous rayon. Part II. Physical activation processes. *Carbon* 2000a;38(3):379–395.

Rodríguez-Reinoso F, Pastor AC, Marsh H, Huidobro A. Preparation of activated carbon cloths from viscous rayon. Part III. Effect of carbonization on CO_2 activation. *Carbon* 2000b;38(3):397–406.

Rodríguez-Reinoso F, Molina-Sabio M, González JC. Preparation of activated carbon-sepiolite pellets. *Carbon* 2001;39(5):776–779.

Rodríguez-Reinoso F, Production and Applications of Activated Carbons. In: *Handbook of Porous Materials*, 3, Eds. Schüth F, Sing KSW, Weitamp J. Wiley-VCH, Chapter 4.8. 2002; 1766–1827.

Rodríguez-Valero MA, Martínez-Escandell M, Molina-Sabio M, Rodríguez-Reinoso F. CO_2 activation of olive stones carbonized under pressure. *Carbon* 2001;39(2):320–324.

Salvador-Palacios F. Process and apparatus for the production of activated carbon. European Patent EP 0974553 B1, 2001.

Shimada T, Yanase H, Morishita K, Hayashi J, Chiba T. Points of onset of gasification in a multi-walled carbon nanotube having an imperfect structure. *Carbon* 2004;42(8–9):1635–1639.

Silvestre-Albero J, Sepúlveda-Escribano A, Rodríguez-Reinoso F. Characterization of microporous solids by immersion calorimetry. *Colloid Surf A: Physico chem Eng Aspect* 2001;187–188:151–165.

Strange JF and Walker Jr PL. Carbon-carbon dioxide reaction: Langmuir-Hinshelwood kinetics at intermediate pressures. *Carbon* 1976;14(6):345–350.

Thomas JM. Microscopic studies of carbon. In: *Chemistry and Physics of Carbon*, 1, Ed. Walker Jr PL. Marcel Dekker Inc., New York, 1965;122–202.

Tomkov T, Siemieniewska T, Czechowski F, Jankowska A. Activation of brown-coal chars with oxygen. *Fuel* 1977;56(1):121–124.

Ubbelohde AR, Lewis FA. In: *Graphite and its Crystal Compounds*, Oxford University Press, 1960.

✓Walker Jr PL. Production of activated carbons: use of CO_2 versus H_2O as activating agent. *Carbon* 1996;34(10):1297–1299.

Walker Jr PL, Rusinko Jr F, Austin LG. In: *Advances in Catalysis*, 11, Eds. Eley DD, Sewood PW, Weisz PB. Academic Press, New York, 1959;133–221.

Wheeler A. In: *Advances in Catalysis*, 3, Eds. Frankenburg WG, Komarewsky VI, Rideal EK. Academic Press, New York, 1951;249–327.

Yang TR, Duan RZ. Kinetics of the reaction between steam and the basal plane of graphite. *Carbon* 1985;23(3):325–331.

Zhang Z, Lussier M, Miller DJ. Stability of hydrogen on Saran char. *Carbon* 2000;38(9):1289–1296.

Zue Z, Lu GQ Max, Finnerty J, Yang RT. Electronic structure methods applied to gas–carbon reactions. *Carbon* 2003;41(4):635–658.

CHAPTER 6

Activation Processes (Chemical)

Chapter 6 describes methods of chemical activation of carbons involving the co-carbonization of a parent feedstock with a chemical. Three chemicals, which have seen frequent use, include zinc chloride, phosphoric acid and potassium hydroxide/carbonate. Section 6.1 is an overview of the comparative behaviors of these three agents, using very similar parent feedstocks, and concluding with the suitability of such activated carbons for methane storage. Section 6.2 is an in-depth analysis of the chemistry of activation by phosphoric acid (H_3PO_4). Section 6.3 is an in-depth analysis of the chemistry of activation by potassium salts, and discusses intercalation chemistry. Section 6.4 reviews additional relevant literature. Enjoy.

6.1 Chemical Activations

6.1.1 Introduction

For some considerable time, activated carbons have been used as adsorbents in applications where impurities, in low concentrations, have to be removed (Von Kienle, 1986; Baker, 1992). Adsorption is the ideal method because it is non-specific (Molina-Sabio and Rodriguez-Reinoso, 2004). Activated carbon must possess a large volume of micropores, with an appropriate pore-size distribution, in order to adsorb molecules of different sizes. In addition, there must be an adequate proportion of mesopores to facilitate access to the micropores. It is therefore understandable that extensive research has been undertaken to provide methods of activation to develop optimum pore-size distributions to meet the broad range of industrial requirements. As the applications become more specific, a more specific control of pore-size distributions is required. For instance, the recovery of gasoline vapor requires an activated carbon with a homogeneous microporosity allowing for the adsorption and desorption of light hydrocarbons found in gasoline vapor. Additional examples are the separation of gas mixtures where the micropore width is able to differentiate between the kinetics of adsorption of two molecules with similar dimensions, and the storage of methane where a well-defined micropore size and capacity are required.

Activated carbons, with microporosity and controlled pore-size distributions, are prepared from: (i) synthetic carbon precursors mainly of polymeric type (polyimide, polyvinyl chloride, resins, etc.) and (ii) the more conventional precursors, such as lignocellulosic materials.

Previous research has shown that meso- and macroporosity in activated carbons, manufactured from lignocellulosic materials, reflect the botanic texture of the precursor (González *et al.*, 1995). Microporosity is developed by activation with gases (mainly steam or carbon dioxide), by widening the rudimentary porosity of the parent char (Rodríguez-Reinoso *et al.*, 1989a; Rodríguez-Reinoso and Linares-Solano, 1989b). Since, in thermal activations, such variables as temperature, pressure, heating rate, etc. do not influence the micropore size distribution greatly (González *et al.*, 1997), the alternative use of well-controlled chemical activations is interesting because, here, the development of porosity is substantially modified.

A typical chemical composition of lignocellulosic materials is 48 wt% C, 6 wt% H and 45 wt% O, the inorganic matter being a minor component. The corresponding atomic ratios are: O/C = 0.73, H/C = 1.5 and H/O = 2.07. As the transformation to the char requires the removal of O and H, the degree of conversion to carbon (carbonization yield) varies widely as a function of the amount of carbon being removed with O and H (as CO_x or hydrocarbons). A typical carbonization process yields around 20–30 wt%, lower than the expected value (up to 48 wt%) if water only was the result of the removal of H and O. This suggests that a chemical activation, where the activating agent is a dehydrating compound, will increase the yield but will also change the thermal degradation of the precursor, leading to a subsequent change in the evolution of porosity.

Several authors have studied the carbonization of lignocellulosic materials (e.g. Caturla *et al.*, 1991; Hu *et al.*, 2001) and found that the main degradation takes place, 200–350 °C, with evolution of H_2O, CO, CO_2, CH_4, aldehydes, etc. Distillation of heavier hydrocarbons (tar) takes place in the range 350–500 °C and above 500 °C, there is little weight loss, thus indicating that the basic structure of the char has already been formed. Chemical activation, using $ZnCl_2$, H_3PO_4, is commonly carried out at 450–600 °C. At these temperatures, the carbonization is not complete, and hence the chemical composition of the activated carbon obtained (after the elimination of the chemical by washing) is between that of the precursor and the char (carbon without activation) (Almansa *et al.*, 2004).

There is a contraction in dimensions of the lignocellulosic precursor during the carbonization process. Studies carried out with almond shells show that the weight loss of around 75 wt% is accompanied by a contraction of around 30 vol%. This change in dimension during carbonization is important in chemical activations, as the reagent has to be incorporated into the interior of the particles where it inhibits the expected contraction with increasing temperature. This means that the reagent may act as a template for the creation of microporosity. Of the many reagents proposed for chemical activation (zinc chloride, phosphoric acid, aluminum chloride, magnesium chloride, potassium hydroxide, sodium hydroxide, etc.), the most commonly used, industrially, are $ZnCl_2$, H_3PO_4 and KOH.

The objective of this section is to present a general view of the chemical activation of lignocellulosic materials for the production of granular activated carbons with a highly developed microporosity, highlighting those processes leading to carbons with uniform microporosity. These analyses of the porosity development allow for a comparison of methods of activation of each reagent. By studying similar parent materials (peach and olive stones) for these chemical activations, a better understanding of chemical activation results and is explained below.

6.1.2 Methodologies and Effects of Impregnations

Chemical activation of lignocellulosic materials is mainly directed toward the preparation of powdered activated carbons (Studies of Caturla *et al.*, 1991; Molina-Sabio *et al.*, 1995, 2003; Almansa *et al.*, 2004). As such, with the precursor being finely divided, the homogeneity of the mixture with the reagent is ensured. However, if the objective is to prepare granular activated carbon the impregnation step has to be carried out with special care to ensure an intimate contact between the precursor and the reagent.

In such studies, the reagent was dissolved in water, mixed with the precursor and the mixture kept at 85 °C without evaporation. In this way, the hydration of the precursor is facilitated, and the swelling of the interior channels of the botanic structure allows for a better access of the reactant to the interior of the particle. Once the impregnation step is complete, carbonization is carried out under a flow of nitrogen and the resulting carbon is washed to eliminate remaining chemical. The same method has also been used to prepare carbon monoliths which includes a conforming step at 130 MPa before carbonization.

Porosity and pore-size distributions were determined by gas adsorption and immersion calorimetry, with the measurement of helium and bulk densities. Volumes of micropores were calculated using the Dubinin–Radushkevich (DR) equation (Section 4.2.3) to interpret the adsorption isotherms of N_2 (77 K), CO_2 (273 K) and n-C_4H_{10} (273 K). Volumes of mesopores were evaluated by subtracting the total volume of micropores from the amount of nitrogen adsorbed at $p/p^0 = 0.95$. The two density values for each carbon were used to calculate the volume of the carbon skeleton and the total volume of pores (including the inter-particle space in monolithic disks). Immersion calorimetry of the carbon into liquids with different molecular dimensions (dichloromethane: 0.33 nm; benzene: 0.37 nm and 2,2-dimethylbutane: 0.56 nm) permits the calculation of the surface area accessible to such liquids and subsequent micropore size distributions. The adsorption of methane has been carried out at 298 K in a VTI high-pressure volumetric adsorption system. Additional techniques such as mercury porosimetry and scanning electron microscopy (SEM) have also been used for the characterization of the carbons.

6.1.2.1 Activation with $ZnCl_2$

An example of the use of $ZnCl_2$ for activation is the work of Caturla *et al.* (1991) who prepared granular activated carbons of around 2 mm particle size by chemical activation of peach stones with $ZnCl_2$. The degree of activation was evaluated from the amount of Zn (X_{Zn}) incorporated to the precursor, this ranging from 0.20 to 1.2 g Zn g^{-1} of precursor. Heat treatment was carried out in a flow of nitrogen at 500 °C. This work is reviewed, with others, by Molina-Sabio and Rodríguez-Reinoso (2004).

Figure 6.1 shows the evolution of porosity with activation. The values plotted for $X_{Zn} = 0$ correspond to the char obtained without impregnation, for which the porosity is low as carbonization is not complete. Small amounts of $ZnCl_2$ induce the initial development of microporosity. At $X_{Zn} > 0.4$ there is further creation of porosity although more slowly and with a more heterogeneous pore-size distribution. Development of porosity continues even at $X_{Zn} = 1$ (equivalent to an impregnation ratio of 2/1 weight of zinc chloride/weight of precursor), mainly corresponding to mesoporosity. The evolution of yield provides

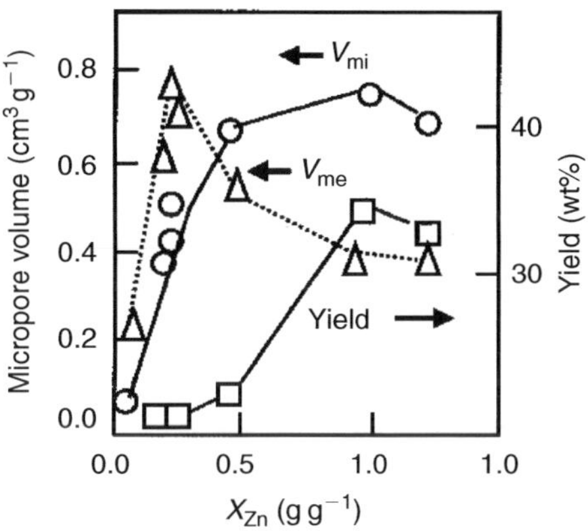

Figure 6.1. Development of micropore volumes (V_{mi}), volume of mesopores (V_{me}) and yield versus degree of activation X_{Zn} (g Zn g^{-1} precursor), prepared from peach stones (Caturla *et al.*, 1991; Molina-Sabio and Rodríguez-Reinoso, 2004).

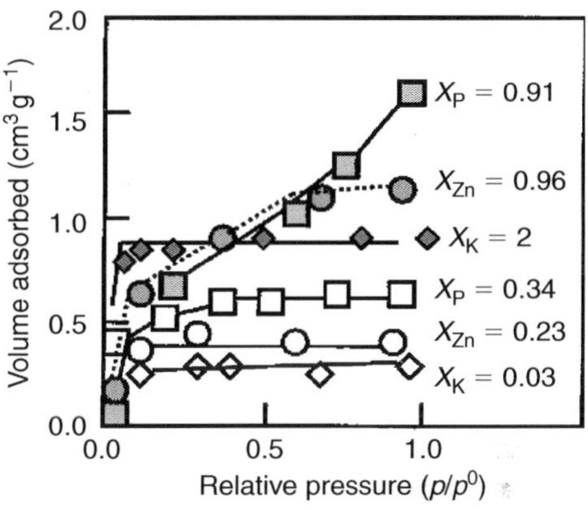

Figure 6.2. Adsorption isotherms for N_2 at 77 K for two series of carbons chemically activated with $ZnCl_2$, H_3PO_4 and KOH. The contents of Zn, P and K (g g^{-1} precursor) are used to distinguish between the carbons (Caturla *et al.*, 1991; Molina-Sabio and Rodríguez-Reinoso, 2004).

additional information as increases in yield are associated with the development of microporosity, this decreasing when mesoporosity starts to develop. It is to be noted that there is a simultaneous increase in yield and microporosity following additions of low values of X_{Zn}.

Adsorption isotherms provide further information on pore-size distributions. Figure 6.2 includes the isotherms (nitrogen at 77 K) for two carbons activated with $ZnCl_2$. The shape of the isotherm at low degree of activation ($X_{Zn} = 0.23$ g g^{-1}) indicates that the granular activated carbon exhibits only micropores with relatively uniform pore size, the contribution of mesoporosity being negligible (this information is confirmed by mercury porosimetry). The shape of the isotherm for $X_{Zn} = 0.96$ g g^{-1} indicates that there is development of small size mesoporosity as there is a plateau at high relative pressures. This evolution of porosity has been described for other lignocellulosic materials (Torregrosa and Martin-Martin, 1991).

The increase in porosity with X_{Zn} suggests that the porosity created by this reactant is due to the spaces left by zinc chloride after washing. In fact, when the volumes of micropores and mesopores are compared with the volume of zinc chloride used for impregnation (using a density for the $ZnCl_2$ solid of 2.9 g cm^{-3}) the micropore volumes are identical for $X_{Zn} < 0.4$, when the carbon only exhibits microporosity. This confirms that the reactant is behaving as template for the creation of porosity. For $X_{Zn} > 0.4$, when the increase in microporosity is less noticeable and the increase in mesoporosity is important, the volume of pores is lower than the volume of reactant. This deviation can be related to the intense chemical attack of the particle by the reactant, in coincidence with the decrease in yield observed for $X_{Zn} > 0.4$.

The effect of the heat treatment (carbonization) on the impregnated precursor has been studied from 500 to 800 °C. There is a slight weight loss, as expected when the carbonization is complete, and accompanied by a slight contraction of the particles, with a subsequent decrease in microporosity with increasing temperature, this being more noticeable when the impregnation ratio is low. It seems then that zinc chloride plays an important role during the thermal degradation of the impregnated particle up to 500 °C, but without further reaction at higher temperatures.

In summary, the information establishing an understanding of the activation mode of zinc chloride in the preparation of activated carbon are: (i) the volume of micropores developed is similar to the volume of zinc chloride introduced into the particle, and the microporosity is uniform and (ii) zinc chloride acts at temperatures <500 °C.

6.1.2.2 Activation with H_3PO_4

In order to study activations by phosphoric acid, Molina-Sabio *et al.* (1995) prepared granular activated carbons (particle size around 2.5 mm) by impregnating peach stones with solutions of phosphoric acid of various concentrations. The degree of activation is defined by the amount of phosphorous incorporated into the particles, X_P, which ranged from 0.09 to 0.91 g (P) g^{-1} precursor. The impregnated material was heat treated under a flow of nitrogen at 450 °C and then washed until no reactant was present. This study, together with other studies, is reviewed by Molina-Sabio and Rodríguez-Reinoso (2004).

Figure 6.3 shows the relationship between porosity and impregnation ratio. There is a rapid development of microporosity as from low concentrations, this being the predominant porosity up to $X_P = 0.3$, with volumes up to 0.6 $cm^3 g^{-1}$. For higher concentrations of H_3PO_4, the development of microporosity is negligible, the main development corresponding to mesoporosity, which reaches values higher than the micropore volume. Phosphoric acid also exhibits a concentration range in which the development of microporosity is coupled with a high yield, the later decreasing when mesoporosity is developed.

The shape of the adsorption isotherms for carbons prepared with low values of X_P are typical of microporous carbons, but increasing degrees of activation produce a heterogeneity of the micropore size and later a mesoporosity. As typical examples, Figure 6.2 (in an earlier page) includes the isotherm for: (i) the carbon activated with $X_P = 0.34\,g\,g^{-1}$, which is of an essentially microporous carbon with wide micropore size distributions and (ii) the carbon activated with $X_P = 0.91$, showing an important contribution of wider mesoporosity. It seems that a characteristic of the carbons activated with phosphoric acid is the heterogeneity of the microporosity (Laine and Lunes, 1992). Such heterogeneity can be evaluated by comparing the volume of micropores (V_{mi}) deduced from the adsorption of *n*-butane and CO_2 at 273 K, as in Figure 6.4. The higher values of (V_{mi} *n*-butane) with respect to ($V_{mi}N_2$) are the consequence of different molecular dimensions of both species (0.43 and 0.36 nm, respectively). It is established that in adsorbents with a wide micropore size distribution the limiting factor for the microporosity is the ratio pore width/molecule diameter rather than the actual pore width (Gregg and Sing, 1982).

The reason for the higher value of ($V_{mi}N_2$) with respect to ($V_{mi}CO_2$) is that both gases measure a different range of microporosity; N_2 (in association with the α_s-plot) measures the total microporosity, whereas CO_2 (in association with DR equation) measures only the

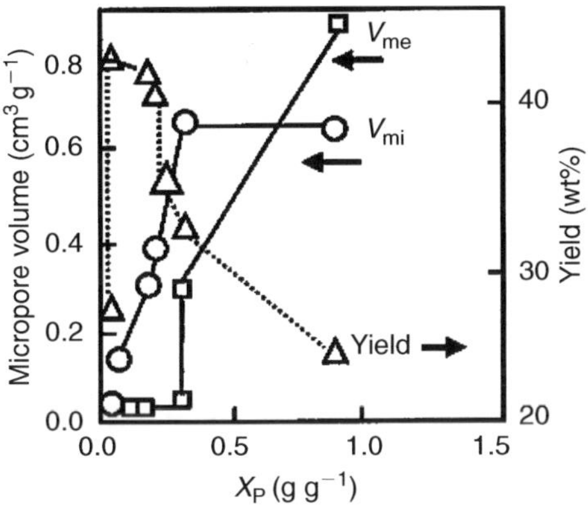

Figure 6.3. Development of the volume of micropores (V_{mi}), volume of mesopores (V_{me}) and yield versus the degree of activation X_P (g (P) g^{-1} precursor) of peach stones (Molina-Sabio and Rodríguez-Reinoso, 2004).

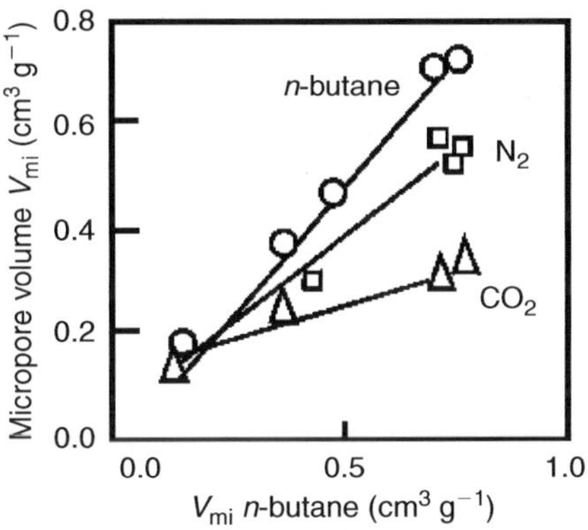

Figure 6.4. Comparison of micropore volume deduced from the adsorption of nitrogen, carbon dioxide and *n*-butane (Molina-Sabio and Rodríguez-Reinoso, 2004).

narrow microporosity (the higher adsorption temperature reduces considerably the relative pressure range reached at atmospheric pressure) (Garrido *et al.*, 1987; Rodríguez-Reinoso and Linares-Solano, 1989b). However, the more important information deduced from Figure 6.4 is the linear relationship between the three values for micropore volume. It seems as if the widening of microporosity induced by phosphoric acid leads to an increase in the volume of micropores, with all pore sizes being developed in a common proportion, thus leading to activated carbons with different volumes of micropores but with the same micropore size distribution.

The volume occupied by the "phosphoric acid" in the interior of the particles can be calculated by considering that at 450 °C the composition is 82 wt% P_4O_{10}–18 wt% H_2O with a density of 2.06 g cm^{-3}. The volume thus deduced is coincident with (V_{mi} *n*-butane) up to values of 0.4 cm^3 g^{-1}. This indicates that there is a wide range of concentration in which the reactant inhibits the contraction of the precursor particle during the heat treatment. The washing of the heat-treated particles leaves behind the porosity. For carbons with higher impregnation ratios (X_P 0.63 and 0.91 g g^{-1}), the volume of the porosity is larger than the volume of the reactant. In these cases, the chemical attack of the highly concentrated phosphoric acid produces the development of meso- and macroporosity.

In summary, it is found that: (i) the volume of micropores developed during activation is similar to the volume of the acid used in the impregnation of the precursor, the heterogeneity of the microporosity being constant in all carbons and (ii) phosphoric acid acts at temperatures <450 °C leading to highly activated carbons.

6.1.2.3 Activation with KOH

Molina-Sabio and Rodríguez-Reinoso (2004) describe a study of the preparation of granular activated carbons using KOH activation without the disintegration of the carbon granules into a powder. They describe the KOH activation of olive stones (particle size around 3 mm) and their chars obtained by their carbonization in nitrogen at 300, 500 and 800 °C, and subsequently impregnated with aqueous solutions of KOH in the ratio 0.005–3 g (KOH) g^{-1} of precursor before being heated under nitrogen. When impregnation ratios above 0.1/1 were used the impregnated olive stone particles sometimes disintegrated to a powder. For ratios above 3/1 $X_K = 2\,g\,g^{-1}$, the impregnated granular char particles always become a powder, after the heat treatment. Further, the temperature of the heat treatment was found to be important for the development of porosity. Porosity in carbons activated at 500 and 700 °C is low, independent of the precursor (original or char) used, whereas temperatures of 850 and 900 °C lead to activated carbons with increasing porosity, thus suggesting a critical temperature range of activation by KOH of carbons. A similar behavior is reported by Hayashi *et al.* (2002) for activation of different lignocellulosic materials with potassium carbonate. In addition, yields around 20–25 wt%, similar to the yield by carbonizing the original precursor, were associated to a low development of porosity, whereas in the more activated carbons the yield is lower, around 10–15 wt% (which in turn is similar to the overall yield in thermal activation of lignocellulosic materials). This suggests that development of porosity by KOH activation may be associated with gasification reactions.

As well as temperature, it is apparent from the above results that the proportion of KOH in the activation system controls extents and dimensions of developed porosity. At low impregnation ratios, for example 0.005/1 (equivalent to $X_K = 0.03\,g\,g^{-1}$) the shape of the isotherm (Figure 6.2) indicates, predominantly, narrow and uniform microporosity. For a high impregnation ratio, for example 3/1 (equivalent to $X_K = 2\,g\,g^{-1}$) microporosity is high. This aspect of activation therefore needs to be explored more deeply and this was done in Figure 6.5.

The discussions of Chapter 4 (Section 4.2.5) conclude that the nitrogen (77 K) and carbon dioxide isotherms (273 K) provide micropore volumes in the wider and narrower of the

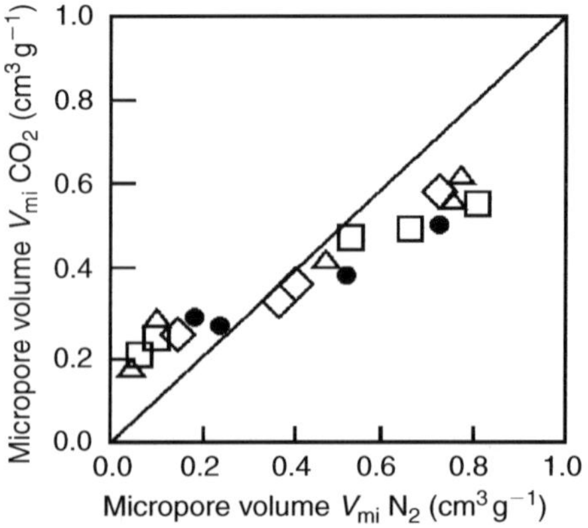

Figure 6.5. Relationship between the volume of micropores deduced from the adsorption of nitrogen and carbon dioxide (all carbons activated with KOH). The precursor is olive stone (●) or the char obtained at 300 (□), 500 (Δ) or 800 °C (◇) (Molina-Sabio and Rodríguez-Reinoso, 2004).

micropores. Figure 6.5 contains the volume of micropores for a range of KOH activated carbons (available to the authors), measured by both nitrogen and carbon dioxide adsorption (a line of slope unity is included as reference). When the impregnation ratio is low the volume of micropores is relatively low, about $0.2\,cm^3\,g^{-1}$, similar to a char ($V_{mi}CO_2$) is larger than ($V_{mi}N_2$) for these carbons, is typical of narrow and uniform microporosity, with values of about 0.35–0.37 nm, for which the access of nitrogen at 77 K is kinetically restricted (it may require a very long time – even weeks – to reach equilibrium). When the volume of micropores is larger, both values become similar ($0.4\,cm^3\,g^{-1}$) and later ($V_{mi}N_2$) > ($V_{mi}CO_2$), when the micropore size distribution is wider. The development shown in Figure 6.5 is independent of the type of precursor (olive stones or chars), although the later require a larger amount of KOH to reach a given volume of micropores. It seems that the interaction with the KOH is more effective for the non-carbonized material than for the char. For medium-to-high impregnation ratios, the KOH does not reach the interior of the char particles, remaining as a solid on the external surface. It was also observed that, independent of the precursor, the maximum value of micropore volume, for granular activated carbons, is $0.4\,cm^3\,g^{-1}$, when ($V_{mi}N_2$) = ($V_{mi}CO_2$). Above this value, the activated carbons obtained are in the form of powder.

The micropore size distribution of some activated carbons, as deduced from immersion calorimetry, is plotted in Figure 6.6. The volume deduced from the adsorption of CO_2 at 273 K has also been included for a comparison. The carbons with up to $0.4\,cm^3\,g^{-1}$ exhibit some molecular sieving effect for molecules above 0.56 nm because the access of 2,2-dimethylbutane is very restricted. This means that in this range, the main pore size developed is around 0.37 nm. No molecular sieving effect is detected above $0.4\,cm^3\,g^{-1}$ when ($V_{mi}N_2$) > ($V_{mi}CO_2$).

In summary, the activation mode of KOH is (i) the reactant acts after the pyrolysis of the precursor, at temperatures above 700 °C; (ii) the development of porosity relates to the extent of impregnation by the KOH; (iii) initially, narrow microporosity is formed followed by the

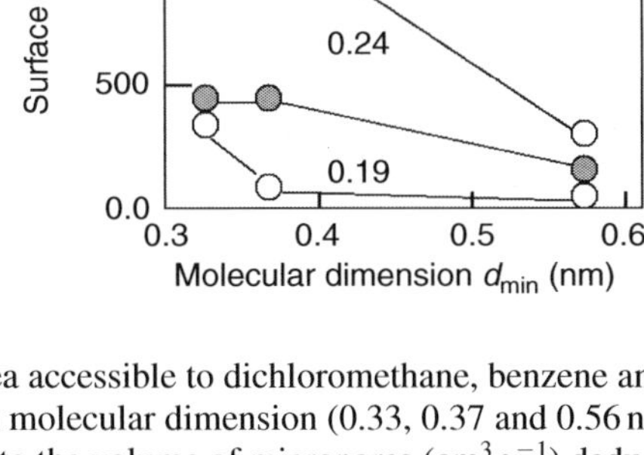

Figure 6.6. Surface area accessible to dichloromethane, benzene and 2,2-dimethylbutane at 30 °C versus the minimum molecular dimension (0.33, 0.37 and 0.56 nm, respectively). The numbers included correspond to the volume of micropores ($cm^3\,g^{-1}$) deduced from the adsorption of carbon dioxide for carbons activated with KOH (Molina-Sabio and Rodríguez-Reinoso, 2004).

wider microporosity at the expense of the narrow microporosity; (iv) activation can proceed without disintegration to form powder and (v) high ratios of KOH to carbon result in the disintegration of the carbon granules to powder.

6.1.2.4 Comparison of mechanisms of activating agents

When chemical activation is considered as a reaction between a solid precursor and the chemical, then concentration, intimacy of mixing, temperature and activation time determine the extent of reaction. For the three activating agents, $ZnCl_2$, H_3PO_4 and KOH, optimized conditions lead to activated carbons with micropore volumes of 0.5, 0.6 and $0.4\,cm^3\,g^{-1}$, respectively.

Impregnation may lead to fragmentation of cellulose and other components of the botanic precursor such as hemicellulose and lignin. Although the three chemicals react with the precursor, some clear differences can be observed at the end of the impregnation step (see Section 6.2).

Particles impregnated with phosphoric acid become elastic and this is explained as follows. The acid separates the cellulose fibers and produces a partial de-polymerization of hemicellulose and lignin (the main components of the matrix) this leading to a decrease in mechanical resistance. Both factors lead to the swelling of the particle. Impregnation also starts the conversion to carbon because a significant amount of tar is observed on the surface of the particles. The tars are the result of the de-polymerization of cellulose catalyzed by phosphoric acid, followed by dehydration and condensation (also catalyzed by the acid), leading to more aromatic and reactive products, with some cross-linking (Solum *et al.*, 1995). Additional cross-linking may be induced by the presence of phosphates.

The de-polymerization and later re-polymerization of cellulose produced by phosphoric acid can be observed in SEM images (Molina-Sabio *et al.*, 1995). The morphology of the carbon impregnated with low impregnation ratios is almost identical to that of the char, but increasing concentrations of the acid show a surface more clearly attacked. At high concentrations, the original cellular morphology of the precursor is lost because a large part of the cellulosic structure has been degraded and extracted from the interior to the exterior of the particle. The intensive re-organization modifies the initial characteristics of the precursor and this modifies the meso- and macroporous structure.

When the degradation of lignin and cellulose is catalyzed by protons, zinc chloride – which at high concentrations gives Brønsted acidity to the solution and dissolves the cellulose – must act in a similar way during impregnation. However, visual changes during impregnation are observed only at high concentrations, because of the lower acidity in comparison with phosphoric acid and the absence of phosphates. The changes are similar to those described above for phosphoric acid including swelling of the particle, lower mechanical resistance and presence of tars on the surface.

No significant visual changes are observed during impregnation with KOH. Only a slight dehydration, with no formation of tar, is seen at low concentrations, but for higher impregnation ratios particles disintegrate to a powder. It is possible that reactions similar to those of a Kraft reaction may take place. The presence of strongly nucleophilic hydroxyl ions produces some fragmentation and solubilization of lignin and hemicellulose because they are more amorphous and less polymerized than the cellulose component.

$ZnCl_2$ and H_3PO_4, in the interior of the particles, produce a dehydrating effect on the cellulose, hemicellulose and lignin components during heat treatment. Dehydration is possible because the chemical is a liquid at the temperature of the process, thus facilitating the bonding to the precursor being thermally degraded. The precursor is able to transfer water to the reactant in the reacting mass to form a hydrated compound which then loses water with increasing temperature. The dehydration produced by both zinc chloride and phosphoric acid is strong, whereas that of KOH does not seem to affect carbonization. The yield reaches values of up to 44 wt% for $ZnCl_2$ (Figure 6.3), and a larger concentration of chemical is required. The values for H_3PO_4 are similar (Figure 6.1) even at low concentrations. For KOH, the maximum yield was 25 wt%, similar to that of the unimpregnated precursor.

Dehydration of the precursor produces a reduction in the dimensions of the particle, although such reduction is partially inhibited because the reactant remains inside during the thermal treatment, thus acting as template for the creation of microporosity. The small size of the $ZnCl_2$ molecule or its hydrates explains the small and uniform size of the micropores created. This does not happen for H_3PO_4 because there are no phosphoric acid molecules, but a mixture of molecules from the small H_3PO_4 and $H_4P_2O_5$ to $H_{13}P_{11}O_{34}$ in the proportion predicted by the liquid–vapor phase diagram for the P_2O_5–H_2O mixtures at the maximum temperature of the treatment, 450 °C. This leads to heterogeneity in the microporosity which is almost independent of the fraction of phosphorous present (X_P).

As KOH does not act as a dehydrating agent on the precursor, it does not react during pyrolysis and it does not inhibit the contraction of the particles upon heat treatment. It starts to react above 700 °C, after the formation of the char and in this sense the activation modes for the original precursor and its char are very similar. It is believed that activation initially consists of a redox reaction, where carbon is oxidized to CO or CO_2, thus, incidentally, creating some porosity and K_2CO_3 as a by-product (Lillo-Ródenas *et al.*, 2003). As a result, KOH is reduced to metallic potassium (Marsh *et al.*, 1984). Since potassium is *intercalated* between the graphene layers of structure the carbon particles break down. In extreme cases, the carbon particles disintegrate into powders.

A comparable study of the activation of *coal-tar pitch fibers*, using CO_2, NaOH and KOH as activating agents, was made by Maciá-Agulló *et al.* (2004). These authors report that NaOH developed the highest value of porosity and KOH developed carbons with a narrow micropore size distribution. Differences in the behavior of NaOH and KOH may be related to differences in the way that sodium and potassium metal atoms (initially as intercalates) are *de-intercalated* (with exfoliation) from the carbon, almost explosively (see Section 6.3.2). In order to compare the results with those obtained by physical activation, the pitch-based fibers were reacted with CO_2, 820–900 °C, flow rate of 100 $cm^3\,min^{-1}$, in a horizontal furnace, for 2.5–26 h. A principal conclusion is that chemical activation produces similar or higher porosity than does physical activation. There are two advantages to the use of chemical activation methods namely (i) higher yields, 27–47 wt% compared with 6 wt% for physical activation and (ii) the surfaces of the activated fibers prepared by chemical activation are less damaged, a factor when considering breakage of fibers.

6.1.3 *Monoliths made from Activated Carbon via Chemical Activation*

The preparation of binderless carbon monoliths by carbonizing conformed cellulose and then activating by reaction with carbon dioxide has already been described (Inomata *et al.*, 2002). Physical activation of such monoliths requires a reaction with carbon dioxide inside the monoliths. Severe diffusion restrictions may apply to such a procedure. Monoliths, by not being a powder or granular, are of high density and have relevance to adsorption processes in limited volumes as, for example in storage of methane.

However, for chemical activation, with the activating agent, from the beginning, being part of the monolith, then activation would proceed from inside the particle, a process quite different from activation, externally, using carbon dioxide. After heating, the observed presence of tars on external surfaces of the precursor particles (previously reported) suggests that conforming under pressure and heating, using impregnated particles would produce enough tar to bind the particles. Furthermore, because the impregnated particles are somewhat elastic (as a consequence of the weakening produced by impregnation), conforming under pressure reduces inter-particle space, thus increasing the bulk density and the adsorption capacity expressed as a volume per volume unit, V/V, (Parkyns and Quinn, 1995; Menon and Komarneni, 1998; Cook *et al.*, 1999).

Molina-Sabio *et al.* (2003), conformed under pressure, 130 MPa, the impregnated precursor (olive stones) with heating to 100 and 150 °C, using phosphoric acid with further heat treatment to 450 °C. The procedure for the preparation of the monolithic disks (20 mm diameter and 10 mm height) uses particle sizes of olive stones of 0.1–0.5 mm. For this H_3PO_4 study, additions were 29, 36, 44 and 52 wt% providing impregnation ratios (X_P g phosphorous:g precursor) of 0.21, 0.28, 0.35 and 0.42, respectively. Results are in Figure 6.7(a, b) which show that a maximum of micropore volume of about 0.8 $cm^3 g^{-1}$ results when the value of

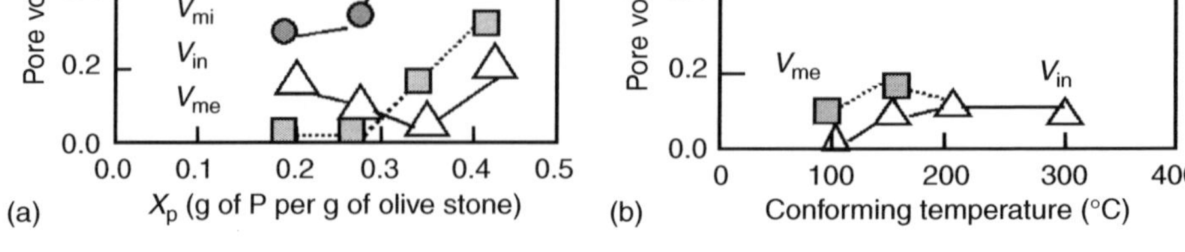

Figure 6.7 (a) The variation of total volumes of microporosity, mesoporosity and inter-particle space with impregnation ratios of phosphorous, for monoliths prepared from olive stones with impregnation by H_3PO_4 (Molina-Sabio *et al.*, 2003). (b) The variation of total volumes of microporosity, mesoporosity and inter-particle space with conforming temperature, for monoliths prepared from olive stones with impregnation by H_3PO_4 (Molina-Sabio *et al.*, 2003).

X_P is about 0.35 ($g\,g^{-1}$). The optimum temperature of conformation is 100 °C. No carbon disks were prepared for the KOH impregnated precursor because of the small effect of this chemical on the pyrolysis stage.

Almansa *et al.* (2004), using olive stones with solutions of $ZnCl_2$, prepared binderless activated carbon disks by pressure compaction at 130 MPa, at 150 and 300 °C, finally heating to 500 °C, using 19, 24, 32, 38 and 48 wt% of $ZnCl_2$. Volumes of micro- and mesoporosities were obtained from isotherms of N_2 at 77 K. Results are in Figures 6.8 and 6.9. The disk 300

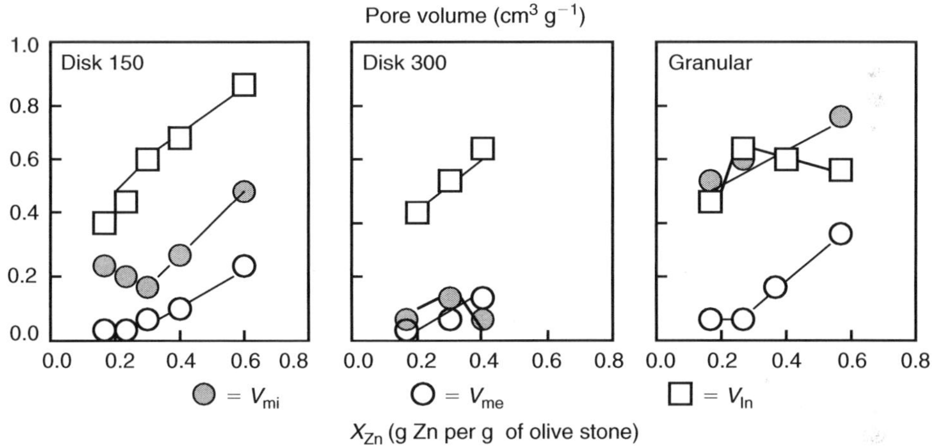

Figure 6.8. The development of volumes of micropores (V_{mi}), mesopores (V_{me}) and inter-particle spaces (V_{in}) with impregnation ratio (X_{Zn}) for the disks compacted at 150 and 500 °C and for granular activated carbon (Almansa *et al.*, 2004).

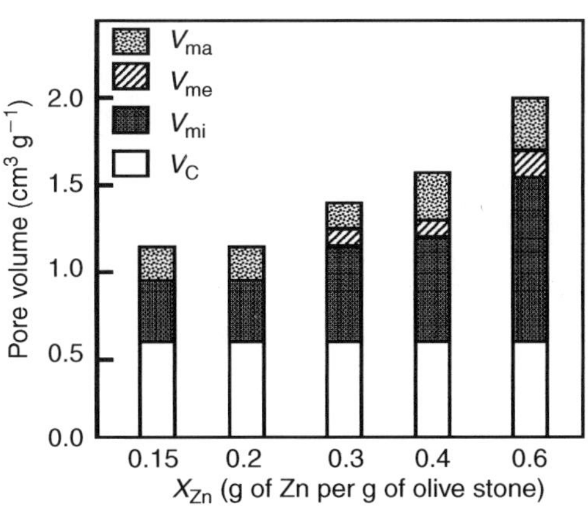

Figure 6.9. Distribution of the different volumes in disks activated with $ZnCl_2$. X_{Zn} is the content in Zn after impregnation. V_C is the volume of the carbon skeleton. V_{mi} is the volume of micropores deduced from the adsorption of nitrogen. V_{me} is the volume of mesopores ($V_{0.95} - V_{mi}$). V_{ma} is the volume of macropores and inter-particle voids. HTT is 500 °C (Almansa *et al.*, 2004; Molina-Sabio and Rodríguez-Reinoso, 2004).

has a dominance of micropores with a minimum of mesoporosity and inter-particle space. Helium densities and methane capacities are also reported (Almansa *et al.*, 2004). Such carbons offer possibilities of methane storage, but require a higher micropore content.

A comparison of Figures 6.7(a, b) and 6.8 indicates that both activating agents lead to disks with similar volumes of carbon skeleton, about $0.6\,cm^3\,g^{-1}$ and a large volume of micropores in respect to other volumes of pores, which are low when $X_P = 0.35$ and $X_{Zn} = 0.30$. For these impregnation ratios, considered above as optimum, the volume of micropores constitutes 86% and 80% of the total volume of pores and voids, the disks exhibiting bulk densities of 0.71 and $0.74\,g\,cm^{-3}$, respectively.

Distributions of the different volumes in disks activated with $ZnCl_2$ are as in Figure 6.9. X_{Zn} is the content of Zn after impregnation, V_C is the volume of the carbon skeleton, V_{mi} is the volume of micropores deduced from the adsorption of nitrogen, V_{me} is the volume of mesopores ($V_{0.95} - V_{mi}$) and V_{ma} is the volume of macropores and inter-particle voids (Molina-Sabio and Rodríguez-Reinoso, 2004). The volume for the carbon skeleton is 50% of the total in disks for $X_{Zn} = 0.15$, and the volume of micropores is 42% of the total in disks for $X_{Zn} = 0.40$.

Should the activated carbon be used for methane storage in a vessel at a pressure of 3.4 MPa, it is necessary to distinguish between: (i) the volume of the carbon skeleton, V_c, which reduces the effective volume of that vessel; (ii) the non-microporous volume, which is the summation of the volumes of mesopores, macropores and inter-particle voids (in this non-microporous volume the density of adsorbed methane is low, similar to that of compressed methane at 3.4 MPa, i.e. $0.023\,g\,cm^{-3}$), and not very useful for storage and (iii) the volume of micropores where the density of the adsorbed methane will be high due to the high adsorption potential within the micropores. The later is the important volume for gas storage when in competition with compressed natural gas (21.0 MPa). Consequently, the carbon for this application must have the maximum volume of micropores and a minimum volume for other pores and voids.

Figure 6.10 describes the development of volumes of methane adsorbed at 25 °C and 3.4 MPa as a function of the impregnation ratio of phosphorous and zinc. In general terms, this development is similar to that seen in Figure 6.9, thus indicating that the microporosity and the amount of methane adsorbed are related, but the correlation is not that good. Thus, the largest amounts adsorbed are 185 and $150\,cm^3\,g^{-1}$ (STP) for $X_P = 0.35$ and $X_{Zn} = 0.40$, whereas the volume of micropores for the disk with $X_{Zn} = 0.6$ was the largest (Figure 6.9). It is more relevant to know the volume of methane adsorbed in a given volume of carbon. Hence, when the values of Figure 6.10 are converted to V/V units, values are obtained of 105, 101, 131 and $88\,cm^3\,cm^{-3}$ for values of X_P of 0.21, 0.28, 0.35 and 0.42, respectively. These values are similar to those of Figure 6.10 for the lower impregnation ratios because the bulk density of the disks is around $1\,g\,cm^{-3}$, the capacity increasing for $X_P = 0.35$. The capacities for the disk activated with zinc chloride are 80, 90, 94, 96 and $70\,cm^3\,cm^{-3}$ for values of X_{Zn} of 0.15, 0.20, 0.30, 0.40 and 0.60, respectively. The tendency is also similar to that of Figure 6.10, although the increase is smoother because of the decrease in bulk density.

Having obtained monoliths by direct activation with H_3PO_4, and having potential for methane storage, a further procedure is to assess if physical activation of these monoliths

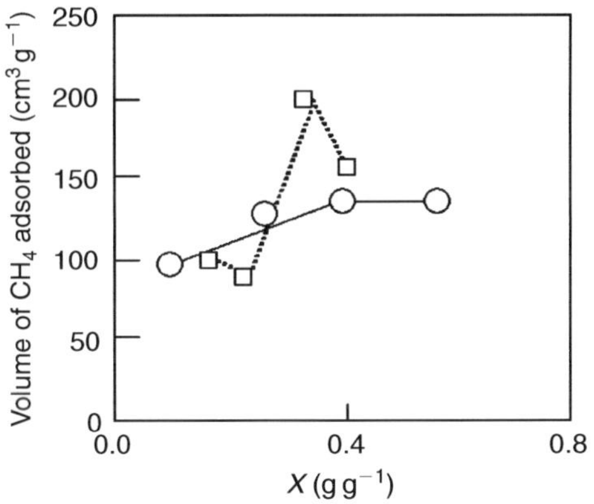

Figure 6.10. Volume of methane adsorbed at 3.4 MPa by the carbon disks activated with H_3PO_4 or $ZnCl_2$. X is the content of P (□) or Zn (○) after impregnation (Molina-Sabio and Rodríguez-Reinoso, 2004).

using carbon dioxide would offer an improvement in values of V/V. Accordingly, Almansa *et al.* (2004) selected the monoliths from $X_P = 0.21, 0.28, 0.35$ and 0.42 and reacted them with carbon dioxide at 725 °C at weight loss rates of 1.0, 1.1, 1.5 and 1.25 wt% h^{-1}. It was found that activation in carbon dioxide produces an increase in the adsorption of methane, this being larger for the two lower impregnation ratios with values reaching up to 150 V/V (20 wt% burn-off, $X_P = 0.21$) to be compared with 131 V/V for direct chemical activation.

6.1.4 Density of Adsorbed Methane

It is convenient to discuss methane storage in this section because of the relevance to microporosity generated by chemical activation, the use of physical activation methods showing much less promise. Interest in the adsorption of methane and hydrogen has re-opened the area of adsorption of gases at supercritical temperatures (Agarwal and Schwartz, 1988; Clarkson *et al.*, 1997; Ming *et al.*, 2003). The established adsorption studies are normally carried out below critical temperatures, as for N_2 at 77 K or benzene at 298 K. Here, it is assumed that the adsorbate fills the micropores at low relative pressures and that the adsorbed molecules are packed as in a liquid, thus allowing for the determination of the volume of micropores from the amount of gas adsorbed. However, for methane storage, adsorption is carried out at room temperature, well above the critical temperature (191 K) where the density of the adsorbed phase is lower than that of the liquid methane. This density is a function not only of the applied pressure but also of the adsorption potential of the porosity, which increases with decreasing width of the pore. For methane storage, a pressure of 3.4 MPa is envisaged where the amount adsorbed has not reached a limiting value. Consequently, amounts adsorbed at 3.4 MPa by two activated carbons with equal volume of micropores but different micropore size distributions will differ. Estimations of the density of methane adsorbed in micropores and its relationship with pore-size distribution are described below.

A first approximation to the experimental density of the adsorbate is to calculate the ratio between the mass of methane adsorbed at 3.4 MPa and the volume of micropores of the carbon, because several authors have shown a relationship between the methane adsorbed at 3.4 MPa, and the volume of micropores or the Brunauer–Emmett–Teller (BET) surface area (Parkyns and Quinn, 1995; MacDonald and Quinn, 1996; Alcañiz-Monge *et al.*, 1997; Lozano-Castelló *et al.*, 2002). By using the adsorption of nitrogen at 77 K for the latter, they found a density ranging from 0.13 to 0.16 g cm^{-3} and hence near the value of 0.16 g cm^{-3} at the critical temperature. There are, however some deviations from this behavior. For instance, when the porosity is narrow (as deduced from the slow adsorption kinetics of nitrogen at 77 K), the carbon may adsorb more methane than predicted from the volume of micropores. It is to be noted that N_2 (77 K) is not filling the narrow microporosity. The relationship is better when using the volume of micropores as obtained from the adsorption of CO_2 at 273 K. In other cases, the presence of larger micropores leads to smaller amounts of methane being adsorbed than expected from the volume of micropores. These two cases illustrate the importance of the appropriate micropore size distribution. Theoretical and simulation studies have shown that the maximum possible value of density is 0.37 g cm^{-3}, when the pore width is equivalent to the molecular dimension of methane (Chen *et al.*, 1997; Ming *et al.*, 2003). This value is quite near that corresponding to the liquid state, 0.42 g cm^{-3}, and decreases rapidly with increasing width of the micropore.

The density of the adsorbed methane has been calculated for the disks activated with $ZnCl_2$ and H_3PO_4 and for carbons obtained by gasification with carbon dioxide (800 °C) of some of the disks. In this way, a relatively large variety of carbon disks covers a wide range of micropore size distributions, wider than when attempting the preparation by only one activation method.

Figure 6.11 includes the data for the volume of methane adsorbed versus the volume deduced from the adsorption of nitrogen and carbon dioxide. In this way, each carbon is represented by two experimental points, one corresponding to (VCH_4) versus ($V_{mi}N_2$) and

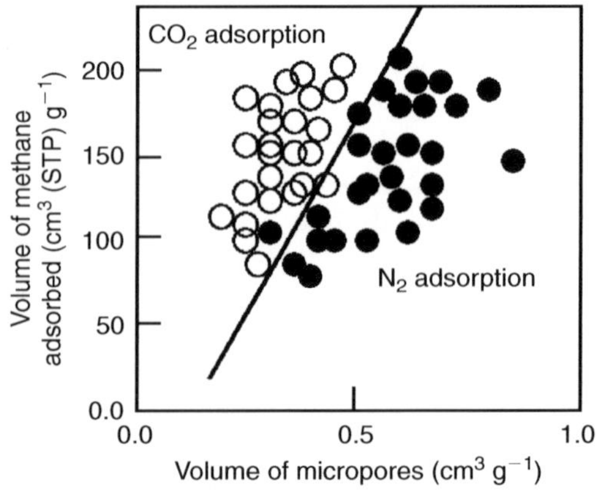

Figure 6.11. Relationship between the volume of methane adsorbed at 3.4 MPa and the volume of micropores deduced from the adsorption of carbon dioxide (○) and nitrogen (●). The line of the plot is for reference (Molina-Sabio and Rodríguez-Reinoso, 2004).

the second to (VCH_4) versus ($V_{mi}CO_2$). Only in carbons in which the microporosity is narrow and uniform – for which ($V_{mi}N_2$) = ($V_{mi}CO_2$) – both points are coincident and the deviation from the reference line is small (the reference line separates the two tendencies: points relating (VCH_4) versus ($V_{mi}CO_2$) are to the left whereas those relating (VCH_4) versus ($V_{mi}N_2$) are to the right of the line). The differences between the points for a given carbon are significant for most carbons (this can be deduced by imagining a horizontal line, common (VCH_4), as the range of values registered by ($V_{mi}CO_2$) is small, between 0.2 and 0.4 $cm^3 g^{-1}$, as compared with the range for ($V_{mi}N_2$) of 0.2–0.9 $cm^3 g^{-1}$). On the other hand, carbons having the same ($V_{mi}CO_2$) adsorb different amounts of methane (this can be deduced by imagining a vertical line), thus indicating that methane is adsorbed not only in narrow micropores but also on wide micropores. However, the adsorbed methane is not retained equally in both types of microporosity, since carbons with very different ($V_{mi}N_2$) may adsorb the same amount of methane. Once again, these results show that the micropore size distribution affects the amounts of methane adsorbed and, consequently the density of the adsorbed phase.

As the adsorption of CO_2 provides the volume of narrow micropores (up to say 0.7 nm) and the adsorption of N_2 the volume of total microporosity (up to 2.0 nm), it is possible to estimate the density of adsorbed methane in each pore size range. The mass of methane adsorbed at a given pressure is the summation of that in narrow micropores ($V_{mi}CO_2$) × density in narrow micropores) and that in wide micropores [($V_{mi}N_2$) − ($V_{mi}CO_2$)] × density in wide micropores):

$$mCH_4 = d_n \times (V_{mi}CO_2) + d_w \times [(V_{mi}N_2) - (V_{mi}CO_2)] \tag{6.1}$$

As the mass of methane adsorbed at 3.4 MPa and the volumes of the micropores are known, both densities can be calculated graphically if a large number of carbons with different micropore size distributions is available. The results are collected in Figure 6.12.

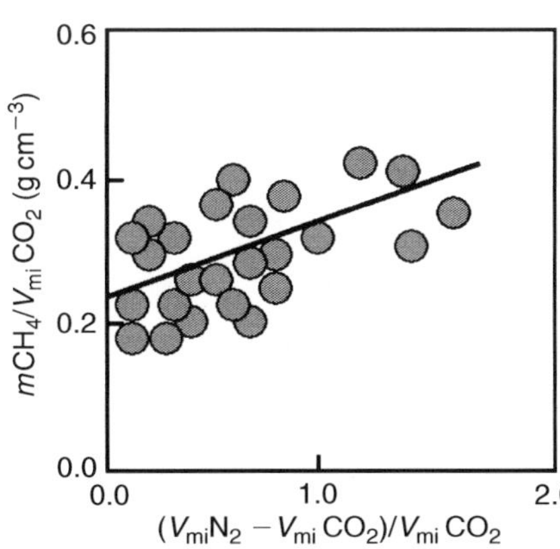

Figure 6.12. Relationship between the methane adsorbed at 3.4 MPa by the carbon disks, from many sources, and the volume of micropores deduced from the adsorption of nitrogen and carbon dioxide. The coordinate and the slope of the straight line (y = 0.09x + 0.25) correspond to the density of methane in narrow and wide micropores, respectively (Molina-Sabio and Rodríguez-Reinoso, 2004).

There is a scatter of data due to the wide range of pore-size distributions used. This makes the calculation of the mass of adsorbed methane from the values of ($V_{mi}N_2$) and ($V_{mi}CO_2$) relatively difficult. Extrapolation of the straight line to the ordinate (it corresponds to a carbon for which ($V_{mi}N_2$) = ($V_{mi}CO_2$) provides a "mean" value for density of 0.25 g cm^{-3} but it can range from 0.41 to 0.19 g cm^{-3}. Chen *et al.* (1997), using Grand Conical Ensemble Monte Carlo (GCEMC) molecular simulation of methane adsorption in slit-shaped carbon pores estimated values for the adsorbed methane of 0.37, 0.27 and 0.20 g cm^{-3} for micropores with dimensions corresponding to 1, 2 and 3 molecular dimensions of methane. These values are slightly higher than the values reported above because they assumed graphitic layers, more dense than layers in activated carbon. In any case, the density is higher than that of methane at the critical temperature, 0.16 g cm^{-3} thus concluding that the adsorbed state is a kind of super-heated liquid, as suggested by Ozawa *et al.* (1976), or an accumulation of compressed gas molecules, as suggested by Dubinin (1960).

Figure 6.12 indicates that if the carbons contain micropores that are detected by the adsorption of nitrogen but not by carbon dioxide (wide micropores) then methane would be adsorbed with a density of 0.09 g cm^{-3} as deduced from the slope of the straight line. Due to the dispersion of experimental points the density could be as high as 0.15 g cm^{-3}. Chen *et al.* (1997) obtained values of 0.14 and 0.09 g cm^{-3} for pores with 4 and 5 molecular dimensions of methane.

The density data discussed above correspond to methane adsorbed at 3.4 MPa. The method has been applied to the adsorption isotherms for methane at different pressures, when plots similar of that of Figure 6.12 are obtained, one for each selected pressure. The values of d_n Equation (6.1) show that the density of adsorbed methane in narrow micropores increases rapidly with pressure up to about 0.25 g cm^{-3} at 2.5 MPa. This means that the maximum capacity for an activated carbon with only narrow micropores is reached at 2.5 MPa. The presence of wide micropores increases the adsorption capacity, but the density in these pores is much lower, and is significantly influenced by pressure. The density of methane adsorbed in these wide micropores at 0.5 MPa is negligible and it increases linearly with pressure up to the value of 0.9 g cm^{-3} at 3.4 MPa, a value much larger than the density for compressed methane at that pressure (around 0.023 g cm^{-3}).

Rodríguez-Reinoso *et al.* (2003) have a patent for carbon monoliths to be used for methane storage.

6.1.5 Summary of Discussions

When a lignocellulosic precursor is activated with increasing concentrations of zinc chloride, phosphoric acid and potassium hydroxide there is a significant development of microporosity. Consequently, there is a degree of activation, specific for each chemical, to maximize micropore volumes with little mesoporosity. These values are 0.4 cm^3 g^{-1} for KOH, 0.5 cm^3 g^{-1} for $ZnCl_2$ and 0.6 cm^3 g^{-1} for H_3PO_4. There is little increase in porosity at higher loadings, the pore-size distribution becoming heterogeneous. KOH, with increasing loading, produces widening of the microporosity to a broader micropore size distribution, with negligible formation of mesoporosity. $ZnCl_2$ develops both wide micropores and small mesopores and H_3PO_4 develops essentially microporosity with some larger mesopores and even macropores.

Note: The carbons obtained from KOH are dependent on the precursor and the experimental conditions used.

Monolithic disks of carbons activated with $ZnCl_2$ and H_3PO_4 have been prepared by simply conforming the impregnated precursor, at 130 MPa pressure, before the heat treatment. The volume occupied by the carbon skeleton is similar in all disks, and is a function of the maximum heat treatment temperature (HTT). The best value reached for methane uptake (185 $cm^3\,g^{-1}$ for $X_P = 0.35$) in microporosity is 131 V/V. Amounts of methane adsorbed are also related to the micropore size distribution. The density of adsorbed methane increases significantly with increasing pressure in the narrow micropores, becoming stabilized at around 2.5 MPa to 0.25 $g\,cm^{-3}$. The density of methane is much lower in wide micropores, but it also increases with pressure up to about 0.09 $g\,cm^{-3}$ when the pressure is 3.4 MPa.

6.2 Chemistry of Activation by H_3PO_4

6.2.1 Methodology

Although the effects of additions of H_3PO_4 are described above, the detail of mechanism was not discussed. This is remedied by summarizing the work of Jagtoyen and Derbyshire (1998) who described, thoroughly, the chemical, physical and morphological changes that occur during activation by phosphoric acid, using white oak and yellow poplar hardwoods which grow in abundance in Kentucky and the southeastern US. These woods differ in their cellular structure, biopolymer composition, density and hardness. The following sections describe, comprehensively, mechanisms of the chemistry of phosphoric acid activation as well as points of correspondence and difference in the behavior of the two hardwoods.

White oak and yellow poplar woods were used in the form of blocks ($22 \times 50 \times 100$ mm) and as a powder (100 mesh (<150 μm)). Thin sections of wood, cut from the blocks, were used to follow dimensional changes during activation. Features such as growth rings survive the process of chemical activation and can be readily seen in the activated carbon. Sections were cut perpendicular to the axial direction of tree growth giving specimens, $22 \times 50 \times 1.5$ mm, for the white oak and $23 \times 71 \times 1.6$ mm for the yellow poplar. The sections contained 6.5 wt% moisture (white oak), and 5.7 wt% (yellow poplar). Other experiments used wood in powder form. Lower H/C and O/C ratios exist for white oak which has lowest content of lignin and lowest content of hemicellulose, and has a higher density than yellow poplar.

Sections of wood (about 1.1 g) or powder (24 g), and used without drying, were soaked in phosphoric acid solution for 1 h at room temperature to allow penetration of acid into the wood. The proportion of H_3PO_4 to wood normally used was 1.45 g H_3PO_4 or 1.5×10^{-2} mol of $H_3PO_4\,g^{-1}$ of wood, providing an excess of reagent. The phosphoric acid was obtained as an 85% solution and diluted to a concentration of 28%. In one series of studies, the ratio of acid to precursor was varied.

The mixtures were heated slowly to 170 °C under flowing nitrogen (0.1 MPa) for 30 min, and then further heated to 650 °C for 1 h. In some experiments the final HTT was <170 °C for 3 h. Carbons were obtained 75–650 °C for white oak, and 150–650 °C for yellow poplar, and leached with water to pH 6 and vacuum dried at 110 °C. Carbons were also prepared without phosphoric acid addition. For white oak, ratios of phosphoric acid to wood of 0.17–1.45 g acid g^{-1} of wood, HTT 500 °C, were used.

An on-line quadrupole mass spectrometer analyzed gases from the wood or a phosphoric acid/wood mixture (rate 2 °C min^{-1} to 600 °C) and included CO (mass 28), CO_2 (mass 44), CH_4 (mass 16, but followed as mass 15, the CH_3 radical, to avoid interference from O) and H_2 (mass 2). Water (mass 18) was in the gas phase at an approximately constant value.

The activated carbons were characterized by (i) chemical analysis for elemental composition, (ii) optical and SEM and (iii) adsorption of N_2 (77 K). Morphological changes were monitored by microscopy and quantified by measurement of dimensional changes using a micrometer to determine distances between growth rings in the radial direction. Pore sizes were obtained from the nitrogen adsorption isotherms (77 K), and surface areas were calculated from isotherms. Micropore volumes, W_o (pores <2 nm diameter) was calculated from the DR equation. Mesopore surface areas and volumes were calculated using the BJH (Barrett, Joyner, Halanda) method (Barrett *et al.*, 1951).

6.2.2 *Porosity Development*

An appropriate ratio of acid to precursor was selected after initial experiments at 500 °C, Figure 6.13. As the ratio of H_3PO_4 to wood was increased, so the micropore volume increased to a maximum at a ratio of about 0.80 g H_3PO_4 g^{-1} of wood and thereon slightly decreased. Mesopore volumes are small at low ratios and then increase as the ratio increases >0.8. In subsequent experiments, the ratio of 1.45 g(H_3PO_4) g^{-1} (dry wood) was used because it gave higher micro- and mesopore volumes, and is typical of ratios used in commercial processes.

A comparison of porosities in activated carbons from white oak and yellow poplar, ratio of 1.45 g (H_3PO_4) g^{-1} (dry wood), is shown as a function of HTT in Figure 6.14, the carbons being very similar. Micropore development begins at about 150 °C, reaching a maximum at about 350 °C. Mesoporosity begin to develop at about 250 °C, reaching a maximum at

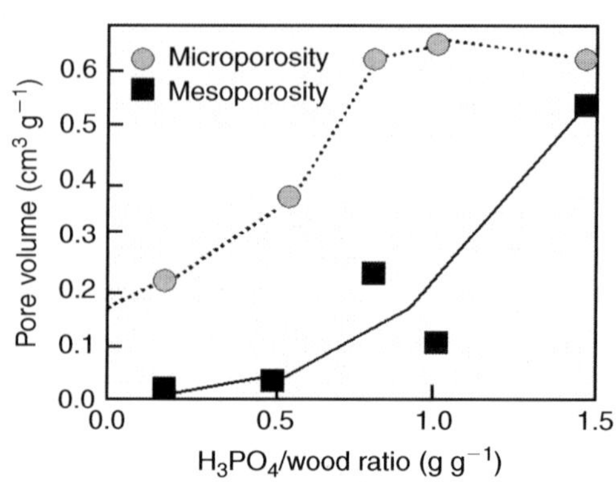

Figure 6.13. Effect of acid:wood ratio on porosity development of white oak (HTT 500 °C, H_3PO_4 strength 28%) (Jagtoyen and Derbyshire, 1998).

500–550 °C. Figure 6.14 also indicates that most of the increase in mesopore volume (>350 °C) equals a decrease in the microporosity, suggesting that pore widening contributes to the increase in mesopore volume. The data of Figure 6.14 also show that the pore volumes are controlled by HTT increasing to maxima between 500–550 °C and then decreasing. Thus, porosity changes are a function of both the ratio of reagent to wood volume and HTT of the system.

6.2.3 Analytical Data

Elemental analyses of the white oak and yellow poplar carbons show that H_3PO_4 treatment accelerates the removal of oxygen and hydrogen, at temperatures as low as 75 °C, Figures 6.15 and 6.16, there being little effect <150 °C, but with a sharp change between 250 and 350 °C. By plotting the atomic ratios H/C versus O/C, all data (with and without additions of H_3PO_4) fall on a single straight line with a gradient of 2.0, suggesting that the net change in the treatments is the elimination of water, Figure 6.17. Thus, the effect of phosphoric acid is to promote dehydration at 150–200 °C, lower than the thermal reaction.

Some differences exist between the two precursors. Although the H and O contents of the carbons from the two hardwoods are similar for thermal treatment, they are not similar for the acid treated products, Figures 6.15 and 6.16. Specifically, the dehydration of white oak has a greater temperature dependence than yellow poplar because, probably, of the higher H and O content of the yellow poplar.

Analyses of evolved gases, in the presence of H_3PO_4, show that CO_2 and CO evolutions begin ~100 °C rising sharply to maxima at ~200 °C. Thereafter, both gases decrease continuously with increasing HTT. Without H_3PO_4, CO_2 and CO evolution begins at a higher temperature (150–200 °C) increasing to maxima between 250 and 350 °C. The concentrations of CO_2 and CO in the gas phase are lower with H_3PO_4. The CO_2 and water originate from both the lignin and cellulose components of the wood.

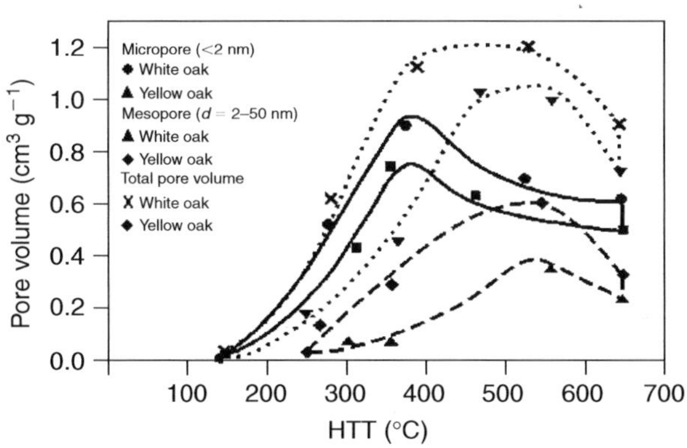

Figure 6.14. Pore volume of activated carbons from white oak and yellow poplar (H_3PO_4:wood = 1.45 g g^{-1}, acid strength 28%) (Jagtoyen and Derbyshire, 1998).

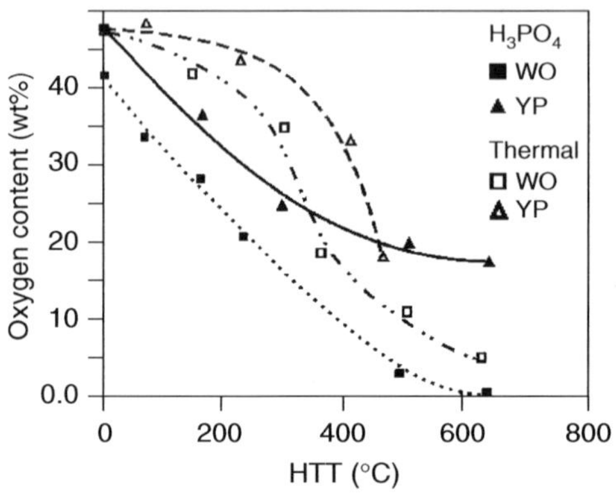

Figure 6.15. Change of oxygen content of carbons from yellow poplar (YP) and white oak (WO) as a function of HTT (H_3PO_4: wood $= 1.45\,g\,g^{-1}$, acid strength 28%) (Jagtoyen and Derbyshire, 1998).

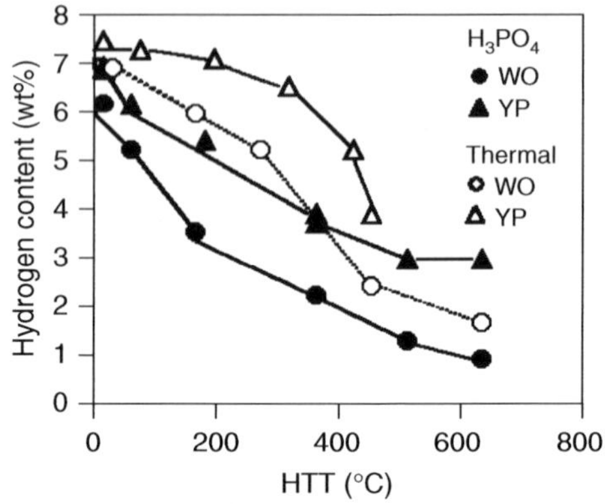

Figure 6.16. Change of hydrogen content of carbons as of Figure 6.15 (Jagtoyen and Derbyshire, 1998).

There is a significant release of CH_4 at about 100 °C rising to a sharp maximum about 250 °C. The generation of CH_4 is a minimum between 350 and 400 °C rising to a second lower maximum between 450 and 550 °C. The release of CH_4 in the thermal case (no acid) is quite different. Here, methane is produced in smaller amounts and is detected above about 250 °C. The production of H_2 starts at about 400 °C in the presence of H_3PO_4, rising to a sharp maximum at about 550 °C. Without the acid, in the gas phase, hydrogen content is always lower and only slightly increases at HTT >450 °C.

Thus, dehydration is only one way that oxygen and hydrogen are removed. In the presence of phosphoric acid, there are also oxides of carbon, with maximum formation at ~200 °C. Hydrogen is also released as light hydrocarbon gases, principally methane, rising to a maximum at about 250 °C; together with small amounts of hydrogen at higher HTT as molecular hydrogen and by secondary methane formation. The constant ratios of Figure 6.17 are therefore somewhat of a coincidence.

6.2.4 Morphological and Dimensional Changes

The microscopy of the fibrous cells with distinct primary (lignin-rich) and secondary (high content of crystalline cellulose) cell walls in the parent wood, showed that, after phosphoric acid treatment to 125 °C, there is evidence of flow and redistribution of cellular material. At 200 °C, this redistribution causes the formerly empty cell lumens to be filled. However, boundaries between the primary and secondary cell walls are still clearly distinguished and the original cell structure is still recognizable even after heat treatment to 350 °C. Thus, it appears that there is relatively little interaction between the crystalline and amorphous regions. In contrast, thermal treatment (no H_3PO_4) causes a massive disruption to the structural integrity of the wood when, at 350 °C, the secondary cell walls have almost disappeared and are indistinguishable from the primary walls.

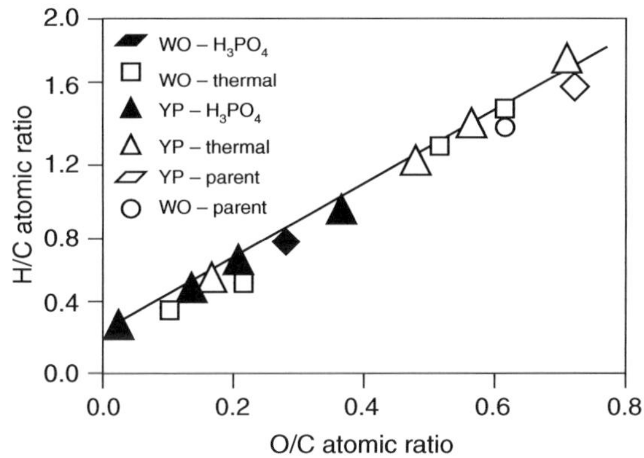

Figure 6.17. The variation of H/C and O/C atomic ratios for carbons produced by activation: (H_3PO_4:wood = 1.45 g g^{-1}, acid strength 28%), the gradient of which indicates elimination of H_2O (Jagtoyen and Derbyshire, 1998).

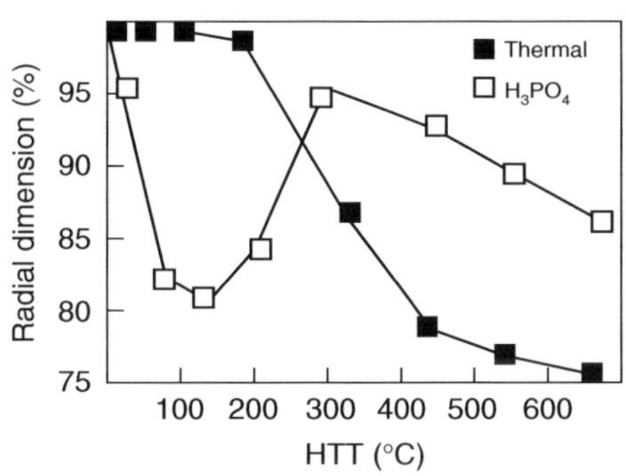

Figure 6.18. Changes in radial dimensions of carbons produced from white oak by H_3PO_4 activation and thermal treatment (H_3PO_4:wood = 1.45 g g^{-1}, acid strength 28%) (Jagtoyen and Derbyshire, 1998).

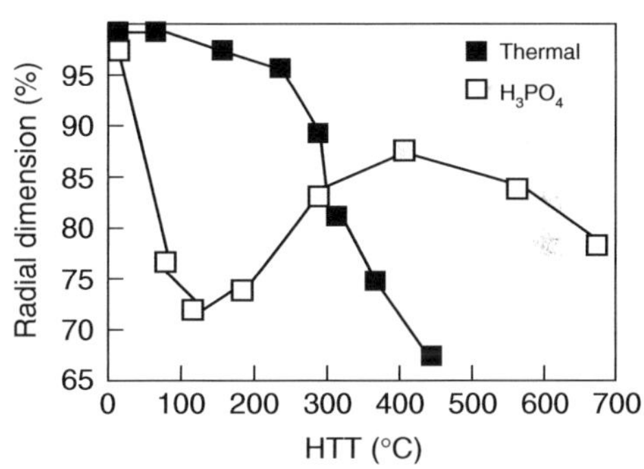

Figure 6.19. Changes in radial dimensions of carbons produced from yellow poplar by H_3PO_4 activation and thermal treatment (H_3PO_4:wood = 1.45 g g^{-1}, acid strength 28%) (Jagtoyen and Derbyshire, 1998).

The transformation of the white oak wood to an activated carbon is accompanied by both dimensional and morphological changes. Upon acid treatment, there is a contraction at low HTT, followed by structural dilation at intermediate HTT and secondary contraction at high HTT. The dilation and secondary contraction are directly related to the development and subsequent decrease in porosity.

The dimensional changes in the radial direction are compared in Figures 6.18 and 6.19. Measurable contraction commences above ~150–200 °C for both hardwoods, and then progresses monotonically with increasing temperature. At 450 °C, the overall contraction in the radial direction is about 20% for white oak and 30% for yellow poplar. When phosphoric acid is present, there is measurable contraction at temperatures as low as 50 °C, consistent with the action of the acid in lowering the temperature threshold for dehydration and weight loss. The contraction sharply increases, 100–150 °C, higher temperatures causing the structure to dilate. This process continues with increasing HTT to a maximum between 300 and 450 °C. With further increases in HTT, a second process of contraction begins. At 650 °C, the overall contraction is 14% for white oak and 20% for yellow poplar, which means that the structures are still dilated relative to the maximum contraction observed at 100–150 °C. For both thermal and acid treatments, the relative dimensional contraction for yellow poplar carbons is always about 10% more than for the white oak series. A possible reason for this difference relates to the much higher lignin content of yellow poplar. If, as suggested, the lignin is readily de-polymerized and migrates into empty structural features, then the contraction of yellow poplar will be appreciably higher than for white oak.

With reference to Figures 6.18 and 6.19, the onset of dilation corresponds closely to the development of microporosity for both hardwoods. Moreover, despite differences in extents of the initial contractions of white oak and yellow poplar, the overall dilation in both cases is about 15%. The micro- and mesopore volumes in the activated carbons follow the same pathway with HTT. Above about 350 °C, the micropore volume decreases while the mesopore volume increases to a maximum between 500 and 550 °C before it also starts to decrease.

6.2.5 Chemistry of Activation by H_3PO_4

Essentially, wood is a complex fiber matrix composite material formed of natural polymers in which the fiber framework consists of crystalline cellulose micro-fibrils, 2–5 nm diameter. The matrix between the micro-fibrils is composed mostly of hemicellulose, and lignin provides the strengthening material that reinforces the surrounding cell wall. Typical contents for the biopolymers of hardwoods are: 42–50% cellulose, 19–25% hemicellulose and 16–25% lignin.

The molecular structures of cellulose, hemicellulose and lignin are in Figure 6.20. The micro-fibrils in the cell walls are formed from cellulose chains that are aligned and held together by hydrogen bonding between the hydroxyl groups on the repeating glucose units. Groups of micro-fibrils are connected by amorphous cellulose (about 10–20% of the total cellulose) and hemicellulose, surrounded by lignin and some hemicellulose. The spaces between the micro-fibrils are of the order of a few nm, and have the same sizes as the larger micro- and mesopores.

6.2.5.1 Reactions at low temperatures <150 °C

The reaction of wood with phosphoric acid begins as soon as the components are mixed. Certainly, after heat treatment to 50 °C, there is evidence of chemical and physical change. It appears that the acid first attacks hemicellulose and lignin, possibly because of easier access to these amorphous polymers than to crystalline cellulose. Cellulose appears to be more resistant to acid hydrolysis than other polysaccharides as seen by the general integrity of the cell structure after reaction with H_3PO_4, even at elevated HTT. Primary effects of acid attack are to hydrolyze glycosidic linkages in the hemicellulose and cellulose and to cleave aryl-ether bonds in the lignin. These reactions are further accompanied by dehydration, degradation and condensation. The primary reactions lead to a reduction in molecular

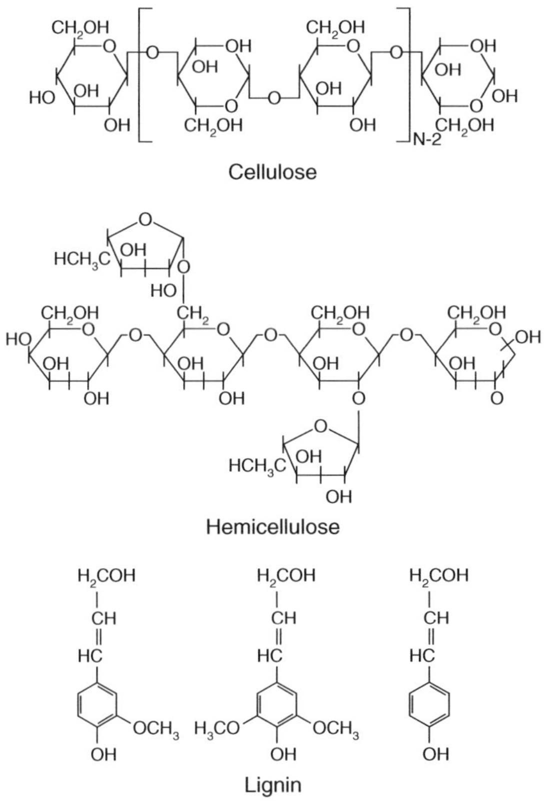

Figure 6.20. Chemical structures of cellulose, lignin and hemicellulose (Jagtoyen and Derbyshire, 1998).

weight, principally of the hemicellulose and lignin, consistent with the ability to form extrudable mixtures of wood and phosphoric acid. Further, the acid-catalyzed hydrolysis of ether linkages in lignin leads to the formation of ketones, as observed by Fourier transform infrared spectroscopy (FTIR) and ^{13}C nuclear magnetic resonance (^{13}C NMR).

These and other bond cleavage reactions that proceed through ionic mechanisms explain the release of CO_2, CO and CH_4. The low temperature evolution of CO_2 and CO is consistent with the observed reduction in C=O functionality found, in part, by esters and carboxylic acid groups in the parent wood (mostly present in hemicellulose and lignin). The substantial release of CH_4 at similar temperatures suggests that the cleavage of aliphatic side chains is relatively facile, and is consistent with a loss of aliphatic character and a corresponding increase in aromaticity.

Other studies have shown that the reaction of phosphoric acid with cellulose, even at low temperatures, produces considerable inter-crystalline and intra-crystalline swelling (Porter and Rollins, 1972; Pandey and Nair, 1974). Further, Pandey and Nair (1974) examined the reaction of phosphoric acid with cotton at temperatures of 10, 29 and 40 °C for 0.5 h using

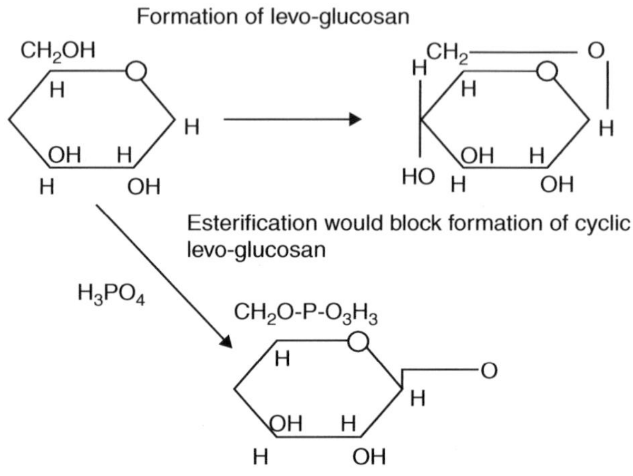

Figure 6.21. Reaction mechanism for flame retardation of cellulose impregnated with phosphoric acid (Jagtoyen and Derbyshire, 1998).

selected concentrations of phosphoric acid (0–81% acid concentration with a constant cotton to liquor ratio of 1:100). They found that the degree of cellulose polymerization decreased with an increase in acid concentration and temperature. This and the other published research (Pandey and Nair, 1974) show that phosphoric acid begins to interact with cellulose at temperatures as low as 10 °C, producing swelling in both crystalline and intercrystalline regions and causing a breakdown of the cellulose polymer chains.

6.2.5.2 Reactions at intermediate temperatures >150 <450 °C

Cross-linking reactions now dominate over bond cleavage and de-polymerization reactions, as indicated by the evolution peak of CO_2 and CO at 200 °C and the first maximum in CH_4 at 250 °C. The high yield of carbon obtained by acid treatment above about 300 °C is due entirely to structural cross-linking that retains the relatively low molecular weight species within the solid phase.

There is evidently a direct connection between porosity development and the process of structural dilation. In attempting to interpret the mechanisms of dilation, a wealth of relevant information is available describing the use of phosphorous compounds as flame retardants for wood and cellulose (Barker and Hendrix, 1979; Hinderschinn and Witschard, 1979; Hindersin and Witschard, 1979) and in another study of the chemical modification of wood (Rowell, 1991). Reaction with phosphoric acid also stabilizes the cellulose structure by inhibiting the formation of levo-glucosan which otherwise offers a route to the substantial degradation of cellulose through its decomposition to volatile products, Figure 6.21. Phosphates also function as flame retardants for synthetic polymers when heated in air (Delobel *et al.*, 1990; Bourbrigot *et al.*, 1995).

The formation of phosphate esters by reaction of cellulose with phosphoric acid is shown in Figure 6.22 which further illustrates how phosphoric acid can be inserted between the

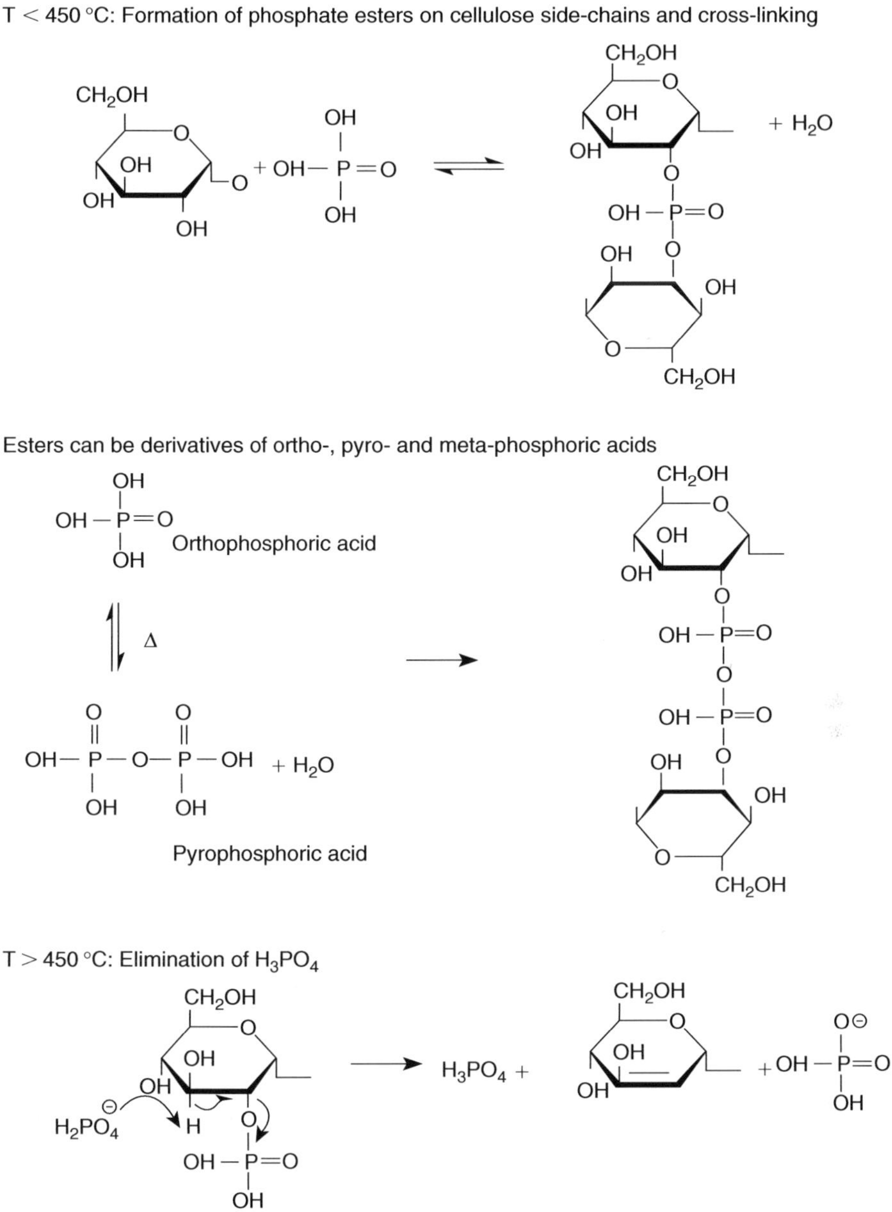

Figure 6.22. Mechanism of phosphate ester formation by phosphorylation of cellulose (Jagtoyen and Derbyshire, 1998).

cellulose chains, disrupting the existing hydrogen bonds, replacing them with other chemical linkages, and simultaneously separating the chains and dilating the structure. The addition or insertion of phosphate groups separates the organic species and effects a dilation of the structure.

Table 6.1. Effect of acid removal for last 50 °C on porosity of product from white oak (Jagtoyen and Derbyshire, 1998).

	Micropore volume ($cm^3 g^{-1}$)		Mesopore volume ($cm^3 g^{-1}$)	
HTT (°C)	H_3PO_4	No H_3PO_4 last 50 °C	H_3PO_4	No H_3PO_4 last 50 °C
250	0.11	0.00	0.03	0.01
300	0.42	0.20	0.06	0.05
350	0.68	0.30	0.10	0.05
450	0.67	0.50	0.36	0.74
500	0.64	0.53	0.55	0.70

An **important distinction** between phosphoric acid activation (to produce a carbon) and the chemical modification of wood by phosphoric acid (fire retardants) is that, in the former case (the commercial case), phosphoric acid is removed after reaction by leaching to recover the reagent and heat treatment is then continued. For this reason, the phosphorous content of the recovered carbons only indicates the amount of residual material that is present, either combined with the organic structure, in the form of insoluble metal phosphates, or physically entrapped.

At 280 °C, the structures are considered to be small polyaromatic units connected mainly by phosphate and polyphosphate bridges, including polyethylene linkages [$(CH_2)_n$]. As the temperature is increased, cyclization and condensation reactions lead to increases in aromaticity and size of the polyaromatic units, enabled by the scission of P—O—C bonds. Between 350 and 500 °C, the char is stable, but at 430 °C and above, the continued cleavage of cross-links leads to an extensive growth in the size of the aromatic units ("embryonic" graphene layers).

Thus, phosphoric acid appears to be able to function in two ways (i) as an acid catalyst in promoting bond cleavage reactions and the formation of cross-links via processes such as cyclization and condensation and (ii) by combining with organic species to form phosphate linkages, such as phosphate and polyphosphate esters, that serve to connect and to cross-link biopolymer fragments. This can be tested by removing (leaching) the phosphoric acid for example from a 250 °C product, and then heating to 300 °C (to be compared with no leaching). Table 6.1 shows the effects of acid removal on the relative porosities of the products, the continued presence of the phosphoric acid maintaining the microporosity, but not the mesoporosity.

Figure 6.23 shows how yield of carbon is a function of HTT for carbons produced with phosphoric acid, compared with runs where the acid was removed before the last 50 °C of heat treatment (H_3PO_4:wood = 1.45 g g^{-1}, acid strength 28%). The H_3PO_4 promotes a higher yield.

6.2.5.3 Reactions at high temperatures >450 °C

At HTT >450 °C, the structure begins to contract, as determined both by the overall dimensions and by the thickness of the secondary cell walls. Micropore volumes decrease steadily with HTT >350 °C, with no apparent effect on the mesopore volume until about 550 °C, when it begins to decrease quite steeply. Between 450 and 550 °C, structural contraction is

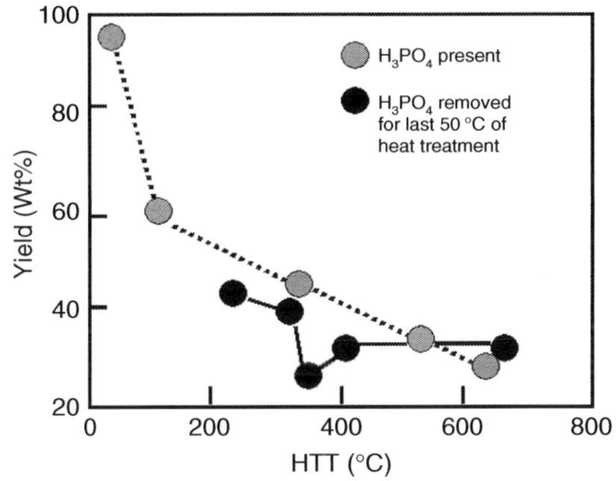

Figure 6.23. Yield of carbon as a function of HTT for carbons produced with phosphoric acid, compared with runs where the acid was removed before the last 50 °C of heat treatment (H_3PO_4:wood = 1.45 g g^{-1}, acid strength 28% (Jagtoyen and Derbyshire, 1998).

accommodated by a narrowing of pore diameters without causing an appreciable shift in the pore size ranges. Above 550 °C, a sharp decline in mesopore volume is paralleled by the secondary contraction of the cell walls. Between 450 and 650 °C, the cell wall thickness is reduced by about 30% of the original dimension, while the overall contraction in the radial direction is only about 11%.

The contraction of the altered crystalline cellulose mirrors its previous extensive dilation. Just as dilation has been explained by the formation of bulky phosphate linkages, its reversal is consistent with their breakdown, due to their attaining the limit of thermal stability at about 450 °C. As the phosphate bridges are considered to be more highly concentrated in the altered cellulose, it is logical that their decomposition would exert the greatest effect on this structure. Relatedly, the FTIR analysis showed a reduction in the intensity of bands attributed to phosphate esters above 450 °C. There is also a significant increase in the aromatic cluster size above 450 °C (as estimated by ^{13}C NMR), indicative of a substantial structural rearrangement that would require a reduction in cross-link density, as found in the studies of flame retardants. The peak in hydrogen evolution at 550 °C is a consequence of aromatic condensation. The corresponding decrease in porosity is also consistent with the occurrence of major structural reordering, where the increased size and alignment of clusters results in a more densely packed (and possibly more anisotropic) structure. Over this same range of temperature, the elimination of residual aliphatic and oxygen-containing groups is consistent with the observed secondary evolution of CH_4 and the disappearance of infrared absorption bands due to ketones and esters.

6.2.6 Summary of Discussions

Phosphoric acid functions both as an acid catalyst to promote bond cleavage and the formation of cross-links via cyclization and condensation reactions, and to combine with organic species to form phosphate and polyphosphate bridges that connect and cross-link biopolymer fragments. The addition (or insertion) of phosphate groups drives a process of dilation that, after removal of the acid, leaves the matrix in an expanded state with an accessible pore structure. Essentially, this is the activation process. At temperatures above 450 °C, a secondary contraction of the structure occurs when the phosphate linkages become thermally unstable. The reduction in cross-link density allows the growth and alignment of polyaromatic clusters, producing a more densely packed structure with some reduction in porosity.

6.3. Chemical Activation: Use of Alkali Metal Salts – K and Na

6.3.1 Historical Introduction

As discussed in Section 6.2 the use of phosphoric acid to prepare activated carbon has origins in a quite different area, namely that of fire retardation. Similarly the use of alkali metal salts, such as the hydroxides and carbonates of potassium and sodium, to prepare activated carbons, is linked to the two industries with major consumptions of carbon, namely the iron and steel industry and the aluminum industry.

The latter half of the last century saw two major changes within the iron and steel industry. First, there was a massive expansion in iron production throughout Europe, the USA and the Far East. As iron production is based on the blast furnace which is dependent on metallurgical coke as the reductant for oxides of iron, so there was an expansion in the need for coal and coal blends to make the metallurgical coke. Although the carbonization characteristics of these coals are of paramount importance, it was also realized that the chemical composition of the mineral matter (which becomes ash) has to be taken seriously into consideration. The mineral matter is much more than an inert diluent of the coke. It became established that certain coals, with mineral matter contents which contained high levels of potassium and sodium, provided metallurgical cokes which did not perform well in the blast furnace. The coke size was found to be reduced and where there exists a reaction bed made up of different sizes of coke pieces, then the permeability of the furnace to the air blast is reduced and, as liquid iron production is a diffusion controlled process, then the efficiency of the blast furnace could fall to unacceptable levels. The question to be solved at that time was how does potassium and sodium cause coke degradation.

To move now to the aluminum industry, a similar question was being asked, although the context of the question was different. Aluminum is produced by the electrolytic reduction of aluminum oxide/fluoride fluxes at temperatures of about 1000 °C, less than the maximum temperatures of a blast furnace (about 1300 °C). The reductant is the carbon anode which is a block of calcined carbon weighing about 1 tonne which has to be inserted into the electrolytic cell. From time to time, these anodes would fail and fall into the molten electrolyte from which they would have to be extracted (a procedure not looked upon with favor within the industry). A major cause of this problem was eventually associated with the presence of sodium (less so with potassium) within the calcined coke. The question to be solved at that time was how does sodium cause degradation and weakening of the carbon anode.

6.3.2 Intercalation Compounds

Attention has now to be directed toward that property of graphite known as *intercalation*. For many decades, intercalation compounds had been studied, mainly because of their interesting electrical properties which could approach the specific electrical conductivity of copper. Reviews are available by Hooley (1969), and Berger *et al.* (1975). However, their extreme sensitivity to air and to moisture precluded any significant applications. Intercalation compounds of graphite are interstitial compounds in which the intercalate is inserted between the graphene layers of graphite resulting in the expansion of the inter-layer spacings. *Isotherms* of uptake of vapor of potassium can be obtained, a process that is essentially reversible. However, the isobars (constant pressure of potassium and increasing temperature) clearly showed the temperature instability of the potassium-intercalation compound. It was inconceivable that degradation to metallurgical coke and anode carbon, by an intercalation mechanism, could occur at temperatures of 1300 and 1000 °C, respectively. So, for some years, this problem of structural damage to carbon artifacts remained unresolved.

Another matter, of relevance to activation of carbons, was the observation that whereas cesium, rubidium and potassium could easily form intercalation compounds, sodium and lithium were quite unreactive, but not totally so. Further, sodium apparently prefers to "intercalate" into less-ordered carbons (not graphitic) and lithium has its own characteristics when it comes to insertion into carbon (electrically driven) for the lithium–carbon battery.

Intercalation is an electron-transfer process, the intercalate atom going to a cation (e.g. $K \rightarrow K^+$) (note that Br_2 forms $2Br^-$). Bonding occurs by transfer of electrons from the alkali metal atom to the conduction band (π-band) of host graphite. There is transfer of electrons from the π-band to bromine with penetration of bromide ions between the graphene layers. In comparing the energy states of atom versus anion, within the graphite lattice, processes of ionization are exothermic for cesium, rubidium and potassium, the sodium being more stable in the liquid state that as an intercalate. However, less organized carbons possess a range of inter-graphene layer distances, unlike graphite. Within these more compact structures, the sodium atom has a reduced ionization energy (the gap is narrower) and so sodium can enter such structures.

A historical story may help to connect the reality of intercalation to the reality of degradation of industrial carbons to the reality of activated carbon. An easily prepared graphite-intercalation compound is graphitic oxide, ($C_8H_2(OH)_2$), made by reacting graphite with a mixture of sulfuric and nitric acids. It is a powdery solid. This material was used in a lecture demonstration by a famous, now deceased, Professor of Inorganic Chemistry. He would heat a sample of graphitic oxide in a porcelain crucible, at the end of his lecture, and then leave the lecture theater. What happened? The graphitic oxide decomposed rapidly and spread throughout the lecture theater eventually to fall over the clean white shirts of his audience. The Professor was never invited back.

The exfoliation process is demonstrated in Figures 6.24 and 6.25. Using the imagination, the upper micrograph could be associations of graphene layers of size about 10 nm. After intercalation of bromine at 298 K for 24 h, followed by rapid heating to about 500 K, the system exploded but did not disintegrate because the "graphene layers" appear to be bonded to each other. The opening up by the heat treatment has created porosity, and the imagination leads to these being micropores. Activation with alkali metal salts (hydroxides and carbonates), certainly, leads to carbons containing microporosity, dominantly. The reasons are apparent from a consideration of Figure 6.25.

Figure 6.24. SEM of natural graphite crystals, as extracted. The width of the micrograph corresponds to 3 mm.

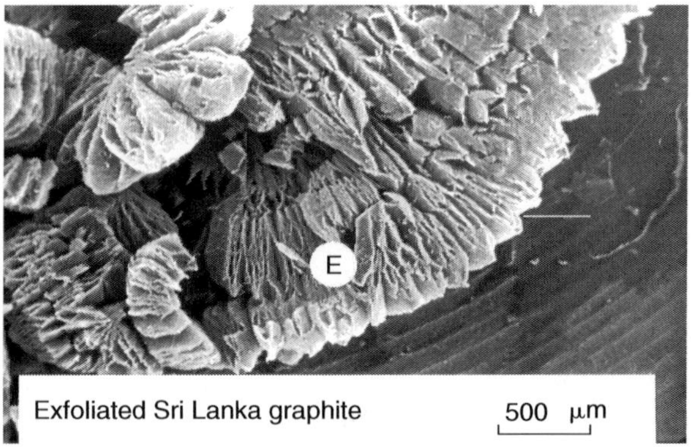

Figure 6.25. SEM of natural graphite crystals, after intercalation with bromine, at 298 K, followed by exfoliation by heating. The width of the micrograph corresponds to 3 mm.

Somewhere down the line a genius saw the connection between these observations and the means to prepare carbons which exposed internal surfaces, provided the products could be suitably contained. And, to cut a long story short, the answer to the degradation problems became clear. The damage to these materials did not occur at the high temperatures of operation but during the heating-up processes in which the alkali metal (when produced) *penetrated* the carbon structures. With increasing temperature, the intercalated compound becomes unstable and with rapid heating (as occurs in these industrial locations) the alkali metal is explosively desorbed and by so doing, destroys the structures in which it was initially located. A further consideration is that, within the electrolytic cell of aluminum production, the electrolyte is a possible source of sodium which diffuses easily throughout the anode.

To return to activated carbons, over the last few decades, research has focused on appropriate procedures to create carbons of controlled porosities, by varying the feedstock, the ratio of alkali salt to carbon and the final HTT and heating rate. With extreme conditions, only powders are formed by the explosive removal of the metal atoms. It should be possible to obtain a granular carbon with a sufficiently enhanced microporosity using more gentle conditions of production (see Section 6.1.2.3).

So far, despite the potential of production of carbons with a wide range of porosities, the adverse economics of such processes have prevented significant commercial production over established methods of physical and H_3PO_4 activations. The extensive microporosity of these carbons has yet to find applications.

6.3.3 Chemistry of Activation using Alkali Salts

The co-carbonization of an organic precursor, such as a resin, coal or carbon with potassium hydroxide or potassium carbonate results in enhanced porosity within the carbon product, a process called chemical activation. Marsh *et al.* (1982) studied carbons from the Amoco process as patented by Wennerberg and O'Grady (1978) which involved heating dry mixes of carbonaceous material with potassium hydroxide. This as done on an industrial scale with rapid heating of these mixes to about 750 °C leading to the explosive removal of the metal. As a result, these Amoco carbons were flocculent, of extremely low density (>0.01 g cm^{-3}) with apparent surface areas as high as 3000 m^2 g^{-1} and had been formed by the total destruction of structure in the parent carbonaceous material. It was as if the graphene layers of structure had been individually separated. Later, Marsh *et al.* (1984) studied activation of green and calcined petroleum cokes with mixes of solid KOH, ratio of 4:1, heating to 750 °C and noted that green cokes produced carbons of highest surface area. Structures within these green cokes (HTT <400 °C) were probably less strongly bonded together compared with the graphitic calcined cokes (extremely hard materials). Again, the destruction of the structure of the green coke was observed.

Marsh *et al.* (1987) reviewed possible mechanisms of activation by potassium and sodium, but did not describe clearly how the potassium was working; this was to come a little later. Chan *et al.* (1993) heated a wide range of carbons with potassium in an inert atmosphere, 773–1273 K, and reported little degradation (comminution) with isotropic porous carbons (PCs) and calcined needle cokes, but metallurgical cokes of very limited graphitizability suffered considerable degradation. The results of this study suggest that size degradation is probably associated with the heterogeneity and varying orientations of anisotropic components in the structure, as well as the presence of mineral matter. Due to these factors expansion (swelling) occurs in non-uniform directions resulting in high stresses and the degradation. These findings offer suggestions for further investigations into chemical activation, using KOH or K_2CO_3, noting that degradation is certainly not wanted by some industries but is welcome for activated carbon production.

To summarize, the intercalation by alkali metal atoms, originating from the reduction of the hydroxide or carbonate, is the initial stage of the degradation process. The expansion of this process introduces strain within the solid carbon. But it requires further heating to drive out the metal from the intercalated system, the vigor of this process determining the porosity of the final carbon.

The reaction pathway for the creation of chemically (KOH, NaOH) activated carbons, involves the intercalation into and removal of these alkali metals from the carbon structures.

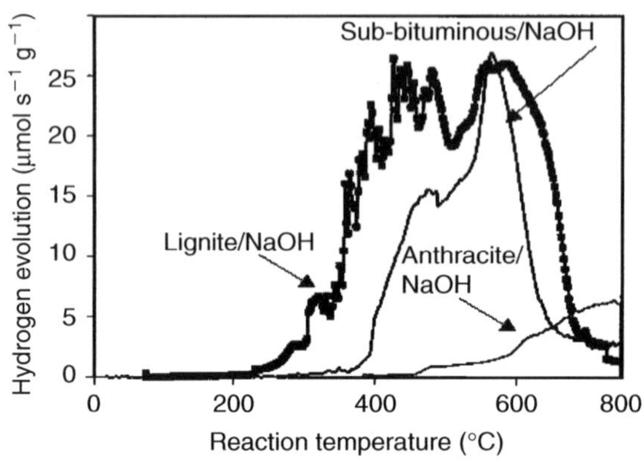

Figure 6.26. Effect of coal rank on the temperature of evolution of hydrogen for reaction with NaOH, using a lignite, a sub-bituminous coal and an anthracite (Lillo-Ródenas *et al.*, 2004b).

What are not so straightforward are the mechanisms which lead from the hydroxides to the metal. These are high-temperature reactions, taking place in the solid phase, where diffusion control over rate is dominant. What is known is interesting because it can be related to discussions of hydrogen inhibition during activation by steam (Sections 5.5 and 6.1.2.3).

Studies of the chemistry of the activation process have been undertaken by the Department of Inorganic Chemistry, University of Alicante (González *et al.*, 1997; Lillo-Ródenas *et al.*, 2001, 2002, 2003, 2004a, b; Lozano-Castelló *et al.*, 2001; Molina-Sabio and Rodríguez-Reinoso, 2004). These latter authors, using a Spanish anthracite, noted that KOH was more effective than NaOH for activation and that impregnation from aqueous solution was more effective than physical mixing, but not for the NaOH system. Increased flow rate of nitrogen during activation enhanced the rate of formation of activated carbon indicating some diffusion control of reaction rate.

Lillo-Ródenas *et al.* (2004b) made a detailed study of the kinetics of activation, using NaOH with the Spanish anthracite. Essential observations for this activation reaction are (a) comminution occurs of the particles of the initial material into smaller sizes, (b) there is enhanced micro- and mesoporosity, (c) there is loss of carbon, (d) below 700 °C the only gaseous product is hydrogen, (e) above ~800 °C carbon dioxide and carbon monoxide appear as gaseous products and (f) a solid product Na_2CO_3 is formed. The evolution of hydrogen is shown in Figures 6.26 and 6.27 (for a lignite, a sub-bituminous coal and an anthracite) indicating that as coal rank increases so the surfaces become less reactive to NaOH, the raw lignin causing evolution of hydrogen as low as 300 °C. On carbonization to 750 °C, the reactivity was reduced with hydrogen evolution not commencing until 450 °C, the overall rate of production of hydrogen by anthracite being only about 10% of that of the carbonized lignin. Lillo-Ródenas *et al.* (2003) report that temperatures of evolution of CO_2 and CO are somewhat higher than noted for the decomposition of surface oxygen complexes formed from molecular oxygen (see Section 4.7.3). Any proposed mechanism of chemical activation has therefore to explain the process of comminution, the loss of carbon, the generation of hydrogen and oxides of carbon and formation of the

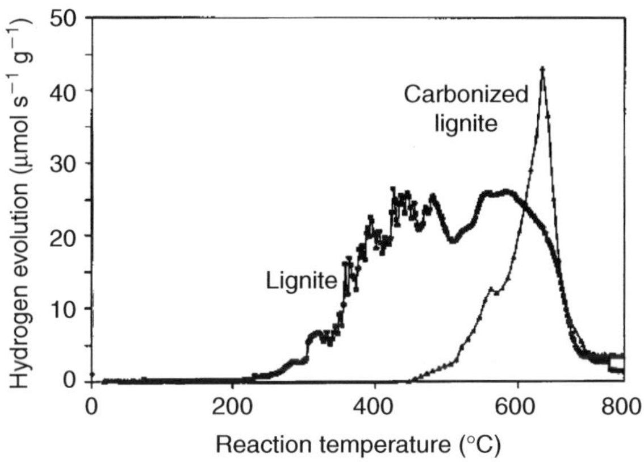

Figure 6.27. Comparison of the evolution of hydrogen versus temperature for the lignite and carbonized lignite (Lillo-Ródenas *et al.*, 2004b).

sodium carbonate. Also, the relative abilities to activate additions of NaOH compared with KOH have to be explained. Experimentally, the NaOH was added, in excess, to the carbon reactant (CR) from aqueous solution in ionic form as Na^+ and OH^-. As the temperature of the CR with added NaOH is increased, residual water is lost leaving the ions deposited on internal surfaces.

By way of explanations, observations (a) and (b) have been already explained, as above, in terms of the metal/intercalation/explosive removal story. Observations (c)–(f) require comment, with attention directed toward the inorganic chemistry involved.

The addition of NaOH to the carbon creates the possibility of reaction with all four elements involved, namely the carbon, the oxygen, the hydrogen and the potassium. The products of these reactions include potassium metal, gaseous hydrogen, gaseous carbon dioxide and monoxide and sodium carbonate. Possible global reactions, for NaOH are as follows:

$$4NaOH + C \leftrightarrow 4Na + CO_2 + 2H_2O \qquad \text{(Reaction 6.1)}$$

$$4NaOH + 2CO_2 \leftrightarrow 2Na_2CO_3 + 2H_2O \qquad \text{(Reaction 6.2)}$$

The mechanistic detail of surface reactions will be much less complicated than indicated by these global reactions which can never occur as written. Rather, the processes will occur via a series of much simpler, simultaneous/consecutive reactions. The initial state of dispersion of the NaOH is important to maximize the contact between the NaOH and the carbon surface. What is uncertain is the chemical state of the NaOH at temperatures of about 350 °C. Has the −OH become bonded to the carbon surface and is the Na-atom/ion free to move in and around the carbon, or is there a form of Na—O complex on the surface. At these temperatures, intercalation is possible. The $(C)_n$—OH appears not to be stable as such but decomposes to $(C)_n$=O and the hydrogen is abandoned. The hydrogen, which must have originated as an atomic species appears to be unable to be chemisorbed to the carbon surface and is rejected from the system as gaseous molecular hydrogen. This observation is in keeping with discussion of reactions between carbon and hydrogen as in Section 5.5. The Na_2CO_3 is probably a secondary product and may not be part of the activation process directly, perhaps indirectly because what is needed to be known is the origin of the carbon in the carbonate.

Expressing this in another way, initially, surface oxygen complexes are formed from the hydroxide; these are removed, thermally, from the surface as CO_2, which is then captured by the excess NaOH, in a solid–gas reaction to form the carbonate. However, it is more than likely that everything takes place via a series of surface reactions. The rather higher thermal stability of the surface oxygen experiments as reported by Lillo-Ródenas *et al.* (2003), could result from the sodium stripping the less stable surface oxygen complexes from the carbon surfaces.

The process of activation, using for example NaOH, is the result of five, critical simultaneous/consecutive processes, that is:

1. The reduction of the hydroxide to free metal.
2. The penetration of the free metal into the lattice (between the graphene layers) of the carbon.
3. The expansion of the structural graphene layers by intercalated sodium.
4. The rapid, "explosive", destructive removal of the intercalate with increasing temperature which forces apart the graphene layers so creating the microporosity. The severity of this process is a function of heating rate and alkali content.
5. *Physical activation* resulting from oxidation by the oxygen of the hydroxide as well as catalytic channeling from mobile clusters of sodium atoms perhaps both from edge carbon atoms and in-plane carbon atoms.

The formation of carbon dioxide, of carbon monoxide, of gaseous hydrogen and of carbonates are the products of secondary reactions and have no role in the mechanism of activation to create enhanced porosity.

Shimada *et al.* (2004), using multi-walled nanotubes, was able to observe, directly by microscopy, the gasification of carbon from within the graphene layers (basal planes), so confirming the possibility of removal of carbon atoms from a surface as distinct from an edge carbon atom.

It is important to consider, in a little more detail, the fact that sodium (with potassium) is well described in the literature as an effective oxidation/reduction (*redox*) catalyst for gasification reactions, and could accelerate the removal of the surface oxides of carbon (Section 5.4.2). This will occur around sites of sodium atoms or clusters of sodium atoms and such preferential removal of carbon atoms from such a region could lead to channeling and pitting and the widening of microporosity leading to the development of the mesoporosity, dependent on the size of the clusters of sodium atoms.

The observations (c)–(f) (above) have now been provided with explanations, perhaps not totally complete, but sufficient to provide for a continuing discussion.

The free energy of formation of surface oxides of carbon (an exothermic reaction) exceeds that of formation of methane (an endothermic reaction) and so it can be assumed that oxides of carbon are formed preferentially to any formation of methane at temperatures less than 700 °C.

The literature reports conflicting behaviors for additions of the potassium carbonate as distinct from the hydroxide. Under quite drastic conditions the carbon material can be totally disintegrated by potassium hydroxide to create a flocculent type carbon with high adsorption capabilities ($>5000\,m^2 g^{-1}$). Under milder conditions, comparable to the use of the potassium hydroxide, no significant activation occurs. One reason for this may be the larger size of the carbonate ion —CO_3^{2-} which prevents entry into the microporosity of the initial reactant. Due to the ionic nature of the solution, or whatever, the potassium ions cannot diffuse into the porosity without the accompanying carbonate (or bicarbonate) ions. As the mechanisms of activation require the presence of both the potassium as well as oxygen species then no significant activation is possible although some may occur on external surfaces. A second reason is that, although the potassium carbonate is able to diffuse into the microporosity of the carbon, the formation of surface oxygen complexes (CO_m) from the carbonate ion is not a favored reaction under these conditions. Continuing studies using carbons of initial wider micro- and mesoporosities at higher temperatures may help to resolve this problem.

The literature also reports that sodium hydroxide is relatively ineffective as an activating agent when compared with the potassium hydroxide (Lillo-Ródenas *et al.*, 2001, 2003). In aqueous solution, the sodium ion is marginally smaller than the potassium ion and so there can be no size exclusion effects. The sodium hydroxide would appear therefore to be as able to enter into the original porosity of the CR as does the potassium hydroxide. So, why is it so inactive? The answer could lie with the higher ionization potential of the sodium atom (or clusters of atoms). It is well established that sodium does not easily form a sodium-intercalation compound with graphitic carbon and this same reluctance is apparent in this activation reaction. As mentioned above, sodium is more effective with the less-ordered carbons which contain graphene interlayer spacings both less and greater than that found in graphite. Sodium enters into the narrower of the spacings between the graphene layers, the percentage of which in a carbon are a function of the carbon precursor and its final HTT.

6.3.4 Summary of Discussions of Alkali Salt Activations

Alkali salts do not react with the precursor – they react with the carbon resulting from the carbonization process. They are equally effective on heating a carbon with the alkali metal salt. Two mechanisms appear to be operating. ***One****, the oxygen atoms of the intimate mixture of hydroxide or carbonate of sodium or potassium gasify carbon atoms to produce* CO_x *so producing porosity. This gasification may be catalyzed by the alkali metal.* ***Two****, the reduction of the alkali metal salt, by carbon, produces sodium or potassium metal atoms which intercalate into the carbon structure so expanding the "lattice". Progressive heating causes the structure to expand further to produce the activated carbon. The characteristics of the developed porosity relate to the rate of heating of the experimental system, rapid heating producing a powder with high levels of porosity. The ratio of salt to carbon is also an important factor.*

6.4 Activation of a Carbon Cloth using Six Chemical Activating Agents

Activated carbon is used mainly in granular or powdered form. Some applications require good control over the containment of the activated carbon, and the use of activated carbon cloths meets this requirement. As well as needing to have adequate micropore volumes, activated carbon clothes must maintain an adequate strength to prevent break-up and dusting. Hence, porosity assessments in conjunction with breaking strengths are a requirement. There are several methods available to produce activated carbon, chemically, in addition to the use of steam and carbon dioxide. A comprehensive study of the activation of a cloth made from viscous rayon is available from Huidobro *et al.* (2001) and this is summarized below.

A viscous rayon cloth was supplied by Moygashel Ltd (Northern Ireland), with a plain weave, a warp density of 9.3 yarn cm^{-1} and a weft density of 8.3 yarn cm^{-1}. Strips of the fabric, 80 × 250 mm, were soaked for 2 min in aqueous solutions, of increasing concentrations (1–12 wt%) of $AlCl_3$, $ZnCl_2$, NH_4Cl, $FeCl_3$, $CuCl_2$, H_3PO_4 and Na_2HPO_3. An additional solution (AZN) containing a mixture of identical concentrations of $AlCl_3$, $ZnCl_2$ and NH_4Cl was used, the total concentration of the three salts ranging from 0.9 to 3.6 wt%. The soaked samples were dried for 90 min (45 min for the H_3PO_4) under a flow of air at 70 °C and later carbonized in an horizontal furnace, under 100 mL min^{-1} flow of nitrogen, heating 5 °C min^{-1} up to 825 °C, soak time of 1 h. Carbonized samples (with H_3PO_4, and Na_2HPO_3) were thoroughly washed in distilled water, or in 1.5 N HCl followed by distilled water (all other samples).

TGA–DTA (thermal gravimetric analysis–differential thermal analysis) analyses (TG-DSC92 Setaram) were carried out on samples of the fabric impregnated with 2 wt% $AlCl_3$ or 3 wt% $ZnCl_2$ using a 5 °C min^{-1} heating rate to 900 °C under a helium flow of 80 mL min^{-1}.

Porosity was evaluated from adsorption isotherms of N_2 (77 K) and CO_2 (273 K) using an automatic equipment (Autosorb 6, Quantachrome). When the kinetics of adsorption of N_2 at 77 K were slow, adsorptions were also carried out at 90 K. Apparent surface areas (from a BET application to the nitrogen isotherms), total micropore volumes (from the α_s-method; Rodríguez-Reinoso *et al.*, 1989a) and narrow micropore volumes (from the DR equation) from the CO_2 isotherms, were calculated.

Sample thickness and breaking load were measured as described by Rodríguez-Reinoso *et al.* (2000a, b) following the American Society for Testing Materials (ASTM) D 1682–64 standard test; using specimens 180 × 260 mm, plus 20 yarns (10 yarns in each side were unraveled to control the width).

Relevant results are reproduced in Figures 6.28 and 6.29. Figure 6.28 shows the variation of breaking load of activated viscous rayon cloth with wt% of several impregnants with Figure 6.29 showing the variation of development of narrow micropore volumes with wt% of several impregnants. Clearly, there are major differences in the way the several activating agents interact with the parent cloth. The use of $AlCl_3$ had the most pronounced effect on breaking load, additions of about 5 wt% increasing the breaking load from 10 to 32 N, with little change in the narrow micropore volumes (~0.26 $cm^3 g^{-1}$). Although Na_2HPO_4 also produced an increase in breaking load, there was an associated reduction in narrow micropore

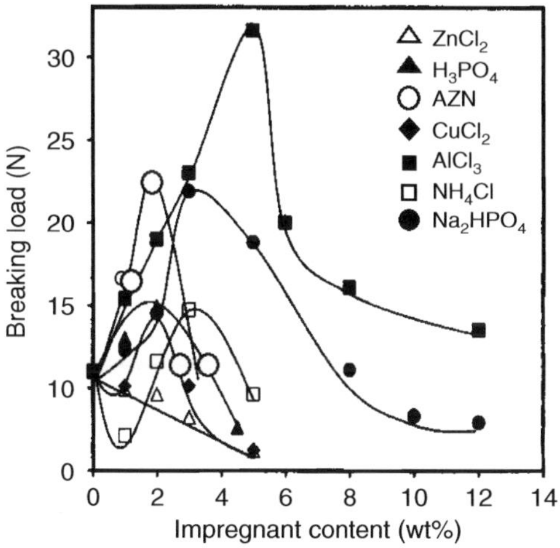

Figure 6.28. The variation of breaking load of activated viscous rayon cloth with wt% of several impregnants (Huidobro *et al.*, 2001).

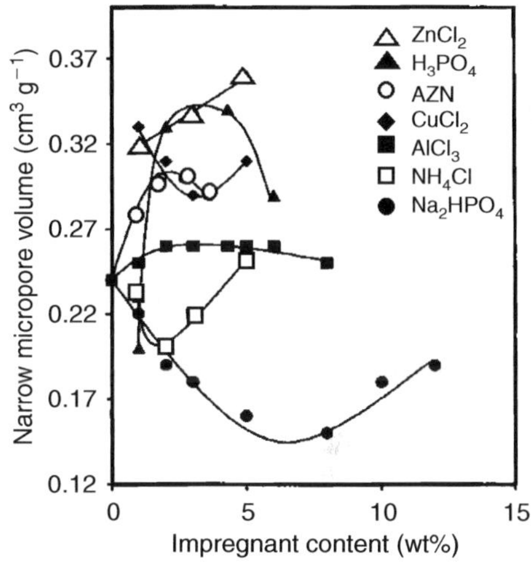

Figure 6.29. The variation of volume of narrow micropores of activated viscous rayon cloth with wt% of several impregnants (Huidobro *et al.*, 2001).

volumes (0.24–0.15 with 5 wt% addition). $ZnCl_2$ enhanced the narrow micropore volumes most (0.38 $cm^3 g^{-1}$) with 5 wt% addition, but not enhancing the breaking load.

The development of porosity is a function of the impregnant with all chlorides producing narrow microporosity with constrictions (causing diffusion problems to the adsorption of nitrogen at 77 K). H_3PO_4 develops wider microporosity, especially at high concentration, with the Na_2HPO_4 producing significant mesoporosity, this being a result of the spaces left by the chemical which remains after carbonization and which is eluted with washing. As for yields, the highest yields were obtained using $AlCl_3$ and H_3PO_4 (36 wt% at 5 wt% of impregnant of both chemicals).

6.5 Information from Selected Publications

Currently, the research journals report many investigations into the production of activated carbons by both physical and chemical methods, and in some cases by combined methods. A short selection of such papers (based on the variety of approaches) is presented below.

Amarasekera et al. *(1998)*: A series of high surface area carbons resulted from heating a brown coal with KOH and NaOH followed by carbonization, HTT 673–1173 K. High resolution nitrogen isotherms were determined for values of p/p^o of 10^{-6}–1 and indicated that pore size development was a function of the KOH:coal ratio. Micropore volumes were reported maximizing at 0.6 $cm^3 g^{-1}$. A replacement of carboxyl groups with K^+ facilitates nitrogen accessibility to the entire micropore space.

Benaddi et al. *(2000)*: Mixtures of wood with either phosphoric acid or diammonium hydrogen phosphate [$(NH_4)_2HPO_4$] were heated under nitrogen or steam/nitrogen flows at temperatures between 300 and 500 °C to obtain activated carbons. The carbons were characterized using potentiometric titration and sorption of nitrogen (77 K). It was found that steam inhibits the incorporation of heteroatoms into the carbon matrix. Activation of wood in the presence of phosphoric acid together with steam gives carbons of high surface area (about 1800 $m^2 g^{-1}$) with well-developed mesoporosity. There is discussion of mechanisms and of surface acidities (see Section 4.7.4).

Gañan et al. *(2004)*: High-porosity carbons were prepared from bituminous coal pitches by combining chemical and physical activation. Chemical activation used KOH impregnation (450 °C for 1 h) followed by carbonization in nitrogen (375–850 °C). The ratio of KOH:pitch of 3:1 optimized the development of porosity. In the carbonization range 800–850 °C, a carbon was produced with a micropore volume of 1.02 $cm^3 g^{-1}$ and a BET surface area of 2350 $m^2 g^{-1}$.

Hayashi et al. *(2002)*: Activated carbons were prepared from phenol–formaldehyde (PF) and urea–formaldehyde (UF) resins by chemical activation with K_2CO_3 with impregnation during the synthesis of the resins. The influence of carbonization temperature (773–1173 K) on surface area and pore volumes and the temperature range at which K_2CO_3 was effective were investigated. The specific surface area and micropore volume of PV-AC and UF-AC increased with increasing HTT (773–1173 K). A carbon from PF possessed a micropore volume of 0.4 $cm^3 g^{-1}$ and a mesopore volume of 0.04 $cm^3 g^{-1}$. The authors deduced an activation mechanism from thermogravimetry and X-ray diffraction (XRD) stating that the K_2CO_3 was reduced by carbon which was removed as CO_2, this causing the activation.

Lyubchik et al. (2004): The approach used was the preparation of activated carbons from Ukrainian anthracites (C_{daf} 92–95%) by combining a chemical pre-treatment and a physical activation. The anthracites were pre-treated with $HClO_4$ or with HNO_3 vapors and then physically activated either with H_2O or CO_2 at 850 °C for 1–8 h. Changes were analyzed using X-ray photoelectron spectroscopy, XRD, elemental analysis and gas adsorption techniques. Chemical pre-treatment with nitric acid of the low-rank anthracite of C_{daf} 92% followed by activation with steam for 2 h at 850 °C produced optimum results. Highly microporous activated carbon of 0.7 $cm^3 g^{-1}$ and estimated surface area of 1600 $m^2 g^{-1}$ was obtained.

Mui et al. *(2004)*: This review presents the current status of research into the production of active carbons from environmental applications using waste tyres. Several activated carbons were produced with BET N_2 surface areas >1000 $m^2 g^{-1}$ using steam activation, with values of 270–980 $m^2 g^{-1}$ for activation with carbon dioxide. Chemical activation with KOH produced a carbon with a surface area of 820 $m^2 g^{-1}$ at 850 °C. Models were developed to describe the conversion of waste tyres to a carbonaceous char. Further research is necessary to lead to successful applications of activated carbons from waste tyres.

Skodras et al. *(2002)*: A technique using the three stages of demineralization, activation and sulfur dispersion was developed for the production of activated carbons from a Greek lignite. Demineralization involved acid treatment and carbons were characterized by XRD. A two-stage activation procedure of pyrolysis under nitrogen, and activation under a carbon dioxide atmosphere was used. Sulfur impregnation of activated carbons involved heating with sulfur under nitrogen flow up to 600 °C. SEM and line-scattering techniques were used to evaluate sulfur distribution in the impregnated activated carbons. Adsorptions of N_2 (77 K) and CO_2 (298 K) were used for characterization. Sulfur impregnated activated carbon samples were proven unreactive and stable at the flue gases temperatures.

Wu et al. *(2004)*: PCs were prepared from PAN-based pre-oxidized cloth with potassium hydroxide (KOH) as active reagent. The PCs were studied by the adsorptions of nitrogen, benzene and iodine. Such process parameters as weight ratio of KOH to the starting material, activation temperature and activation time are crucial for preparing high quality PCs. A series of PCs with high BET surface area and well-developed porous structure in which micropores are dominant (0.7 nm) were obtained with less KOH and shorter activation time in comparison to the traditional methods. The optimum conditions for preparing PCs with high BET surface area from PAN-based pre-oxidized cloth were given, and the relationships between pore structure and adsorption property of PCs were explored.

Yoon et al. *(2004)*: Catalytically grown carbon nanofibers (CNF) were activated with KOH. The structural changes of CNFs were investigated using SEM, transmission electron microscopy (TEM) and XRD. The observed increase in surface area in KOH activation of CNF is ascribed to local broadening of graphene interstices or local burn-off of graphene layers. Activation under severe conditions destroyed the fiber structure of CNF. Activated CNFs were applied as an electrode for electrical double layer capacitors to be compared to active carbon fibers in terms of the surface properties.

References

Agarwal RK, Schwartz JA. Analysis of high-pressure adsorption of gases on activated carbon by potential theory. *Carbon* 1988;26(6):873–887.

Alcañiz-Monge J, Casa-Lillo MA, Cazorla-Amorós D, Linares-Solano A. Methane storage in activated carbon fibres. *Carbon* 1997;35(2):291–297.

Almansa C, Molina-Sabio M, Rodríguez-Reinoso F. Adsorption of methane into $ZnCl_2$-activated carbon derived discs. *Micropor Mesopor Mat* 2004;76(1–3):185–191.

Amarasekera G, Scarlett MJ, Mainwaring DE. Development of microporosity in carbons derived from alkali digested coal. *Carbon* 1998;36(7–8):1071–1078.

Baker FS. Activated carbon. In: *Kirk-Othmer Encyclopedia of Chemical Technology*, 4th Edition, Vol. 4, Eds. Miller CE, Repik AJ, Tolles ED. Wiley, New York, 1992, pp. 1015–1037.

Barker RH, Hendrix JE. Cotton and other cellulose polymers. In: *Flame Retardancy of Polymeric Materials*. Eds. Kuryla H, Papa AJ. Marcel Dekker Inc, New York, 1979, pp. 35–65.

Barrett EP, Joyner LG, Halenda PH. The determination of pore volume and area distributions in porous carbons. I. Computation of nitrogen isotherms. *J Am Chem Soc* 1951;73:373–380.

Benaddi H, Bandosz TJ, Jagiello J, Schwarz JA, Rouzaud, Legras D, Béguin F. Surface functionality and porosity of activated carbons obtained from chemical activation of wood. *Carbon* 2000;38(5):669–674.

Berger D, Carton B, Métrot A, Hérold A. Intercalation of potassium and sodium with carbons. In: *Chemistry and Physics of Carbon*, Vol. 12, Ed. Walker Jr PL. Marcel Dekker Inc, New York, 1975, pp. 1–37.

Bourbrigot S, Le Bras M, Deobel R, Breant P, Tremillon JM. Carbonization mechanisms resulting from intumescence. Part II. Association with an ethylene terpolymer and the ammonium polyphosphate–pentaerythriol fire retardant system. *Carbon* 1995;33(3):283–294.

Caturla F, Molina-Sabio M, Rodríguez-Reinoso F. Preparation of activated carbon by chemical activation with $ZnCl_2$. *Carbon* 1991;29(7):999–1007.

Chan BKC, Thomas KM, Marsh H. The interactions of carbons with potassium. *Carbon* 1993;31(7):1071–1082.

Chen XS, McEnaney B, Mays TJ, Alcañiz-Monge J, Cazorla-Amorós D, Linares-Solano A. Theoretical and experimental studies of methane adsorption on microporous carbon. *Carbon* 1997;35(9):1251–1258.

Clarkson CR, Bustin RM, Levy JH. Application of mono/multilayer and adsorption potential theories to coal–methane adsorption isotherms at elevated temperature and pressure. *Carbon* 1997;35(12):1689–1705.

Cook TL, Komodromos C, Quinn DF, Ragan S. Adsorbent storage for natural gas vehicles. In: *Carbon Materials for Advanced Technologies*, Ed. Burchell TD. Pergamon Press, Amsterdam, 1999, pp. 269–302.

Delobel R, Le Bras M, Ouassou N. Fire retardance of polypropylene by diammonium pyrophosphate pentaerythriol: spectroscopic characterization of protective coatings. *Polym Degrad Stabil* 1990;30:41.

Dubinin MM. The potential theory of adsorption of gases and vapours for adsorbents with energetically non-uniform surfaces. *Chem Rev (London)* 1960;60:235–241.

Gañan J, González-García CM, González JF, Sabio E, Macías- García A, Díaz-Díez MA. Preparation of activated carbons from bituminous coal pitches. *Appl Surf Sci* 2004;238:347–354.

Garrido J, Linares Solano A, Martín JM, Molina M, Rodríguez-Reinoso F, Torregrosa R. Use of N_2 vs. CO_2 in the characterization of activated carbons. *Langmuir* 1987;3:76–81.

González JC, González MT, Molina-Sabio M, Rodríguez-Reinoso F, Sepúlveda-Escribano A. Porosity of activated carbons prepared from different lignocellulosic materials. *Carbon* 1995;33(8):1175–1177.

González JC, Sepúlveda-Escribano A, Molina-Sabio M, Rodríguez-Reinoso F. Production, properties and applications of activated carbon. In: *Characterization of Porous Solids*. Vol. IV, Eds. McEnaney B, Mays TJ, Rouquerol F, Rodríguez-Reinoso F, Sing KSW, Unger KK. The Royal Society of Chemistry, Cambridge, 1997, pp. 9–16.

Gregg SJ, Sing KSW. *Adsorption, Surface Area and Porosity*, 2nd Edition. Academic Press, London, 1982.

Hayashi J, Uchibayashi M, Horikawa T, Muroyama K, Gomes VG. Synthesizing activated carbons from resins by chemical activation with K_2CO_3. *Carbon* 2002;40(5):2747–2752.

Hinderschinn RR, Witschard G. Intumescence and char. In: *Flame Retardancy of Polymeric Materials*, Eds. Kuryla H, Papa AJ. Marcel Dekker Inc, New York, 1979, pp. 67–78.

Hindersin RR, Witschard G. In: *Flame Retardancy of Polymeric Materials*, Eds. Kuryla H, Papa AJ. Marcel Dekker Inc, New York, 1979, pp. 1–34.

Hooley JG. Intercalation isotherms on natural and pyrolytic graphites. In: *Chemistry and Physics of Carbon*, Vol. 5, Ed. Walker Jr PL. Marcel Dekker Inc, New York, 1969, pp. 321–373.

Hu Z, Srinivasan MO, Ni Y. Novel activation process for preparing highly microporous and mesoporous carbons. *Carbon* 2001;39(6):877–886.

Huidobro A, Pastor AC, Rodríguez-Reinoso F. Preparation of activated carbon cloth from viscous rayon Part IV. Chemical activation. *Carbon* 2001;39(3):389–398.

Inomata K, Kanazawa K, Urabe Y, Ozono H, Araki T. Natural gas storage in activated carbon pellets without binder. *Carbon* 2002;40(1):87–93.

Jagtoyen M, Derbyshire F. Activated carbons from yellow poplar and white oak by H_3PO_4 activation. *Carbon* 1998;36(7–8):1085–1097.

Laine J, Lunes JLS. Effect of the preparation method on pore size distribution of activated carbon from coconut shell. *Carbon* 1992;30(4):601–604.

Lillo-Ródenas MA, Lozano-Castelló D, Cazorla-Amorós D, Linares-Solano A. Preparation of activated carbons from Spanish anthracite. Part II. Activation by NaOH. *Carbon* 2001;39(5):751–759.

Lillo-Ródenas MA, Carratalá-Abril J, Cazorla-Amorós D, Linares-Solano A. Usefulness of chemically activated anthracite for the abatement of VOC at low concentrations. *Fuel Process Technol* 2002;77–78:331–336.

Lillo-Ródenas MA, Cazorla-Amorós D, Linares-Solano A. Understanding chemical reactions between carbons and NaOH and KOH: an insight into the chemical activation mechanism. *Carbon* 2003;41(2):267–275.

Lillo-Ródenas MA, Cazorla-Amorós D, Linares-Solano A, Clinard C, Rouzaud JN. HRTEM study of activated carbons prepared by alkali hydroxide activation of anthracite. *Carbon* 2004a;42(7):1305–1310.

Lillo-Ródenas MA, Juan-Juan J, Cazorla-Amorós D, Linares-Solano A. About reactions occurring during chemical activation with hydroxides. *Carbon* 2004b;42(7):1371–1375.

Lozano-Castelló D, Lillo-Ródenas MA, Cazorla-Amorós D, Linares-Solano A. Preparation of activated carbons from Spanish anthracite. Part I. Activation by KOH. *Carbon* 2001;39(5):741–749.

Lozano-Castelló D, Cazorla-Amorós D, Linares-Solano A, Quinn DF. Influence of pore size distribution in methane storage at relatively low pressures; preparation of activated carbons with optimum pore size. *Carbon* 2002;40(7):989–1002.

Lyubchik SB, Galushko LY, Rego AM, Tamarkina YV, Galushko OL, Fonseca IM. Intercalation as an approach to the activated carbon preparation from Ukrainian anthracites. *J Phys Chem Solids* 2004;65(2–3):127–132.

MacDonald JAF, Quinn DF. Adsorbents for methane storage made by phosphoric acid activation of peach pits. *Carbon* 1996;34(9):1103–1108.

Maciá-Agulló JA, Moore BC, Cazorla-Amorós D, Linares-Solano A. Activation of coal tar pitch fibres: physical activation vs chemical activation. *Carbon* 2004;42(7):1367–1370.

Marsh H, Crawford D, O´Grady TM, Wennerberg A. Carbons of high surface area. A study by adsorption and high resolution electron microscopy. *Carbon* 1982;20(5):419–426.

Marsh H, Yan DS, O´Grady TM, Wennerberg A. Formation of active carbons form cokes using potassium hydroxide. *Carbon* 1984;22(6):603–611.

Marsh H, Murdie N, Edwards IAS, Boehm H-P. Interactions of carbons cokes, and graphites with potassium and sodium. In: *Chemistry and Physics of Carbon*, Vol. 20, Ed. Thrower PA. Marcel Dekker Inc, New York, 1987; pp. 213–272.

Menon VC, Komarneni S. Porous adsorbents for vehicular natural gas storage: a review. *J Porous Mat* 1998;5(1):43–58.

Ming L, Anzhong GD, Xuesheng L, Rongshun W. Determination of the adsorbate density for supercritical gas adsorption equilibrium data. *Carbon* 2003;41(3):585–588.

Molina-Sabio M, Rodríguez-Reinoso F. Role of chemical activation in the development of carbon porosity. *Colloid Surfaces A: Physicochem Eng Aspects* 2004;241:15–25.

Molina-Sabio M, Rodríguez-Reinoso F, Cartula F, Sellés MJ. Porosity in granular carbons activated with phosphoric acid. *Carbon* 1995;33(8):1105–1113.

Molina-Sabio M, Almansa C, Rodríguez-Reinoso F. Phosphoric acid activated carbon discs for methane adsorption. *Carbon* 2003;41(11):2113–2119.

Mui ELK, Ko DCK, McKay G. Production of active carbons from waste tyres – a review. *Carbon* 2004;42(14):2789–2805.

Ozawa S, Kusumi S, Ogino Y. Physical adsorption of gases at high pressures. IV. An improvement of the Dubinin–Astakhov adsorption equation. *J Colloid Interf Sci* 1976;56(1):83–91.

Pandey SN, Nair P. Effect of phosphoric acid treatment on physical and chemical properties of cotton fiber. *Text Res J* 1974;December:927.

Parkyns ND, Quinn DF. Natural gas adsorbed on carbon. In: *Porosity in Carbons*, Ed. Patrick JW. Edwards Arnold, London, 1995, pp. 291–325.

Porter BL, Rollins ML. Changes in porosity of treated lint cotton fibers. I. Purification and swelling treatments. *J Appl Polym Sci* 1972;16:217.

Rodríguez-Reinoso F, Linares-Solano A. Microporous structure of activated carbons as revealed by adsorption methods. In: *Chemistry and Physics of Carbon*, Vol. 21, Ed. Thrower PA. Marcel Dekker Inc, New York, 1989b, pp. 1–146.

Rodríguez-Reinoso F, Garrido J, Martín-Martínez JM, Molina-Sabio M, Torregrosa R. The combined use of different approaches in the characterization of microporous carbons. *Carbon* 1989a;27(1):23–32.

Rodríguez-Reinoso F, Pastor AC, Marsh H, Martínez MA. Preparation of activated carbon cloths from viscous rayon. Part II. Physical activation processes. *Carbon* 2000a;38(3):379–395.

Rodríguez-Reinoso F, Pastor AC, Marsh H, Huidobro A. Preparation of activated carbon cloths from viscous rayon. Part III. Effect of carbonization on CO_2 activation. *Carbon* 2000b;38(3):397–406.

Rodríguez-Reinoso F, Molina-Sabio M, Almansa C. Procedimiento para la obtención de monolitos de carbón activado, monolitos obtenidos y su empleo en el almacenamiento de gas natural/metano. *Patente española*, 2003, 2.165.784.

Rowell RM. In: *Word and Cellulosic Chemistry*, Vol. 15, Eds. Hon DNS, Shirashi N. Marcel Dekker Inc, New York, 1991, p. 703.

Shimada T, Yanase H, Morishita K, Hayashi J, Chiba T. Points of onset of gasification in a multi-walled carbon nanotube having an imperfect structure. *Carbon* 2004;42(8–9):1635–1639.

Skodras G, Orfanoudaki Th, Kakaras E, Sakellaropoulos GP. Production of special activated carbon from lignite for environmental purposes. *Fuel Process Technol* 2002;77–78:75–87.

Solum MS, Pugmire RJ, Jagtoyen M, Derbyshire F. Evolution of carbon structure in chemically activated wood. *Carbon* 1995;33(9):1247–1254.

Torregrosa R, Martín-Martín JM. Activation of ligno-cellulosic materials: a comparison between chemical, physical and combined activation in terms of porous texture. *Fuel* 1991;70(10):1173–1180.

Von Kienle H. In: *Ullmann´s Encyclopedia of Industrial Chemistry*, 5th Edition, Vol. A5, Eds. Gerhatz W, Yamamoto YS, Campbell FT. VCH, New York, 1986, pp. 124–140.

Wennerberg AN, O´Grady TM. Active carbon process composition. US Patent 4082694; Amoco Corporation, 1978.

Wu M, Zha Q, Qiu J, Guo Y, Shang H, Yuan A. Preparation and characterization of porous carbons from PAN-based pre-oxidized cloth by KOH activation. *Carbon* 2004;42(1):205–210.

Yoon S-H, Lim S, Song Y, Ota Y, Qiao W, Tanaka A, Mochida I. KOH activation of carbon nanofibers. *Carbon* 2004;42(8–9):1723–1729.

CHAPTER 7

SEM and TEM Images of Structures in Activated Carbons

7.1 Introduction

The charisma of activated carbon is the uniqueness of its porosity which takes over the entire particle of the carbon with its overwhelming number of shapes and sizes of microporosities. A human wish is to know the structure and interrelationships of these microporosities. However, the wish is very difficult to realize, the very smallness of the microporosities, <2 nm entrance dimension, just coming within reach of modern electron microscopes. Chapter 3 is devoted to a critical assessment of the methods of structural analysis and of model generation to be found in the literature, presenting about 30 models representing a wide range of approaches. It is a manifestation of the problem facing the modelers that there is little in common between these models. The computer-generated models depend upon the quality and quantity of information supplied. The X-ray and neutron diffraction techniques (high- and low-angle studies) require theoretical models to translate diffraction information to structural information.

However, high-resolution transmission electron microscopy (HRTEM) produces images of structure, of which Figure 3.28(f) is a very good example. An analysis of structure within Figure 3.28(f) shows how complex is this structure with defective graphene layers, of different dimensions and shapes all co-bonded in close proximity to each other to create the spaces between the graphene layers, called the microporosity.

So complex is this structure which encapsulates microporosity that its accurate prediction from diffraction studies and computer modeling may never be possible.

Analyses of fringe-images derived from transmission electron microscopy (TEM) (Section 3.4.21) give the impression that all of the detail is in one plane within the photographed image. This may not be so, but this qualification may not be important as it is the general structural detail which is looked for, not accuracy of position of structure within 15 nm^3 of activated carbon. The validity of the fringe-images can be assessed from the publications of Sharma *et al.* (2000) and Aso *et al.* (2004) who undertook X-ray diffractions (XRD) determinations of the so-called *crystallite sizes* as well as fringe-images from a graphitizable

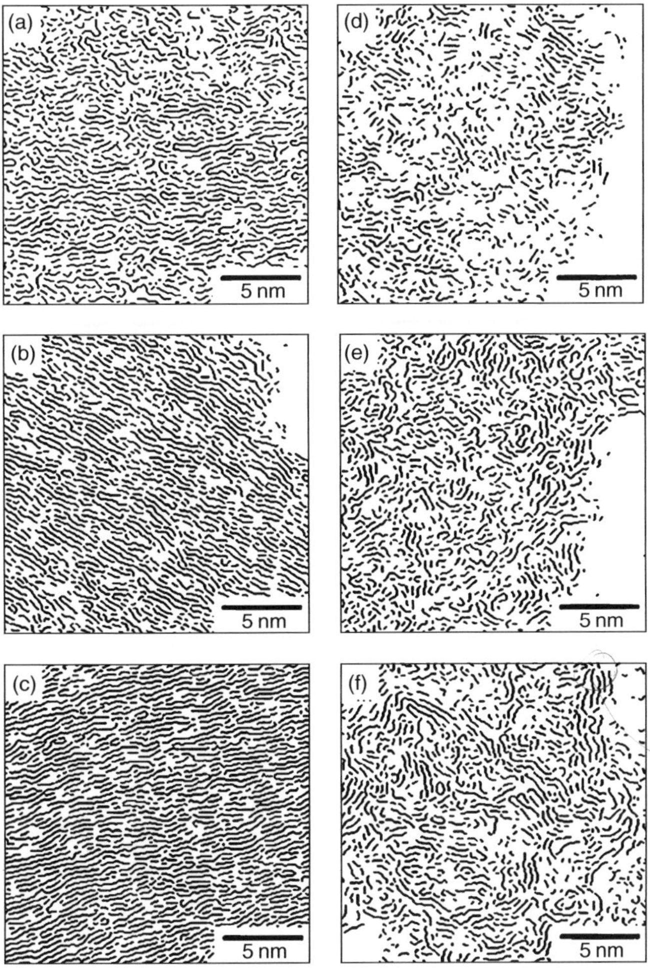

Figure 7.1. (a–f) Fringe-images derived from PVC-1000 °C and PFA-1000 °C (Aso *et al.*, 2004).

carbon from polyvinyl chloride (PVC) and a non-graphitizable carbon from polyfurfuryl alcohol (PFA). The fringe-images derived from PVC-1000 °C, and PFA-1000 °C are shown in Figure 7.1(a–f). Figure 7.1(a–c) indicates the parallelism of graphene layers associated with a graphitizable carbon, with Figure 7.1(d–f) indicating the much more random relative orientations of layers associated with non-graphitizable carbons. An associated XRD assessment of structure quote found that values of the crystallite size (L_c (nm)) decreased from 1.7 to 1.4 nm (heat treatment temperature (HTT) 600–800 °C) increasing to only 1.6 nm (HTT 1000 °C) for the PVC carbons. These data are in contradiction to the appearance of the fringe-images, see Figure 7.1(a–c). The serious limitations of the XRD data and the explanation of this anomaly are available in Section 2.10.

There is close resemblance between what is seen in Figure 3.28(f) (Rouzaud *et al.*, 2002), and the labyrinths of Figure 3.28(c, d). The labyrinth is too symmetrical and if drawn in a more disordered way, could create a model which would meet most of the requirements for a model as listed in Section 3.2.

When one is asked to describe porosity, using paper and pencil, first attempts are as seen in Chapter 3, as Figures 3.1–3.3. The approach of textbooks when describing adsorption in porous solids is the thermodynamics of adsorption processes. Although high-resolution transmission and scanning electron microscopy (HRTEM and SEM, respectively) have been available for at least 20 years, this facility has probably been under-exploited in analysis of structure in activated carbons.

This approach of microscopy has its critics who say that micrographs can be obtained to prove the existence of any suggested structure. There is some truth in such comments, but the critics are not totally correct. There is no doubt that great care has to be taken not to misrepresent structure. This can be easily done by focussing on an image which looks most interesting but which is not representative. The microscopist loves to photograph the rare and the unusual. HRTEM works routinely at magnifications of about one million times. This is equivalent to the inspection of an area 1 mm^2 in an original area of 1 km^2. The need for representation is therefore obvious.

Chapter 7 reports on a study of the microscopy of some activated carbons, and the adsorption properties of some being contained within Chapter 5. Greatest care is taken in this study to ensure that what is photographed is representative of the carbon material, facilitated by the *extensive homogeneity* of structure which exists within such carbons.

Chapter 7 explores the possibilities suggested by Rouzaud *et al.* (2002) of using HRTEM (not fringe imaging) to study a range of activated carbons which includes: (a) samples of wood; (b) four commercial activated carbon, activated with phosphoric acid and steam; (c) activated carbon from olive stones, using carbon dioxide and steam and (d) activated carbon from olive stones using potassium hydroxide as the activating agent.

7.2 Experimental Microscopy

The SEM was a JEOL, JSM-840. For specimen examination, small fragments of the carbon, about 1 mm^3 were mounted on stubs and coated with silver to obtain maximum contrast with the photography. The SEM is useful for imaging the morphology of the carbon specimen (e.g. original botanical structures), but not porosity.

A JEM (JEOL) 2010 TEM was used, this having a maximum magnification of 1.2 million, with a resolution between planes of 0.14 nm, and between points of 0.25 nm. The carbon specimen was ground in a pestle and mortar and a suspension made in acetone of the powder. Small volumes of this suspension were placed over a "holey" carbon film supported on a copper grid so allowing direct transmission of the electron beam. HRTEM provided images of the effects of activation with resolutions of about 1–2 nm.

This project was co-ordinated by Dr. Cristina Almansa, University of Alicante.

7.3 Selected Micrographs

7.3.1 Photography and SEM

As indicated in Chapter 1, porous carbon made from coconut shell, indicated to the world the potential of activated carbon for purification and separation. World War I was a turning point. Figure 7.2(a) is a photograph of a humble coconut with its fibrous outer covering. Figure 7.2(b) shows the interior of the coconut, the white interior being the copra to

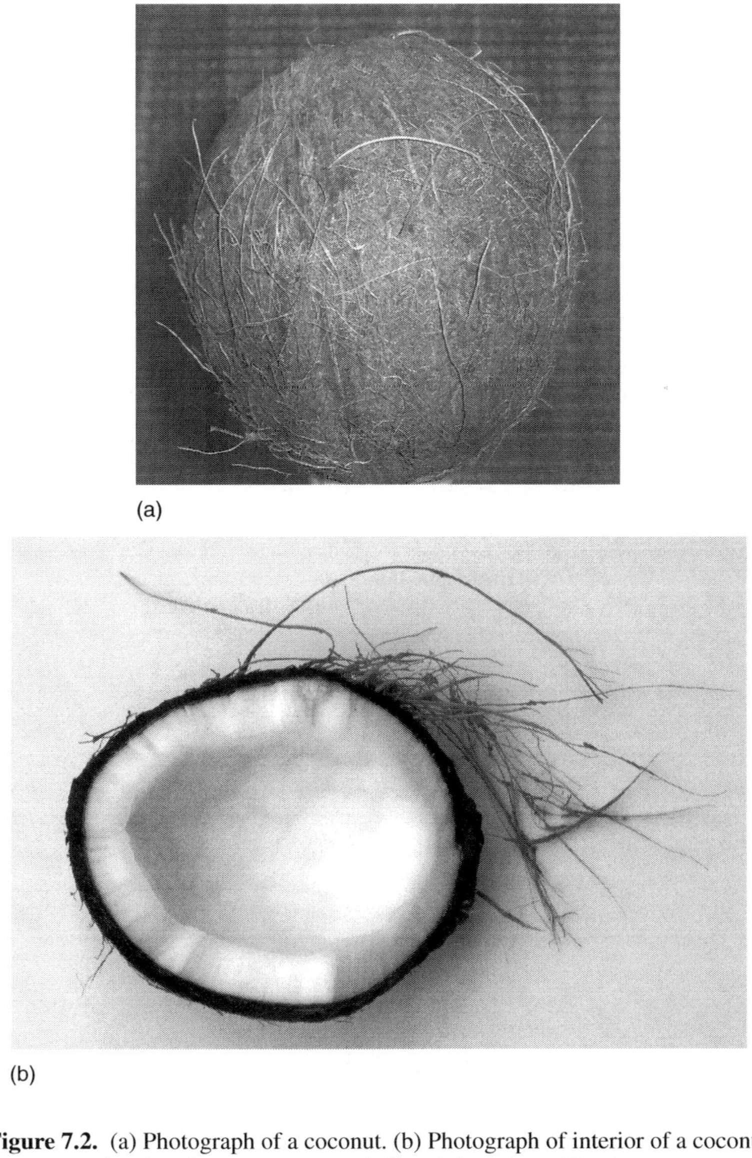

Figure 7.2. (a) Photograph of a coconut. (b) Photograph of interior of a coconut showing the outer coconut shell and the inner white copra.

be used in cake-making and for its oil. It is the hard, dense shell of the coconut which is carbonized/activated to make the activated carbon. Although porous carbons can be made from most lignocellulosic parent materials, it is those with hardness and a high density which find commercial application as activated carbon. Figures 7.3 and 7.4 are SEM micrographs of fracture surfaces of coconut shell. The compactness (high density) of the shell structure is to be seen.

Figures 7.5–7.8 are SEM micrographs of carbonized woods. Wood anatomy is a developed science with an extensive literature. These four micrographs are included to show the

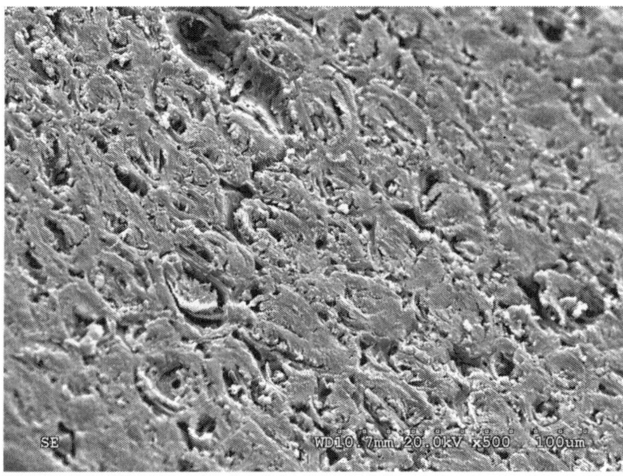

Figure 7.3. SEM micrograph of fracture surface of coconut shell (the width of micrograph, $w = 200\ \mu m$).

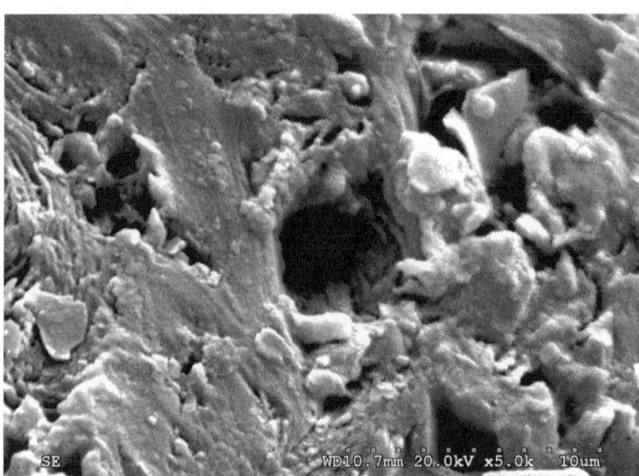

Figure 7.4. SEM micrograph of a fracture surface of coconut shell ($w = 25\ \mu m$).

diversity of structures which is exhibited, but also to make another point. The vesicles and spaces within these micrographs are of micrometer size and contribute nothing to the adsorption capacity of an activated carbon (V_{micro} ($cm^3\,g^{-1}$)).

It may be obvious, but pieces of activated carbon, in an adsorption system, do not act as a "magnet" to an adsorbate, behaving essentially as a "getter". These pieces of carbon are essentially passive within an adsorbing system. They do not attract the adsorbate molecules towards them. But once an adsorbate molecule makes contact with surface of microporosity, then the forces of adsorption come into play and the adsorbate molecule enters the micropore network.

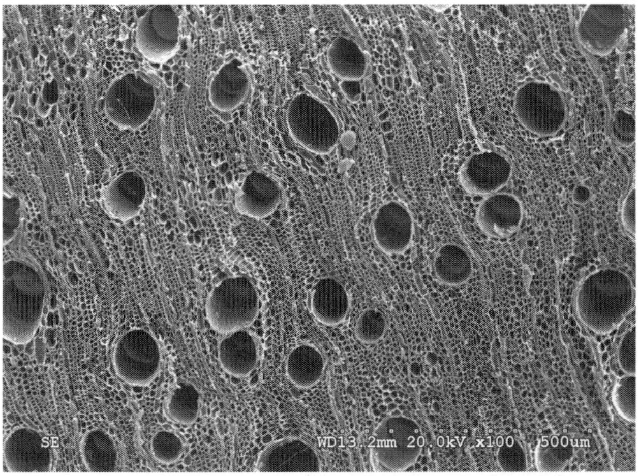

Figure 7.5. SEM micrograph of a fracture surface of a wood (w = 1000 μm).

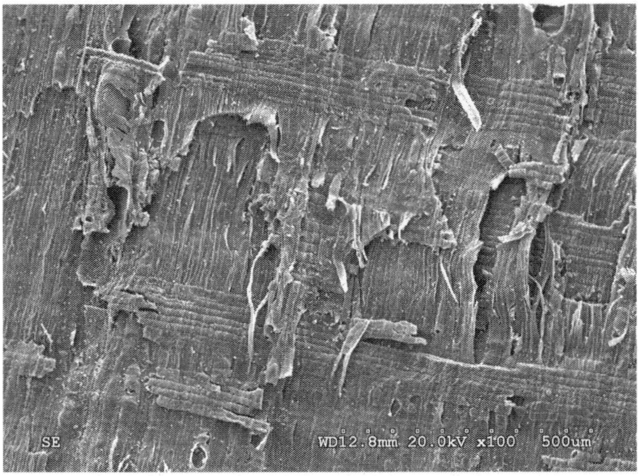

Figure 7.6. SEM micrograph of a fracture surface of a wood (w = 1000 μm).

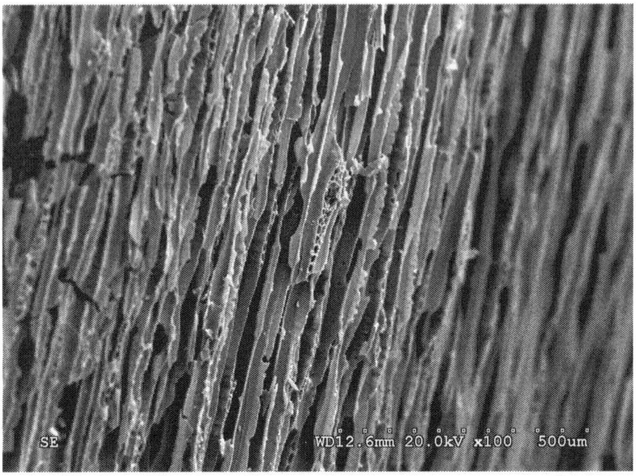

Figure 7.7. SEM micrograph of a fracture surface of a wood (w = 1000 μm).

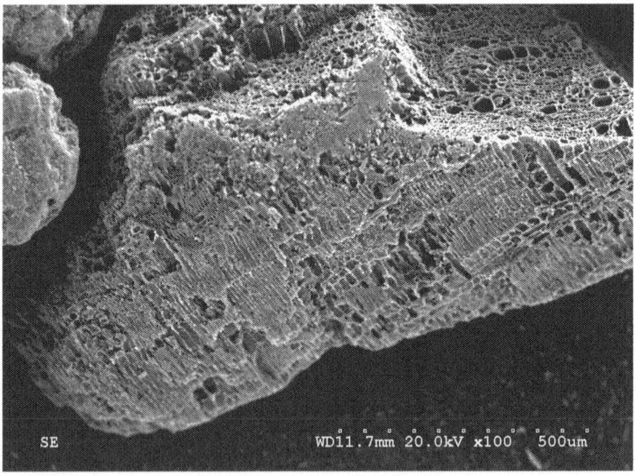

Figure 7.8. SEM micrograph of a fracture surface of a wood (w = 1000 μm).

Figures 7.9–7.12 are SEM micrographs of surfaces of carbonized olive stones (D19 $V_{micro} = 0.31\,cm^3\,g^{-1}$), activated in carbon dioxide to 19 wt% burn-off (Rodríguez-Reinoso *et al.*, 1995). Examination of this carbon failed to show external evidence of gasification. The cracks of Figures 7.11 and 7.12 are considered to be shrinkage cracks produced on carbonization. The burn-off (19 wt%) has created the micropore volumes as measured in adsorption studies. Figures 7.13 and 7.14 are of surfaces of olive stones (D52 $V_{micro} = 0.55\,cm^3\,g^{-1}$), activated in carbon dioxide to 52 wt% burn-off, and show evidence of surface erosion (ablation) creating extensive macroporosity. Figure 7.15 is of the surface of carbon from olive stones (H22 $V_{micro} = 0.33\,cm^3\,g^{-1}$), activated in steam to 22 wt% burn-off. Overall, it is reported that these surfaces exhibited more surface ablation

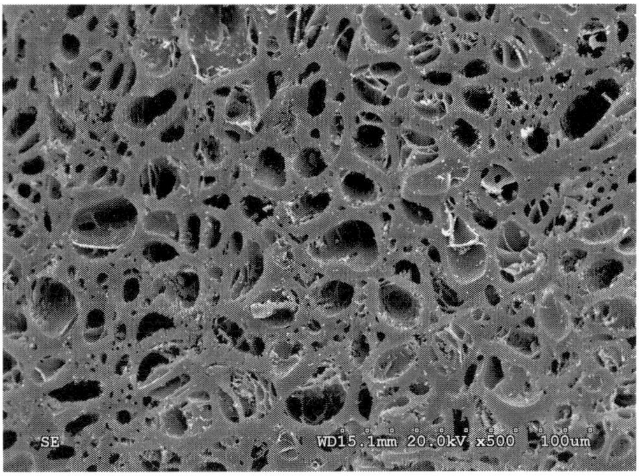

Figure 7.9. SEM micrograph of a surface of activated carbon from olive stones D19 ($w = 250\,\mu m$).

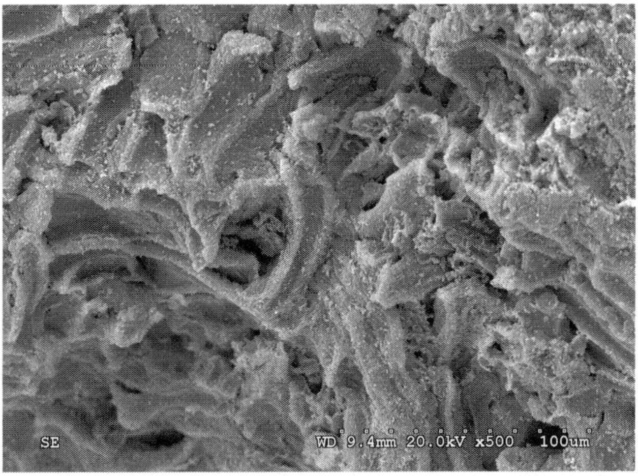

Figure 7.10. SEM micrograph of a surface of activated carbon from olive stones D19 ($w = 250\,\mu m$).

than the corresponding carbon activated with carbon dioxide, as noted with the activation of carbon fibres (Figure 5.31).

7.3.2 Transmission Electron Microscopy

Figure 7.1(a–f), being fringe-images produced by diffraction interference effects, is essentially a virtual image. Another approach to a structure analysis is to use transmission methods in which the image is created by absorption effects (thickness contrast). This

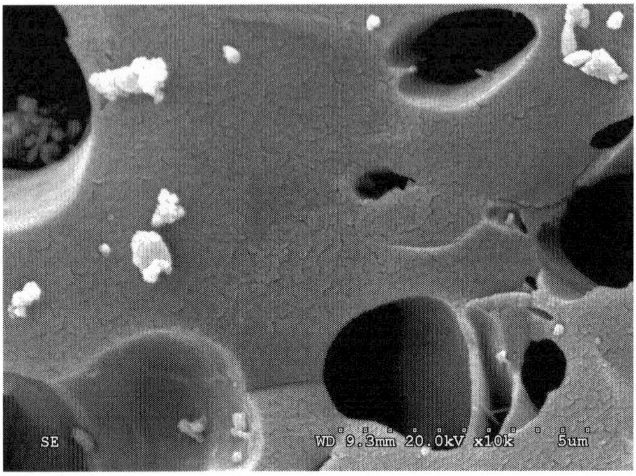

Figure 7.11. SEM micrograph of a surface of activated carbon from olive stones D19 ($w = 10\,\mu m$).

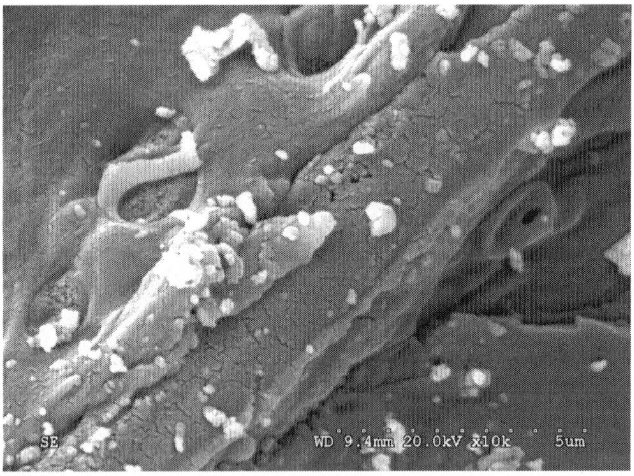

Figure 7.12. SEM micrograph of a surface of activated carbon from olive stones D19 ($w = 10\,\mu m$).

method has its own limitations in that the graphene layers of structure will be orientated in all directions relative to the axis of the electron beam. Porosity can be identified because of zero absorption. Maximum contrast results when the graphene layers are vertical.

Figures 7.16–7.20 are TEM micrographs from four commercial activated carbons. The effects of catalytic gasification have been discussed already in Chapter 5, Section 5.4. These effects are further seen in Figure 7.16 which shows gasification etching and channeling of mobile catalytic particles. Figure 7.17 shows etching by a larger mobile catalytic

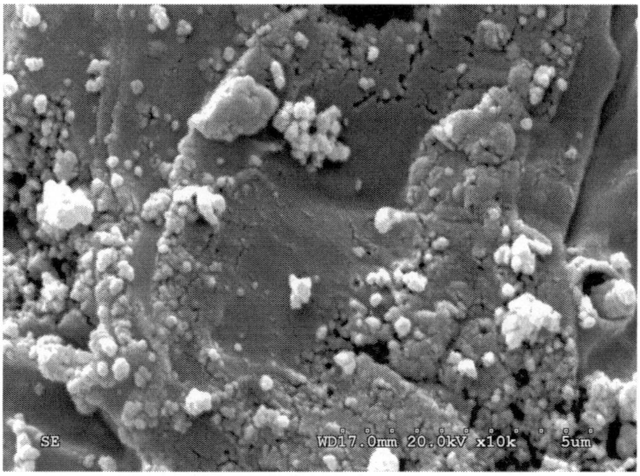

Figure 7.13. SEM micrograph of a surface of activated carbon from olive stones D52 ($w = 10\,\mu m$).

Figure 7.14. SEM micrograph of a surface of activated carbon from olive stones D52 ($w = 10\,\mu m$).

particle, size about 1 μm at the edge of the carbon particle. Channeling is also observed, of size about 50–100 nm. The active metal in these particles appears to be iron. This channeling has the size of macroporosity (>50 nm) and makes no contribution to adsorption capacity although it may assist with transport of the adsorbate into the interior of the particle.

Figures 7.18–7.20 are TEM micrographs of higher magnification. Commercial activated carbons (3) and (6) are prepared by chemical activation using phosphoric acid (Figures 7.18(a, b) and 7.19(a, b)) with activated carbon (5) being prepared by thermal activation.

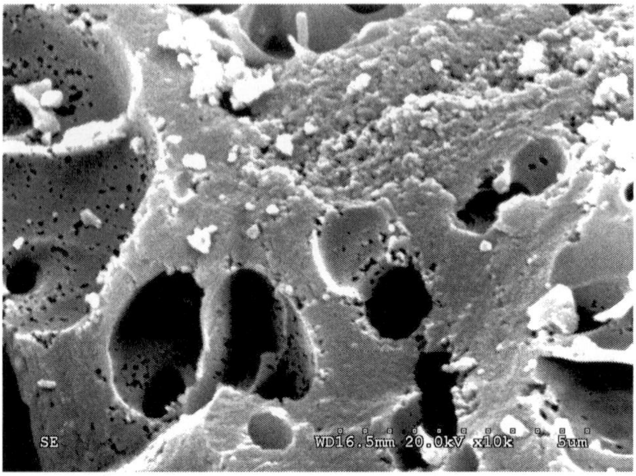

Figure 7.15. SEM micrograph of a surface of activated carbon from olive stones H22 ($w = 10\,\mu m$).

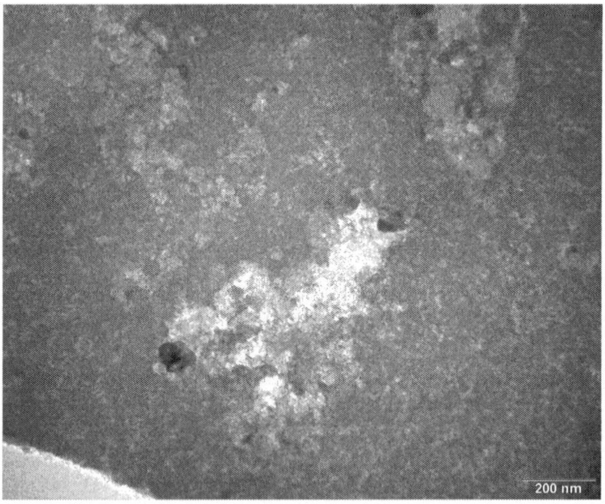

Figure 7.16. TEM micrograph of section of commercial activated carbon (6) ($w = 1.6\,\mu m$).

There are differences to be seen between these two methods of activation with activated carbon (5) appearing to be more homogeneous. The three carbons, at the higher magnifications, show the vertical projection of the graphene layers, indicating their defective nature (non-planarity, and between which is contained the microporosity.

Figures 7.21–7.27 are TEM micrographs of activated carbons from olive stones, D19, using carbon dioxide and H22 using steam (Rodríguez-Reinoso *et al.*, 1995). Figure 7.21, as well as showing the lace-like assemblies of graphene layers (as in Figure 7.20(b)), indicates that spherules of material (20–50 nm in size) appear to be resistant to gasification.

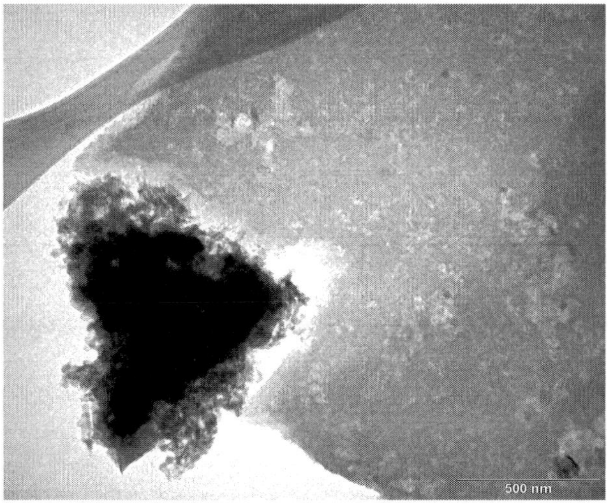

Figure 7.17. TEM micrograph of section of commercial activated carbon (10) ($w = 2.0\,\mu m$).

(a) (b)

Figure 7.18. (a,b) TEM micrograph of section of commercial activated carbon (3) ($w = 1000\,nm$ (a)).

Figures 7.22(a, b) and 7.23(a, b), at higher magnifications, illustrate the three-dimensional *lace-work which is structure in activated carbons*. It is the differences in the pattern of this lace-work which creates differences in the adsorption characteristics of activated carbons.

Moving now to Figures 7.24–7.27, for H22, Figures 7.24 and 7.25 show that residual mineral matter in the activated carbon produces channeling of size 5 to >50 nm, coming into the range of mesoporosity. Figure 7.26 indicates that, in some parts of the carbon H22, gasification has produced cavities of mesopore size. Finally, Figure 7.27(a, b) again shows the three-dimensional *lace-work which is structure in activated carbons*. It is not possible to distinguish between D19 and H22 using HRTEM. The last word, here, is the sensitivity of the adsorption isotherm, which is where this story began.

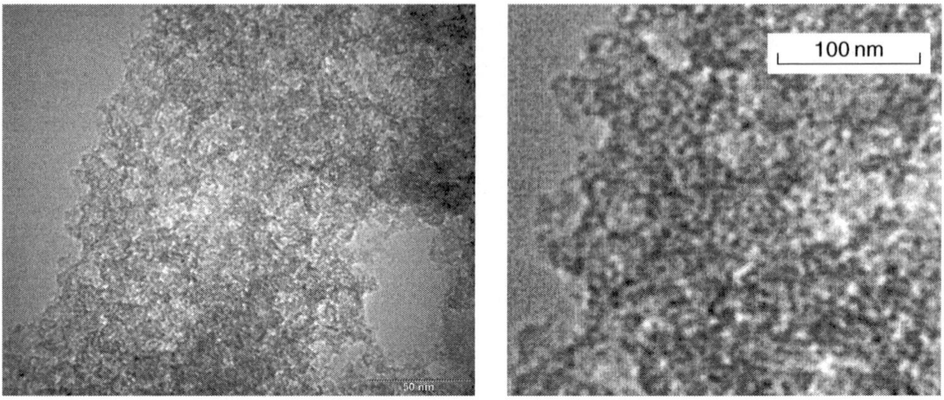

(a) (b)

Figure 7.19. (a,b) TEM micrograph of section of commercial activated carbon (6) (w = 100 nm (a)).

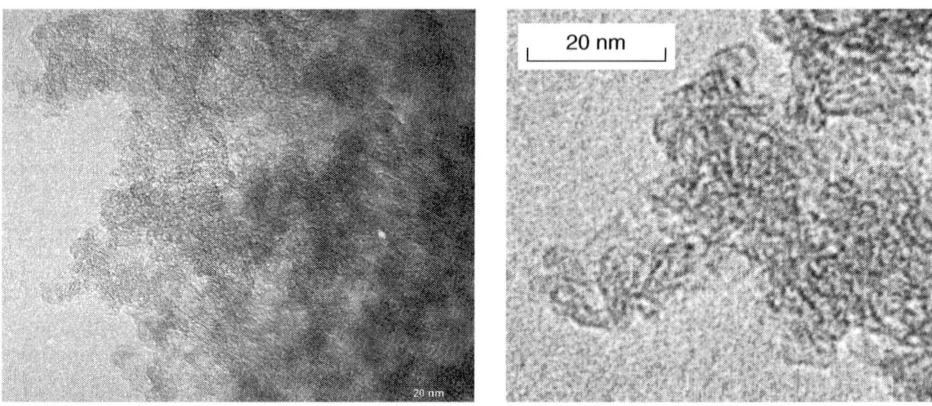

(a) (b)

Figure 7.20. (a,b) TEM micrographs of section of commercial activated carbon (5) (w = 225 nm (a)).

The activated carbons prepared by chemical activation of olive stones and mesophase spheres could not be distinguished from each other by HRTEM. Nor could they be distinguished from olive stones activated thermally.

7.4 Conclusions

The technique of HRTEM appears to be able to indicate the presence of microporosity in activated carbons. Micrographs suggest a *three-dimensional lace-work to model structure.* The model of the labyrinth (Figure 3.28(c)), with all elements of symmetry removed, is appropriate.

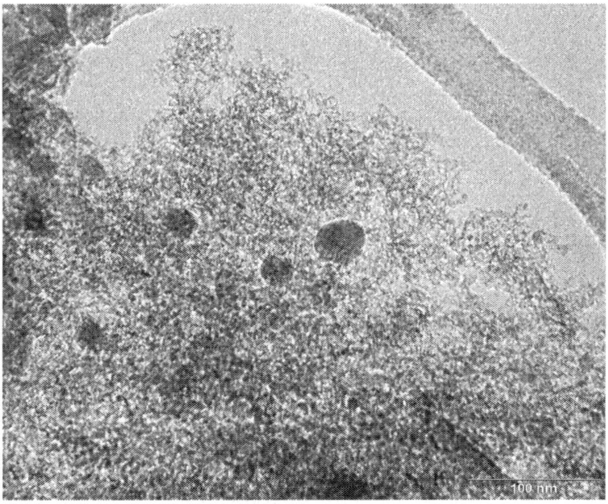

Figure 7.21. TEM micrograph of section of activated olive stones D19 (w = 500 nm).

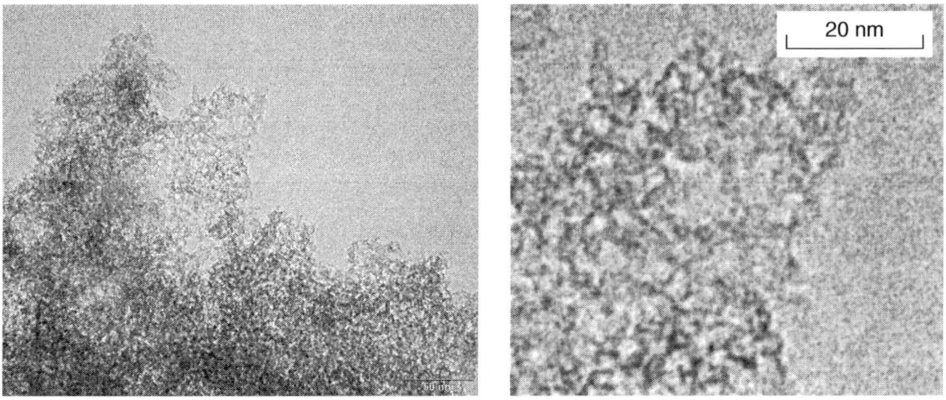

(a) (b)

Figure 7.22. (a,b) TEM micrographs of section of activated olive stones D19 (w = 250 nm (a)).

A more comprehensive study of activation processes, by HRTEM and SEM is suggested from these results. The removal of background noise, according to Rouzaud *et al.* (2002) would facilitate this analysis.

Of the other currently available techniques of structural analysis of activated carbons the following comments are possible.

Computer-generated models need significant development if they will ever be able to predict the detail of this three-dimensional lace-work.

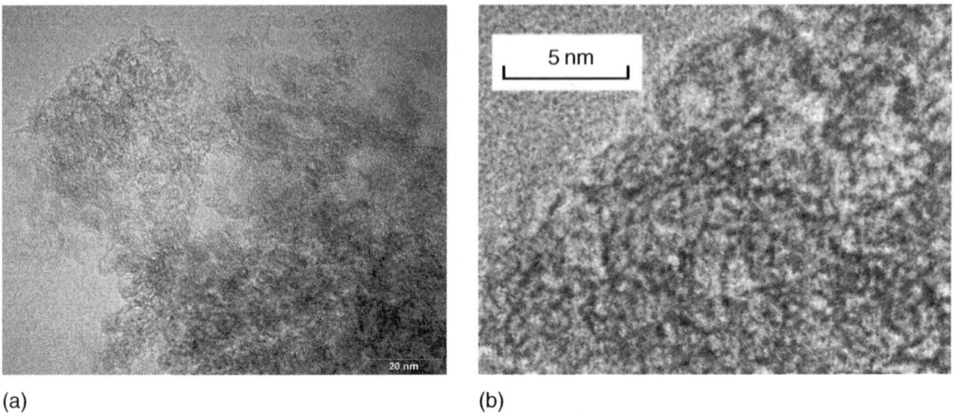

Figure 7.23. (a,b) TEM micrographs of section of activated olive stones D19 (w = 100 nm (a)).

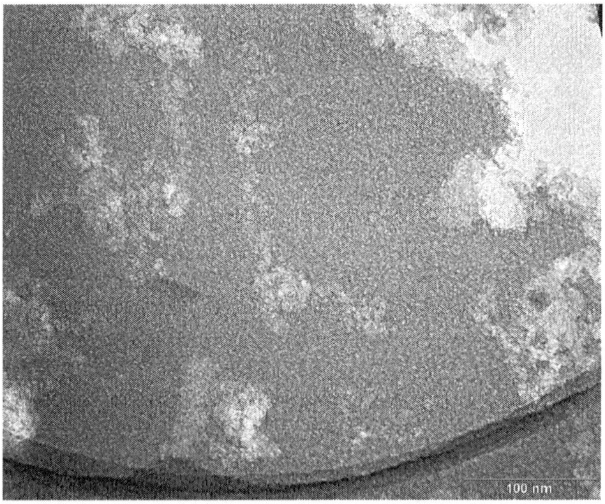

Figure 7.24. TEM micrograph of section of activated olive stones H22 (w = 500 nm).

The use of XRD with small-angle X-ray scattering (SAXS) and small-angle neutron scattering (SANS) provides bulk structural properties and cannot contribute to the detail of structure in activated carbons.

Vibrational spectroscopy, including infrared (IR), Fourier transform infrared spectroscopy (FTIR) and Raman spectroscopy, provides detail of structure of the constituent graphene layers, but cannot contribute to the relative positions of these graphene layers in the way they create the microporosity (Centrone *et al.*, 2005).

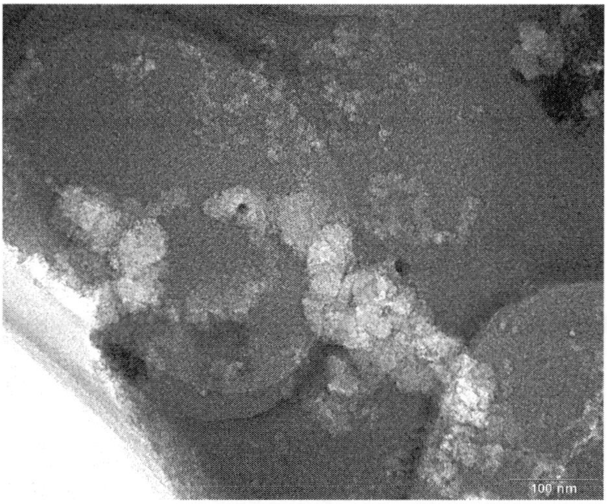

Figure 7.25. TEM micrograph of section of activated olive stone H22 (w = 500 nm).

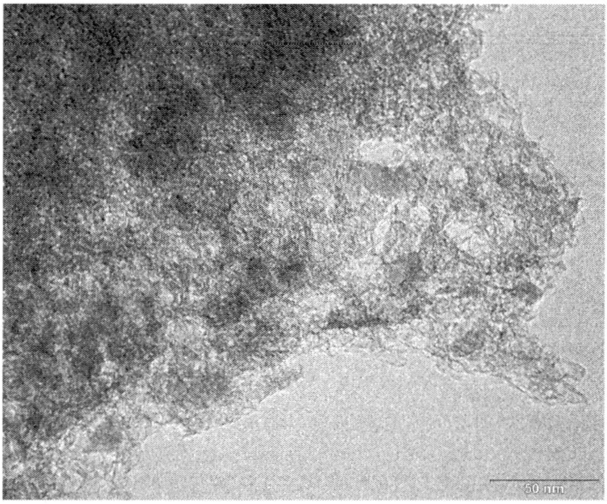

Figure 7.26. TEM micrograph of section of activated olive stone H22 (w = 250 nm).

Phase-contrast, high-resolution, TEM is able to provide meaningful information of the size and shape of constituent graphene layers and their relative positions, in virtual images. A systematic study of such images in association with adsorption properties is still awaited.

The adsorption of gases into activated carbons, to create the adsorption isotherm, provides, totally, the frequency distributions of adsorption potentials (each adsorbate/adsorbent) of

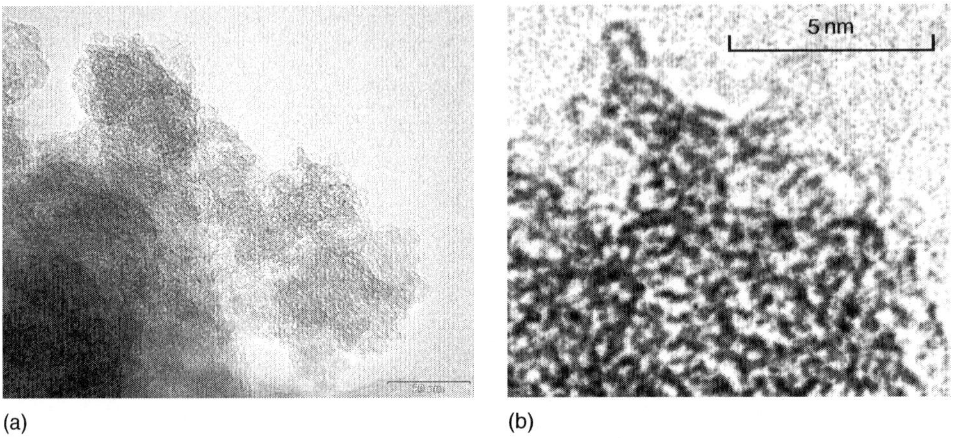

Figure 7.27. (a,b) TEM micrographs section of activated olive stone H22 (w = 100 nm (a)).

the adsorption sites (locations). It is this information which is of paramount importance in the characterization of activated carbon. The relative positions of these adsorption sites become discussion items for the theoretical chemist.

References

Aso H, Matsuoka K, Sharma A, Tomita A. Structural analysis of PVC and PFA carbons prepared at 500–1000 °C based on elemental composition, XRD and HRTEM. *Carbon* 2004;42(14):2963–2973.

Centrone A, Brambilla L, Renouard T, Gherghel L, Mathis C, Müllen K, Zerbi G. Structure of new carbonaceous material. The role of vibrational spectroscopy. *Carbon* 2005;43(8):1593–1609.

Rodríguez-Reinoso F, Molina-Sabio M, González MT. The use of steam and CO_2 as activating agents in the preparation of activated carbons. *Carbon* 1995;33(1):15–23.

Rouzaud J-N, Clinard C. Quantitative high-resolution transmission electron microscopy: a promising tool for carbon materials characterization. *Fuel Process Technol* 2002;77–78:229–235.

Sharma A, Kyotani T, Tomita A. Comparison of structural parameters of PF carbon from XRD and HRTEM techniques. *Carbon* 2000;38(14):1977–1984.

CHAPTER 8

Applicability of Activated Carbon

8.1 Liquid-Phase Adsorptions

Applications of activated carbons (AC) in liquid-phase adsorptions are extensive, the number running into thousands. This Chapter makes no attempt to summarize such involvements, but concerns itself with explanations of mechanisms of adsorption of inorganic and organic species from the aqueous phase. In this way, an understanding of the factors which control extents of adsorption is made available and can be extended to other systems. This Chapter also highlights applications of major industrial importance.

8.1.1 Introduction

AC is dominantly used for purposes of adsorption, a task for which it is well designed. Essentially, adsorption is restricted to working in one of two phases, from the gas/vapor phase (usually air) or from the liquid phase (usually water). Fortunately, there is only one gas/vapor phase, but for liquids, two distinct phases need to be considered, namely adsorption from aqueous systems and adsorption from non-aqueous systems.

From the point of view of research investment, studies into the properties of AC have massively been concerned with one-component systems (gas and liquid phases). To go to two-component, gas-phase systems requires a significant upgrading of experimental complexity requiring analysis of the gas phase as well as pressure changes. To go to three-component systems, as may be found in industrial environments, indeed, will be a challenging experiment.

Adsorption from solution, in comparison, is relatively simple to do experimentally, as no volume changes are involved and modern analytical techniques can be easily adapted to measure concentration changes in solutions as complex as you wish. But, do not be deceived by this apparent simplicity. Competitive adsorption may occur between the solvent and the solute. Adsorption from solution may be further complicated because the solute may change chemically, or its concentration may change in the solution. Thus, concentration of an acidic molecule is a function of the pH of the solution and it is to be noted that both the non-dissociated molecule and an associated ion may be adsorbed.

AC, as well as being a unique material in terms of its variable microporosities, as well as being amphoteric, are also essentially hydrophobic and adsorb species from aqueous solution, apparently ignoring the presence of the water, provided the surfaces are relatively free of surface oxygen complexes (see below for some qualifications to this statement).

8.1.2 Adsorption of Iodine and Acetic Acid from Aqueous Solution

The need to know the chemical nature of the adsorptive is well illustrated by two simple studies, that of iodine (I_2) from an aqueous solution in potassium iodide (KI) and of acetic acid (CH_3COOH) from an aqueous solution. The KI acts as a reservoir for the iodine by forming KI_3 (K^+ and I_3^-). It is necessary to do this because of the limited solubility of iodine, as such, in water. What is needed to be known, for the construction of appropriate isotherms is the chemical nature of the adsorptive, be it (I_2) or (I_3^-) for the iodine system, or CH_3COOH or $CH_3COOH(H_2O)_n$ for the acetic acid system. Further, there is the need to elucidate if (a) the iodine is adsorbed via pore filling or via capillary condensation mechanisms and (b) to demonstrate uncertainties which exist when adsorbing species can be associated with water molecules in aqueous solution, as for adsorption of acetic acid (Hill and Marsh, 1968).

The carbons used, for these two studies, were prepared from polyvinylidene chloride (PVDC) by carbonizing to 850 °C in nitrogen and activating in carbon dioxide at 850 °C to 24, 41, 70 and 85 wt% burn-off. A commercial charcoal (Sutcliffe Speakman & Co. Limited, No. 112 (UVD/20) prepared for adsorption from the liquid phase) was also used. Characterizations were made by adsorptions of nitrogen at 77 K and of carbon dioxide at 273 and 195 K, using the Brunauer–Emmett–Teller (BET) and Dubinin–Radushkevich (DR) equations of adsorption. The saturation vapor pressures of CO_2 at 273 and 195 K are taken as 3.44 and 0.186 MPa, respectively. The carbons are described in Table 8.1. Thus, a

Table 8.1. Estimated (apparent) surface areas ($m^2 g^{-1}$) for charcoal and 850 °C PVDC AC (Hill and Marsh, 1968).

Method of measurement	Carbon					
	Charcoal	PVDC 850 °C 0 wt% burn-off	PVDC 850 °C 24 wt% burn-off	PVDC 850 °C 41 wt% burn-off	PVDC 850 °C 70 wt% burn-off	PVDC 850 °C 85 wt% burn-off
CO_2 – 273 K DR equation	650	1050	900	1000	950	800
CO_2 – 195 K $\sigma = 0.170\,nm^2$	1350	1400	2000	2000	3600	5300
N_2 – 77 K $\sigma = 0.158\,nm^2$	1100	1550	2050	2100	3600	4400
Iodine – 289 K $\sigma = 0.40\,nm^2$	800	1350	1950	3000	3000	3050
Acetic acid – 289 K $\sigma = 0.205\,nm^2$	–	625	–	1250	1250	–
	Gradients of Dubinin–Polanyi plots for CO_2/273 K isotherms $\times 10^2$					
	21.7	11.6	13.7	13.9	13.8	16.7

direct comparison with gas-phase adsorptions is available. Note: The use of the parameter "surface area" is one of convenience for comparative purposes. Nothing fundamental is suggested by its use.

A range of iodine concentrations was used with controlled potassium iodide concentrations and equilibrated over known weights of carbon. To know extents of adsorption of iodine by the carbon, the latter was centrifuged from the solution and the equilibrated iodine **desorbed** from the carbon by shaking the carbon with benzene and titrating the benzene with sodium thiosulfate solution. Total iodine concentrations ($[I_2] + [I_3^-]$) were measured by direct titration, with sodium thiosulfate, of the solution as separated from the carbon. Concentrations of I_2 were calculated from the equilibrium constant for the reaction $[I_2 + I^- \leftrightarrow I_3^-]$ taken as 870 L mol^{-1} at 298 K.

By adding the isotope, ^{131}I, (β-emitter) as iodide, and monitoring the radioactivity of the adsorbed iodine, complete exchange of iodine molecules occurred in less than 15 s. This therefore indicates complete mobility and exchange, in solution, of free iodine, of iodide and of tri-iodide, as well as adsorbed iodine.

The adsorption of iodine was expressed in two ways, by plotting the amount of I_2 adsorbed (mmol) per gram of carbon against either the calculated free iodine concentration $[I_2]$, as in Figure 8.1, or as total titratable iodine in solution, ($I_2 + I_3^-$), as in Figure 8.2.

The graphs of extent of adsorption of iodine with **total iodine** equilibrium concentration, Figure 8.2, for the charcoal and the 850 °C PVDC carbon exhibit **areas** within which any point is a position of equilibrium in the system. This area is restricted in the upper region by the limiting solubility of iodine in the iodide solution and in the lower region by the

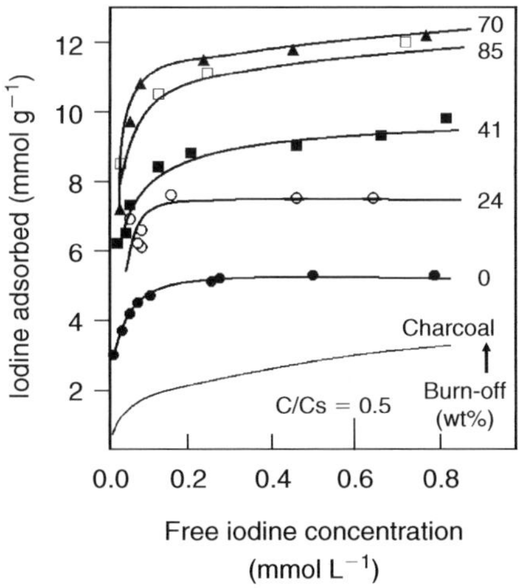

Figure 8.1. Adsorption of iodine at 298 K using **free** iodine concentrations on charcoal, un-activated and activated PVDC (850 K in CO_2) carbons (Hill and Marsh, 1968).

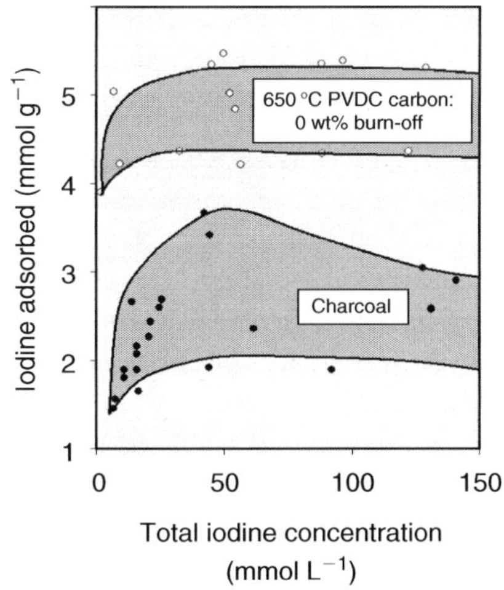

Figure 8.2. Adsorption of iodine at 298 K using **total** iodine concentrations on charcoal, un-activated and activated PVDC (850 K in CO_2) carbons (Hill and Marsh, 1968).

effect of the high iodide to iodine ratios which repress the concentration of free iodine and hence the amount of iodine adsorbed. A vertical line through this area is a line of an infinite number of positions of equilibrium of extent of adsorption with one total iodine concentration. Thermodynamically such a concept is completely unacceptable – the ordinates used in this graph must therefore be incorrect.

The graphs of extent of adsorption of iodine (as I_2) with free iodine concentration (calculated as shown above) for the six carbons of this study (Figure 8.1) show that all the points contained in the areas of Figure 8.2 now fall on single lines which represent true adsorption isotherms.

Concentration changes of acetic acid were measured by titration with a standard sodium hydroxide solution. Extents of adsorption of acetic acid on 850 °C PVDC carbons (0, 41 and 70 wt% burn-off) are plotted against acetic acid concentration in Figure 8.3. Extents of adsorption of acetic acid by the carbons reached a maximum in about 4 M acetic acid (mole fraction = 0.09). The maximum amounts of acetic acid plotted in these isotherms (Figure 8.3) correspond more to capillary condensation than pore filling mechanisms (Table 8.1) with the possibility of capillary condensation of hydrated acetic acid molecules. Thus, no unequivocal conclusions are possible. The conclusions of this study, relevant to understanding adsorptions from solution are as follows:

1. Adsorption of iodine from aqueous I_2/KI solution resembles adsorption of CO_2 at 195 K and N_2 at 77 K. That is, adsorption by iodine proceeds in the microporosity by a pore filling mechanism, and by capillary condensation in the mesoporosity, as does nitrogen at 77 K. The adsorption of iodine is apparently not influenced by the solvent, in this case, water which is an inert diluent.

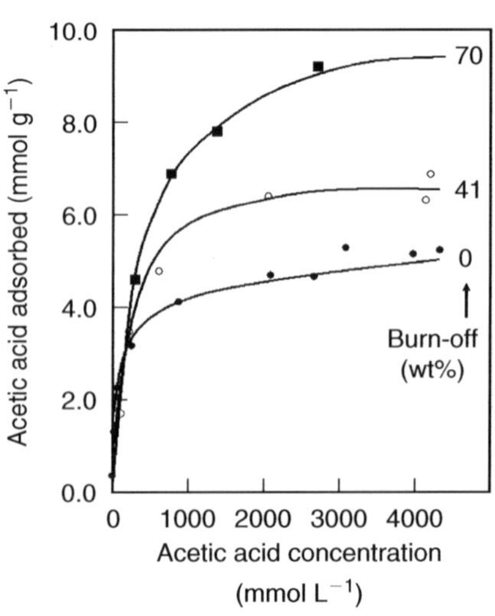

Figure 8.3. Adsorption of acetic acid at 298 K on charcoal, un-activated and activated PVDC (850 K in CO_2) carbons (Hill and Marsh, 1968).

2. The correct isotherm to represent iodine adsorption uses the concentration of free iodine in solution and not total iodine concentration. No adsorption of I^- or I_3^- could be detected at the surface coverages studied.
3. Adsorption of acetic acid from aqueous solution most probably involves competition with the water molecules. Hence, it is difficult to distinguish between adsorption restricted to the microporosity (micropore pore filling), from the adsorption of hydrated acetic acid molecules (capillary condensation).

8.1.3 Adsorption of Inorganic Solutes from Aqueous Solution

The literature describing adsorption from aqueous solutions, involving of course, major water treatment programs is, indeed, massive with Radovic *et al.* (2000) quoting 777 references, many of these, themselves, being further reviews with their own collection of references. These authors do not discuss desorption (regeneration) which is important to carbon recovery (see Chapter 9). Equilibrium states are not often studied, it being more economical to work with initial, rapid extractions. Adsorptions of the following species are reviewed. Ions of: arsenic, cadmium, chromium, cobalt, copper, gold, lead, molybdenum, nickel, phosphates, silver, zinc and uranium.

In wastewater treatments, special attention is given to removal of dangerous inorganic material such as the heavy metals, including mercury, chromium, molybdenum, cobalt, nickel, copper, cadmium, lead, uranium, gold, arsenic, barium, iron and vanadium. Whereas the forces of adsorption from the gas phase are well understood, adsorption from solution of inorganic species is not so well understood. So, it is most pertinent to enquire about the sites which are responsible for adsorption of inorganic species.

However, the central issue for adsorption of ions from an aqueous medium is the understanding of the mechanisms by which ionic species become attached to a carbon surface. This phenomenon is more complex than the relatively simple adsorption of carbon dioxide from the gas phase or of molecular iodine (I_2) from the aqueous phase.

There are three schools of thought which discuss mechanisms by which metallic ions (or other ions) are removed from an aqueous solution. The **first** states that the process is of electrostatic adsorbate–adsorbent interactions and is totally dependent on the existence of functionality on the carbon surfaces, especially oxygen surfaces complexes, that is, it is a process of ion-exchange. The **second** school of thought suggests that enhanced adsorption potentials, as in the narrowest of microporosity, may be strong enough to "adsorb" and retain ions. The **third** mechanism is based on the hard and soft acids and bases (HSAB) concept as suggested by Alfarra *et al.* (2004) and is a response to the amphoteric nature of carbon surfaces.

There is no doubt, as shown below, that the presence of oxygen functionality on carbon surfaces is critically important to the removal of ionic species. In fact, this mechanism may dominate all adsorptions of metal ions. Carboxylic (—COOH) and phenolic (—OH) have a major role in which the ion-exchange produces (—COOMOOC—) and (—OMO—) with protons ($(H_2O)_n(H^+)$) being formed within the aqueous medium. The specific inorganic chemistry of each metallic ion controls the conditions under which ion-exchange (adsorption?) takes place. The detail of the chemistry of adsorption of the many ions is the basis of the review of Radovic *et al.* (2000).

There is little convincing evidence that dispersion forces are responsible for ion uptake. The matter of electrostatic charge has to be considered because, overall, there has to be electrostatic neutrality. Further, if adjacent micropores accepted only cations (M^{n+}) then considerable forces of repulsion would be set up. Such systems would never be created because of this inherent instability.

The HSAB approach originates from the concepts of Lewis of acid–base theory. *Soft bases* are easy to oxidize and hold their valence electrons loosely, for example (I^-). *Hard bases* hold their valence electrons tightly and are difficult to oxidize, for example (F^-). *Soft acids* are associated with large acceptor atoms; they contain unshared electron pairs in their valance shells, and have high polarizability and low electronegativity, examples being Cu^+ and Ag^+. On the other hand, *hard acids* are small ionic species with low polarizability and low electronegativity, examples being H^+, Li^+ and Mg^{2+}.

Alfarra *et al.* (2004) applied this concept of HSAB to AC. The surfaces of such carbons are composed not only of carbon atoms, but also contain hydrogen, oxygen, nitrogen, sulfur, perhaps phosphorous and possibly chlorine. It is considered that the π-orbitals of heteroatoms overlap with those of carbon and form bonding and anti-bonding molecular orbitals extending throughout a graphene layer. Therefore, bonding molecular orbitals have a high contribution from heteroatoms so creating a partial negative charge (δ^-) at the site. The positive charge (δ^+) is delocalized over the graphene layer. The softness of benzene ($\eta = 5.3$) is higher than that of anthracene ($\eta = 3.3$) suggesting to the authors that it is reasonable to attribute a strong softness to the graphene layers, particularly in the centers of the graphene layers. This approach predicts that hard acids (Li^+, Na^+, K^+, Mg^{2+}, Ca^{2+}, Al^{3+}, Cr^{3+}, Fe^{3+}) would be "adsorbed" on surface functional groups, usually on edge atoms. Soft acids (Cu^+, Ag^+, Pd^{2+}, Pt^{2+}, Hg^{2+}, I_2) would be "adsorbed" over the surfaces of graphene layers, extents of adsorption being a function of surface area.

Alfarra *et al.* (2004) report that formation of surface oxygen complexes promotes the ionic exchange adsorption of transition metal ions such as Zn^{2+}, Cu^{2+}, Cr(III) and (VI) and Ni^{2+}. Under these conditions the effect of specific surface area is almost negligible (except that it controls extents of complex formation). When using an oxidized charcoal with a BET surface area of about $5\,m^2\,g^{-1}$, cation adsorption for the above four ions was effective. The HSAB concept explains this because hard acids, such as Cr^{3+}, are trapped by the dissociated surface oxygen groups that are considered to be hard bases. (*Note: the use of N_2 (77 K), as an adsorbate for microporous carbons, may provide seriously misleading information. In fact, the surface area of $5\,m^2\,g^{-1}$ is impossibly low. The use of CO_2 (273 K), as an adsorbate, could indicate a surface area of $>300\,m^2\,g^{-1}$ for this type of carbon*).

The adsorption of cadmium (Cd^{2+}) further illustrates the complexity of these processes. Cadmium ions are removed from aqueous solution (pH of 3 to 1.1) by an AC. The adsorption mechanism is considered to involve reaction between the (Cd^{2+}) and the graphene layers, the resultant bonding having a strong covalent nature. On increasing the values of pH, the acidic groups of the complexes dissociate so facilitating the formation of ionic bonds with the (Cd^{2+}). These conclusions are supported from isotopic cadmium exchange studies which show a much slower rate of exchange, at this pH of 1, between cadmium adsorbed and cadmium in solution (see also Section 8.1.3.3).

Chemical adsorption–reduction reactions also occur on carbon surfaces resulting in uptake of ions (cations) from aqueous solutions. This occurs when the redox potential of metal ions is high compared with that of the carbon, for example for gold, $E^{\circ}(AuCl_4^-/Au) = 1.00$ V, for silver, $E^{\circ}(Ag^+/Ag) = 0.80$ V and for platinum, $E^{\circ}(PtCl_4^{2-}/Pt) = 0.755$ V. These metal ions are "adsorbed" as a result of the following type of Reaction 8.1:

$$(PtCl_4)^{2-} + 2e^- \rightarrow Pt^+ + 4Cl^- \qquad \text{(Reaction 8.1)}$$

The gold and silver ions behave differently with respect to the carbon surface. The *soft* gold ions become attached to the graphene layers of the AC and become extensively reduced because of their high redox potential, a process which is independent of the porosity of the AC and of its surface functionality. On the other hand, the $Ag(NH_3)^{2+}$ ion is weakly trapped and reduced by, in this case, carbon fibers, this being promoted by surface oxygen groups introduced from nitric acid. The ion $(PtCl_4)^{2-}$ behaves as a *soft* base with adsorption on the carbon surface. Thus, extents of adsorption are a function of surface area and are independent of oxygen surface functionality. From the practical point of view of recovery of gold and silver, the so-deposited metal is difficult to retrieve from the carbon surface and cyanide complexes are used instead because of their lower redox potentials (−0.58 and −0.31 V, respectively) which facilitate adsorption without reduction. Adsorption is considered to occur on the surfaces of the graphene layers via π–π interactions.

Summarizing, adsorption of metallic ions from aqueous solution is far from being a straightforward process. Factors controlling extents of adsorption may be listed as follows.

1. The chemistry of the metal ion or metal ion complex which is in solution.
2. The concentration of the solute and the derived isotherm.
3. The pH of the solution containing the ionic species and the point of zero charge (PZC) on the surface, $\mathbf{pH_{PZC}}$, and isoelectric point (IEP) $\mathbf{pH_{IEP}}$ (see Section 8.1.4.4).
4. The origins of the AC, its surface area and porosity including volumes in narrow and wider microporosity.
5. The surface composition of the AC, including heteroatoms bonded within the graphene layer (in-plane) as well as edge-bound oxygen functionality.
6. The chemical composition of oxygen functionality which is dependent on its method of formation, for example from molecular oxygen, carbon dioxide, nitric acid, phosphoric acid, hydrogen peroxide, ozone, hydrogen peroxide and ammonium peroxodisulfate.
7. The size of the adsorbing species, that is the hydrated ion, in the range 1.0–1.8 nm, is of importance in carbons with significant volumes of narrow microporosity.
8. Temperature.

Experimentally, to ascertain the role, if any, of surface oxygen complexes in the uptake of metallic ions, a correlation is looked for between amounts of ions adsorbed (N_A) and the amounts of participating oxygen functionality ($\boldsymbol{N}_{\mathbf{F}}$). The latter requires the use of Diffuse Reflectance Infrared Fourier Transform Spectroscopy (DRIFT) (Figure 4.28) and X-ray photoelectron spectroscopy (XPS) (Figure 4.29) to identify the surface species associated with temperature-programmed desorptions (TPD) (Figure 4.30) and Boehm titrations to assess, quantitatively, amounts of surface species (Section 4.3.3). The difficulties of assigning a number to the availability of sites capable of adsorbing a metallic ion can be envisaged. The experimentation would also involve studying carbons with increasing contents of surface functionality at several pH values of the solution.

A numerical approximation between ($\boldsymbol{N}_{\mathbf{A}}$) and ($2\boldsymbol{N}_{\mathbf{F}}$) would indicate a dependency of adsorption upon surface functionality. The uptake of M^{2+} on say carboxylic groups releases two protons (H^+) into the solution. If ($\boldsymbol{N}_{\mathbf{A}}$) >> ($\boldsymbol{N}_{\mathbf{F}}$) then this would be a clear indication of the role of the *soft* sites of the graphene layer. If ($\boldsymbol{N}_{\mathbf{A}}$) << ($\boldsymbol{N}_{\mathbf{F}}$) then this could indicate the use of an inappropriate pH for the solution, or an inability of the metallic ion (or ion complex) to enter microporosity in which the functionality was bonded. Very few studies have been made of such changes. Some authors use the ratio (***R***) which equals $N_A/(2N_F/2)$, using equivalents.

A sustained and comprehensive experimental program is needed to provide unequivocal results for each cation studied.

The concepts and origins of surface acidity and basicity are discussed in Section 4.3.4 and it is to be noted that these concepts parallel closely the approach of HSAB, differences being mainly in the descriptions of the terms used, with acidic and basic behavior of a graphene layer being equated to hard and soft properties.

A comprehensive survey of adsorption of 11 metallic ions by an AC (surface area of $417\,m^2\,g^{-1}$) prepared by the phosphoric activation of pecan shells is reported by Dastgheib and Rockstraw (2002). Adsorptions were made, in the concentration range 0.0–10.0 $meq\,L^{-1}$ (same as $meq\,L^{-1}$) with pH values close to 3.0, at 25 °C. No data are provided for analyses of surface oxygen complexes, although it is known that phosphoric acid, as an activating agent provides oxygenated surfaces. Significant differences in extents of adsorption of the cations are reported, varying from 0.4 $meq\,g^{-1}$, for magnesium (Mg(II)), to 1.9 $meq\,g^{-1}$ for Hg(II), from solutions of 5.0 $meq\,L^{-1}$. Results are summarized in Table 8.2 indicating

Table 8.2. Metal ion electro-negativities, first stability constants of metal ion hydroxides and metal ion adsorption (Dastgheib and Rockstraw, 2002).

Metal ion	Electronegativity	Log K_1	Metal ion adsorption ($meq\,g^{-1}$)
Hg(II)	1.9	10.6	1.9
Cr(III)	1.6	10.1	1.6
Cu(II)	2.0	6.5	1.0
Ni(II)	1.8	4.1	0.9
Cd(II)	1.7	3.9	0.9
Ca(II)	1.0	1.3	0.9
Sr(II)	1.0	0.8	0.6
Co(II)	1.8	4.3	0.5

(somewhat) that higher levels of metal ions adsorbed on to AC correspond to higher electronegativities and stability constants, the latter determining the speciation of metal ion species. The correlations are not good, however, indicating that other factors are operating.

From changes in pH values of the metal ion solutions, before and after adding the pecan carbon, together with the values of metal ion adsorbed, it was possible to calculate the number of protons released per gram of carbon during the adsorption process. The ratios (R) of equivalent metal ions adsorbed to protons released were calculated for all concentrations. For most of the data of Table 8.2 this ratio is close to unity, more so at low concentrations. Jia and Thomas (2000) report similarly for the adsorption of cadmium.

At low concentrations, all metal ions are present totally in the form of M^{n+} when values of (R) are close to unity indicating the exchange of one proton for one equivalent metal ion. At higher concentrations, values of (R) increase to 1.5 suggesting the formation of other surface complexes in addition to the release of one proton by the metal ion. For Hg(II) values of (R) increase to 2 suggestive of formation of one $HgCl_2$ complex. The formation of surface complexes has been confirmed for copper ion adsorption by Biniak *et al.* (1999).

Equilibrium uptakes of metallic cations, dyes, and even gold and anionic adsorbates are largely governed by electrostatic attraction or repulsion. It is suggested that further studies of the role of the carbon surface are needed to show whether this is indeed the case. If not, only then does the role of specific interactions (e.g. complex formation), or even nonspecific van der Waals interactions, become a reasonable alternative or complementary argument. Indeed, in the adsorption of inorganic solutes, the main fundamental challenge remains how to "activate" the entire carbon surface in order to achieve maximum removal efficiencies. The solution conditions, at which maximum uptake can be achieved, depend on the surface chemistry of the adsorbent.

8.1.3.1 Adsorption of mercury species

One example (out of many) to illustrate the complexity of adsorption from solution (as compared with gas-phase adsorptions), is the removal of mercury, an unacceptable toxic pollutant in aqueous systems. It is found in wastewaters (before treatment) from such manufacturing industries as chloroalkali, paper and pulp, oil refining, plastic and batteries, and can exist as free metal, as Hg(I) and Hg(II). Mercury adsorption capacity, on AC, increases as the pH of the aqueous systems decreases. Carbons with different activation methods have widely different capacities. Sulfurization of the carbon, loading the carbon with zirconium, as well as the dispersion of FeOOH species over the carbon, enhanced Hg(II) uptake. Mercury vapor can be taken up using AC which have been pre-treated with sulfur, the effect of chemisorbed oxygen being to retard (not prevent) the uptake of mercury, López-González *et al.* (1982).

Yardim *et al.* (2003) studied the removal of mercury(II) from aqueous solutions of $HgCl_2$. The carbon was prepared by carbonizing furfural to 800 °C and activating in steam at 800 °C for 1 h. Presumably, surface oxygen complexes would be formed. The resultant carbon had a BET surface area of $1100\,m^2\,g^{-1}$ with a micropore volume of $0.425\,cm^3\,g^{-1}$. Using solutions with a pH of 8.6 the maximum amount of adsorption of mercury was $174\,mg\,g^{-1}$, an amount comparable to that reported in Table 8.2. Amounts adsorbed increased with changes in pH from 1.0 to 8.0. Desorption experiments were carried out with hot water when it was found difficult to remove more than about 6 wt% of the adsorbed mercury. Yardim *et al.* (2003) state their studies confirm the formation of stable chemical bonds between Hg(II) and surface oxygen complexes, a process of chemisorption, with only about 6 wt% being physically adsorbed, presumably on the graphene layers. This result does not contradict that of López-González *et al.* (1982) who observed that oxygenated surfaces still adsorbed some mercury.

Goel *et al.* (2005) studied the removal of mercury(II) from aqueous solutions of $HgCl_2$. The carbon used was a carbon aerogel with a BET surface area of 700 $m^2 g^{-1}$ with a micropore volume of 0.10 $cm^3 g^{-1}$ and a mesopore volume of 1.36 $cm^3 g^{-1}$. Using solutions with a pH of 8.32 the maximum amount of adsorption of mercury was 35 $mg g^{-1}$ at 20 °C. Amounts of mercury adsorbed increased with increasing pH. At pH values >5.0 it is suggested that $Hg(OH)_2$ is retained within the porosity. However, on increasing the temperature to 70 °C, amounts of Hg adsorbed increased by ~40 wt% indicating the occurrence of a chemical reaction or of a kinetic activated diffusion restriction of the mercury into the microporosity. At pH >7.0 it is considered that Hg^{2+} or $Hg(OH)^+$ exchange with the protons of the acids of surface oxygen complexes.

Maroto-Valer *et al.* (2005) showed that those properties of carbon which promote adsorption of mercury from aqueous solutions are equally effective in the removal of mercury found in flue gases. Coal-fired utility boilers are the largest source of mercury resulting from human activity and account for 33% of all mercury emissions. For the removal of flue gas mercury, a promising technological approach is the location of an adsorbent carbon upstream of the electrostatic precipitators (baghouse) used for dust suppression, Pavlish *et al.* (2003).

The mechanism of this mercury removal has been in dispute within the literature, with, on the one hand, Li *et al.* (2003) concluding that surface oxygen complexes are the active sites for entrapment of Hg^0, and with Kwon *et al.* (2002), on the other hand, stating that surface oxygen complexes are in no way beneficial. The purpose of the study of Maroto-Valer *et al.* (2005) was to resolve this dilemma (if possible).

The fly ash was from a burner using pulverized high volatile bituminous coal with a carbon content of ~58 wt%. The carbon content was enriched by a float-sink procedure followed by extraction of mineral matter by $HCl/HNO_3/HF$, called DEM-PCC1. This carbon was activated with steam at 850 °C for 60 min in a horizontal furnace, called AC-PCC1. Analytical data for these two carbons and a commercial carbon, Darco Insul, are in Table 8.3.

Table 8.3. Porosity and mercury capacity of fly ash carbons and Darco Insul charcoal (Maroto-Valer *et al.*, 2005).

Carbon	Ash content (wt%)	BET surface area ($m^2 g^{-1}$)	$V_{0.95}$ ($mL g^{-1}$)	Average pore diameter D_a (nm)	Mercury capacity ($mg g^{-1}$)
DEM-PCC1	3.6	53	0.04	1	1.85
AC-PCC1	11.2	863	0.49	2.3	0.23
Darco Insul	–	700	–	–	2.77

The three carbons were tested for mercury uptake using a fixed-bed system, at 138 °C with a simulated flue gas composed of 16% CO_2, 5% O_2, 2000 ppm SO_2, 270 ppm Hg and balanced with nitrogen. The AC-PCC1, despite its larger surface area, was less effective than the parent DEM-PCC1 for mercury capacity. An XPS study revealed a higher oxygen content on the surfaces of DEM-PCC1. There appears to be some inconsistency within these data. However, it is more than likely that the low surface area of the DEM-PCC1 is the result of activated diffusion effects during the adsorption of nitrogen at 77 K (it may be as high as 500 $m^2 g^{-1}$ if carbon dioxide had been used as the adsorbate at 273 K). The authors conclude that surface oxygen complexes provide sites for mercury bonding where lactone and carbonyl groups are active sites for the Hg^0 capture.

8.1.3.2 Adsorption of chromium species

Chromium is one of the elements in the top priority of toxic pollutants as defined by the United States Environmental Protection Agency (US-EPA) because of the extreme toxicity of Cr(VI). The acceptable limit of the EPA is 0.05 $mg L^{-1}$. Cr(VI) is a strong oxidizing agent at room temperature and destroys biological tissue with which it comes into contact. The concern over chromium stems from its widespread use in the metallurgical industries (steel and ferro- and non-alloys), the refractories industries (chrome and chrome magnesite) and the chemical industries (pigments, electro-plating, tanning).

Chromium, in the environment, commonly exists as Cr(III) and Cr(VI). The Cr(III) species is a trace element and supports the chemistry of living organisms. Conversely, Cr(VI) is a most unpleasant species inducing serious medical conditions such as asthma and dermatitis. Cr(VI) species are highly soluble and this ensures a wide dispersion in water, in soils and the atmosphere, whereas Cr(III) species are sparingly soluble.

A comprehensive commentary of the speciation of chromium compounds and methods of analysis is published by Kotaś and Stasicka (2000). Radovic *et al.* (2000) present a historical review of the literature of the characteristics of chromium as an adsorbate. A reading of such literature is not that rewarding because of the way much of the research is faceted and uncoordinated. As indicated above, the need today is to consider, in the one study, the speciation of the adsorbate, the characterization of several carbon adsorbents (varying porosity and surface chemistry, including its amphoteric nature) and pH, and the effects of temperature.

A study of Babel and Kurniawan (2004), into the suitability of using carbons from coconut shell (CSC) as adsorbents for chromium removal from wastewater, incorporates considerations of speciation, surface chemistry of the carbon and pH, as well as the economics of the process. A CSC and a commercial activated carbon, from Thailand, (CAC) were used, the former having a surface area of 5–10 $m^2 g^{-1}$ and the latter 900–1100 $m^2 g^{-1}$. Pore volumes were 0.06 and 0.73 $mL g^{-1}$, respectively. The carbons were oxidized with nitric acid (65 wt%) at 110 °C for 3 h and sulfuric acid (2 vol%) at 110 °C for 2 h. The as-received CSC adsorbed 25% of chromium from a solution containing 10 $g L^{-1}$ of Cr(IV), that oxidized with nitric acid 73%, and that oxidized with sulfuric acid 40%. Considering the effect of solution pH, adjusted using 0.1 N NaOH and 0.1 N H_2SO_4, and the nitric acid treated CSC, the adsorption increased a little from pH 2 (78%) to 88% at pH 5, to decrease to 33% at pH 9. The maximum adsorption capacity of the CSC-nitric acid was 10.88 mg of Cr(VI) per gram of carbon compared with 15.47 mg of Cr(VI) per gram of CAC-nitric acid.

Mechanistically, it is suggested that oxidation with nitric acid introduced carboxylic, hydroxyl and carbonyl groups to the carbon surface and these groups, at low pH, are oxidized as follows:

$$Cr^{n+} + n(\text{—COOH}) \leftrightarrow (\text{—COOH})_nM + nH^+ \qquad \text{(Reaction 8.2)}$$

The oxyanion of dichromate ($Cr_2O_7^{2-}$), used in synthetic wastewater by Babel and Kurniawan (2004), is Cr(IV) but more commonly existing as $HCrO_4^-$ in wastewaters, etc. This is reduced by the carbon to Cr(III) as follows:

$$HCrO_4^- + 7H^+ + 3e^- \leftrightarrow Cr^{3+} + 4H_2O \qquad \text{(Reaction 8.3)}$$

As the pH of wastewater is generally <6.0, the Cr(III) species mostly exists as $[Cr(OH)]^{2+}$ and thus the overall adsorption reaction is as follows, where A represents an adsorption site on the acidic surface of CSC:

$$Cr^{3+} + H_2O + HA \leftrightarrow ([Cr(OH)]^{2+} \cdots A^-) + 2H^+ \qquad \text{(Reaction 8.4)}$$

It is further suggested in the review of Babel and Kurniawan (2004) that a Cr(III) ion may attach itself to three adjacent hydroxyl, carboxylic and oxyl groups, which donate some lone pairs of delocalized π-electrons on the surface to the metal ion for formation of surface oxide compounds ($[Cr(OH)]^{2+} \cdots A^-$).

8.1.3.3 Adsorption of cadmium species

The relationship between surface functionality and extents of adsorption of cadmium from aqueous solution is well described by Jia and Thomas (2000) who attempted to establish a relationship between (N_A) and (N_F). The commercial carbon used (carbon C) was derived from coconut shell and activated with steam at ~900 °C. Oxygen functionality was introduced by reaction with nitric acid (100 °C) for 4 and 48 h (carbons CN1, CN2). A range of carbons, of decreasing surface oxygen contents, was prepared by progressive heat treatment to 800 °C. These carbons were subsequently analyzed using the surface analytical techniques of (1) Temperature programmed Desorption (TPD), (2) Fourier-transform infrared spectroscopy (FTIR), (3) X-ray absorption near edge structure (Xanes), (4) Titration studies using the Boehm method, (5) Adsorption and desorptions of cadmium from aqueous solutions of cadmium nitrate, (6) Flow-microcalorimetry (Section 4.6.4), (7) Adsorption of carbon dioxide (273 K) and nitrogen (77 K) and (8) Energy dispersive X-ray analysis (EDA).

The surface analyses indicated that the nitric acid oxidation introduced, principally, carboxylic acids with some phenol and quinone groups. Lactone groups were also found in the heat-treated carbons. The carboxylic acids possessed the least thermal stabilities. Changes observed in pore volumes and surface areas did not correlate with extents of cadmium uptake from solution. However, these extents followed closely the introduction of surface functionality and its subsequent removal by heat treatment. Of critical importance was the observed ratio value of 2 of (N_A)/(N_F), for the initial stages of adsorption, indicating the formation of a cadmium link with two carboxylic groups. The ratio of (N_A)/(N_F) decreased slightly with further cadmium uptake.

$$\cdots COOH + HOOC \cdots + M^{2+} \leftrightarrow \cdots COO\text{-}M\text{-}OOC \cdots + 2H^+ \qquad \text{(Reaction 8.5)}$$

In addition, Jia and Thomas (2000) observed that the adsorption of cadmium from aqueous solution was both irreversible and reversible, the irreversibility occurring with the initial adsorptions of the cadmium, becoming reversible with increasing up-take of cadmium. Removal of carboxylic acids caused a significant reduction in extents of irreversibility. The calorimetric data indicated a wide range of adsorption mechanisms for the cadmium associated, not only with carboxylic groups but also with hydrogen bonding of hydrated cadmium ions and other functional groups.

8.1.3.4 Adsorption of gold and silver species

A major use of AC in the metallurgical industries is the extraction of gold and silver from low-grade ores. Three processes have been used to extract these metals. In the first, the Carbon-in-pulp (**CIP**) method, the prepared ore is conditioned with air and lime and mixed with an oxygenated solution of sodium cyanide plus caustic soda and then contacted with AC (coarser in size) which retains the gold–cyanide complex, the carbon being then screened from the system. The carbon-in-leach (**CIL**) method is a variant of the CIP method and involves making a pile of crushed ore together with lime or sodium hydroxide. This pile is sprayed with oxygenated alkaline sodium cyanide solution, the emerging solution then being passed through a carbon adsorption circuit, the carbon adsorbing up to 3 wt% of metal. The carbon-in-tower (**CIT**) method involves contacting the solution of silver and gold in vertical columns in which carbon is circulated counter-currently. The gold is recovered electrolytically, by alcohol or glycol stripping or zinc precipitation. The AC is regenerated by heating 600–800 °C in an inert atmosphere or by using superheated steam. Coconut shell is the major source of AC because of its hardness with Norit producing carbons from peat and claiming a gold capacity of 4.5–5.0 wt%.

Gold complexes with many ligands of which the $[Au(CN)_2]^-$ complex is the most stable and is involved in the CIP and similar gold extraction processes. The nature of the species adsorbed on the carbon surface has proved difficult to resolve. Complexes with sodium, potassium and calcium salts of $[Au(CN)_2]^-$ have been suggested with formation of electrical double layers. The possibility of simple ion-exchange appears not to have been favored. The absence of oxygen from the extraction systems reduces the adsorption capacity of the carbon for gold. However, this oxygen effect was found to be negligible under conditions of low and high ionic strength as existing in the CIP process.

Radovic *et al.* (2000) review a selection of the considerable available literature which attempts to explain the processes of adsorption of gold and silver complexes on carbon surfaces. No unequivocal message emerges from this review. Radovic *et al.* (2000) report that, in the presence of oxygen, the species adsorbed was $M^{n+}[Au(CN)_2^-]_n$. At low ionic strengths adsorption was considered to involve ion-exchange sites.

However, an earlier study of Jia *et al.* (1998), although not having available the surface analytical techniques used in their later cadmium paper (Jia and Thomas, 2000), most strongly indicates where adsorption is taking place. The carbons used were from coal, coconut shell and polyacrylonitrile (a source for nitrogen in a carbon) and activated progressively to ~60 wt% burn-off in steam at 900 °C. They were oxidized in boiling nitric acid (25%) for 48 h. The CSC was reacted with ammonia for 1 h at 700 °C. Carbons were characterized by adsorption of nitrogen (77 K) and carbon dioxide (273 K), Raman spectroscopy and infrared spectroscopy. Solutions for gold adsorption were prepared from solutions of $KAu(CN)_2$ and NaCN (using $KAg(CN)_2$ for silver studies). Adsorptions were studied in the presence of alcohols and excess cyanide.

There was no evidence within the results to show any influence of oxygen surface complexes. The data from Raman spectroscopy indicated that only the species $Au(CN)_2^-$ exists in gold solution, with $Ag(CN)_3^{2-}$ existing in the silver solution but changing to the larger $Ag(CN)_3(H_2O)^{2-}$ and $Ag(CN)_4^{3-}$ species in excess cyanide solution. As a result of these differences in size, gold exhibited preferential adsorption over silver. The addition of alcohols to the gold solution reduced gold adsorption and this is attributed to the preferential adsorption of alcohols, by dispersion forces, on to the graphene layers which constitute the walls of the porosities of these carbons, the adsorption being related to electron donation via the π-orbitals of the graphene layer. A comparison of the porosities and adsorption abilities of the three carbons of the study indicated that correlations with surface area and total pore volumes were not sufficiently strong leading to the conclusion that *pore-size distributions* (PSD) within the total pore volumes were responsible. No further work in this area seems to have been done.

The details of the mechanism(s) of gold adsorption remain uncertain, mainly because of the many variables involved with the carbon itself. There is a need to characterize fully porosity and PSD. There is the question of extents of surface oxygen complex formation, the chemistry of these complexes and their location in porosities, some of which may not be accessible to gold complex species. At the same time, the pH and the ionic strengths of the solutions have to be controlled. Such a series of diagnostic experiments would be expensive to carry out and may not now be necessary if the use of cyanide solution is replaced by less toxic materials, such as thiocyanates. Already, there is experience with this lixiviant, as it is easily available, it is of low toxicity, stable in acid solutions, carbons can be easily regenerated and solutions can be used repeatedly.

For example, Kononova *et al.* (2005) studied the removal from solution of gold and silver thiocyanates in order to evaluate the use of a new Russian carbon made from Siberian coals, at the same time assessing the importance of surface oxygen functionality.

The carbon from the Siberian anthracite is the most attractive of the suite of carbons studied. Its high surface area ($1643\,m^2\,g^{-1}$) indicates that it must be an AC and, therefore, probably, will possess oxygen surface complexes (Table 8.4). The cocoa stone carbon (surface area $1450\,m^2\,g^{-1}$) adsorbs very little in the way of gold. This carbon may contain little in the way of surface oxygen complexes. The authors conclude that a reduction of the gold ions takes place with the formation of surface complexes with the carbon π-system at the expense of partial transfer of charge from the graphene layers to the adsorbed ion. That anthracite possesses well-developed graphene layers, with relatively high conductivities, may promote this process. The best selectivity to gold thiocyanate ions is at a pH of 2.6, >90% recovery. At a pH of 8–10, the gold recovery decreases to 70%. For regeneration, Kononova *et al.* (2005) used alkaline solutions of thiocarbamide at 145 °C obtaining ~95% recovery rates.

In summary, the chemistry of the adsorption processes for soluble gold complexes in terms of the surface characteristics of the adsorbing AC remains unresolved but with strong indications of adsorption of metal complexes on graphene layers, rather than ion-exchange of metal cations.

Table 8.4. Characteristics of the four carbon adsorbents used for gold adsorption (Kononova *et al.*, 2005).

Carbon	Carbon origin	BET surface area ($m^2 g^{-1}$)	Total pore volume ($cm^3 g^{-1}$)	Total pore volume ($cm^3 g^{-1}$)			Gold adsorption ($mmol\,g^{-1}$)
				Macropores	Micropores	Mesopores	
BAU	Charcoal	760	0.70	0.37	0.20	0.13	2.3
ABG	Brown coal	420	0.56	0.32	0.17	0.07	1.3
UC	Cocoa stones	1450	0.52	0.29	0.14	0.14	0.5
LK4	Siberian anthracite	1643	0.65	0.33	0.18	0.18	3.0

8.1.3.5 Adsorption of zinc species

Carrott *et al.* (1997), in a study of the adsorption of zinc species from aqueous solution appear to have arrived at a quite unequivocal understanding of the phenomena involved. They used five carbons, derived from peat, lignin and pitch by activation with molten KOH (the Anderson process). Surface areas (BET, N_2 at 77 K) were 644, 860, 1138, 1603 and 3085 $m^2 g^{-1}$, respectively, with *PZC* values of 8.3, 11.8, 7.5, 6.9 and 6.2, respectively. Attention was given to the speciation of the zinc in the pH range 0.0–14, and involving Zn^{2+} (pH < 8) and $Zn(OH)^+$, $Zn(OH)^{3-}$ and $Zn(OH)_4^{2-}$. Zinc was adsorbed from solutions of zinc nitrate, the pH of which were adjusted by addition of nitric acid or sodium hydroxide.

For all carbons, with the exception of the peat carbon of $PZC = 11.8$, the adsorption of zinc is attributed to reaction with the ionized acid sites of the carbon surface, calculated equilibrium constants being in the range 10^6–10^7. For the exceptional peat carbon, which has no acid sites, but a high concentration of basic sites, the adsorption of zinc is attributed to negatively charged hydroxyl complexes on protonated basic sites, with an equilibrium constant of 10^8. The authors state extents of adsorption do not correlate with surface area or porosity, but only to concentrations of acidic and basic sites. Some of the surface areas reported (N_2 at 77 K, BET equation) appear to be too high, and contain significant contributions from nitrogen in mesoporosity filled by capillary condensation. Correlations with these surface area values are therefore not to be expected.

8.1.3.6 Adsorption of Zn(II), Cu(II), Fe(III) from Mn solutions

The molecular sieve properties of AC (Section 8.2.3) enable components of gas mixtures to be separated, the process utilizing differences in pore dimensions or differences in rates of adsorption (kinetic effects). The same general phenomenon of separation (the detail being different) is possible by adsorptions from solutions. Kononova *et al.* (2001) purified solutions of manganese sulfate from the contaminants of Zn(II), Cu(II), Fe(III) using AC from anthracite and brown coal (activated in steam 830–850 °C, reacted with air at 400–410 °C), and other cation exchangers.

The production of active manganese dioxide from manganese ores has, as an end-product, a solution of $MnSO_4$ containing a few percent of zinc, copper and iron. The industrial solutions contained 35 mmol L^{-1} of manganese with concentrations of Zn^{2+}, Cu^{2+}, Fe^{3+} of about 0.2 mmol L^{-1}. Initial batch studies established that amounts of the three metals maximized in the pH range 3.5–4.5. The extents of adsorption were in the order:

$$Cu^{2+} > Fe^{3+}/Fe^{2+} > Zn^{2+} >> Mn^{2+}$$

This sequence is attributed to differences in the stabilities (dissociation constants) of the complexes formed between the functional groups of the carbon surface introduced by the activation and oxidation procedures.

A study of the multi-component system under dynamic conditions of flow through a bed of the AC (to determine the breakthrough curves) indicated the complete separation of the metal-ion impurities from the manganese sulfate solution. The breakthrough for the three metals occurred almost simultaneously. Kononova *et al.* (2001) report that the purified solution of manganese sulfate can then be used, directly, for the electrochemical precipitation of manganese dioxide.

Desorption (regeneration) of the carbon is carried out by washing with 0.5 M H_2SO_4.

8.1.4 Adsorption of Organic Solutes from Aqueous Solution

8.1.4.1 Introduction

Moreno-Castilla (2004) reviewed adsorption from aqueous solution with special attention to the properties of the carbon under investigation. (*These days, a publication which does not fully characterize its carbons has little value. Editors of appropriate journals should insist that adequate characterization details be provided in their papers.*) As the overall emphasis of this book concerns the properties of carbons, it is essential to cite, rather fully, this paper by Moreno-Castilla (2004).

Dabrowski *et al.* (2004), a little later, reviewed the adsorption of phenols from aqueous systems on to AC. The emphases of these two reviews are different, but with both being timely and complementary to each other.

Adsorption of organic solutes covers a wide spectrum of systems such as drinking water and wastewater treatments, and applications in the food, beverage, pharmaceutical and chemical industries. AC adsorption has been cited by the US-EPA as one of the best available environmental control technologies.

In spite of the large market for AC, the specific mechanisms by which the adsorption of many compounds, especially organic compounds, take place on this adsorbent are still uncertain. Thus, current knowledge about the fundamental factors controlling adsorption processes in aqueous phases (for which there is some certainty) needs to be summarized, quoting examples.

Adsorption of organic solutes is just as important as inorganic solutes, from the point of view of environmental control. Here, adsorption of organic compounds and of aromatics in particular, is a complex interplay of electrostatic and dispersive interactions. This is particularly true for phenolic compounds. The following arguments appear to have experimental and theoretical support:

1. Adsorption of phenolic compounds is partly physical and partly chemical.
2. The strength of π–π-interactions (adsorbent–adsorbate) can be modified by ring substitution on the adsorbent, leading to unfavorable situations affecting the feasibility of thermal regeneration (recovery of the AC).

8.1.4.2 Adsorption from dilute solutions

Contaminants to water supplies are in low concentrations and knowledge of their behavior toward adsorbents is critical, hence the importance of the dilute solution. The main differences between adsorption from the gas phase and the liquid phase are as follows:

(a) A solution is typically a system of more than one component. In experimental systems, there are at least two substances that can adsorb. For a dilute solution, adsorption of one type of molecule (M_1) involves replacement of the other (M_2). Thus, molecules adsorb not only because they are attracted to solid surfaces but also because the solution may reject them. A typical illustration is of the attachment of hydrophobic molecules on hydrophobic adsorbents from aqueous solutions which is mainly driven by their dislike of water and not by their attraction to the surface.

(b) Isotherms from solution may exhibit non-ideality, not only because of lateral interactions between adsorbed molecules but also because of non-ideality in the solution.

(c) Capillary condensation processes from solution appear to be less common than from the gas phase, because of the stronger screening interaction forces in condensed fluids.

Adsorption isotherms are normally developed to evaluate the capacity of AC for the adsorption of a particular molecule. The shape of the isotherms is the first experimental tool (of considerable utility) to diagnose the nature of a specific adsorption phenomenon, and it is expedient to classify the most common types, phenomenologically. There are several types of isotherms (of adsorption from solution) but those mainly found in carbon materials are the five depicted in Figure 8.4. (Note that this classification of Moreno-Castilla (2004) presents isotherm shapes in a way which differs from that of Figure 4.8 (for gas-phase adsorption)).

Long-linear isotherms are not common in adsorption on carbons but are to be found in the initial part of isotherms on homogeneous surfaces. The Langmuir type (L) frequently occurs, even when the premises of the Langmuir theory are not satisfied. Type F, typical for heterogeneous surfaces, is perhaps the most common. High-affinity isotherms are characterized by a very steep initial rise, followed by a pseudo-plateau. Sigmoidal isotherms have been obtained with homogeneous surfaces such as the graphitized carbon blacks Graphon and V3G.

Statistically, adsorption from dilute solutions is simple because the solvent can be interpreted as primitive, that is to say as a structureless continuum. Therefore, all the equations derived from monolayer gas adsorption remain valid after replacing pressure by concentration and modifying the dimensions of some parameters.

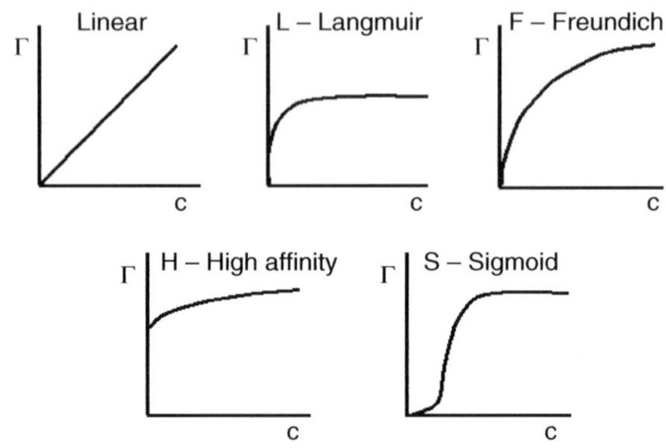

Figure 8.4. Most common adsorption isotherms found from dilute aqueous solutions on carbon materials (Moreno-Castilla, 2004).

Batch equilibrium tests are often complemented by dynamic column studies to determine system size requirements, contact time and carbon usage rates. These parameters can be obtained from the breakthrough curves (Section 4.5.2).

Adsorption measurements should preferably be supplemented by microcalorimetry, such as immersion and flow-microcalorimetry (see Figure 4.8. The IUPAC classification of adsorption isotherm shapes; see also Figure 8.4). These techniques give additional information about the nature of surfaces of the adsorbent and the mode or mechanism of adsorption.

8.1.4.3 Factors that control the adsorption process

Adsorption is a spontaneous process that takes place if the free energy of adsorption, ΔG_{ads}, is negative. There is a wide range of energies contributing to the free energy of adsorption, which can be grouped into non-electrostatic and electrostatic (Equation (8.1)).

$$\Delta G_{\text{ads}} = \Delta G_{\text{non-elect}} + \Delta G_{\text{elect}} \tag{8.1}$$

Although at atomic levels, all ionic and molecular interactions can be interpreted as "electric", this term is restricted to coulombic interactions and all other interactions are termed non-electrostatic, whatever their origin.

Electrostatic interactions appear when the adsorptive is an electrolyte that is dissociated or protonated in aqueous solution under the experimental conditions used. These interactions, either attractive or repulsive, are strongly dependent on the charge densities for both the carbon surface and the adsorptive molecule and on the ionic strength of the solution. The non-electrostatic interactions are always attractive, and include van der Waals forces, hydrophobic interactions and hydrogen bonding.

8.1.4.4 Characteristics of the adsorbent

The important characteristics of carbon are PSD, surface chemistry (functionality) and mineral matter content. The adsorption capacity depends on the accessibility of the organic molecules to the microporosity which is dependent on their size. Thus, under appropriate

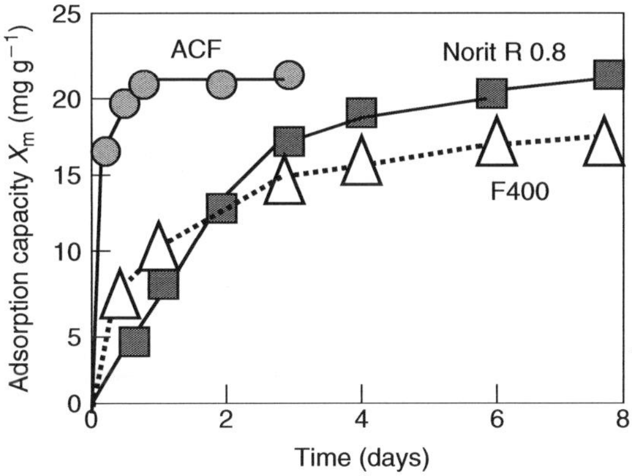

Figure 8.5. Adsorption kinetics of aminetriazol on an ACF and AC (F400 and Norit R 0.8) (Moreno-Castilla, 2004).

experimental conditions, small molecules such as phenol can access micropores; natural organic matter (NOM) can access mesopores, with bacteria having access to macropores. Activated carbon fibers (ACFs) may be a better adsorbent than granular activated carbons (GAC), because of their higher adsorption kinetics and adsorption capacity (Figure 8.5). But, there are difficulties associated with the use of carbon fibers which include (a) difficulty in handling, (b) health problems (to be compared with asbestos) and (c) regeneration is more difficult in comparison with GAC.

The surface chemistry of AC is dominated by their amphoteric nature. That is an ability to demonstrate both acidic and basic functionality. To a large extent, the surface chemistry essentially depends on heteroatom content, mainly on their surface oxygen complex contents. These determine the charge of the surface, its hydrophobicity and the electronic density of the graphene layers. Thus, when a solid such as a carbon material is immersed in an aqueous solution, it develops a surface charge that comes from the dissociation of surface groups or the adsorption of ions from solution. This surface charge (Figure 8.6) depends on the solution pH and the surface characteristics of the carbon. A negative charge results from the dissociation of surface oxygen complexes of acid character such as carboxyl and phenolic groups (Brönsted type). The origin of the positive surface charge is more uncertain because, in carbons without nitrogen functionalities, it can be due to surface oxygen complexes of basic character like pyrones or chromenes, or to the existence of electron-rich regions within the graphene layers acting as Lewis basic centers, which accept protons from the aqueous solution (see Section 4.7).

The surface charge can be determined by electrokinetic or titration methods. These are complementary methods, especially in the case of granular porous carbons. The first method primarily measures the surface charge at the more external surface of the particles whereas the second one provides a measure of the total surface charge. The pH at which the external surface charge is zero is called the point of zero charge, $\mathbf{pH_{PZC}}$. For typical amphoteric carbons, the surface is positively charged at pH $<\mathbf{pH_{PZC}}$ and negatively charged at pH $>\mathbf{pH_{PZC}}$, as determined by the use of titrations. The so-called IEP, also

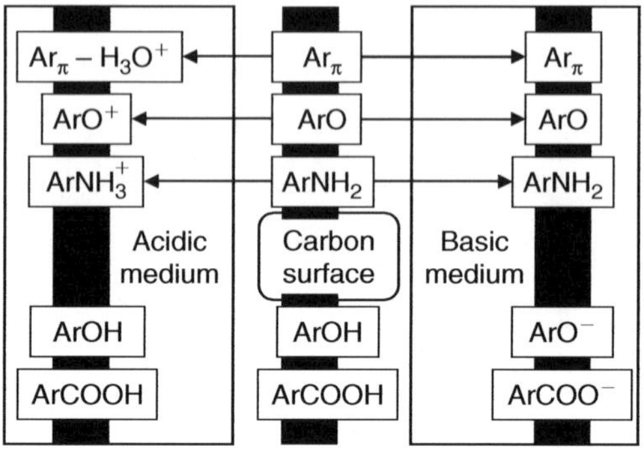

Figure 8.6. Macroscopic representation of the features of carbon surface chemistry (Radovic *et al.*, 2000; Moreno-Castilla, 2004).

denotes the pH at which the external surface charge is zero, $\mathbf{pH_{IEP}}$, and is determined electrokinetically, not by titrations that is, the position of zero zeta-(ξ) potential. In practice, $\mathbf{pH_{IEP}}$ is usually lower than $\mathbf{pH_{PZC}}$ for AC. The larger the difference between the two values, then the more heterogeneous is the distribution of surface charges.

According to Garten and Weiss (1957), the H-carbons (a description of these authors) are those with a high IEP where $pH_{IEP} > 7.0$ and the L-carbons have a low IEP of <7.0. It is observed that absolute values of surface charge are larger for the H-carbons. The amphoteric nature of carbons is also described in terms of HSAB, Alfarra *et al.* (2004).

In addition, surface oxygen complexes affect the surface hydrophobicity, which determines the hydrophobic interaction. Hydrophobic interaction or, more specifically, hydrophobic bonding, describes the unusually strong attraction between hydrophobic molecules and hydrophobic carbon surfaces. The surface hydrophobicity of carbon materials can be determined, for instance, by measuring their contact angle with test liquids of known surface tension, by inverse gas chromatography or by water adsorption.

Hydrophobic bonding occurs exclusively in aqueous solution, mainly coming from the strong tendency of water molecules to associate with each other by hydrogen bonding and from the characteristic structure of water molecules. Hydrophobic bonding plays an important role in interface and colloid science. For instance, the increasing adsorbability of aliphatic acid molecules with increasing hydrocarbon length, also known as Traube's rule, is essentially due to the increasing hydrophobic effect. From this, it is important to plot the adsorption isotherms obtained for adsorptives with different solubilities against the concentration relative to that in saturated solution. This normalization eliminates the differences in hydrophobicity between the adsorptive molecules, so that the normalized isotherms more truly reflect the affinity for the surface.

In general, an increase in the oxygen content of carbon brings about a decrease in its hydrophobicity. Thus, Pendleton *et al.* (2002) showed (Figure 8.7) that the adsorption of dodecanoic acid on different AC decreased linearly when the oxygen content of the

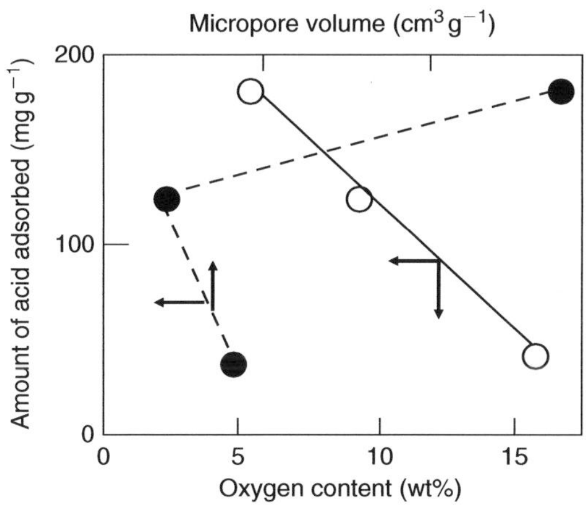

Figure 8.7. Relationship between the maximum amount of dodecanoic acid adsorbed on different carbons with their oxygen content and micropore volume (Pendleton *et al.*, 2002; Moreno-Castilla, 2004).

carbonaceous adsorbent increased and that there was no relationship with the micropore volume of the adsorbent. Water molecules are bound to the surface oxygen complexes by H-bonds, reducing the accessibility of the hydrophobic aliphatic chain of dodecanoic acid to the hydrophobic parts of the carbon surface.

This model explains the decrease in methyl isoborneol (MIB) adsorption when the oxygen content of carbons increased, Pendleton *et al.* (1997). In this case the average enthalpy of displacement of water by MIB changed (-40 to $-20\,kJ\,mol^{-1}$) when the concentration of hydrophilic sites on the surface increased from 0.3 to 1.1 $mmol\,g^{-1}$ suggesting that it is more difficult for MIB molecules to displace water molecules from carbon surfaces when their oxygen content increases. Results obtained show that the enthalpy of displacement reaches a limiting value when the number of hydrophilic sites is higher than about 0.7 $mmol\,g^{-1}$.

Finally, surface oxygen complexes also affect the electronic density of the graphene layers, Coughlin and Ezra (1968), so affecting the dispersion interactions between the carbon surface and adsorptive molecules. Carboxyl groups fixed at the edges of the graphene layers have the ability to withdraw electrons, whereas phenolic groups release electrons.

On the theoretical side, Tamon and Okazaki (1996), using a semi-empirical quantum chemical method, determined the highest occupied molecular orbital (HOMO) and lowest unoccupied molecular orbital (LUMO) levels of a non-substituted phenanthroperylene cluster (AC), and the same cluster substituted with two phenolic groups (AC OH) and with two carboxyl groups (AC COOH). Results indicated that the withdrawing groups, such as carboxyl groups, decreased HOMO and LUMO levels with respect to the original carbon cluster, whereas the donating groups increased these levels.

Finally, the adsorption process can be influenced by the mineral matter content of the carbon. In general, this has a deleterious effect on the adsorption because it can block the porosity of the carbon matrix and can preferentially adsorb water because of its hydrophilic nature, so reducing the adsorption of the adsorptive.

8.1.4.5 Characteristics of the adsorptive

The dominant properties of the adsorptive in solution are its molecular size, solubility, pK_a and substituents to the molecule, should it be aromatic. This is because (a) the molecular size controls the accessibility into the porosity, (b) solubility controls the hydrophobic interactions, (c) the pK_a controls the dissociation of the adsorptive (if it is an electrolyte) and is closely related to the pH of the solution and (d) the substituent of the aromatic ring of the adsorptive withdraws or releases electrons from the ring and this affects the dispersion interactions between the aromatic ring of the adsorbate and the graphene layers of the adsorbent.

Currently, because of the frequency of their occurrence in wastewaters, the adsorption of phenolic compounds on carbons and the influence of surface oxygen complexes on their uptake are the most frequently studied, Radovic *et al.* (1997). It is well established that an increase in surface acidity of AC, after an oxidation, causes a decrease in phenol adsorption from dilute aqueous solution. For example, Figure 8.8 shows the adsorption isotherms of phenol on oxidized carbons, Mahajan *et al.* (1980). There was a large decrease in phenol uptake after oxidation, the effect of oxidation not being trivial. This phenol uptake progressively increased as the surface acidity decreased, and the oxidized sample (heat treatment temperature (HTT) 950 °C) had the same adsorption capacity as the original carbon (about 1.5 $\mu mol\, g^{-1}$).

Three mechanisms have been proposed to explain this behavior, namely (a) the π–π-dispersion interaction mechanism, (b) the hydrogen-bond formation mechanism and (c) the electron-donor–acceptor complex mechanism. To give these ideas a historical

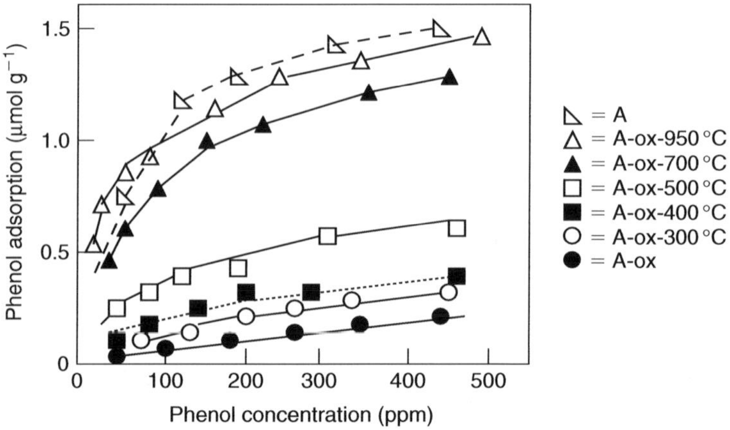

Figure 8.8. Adsorption isotherms of phenol on AC with different degrees of oxidation (Mahajan *et al.*, 1980; Moreno-Castilla, 2004).

perspective, mechanisms (a) and (b) were suggested by Coughlin and Ezra (1968), and mechanism (c) was suggested by Mattson *et al.* (1969). At that time, it was known that phenol adsorbed in a flat position on the graphene layers, and here adsorption was driven by the π–π-dispersion interactions between the aromatic ring of phenol and the aromaticity of the graphene layers.

Coughlin and Ezra (1968) proposed that acidic surface oxygen groups, located at the edges of the graphene layers, remove electrons from the π-electron system, so creating positive holes in the conducting π-band of the graphene layers, so weakening interactions between the π-electrons of the phenol aromatic ring and the π-electrons of the graphene layers, so reducing the phenol uptake. Coughlin and Ezra (1969) also suggested that the bonding of water molecules to the oxide functional groups by H-bonding plays a role in the uptake of phenolic compounds, assuming that water molecules, adsorbed on oxygen groups, become secondary adsorption centers (complexes) retaining other water molecules by means of H-bonds. These complexes then block the entry of organic molecules to significant parts of the surface. However, Coughlin and Ezra (1968) further observed that surface oxygen complexes did not influence phenol uptake from concentrated solutions. But, this may be another story.

Mattson *et al.* (1969), alternatively, suggested that aromatic compounds adsorb on carbons by a donor–acceptor complex mechanism, with the carbonyl oxygen of the carbon surface acting as the electron donor and the aromatic ring of the adsorbate acting as the acceptor. With progressive uptake of, for example, aromatic compounds, once the carbonyl groups had all reacted, the aromatic compounds then form donor–acceptor complexes with the graphene layers. Thus, Mattson and co-workers explained a decrease in phenol uptake, after carbon oxidation, as being due to the oxidation of carbonyl groups to carboxyl groups. As a result, the electron-donor–acceptor complexes could not be formed. Figure 8.8 illustrates this effect where the higher HTTs cause reduction in surface oxygen contents.

Since these initial proposals (differing) by Coughlin and Ezra (1968) and Mattson *et al.* (1969), and after many papers have been published, significant clarification has not been forthcoming. The publication of Haydar *et al.* (2003) describes the adsorption of *p*-nitrophenol on an AC with different levels of oxidation, the authors providing experimental evidence for the π–π-dispersion interaction mechanism.

There is further good experimental evidence for the π–π-dispersion interactions in this publication by Mahajan *et al.* (1980) of phenol adsorption on both graphite and a boron-doped graphite, where the substitutional boron removes π-electrons from the graphene layers, resulting in a lowering of the phenol adsorption from water. These authors also indicated that phenol and water both compete to form H-bonds with surface oxygen complexes, such as carboxyl groups. In this competition, the water molecules are preferentially bonded, by H-bonds. Of course, a definitive experiment is to adsorb phenol from a solution in cyclohexane so removing any competitive adsorption and, here, Mahajan *et al.* (1980) found that, for two heat-treated carbons, the carbon with higher surface acidity exhibited a higher phenol uptake.

Franz *et al.* (2000) studied the adsorption of phenol, aniline, nitrobenzene and benzoic acid from both aqueous and cyclohexane solutions. Although the interpretations of the authors of the π–π-dispersion may not be that clear, their experimental results apparently indicate that the adsorption mechanism is by both water H-bonding with carboxyl groups and by π–π-dispersion interactions between the aromatic ring of the adsorptive and the graphene layers. These authors concluded that Mattson's mechanism, involving electron-donor–acceptor complexes, is not the driving force for the adsorption of aromatics on AC. In view of the now known properties of carbons this is probably correct. A weak point in the Mattson mechanism is that, although there is much experimental evidence that oxidation of carbons increases their concentration of CO_2-evolving groups, the CO-evolving groups also increase or remain essentially unchanged.

Additional experimental evidence of the influence of water adsorption comes from the enthalpy of immersion into water of AC, which reflects the specific and non-specific interactions between the liquid and the solid, López-Ramón *et al.* (2000) (see Figure 4.8. The IUPAC classification of absorption isotherm shapes; see also Figure 8.4; Figure 8.9). This revealing study of many carbons oxidized to different levels indicates a simple correlation between the enthalpy of immersion, the oxygen content of the surface, the basic groups as titrated by HCl, micropore filling and the wetting of the external surface. Thus, the specific interaction between the surface oxygen groups and water is 12.1 kJ mol^{-1} (oxygen), involving an average of two water molecules per oxygen atom.

Using the same technique, MacDonald and Evans (2002) measured the enthalpy of exchange of phenol–water from dilute aqueous solutions on commercial BPL carbons with different oxygen contents. They found that a strong preference of water to H-bond to surface oxygen complexes lowers the phenol exchange enthalpy from aqueous solution because the phenol exchange enthalpy was inversely proportional to the net molar enthalpy of immersion into water.

The two mechanisms proposed by Coughlin and Ezra (1968) explain much of the experimental results available to date. However, some phenol is *chemisorbed* on to carbon surfaces and cannot be desorbed by treating with different solvents or by heat treatment in an

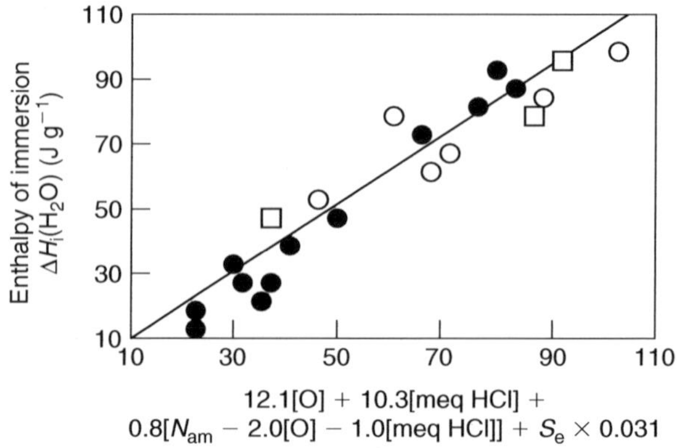

Figure 8.9. Correlation between the enthalpy of immersion into water, the total surface oxygen, the basic groups titrated with HCl, the micropore filling and the wetting of the non-porous surface area (López-Ramón *et al.*, 2000; Moreno-Castilla, 2004).

inert atmosphere. Rather, the chemisorbed part is evolved as light gases and heavy products from the surface of the carbon and a polymeric carbon residue remains on the surface. This residue reduces the adsorption capacity of thermally regenerated AC. However, as indicated earlier in this book, carbon surfaces sometimes exhibit sites of exceptional activity and this chemisorption of phenols is yet another manifestation of this property (see Section 8.1.4.9).

8.1.4.6 Chemistry of solutions: the importance of pH and ionic strength

Factors that significantly influence adsorption processes include the solution pH and ionic strength, pH being a key factor that controls the adsorption processes of organic weak electrolytes and polyelectrolytes because it controls the electrostatic interactions between the adsorbent and the adsorbate. Thus, solution pH determines the carbon surface charge and the dissociation or protonation of the electrolyte. At a solution pH lower than the pH_{PZC} or the pH_{IEP}, the total or external surface charges, respectively, are on average positive, whereas at a higher solution pH, they are negative. Further, the solution pH also controls the dissociation or ionization of the electrolyte through its pK_a. Thus, for instance, acidic electrolytes are dissociated at pH $>pK_a$. Therefore, the solution pH controls the adsorptive–adsorbent and adsorptive–adsorptive electrostatic interactions, which have profound effects on the adsorption process.

Also, the adsorption of substituted phenols on AC depends on the solution pH, as illustrated by Moreno-Castilla (2004) in Figure 8.10. At acidic pH values (<7.0), the uptake maximized because the phenols were undissociated and dispersion interactions predominated. At basic pH values (>7.0), uptakes become lower because of electrostatic repulsions between the negative surface charge and the phenolate anions and between phenolate–phenolate anions in solution. The pH at which the uptake decreased was found to depend

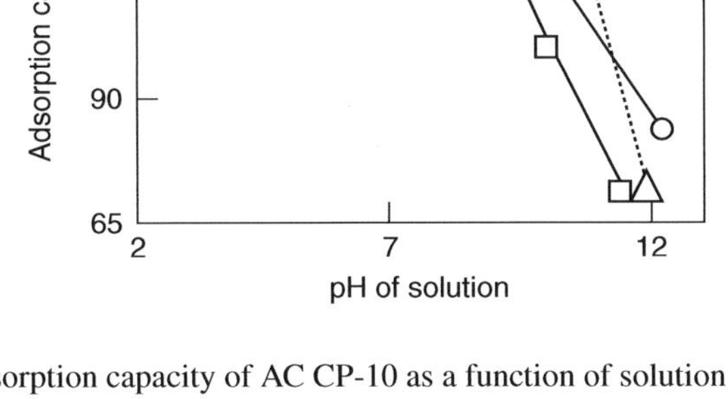

Figure 8.10. Adsorption capacity of AC CP-10 as a function of solution pH. ○ = phenol (pK_a = 9.96); □ = *m*-chlorophenol (pK_a = 8.80); △ = *p*-nitrophenol (pK_a = 7.13) (Moreno-Castilla *et al.*, 1995; Moreno-Castilla, 2004).

Table 8.5. Characteristics of AC beds when removing *ortho*-chlorphenol at pH values of 2 and 10. Particle size is between 0.15 and 0.25 mm (Rivera-Utrilla *et al.*, 1991; Moreno-Castilla, 2004).

Sample	pH = 2		pH = 10	
	V_B (cm^3)	H_{MTZ} (cm)	V_B (cm^3)	H_{MTZ} (cm)
C-2	4450	1.02	550	2.02
C-13	6500	0.95	1820	1.54
C-24	10,650	0.48	2800	1.32

on the adsorptive pK_a and the difference between the pH_{PZC} and the pH_{IEP}, Moreno-Castilla *et al.* (1995).

Solution pH also affects the characteristics of carbon beds, as shown in Table 8.5. Thus, adsorption of *ortho*-chlorophenol (OCR) is favored at acidic pH, as is further noted by their higher values of breakthrough volumes, V_B **(cm^3)**, and their lower values of the height of the mass transfer zone (MTZ), H_{MTZ} (cm), found at this pH (Section 4.5 – Breakthrough Curves). The lower adsorption at basic pH is due, as explained above, to the repulsion between the negatively charged carbon surface and the *ortho*-chlorophenolate anions, Rivera-Utrilla *et al.* (1991).

The effect of pH and the nature of functional groups of both the aromatic adsorptive and the adsorbent on the adsorption process were studied by Radovic *et al.* (1997) who used an AC oxidized with nitric acid and nitrided with ammonia to study the adsorption of aniline and nitrobenzene, which are, respectively, electron-donating and electron-withdrawing groups. At pH = 2, anilinium cations were predominant in solution and the surface charge was positive. Carbon oxidation enhanced adsorption relative to the original carbon, whereas nitriding decreased it, the authors indicating that the electrostatic adsorbent–adsorbate interactions were more important than the dispersion interactions. Thus, oxidation shifted the pH_{PZC} to a value close to the solution pH, whereas nitriding increased the pH_{PZC}, so enhancing the adsorbate–adsorbent interactions.

At pH = 11, the surface charge was predominantly negative and aniline was not protonated in solution, with oxidation and nitriding of the carbon, both being detrimental to adsorption because dispersion interactions predominate and are not favored because the oxygen- and nitrogen-containing functional groups withdraw electrons from the graphene layers. However, in this case, some preferential H-bonding of water to oxygen and nitrogen surface functional groups may also be contributory to the decrease. At a solution pH equal to the pH_{PZC}, aniline uptake was relatively high on all adsorbents. According to the authors this is consistent with minimized electrostatic repulsions.

Results obtained with nitrobenzene as the adsorbate indicate that solution pH has little effect on equilibrium uptake because nitrobenzene molecules are the dominant species across the entire range of solution pH. However, as commented above, the H-bonding of water to some surface functionalities also plays a role. A maximum uptake is obtained at pH = pH_{PZC} because it is here that dispersion interactions are maximized. These results

Table 8.6. Surface areas and pore volumes of carbons W and C (Newcombe and Drikas, 1997).

Carbon	Surface area ($m^2 g^{-1}$)	Pore volume ($cm^3 g^{-1}$)		
		Micropore width <0.8 nm	Micropore width >0.8–2.0 nm	Mesopores 2–50 nm
Carbon W	1197	0.30	0.30	0.49
Carbon C	1018	0.30	0.17	0.07

indicate that functionalization of either the adsorptive or the adsorbent, which increases the π-electron density, leads to enhanced or stronger adsorption when adsorption processes are governed by dispersion forces. The converse is also true.

Ionic strength is the other key factor that controls the electrostatic interactions. Thus, these interactions, either attractive or repulsive, can be reduced by increasing the ionic strength of the solution. This is due to a screening effect of the surface charge produced by the added salt. Therefore, when the electrostatic interaction between the surface and the adsorptive is repulsive, or the surface concentration is sufficiently high, an increase in ionic strength increases extents of adsorption. Conversely, when the electrostatic interactions are attractive, or the surface concentration is sufficiently low, an increase of the ionic strength diminishes extents of adsorption. These effects take place with weak organic electrolytes and polyelectrolytes, for instance, in the case of bisphenol A and NOM.

NOM is found in varying concentrations in all natural water sources. It is a complex mixture of compounds varying from small hydrophilic acids, proteins and amino acids to larger humic and fulvic acids. Newcombe and Drikas (1997) studied the adsorption of NOM (nominal molecular weight between 500 and 3000 Da) on two AC. These were Carbon C, Calgon F-400, a bituminous coal-based carbon and Carbon W, (Picazine), produced by PICA, a chemically activated wood-based carbon. Details of these carbons are in Table 8.6.

Figure 8.11 illustrates the adsorption of NOM at pH = 7 on carbon W. At this pH, both the carbon surface and NOM are negatively charged. At very low surface coverage direct surface adsorbate interactions predominate. An increase in salt concentration effectively screens repulsive interactions, resulting in increased adsorption. For carbon C, Figure 8.12, the adsorption of NOM was carried out at pH = 4 when the carbon surface is positively charged, whereas NOM will be negatively charged. The adsorption isotherms show an intersection point indicating a transition from a screening-reduced to a screening-enhanced regime. At low surface concentrations, below the intersection point of the isotherms, attractive interactions between the adsorbent and the adsorbate are screened by an increase in salt concentration, decreasing the adsorption. Above the intersection point, where the surface concentration is higher, salt screens the repulsion between charged extents of adsorption.

When the adsorption of organic molecules is governed by non-electrostatic interactions, such as π–π-dispersion or hydrophobic interactions, the area of the adsorbent occupied by the adsorbate depends on the porosity of the former and the molecular size of the latter. Thus, it has been shown by Stoeckli *et al.* (2001), using immersion calorimetry measurements, that

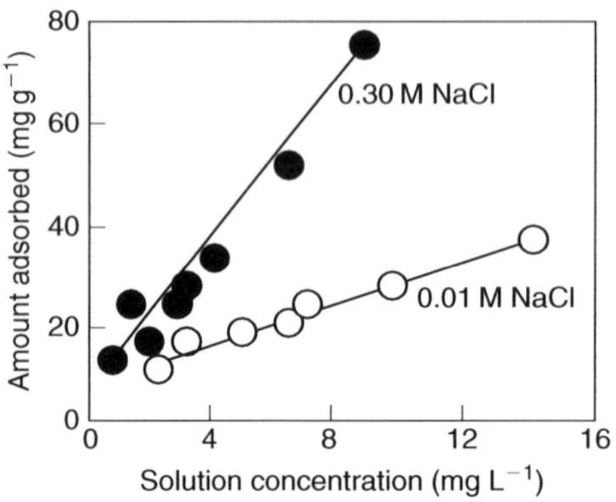

Figure 8.11. Adsorption of NOM on carbon W ($pH_{PZC} = 4.5$) at two ionic strengths and at solution pH = 7 (Moreno-Castilla, 2004).

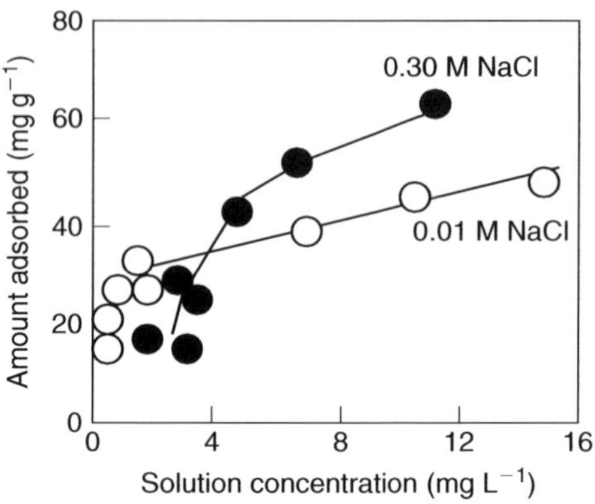

Figure 8.12. Adsorption of NOM on carbon C ($pH_{PZC} = 7.5$) at two ionic strengths and at solution pH = 4 (Moreno-Castilla, 2004).

adsorptions from dilute aqueous solutions of phenol and meta-chlorophenol are as monolayers by both porous and non-porous carbons with basic surface properties, provided that the adsorptive is undissociated at the solution pH. (This does not apply where molecular sieve effects reduce accessibility to the micropore system, a point that must never be overlooked.)

With regard to the effect of temperature on adsorption processes, an increase in uptake of organic molecules is expected when the adsorption temperature decreases (adsorption is a spontaneous process). However, some examples have been reported where the amount adsorbed increased with temperature. Thus, for instance, Terzyk *et al.* (2003) reported that the adsorption of paracetamol from diluted aqueous solution increased with adsorption temperature. This effect was independent of the type of carbon, its surface chemistry or solution pH. The authors explained this behavior as due to phase changes in the crystal form of the adsorptive. Alternatively, diffusion restrictions may be operating because of the large size of the paracetamol molecule when an enhanced temperature increases rates of adsorption toward an equilibrium position.

8.1.4.7 Summary of review of Moreno-Castilla (2004)

Adsorption of organic molecules from dilute aqueous solutions on carbon materials is a complex interplay between electrostatic and non-electrostatic interactions. Electrostatic interactions appear with electrolytes, essentially when they are ionized at the experimental conditions used. Non-electrostatic interactions are essentially due to dispersion and hydrophobic interactions. The surface chemistry of the carbon has a major influence on both electrostatic and non-electrostatic interactions, and can be considered as the main factor in the adsorption mechanism from dilute aqueous solutions.

Aromatic compounds are physisorbed on carbon materials essentially by dispersion interactions between the π-electrons of the aromatic ring and those of the graphene layers. Functionalization of either the adsorbent or the adsorptive profoundly affects these dispersion interactions. In addition, the presence on carbon surfaces of chemical functionalities that can give rise to H-bonds with water can effectively reduce the adsorption of the adsorptive molecules.

Strongly adsorbed aromatic compounds, which cannot be desorbed even at high temperatures, probably involve an electron-donor–acceptor or charge-transfer mechanism. It has been shown that the surface of a carbon is more effectively used if non-electrostatic interactions are the driving force for the adsorption. This condition can be reached by controlling the surface chemistry of carbon, the pH and ionic strength of the solution.

Finally, research trends on this topic include the study of competitive adsorption, for instance, between NOM and micro-pollutants such as pesticides; adsorption of bacteria which in turn can catalyze the decomposition of the adsorptive; adsorption/desorption of drugs for medical applications and modifications of the adsorptive depending on the solution and carbon surface chemistry.

8.1.4.8 Summary of review of Dąbrowski et al. *(2004): adsorption mechanisms*
Phenolic compounds are to be found in contaminated water sources. At low concentrations, drinking water can be rendered unpalatable and contaminants may be carcinogenic to humans. Phenols are by no means rare on the industrial scene and are associated with oil refineries, with sites of coal gasification and coal carbonization, as well as the petrochemical industry. The production of plastics, colours, pesticides and insecticides are other examples of sources of phenols. Degradation of waste and used plastics produces phenols.

AC provide one of the most efficient materials for the removal of phenols from aqueous media. Traditionally, powdered AC and GAC have been used with new attention being given to the use of ACFs, Rodríguez-Reinoso (1997). It has been known for some time that adsorptions of phenols exhibits irreversibility and Dąbrowski *et al.* (2004) review this topic. Applications of AC to water purification must produce a pollution-free water supply, with little interest in phenol recovery. Hence, irreversible adsorption ensures an efficient way of trapping the phenols.

In adsorption process of phenols, Dąbrowski *et al.* (2004) emphasize (a) the importance of the porosity of the AC and its functionality, (b) the molecular structure of the adsorbate, including pK_a, polarity, shape and size, as well as (c) the pH of the solution, its ionic strength and concentration. The amphoteric nature of the carbon surface must not be overlooked, the nature of the carbon surface being directly related to the pH of the solution. The work of Giles *et al.* (1974a, b) is quoted where 17 different shapes of isotherms are described indicating weak adsorptions (convex shape), strong adsorptions (concave shape) to extremely strong adsorptions where the initial sections of the isotherm are vertical.

Unlike adsorptions of metal ions (M^{n+}) a phenol, in aqueous solution, can exist as a non-dissociated and as an anionic phenolate, dependent on the pH, the more acidic the solution, the higher the concentration of the phenol molecule. Thus, the mechanisms of adsorption of a phenol change drastically with pH. In the molecular state, adsorption is more a function of the porosity of the carbon, principally the PSD of the microporosity. Extents of adsorption of a phenol are a maximum at neutral pH decreasing in both alkali and acidic solutions. At low values of pH, increased adsorption of protons on to the carbon surface reduces the facility of adsorption by the phenol. Also, the more polar is the surface of the carbon, the lower is the extent of adsorption of the phenol (see Terzyk, 2003; Terzyk *et al.*, 2003).

8.1.4.9 Summary of review of Dąbrowski et al. *(2004): irreversible adsorption*
Irreversibility of phenol adsorption is important because of the need (if possible) for regeneration of the AC, the adsorbed (trapped) phenol having little (or zero) commercial value.

The irreversibility of adsorption of phenols has been established since the late 1960s. The dominant features of irreversibility (chemisorption of phenol) were summarized by Magne and Walker (1986) and include (a) irreversibility increases with time spent by the phenol on the carbon surface, (b) irreversibility increases with increasing temperature of the system, (c) irreversibility is inhibited by the presence of surface oxygen functionality on the carbon surface, (d) the sites of formation of chemisorbed phenol are oxygen-free sites at the edges of graphene layers and (e) physisorption occurs over the entire surface of the carbon.

Grant and King (1990) provided an insight into the causes of irreversibility. They were able to extract the chemisorbed phenol by acetone and to examine the solute by mass spectrometry, noting the formation of polymers of phenol, usually dimers with some higher polymers. The effect of increasing pH was to increase the fraction of reversible adsorption and to reduce the fraction of irreversible adsorption almost to zero (pH 1.8). Increasing adsorption temperature increased the amounts of phenol taken up by the carbon, an effect suggestive of a chemical reaction rather than a physical adsorption effect. Increasing the contents of surface oxygen complexes by treatment with nitric and sulfuric acids enhanced phenol adsorptions at pH = 7.0, this being attributed to the presence of lactones. At the same time, extents of irreversibility increased because of the formation of strong bonds with adsorbed molecules.

The polymerization reactions, essentially oxidation reactions which remove hydrogen, involve the adsorption of both the phenol and oxygen to the carbon surface. More adsorption of phenol occurs under **oxic** (with surface oxygen) conditions than under **anoxic** (oxygen-free surfaces) conditions. However, the presence of **acidic** surface oxygen complexes inhibits the adsorption of phenols by reducing the ability of the carbon to promote oxidative coupling reactions leading to polymerization. Such basic groups as chromenes, lactones and pyrones promote polymerization.

8.1.5 Other Studies: Examples

Other examples from the literature are summarized below to indicate the many considerations necessary to analyze adsorptions from solutions.

González-García et al. *(2001)* describe the adsorption isotherms at 20 °C of four AC of the *non-ionic surfactant* Triton X-100 from aqueous solution over a wide concentration range. The adsorption was explained using one or a combination of two Langmuir equations, depending on the equilibrium concentration range studied. The results indicate that there are at least two kinds of interactions, the first related to a direct interaction between the AC surface and adsorbate molecules, and the second mainly due to the interaction between surfactant molecules at the adsorbent–solution interface leading to the formation of interfacial aggregates.

Ismadji and Bhatia (2001) discuss a modification of the DR pore filling model by incorporation of the repulsive contribution to the pore potential, and of bulk non-ideality, for the characterization of AC using liquid-phase adsorption. For this purpose, experiments were performed using ethyl propionate, ethyl butyrate and ethyl isovalerate as adsorbates and the microporous–mesoporous AC – Filtrasorb 400, Norit ROW 0.8 and Norit ROX 0.8 as adsorbents. The repulsive contribution to the pore potential is incorporated through a Lennard-Jones intermolecular potential model, and the bulk-liquid-phase non-ideality through the Universal quasichemical Functional group Activity Coefficient (UNIFAC) activity coefficient model. For the characterization of AC, the generalized adsorption isotherm is utilized with a bimodal gamma function as the PSD function. It is found that the model represents the experimental data very well, and significantly better than when the classical energy-size relationship is used, or when bulk non-ideality is neglected. Excellent agreement between the bimodal gamma PSD and DFT-cum-regularization-based PSD is also observed, supporting the validity of the proposed model.

In their study, Ismadji and Bhatia (2001) made use of a modified DR equation which takes into account the repulsive part of the adsorption potential within the micropores. Consideration is also given to the non-ideality of the bulk-liquid phase of the adsorbate. To describe the PSD, a bimodal gamma function is used. These developments of the DR equation yielded better agreement with experimental data for adsorption of esters on AC, than the original DR equation.

Lu and Sorial (2004) discuss the importance of the micropore size distributions, and the electrostatic status of the surface. They studied the impact of adsorbent PSD on adsorption mechanisms from multi-solute systems. **Oxic** (with oxygen present) and **Anoxic** (without oxygen present) adsorption equilibria were obtained for the single solute (phenol), binary solute (phenol/2-methylphenol) and ternary solute (phenol/2-methylphenol/2-ethylphenol) systems on a GAC F400, and on types of ACFs, namely activated carbon cloth-10 (ACC-10) and ACC-15. F400 has a wide PSD, while ACC-10 and ACC-15 have narrow PSD and their critical pore diameters are 0.80 and 1.28 nm, respectively. In single solute adsorption, the higher adsorptive capacity under **Oxic** conditions, when compared with **Anoxic** ones, was related to the PSD of the adsorbent. Binary solute adsorption on ACC-10 and ternary solute adsorption on ACC-15 indicated no impact of the presence of molecular oxygen on the adsorptive capacity and the adsorption isotherms were well predicted by the ideal adsorbed solution theory (IAST). Significant differences between **Oxic** and **Anoxic** isotherms were noticed for other multi-component adsorption systems. The narrow PSD of ACFs was effective in hampering the oligomerization of phenolic compounds under **Oxic** conditions. Such a phenomenon provides accurate predictions of fixed-bed adsorbers in water treatment systems.

Nagoka and Yoshino (1986) studied the surface properties of electrochemically pre-treated glassy carbon by cyclic voltammetry and adsorption of catechol (1,2-dihydroxybenzene). The authors concluded that adsorption of quinones (on glassy carbon and pyrolytic graphite) depends on an electronic effect such as an electrostatic attraction between the adsorbate and partial surface charges, rather than a specific chemical effect.

Tanada et al. *(1997)* studied the aqueous solution removal of chloroform by a surface-modified AC. They report a monotonic decrease in the amount adsorbed with increasing acidity of the carbon, although the reported surface areas are essentially the same. These authors state that the adsorption capacity is dominated by pore structure and that heteroatoms (such as oxygen, nitrogen and hydrogen) have to be considered. They conclude by stating that aggregates of water molecules on AC prevent the invasion of chloroform into the adsorption sites.

Xiao et al. *(2005)* studied adsorption of cationic–anionic surfactant mixtures on AC. Surfactants have applications on oil–water interfaces and on rock and soil interfaces. It is reported that mixtures of surfactants have enhanced surface activity when compared with the individual surfactant. These authors adsorbed two mixtures of surfactants on an AC (as a model surface) of $1050\,m^2\,g^{-1}$. Results were dependent on the ionic strength of the solution. When using deionized water, this synergistic effect was observed, but when using mineralized water (containing e.g. $CaCl_2$, $MgCl_2$, $NaHCO_3$, Na_2SO_4, NaCl) the opposite effect was observed, that is, the addition of the second surfactant reduced the adsorption of the first surfactant. This is attributed to a screening effect by the ions from the salts around the existing adsorption of the surfactant so preventing its further adsorption following addition of the second surfactant (an effect due to electrostatic interactions between the oppositely charged surface surfactant ions).

8.2 Gas-Phase Applications

8.2.1 Introduction

It is impossible, in a single text, to describe adequately the large number of gas-phase applications allocated to AC. The following list indicates the wide-ranging scenarios for AC. Selected topics are dealt with more fully below:

(a) Personal protection.
(b) Cigarette filters.
(c) Industrial gas masks. Chemical warfare agent protection, including clothing, gas masks, and atmospheres in warships, submarines, tanks and aircraft.
(d) Effluent gas purification.
(e) Industrial off-gas purification, removal of SO_2, H_2S, CS_2, etc.
(f) Petroleum refineries.
(g) Sewage and geothermal plants.
(h) Vinyl chloride monomer (VCM) plants and solvent recovery in general.
(i) Separation of gas mixtures using carbon molecular sieves (CMS).
(j) Organic and inorganic process catalysis, both as a support material and as a catalyst.
(k) Adsorption of radionuclides.
(l) Natural gas storage and purification.

(m) Automobile/gasoline recovery.

(n) Odor control generally.

Rodríguez-Reinoso and Sepúlveda-Escribano (2001) and Rodríguez-Reinoso (2002) reviewed the applications of porous carbons in adsorptions and catalysis. For gas-phase applications, carbon adsorbents, generally, are used in the form of hard granules, hard pellets, fibers, cloths and monoliths, because these avoid an excessive pressure drop when the gas to be treated passes through the adsorbent bed (see Chapter 9 for further details). They usually have a well-developed microporosity to provide a high adsorptive capacity and selectivity for gases and organic vapors. Their surface areas range between 1000 and 2000 $m^2 g^{-1}$, and have a high adsorptive capacity per unit volume, high retention capacity, high preferential adsorption of gases in the presence of moisture, low resistance to gas flow and complete release of adsorbates at increasing temperatures and decreasing pressures.

The production of AC, in the USA, for the year 2002, was 350 million pounds (about 150×10^6 kg or 150,000 tonnes). Of this production, 79% is for liquid-phase applications, the other 21% being used for air purity projects, the spent AC having to be regenerated. The carbons for air quality control are generally somewhat more expensive than liquid-phase carbons, the prices averaging out at about $1.15 per pound. The US production has a capacity of about 435 million pounds.

From a practical point of view, although the equilibrium isotherm provides information about equilibrium adsorption capacities across the PSD range, the isotherm cannot be used to predict performance (kinetics) in an operational bed. To do this, use is made of the breakthrough curve (Section 4.5). This consists of a bed of carbon through which can be passed the gas (or liquid) containing of two or more components which need to be separated from each other.

For this, test bed breakthrough curves are needed and as the gas stream passes through the adsorbent bed, the composition of the latter can be divided into three different zones, as schematized in Figure 8.13. **Zone I** is where the adsorbent is already saturated with the adsorbate to its full equilibrium capacity. **Zone II** is the part where the transfer of the adsorbate from the gas phase to the solid is taking place. It is called the **MTZ**. **Zone III**, finally, is the part where the adsorbent is still free from adsorbate.

As represented in Figure 8.14, the MTZ moves through the adsorbent as the gas stream passes through. When the adsorption time is small, the adsorbate concentration in the effluent (the gas that has passed through the adsorbent bed) is nil, but it starts increasing when the MTZ reaches the end of the bed. The graph of the evolution of the adsorbate concentration in the effluent with the operation time is called the breakthrough curve (see Section 4.5). When the adsorbate concentration at the exit of the bed reaches the acceptable emission limit, adsorption must be stopped and the adsorbent should be replaced or regenerated. It can be seen in Figure 8.13 that the average adsorption capacity in the MTZ is about 50% of the equilibrium capacity. As the adsorption process must be stopped before the whole MTZ has passed through the end of the bed, it follows that some of the adsorbent will not be fully used. Therefore, the length of the MTZ plays a very important role in determining the amount of component that can be adsorbed before reaching a too

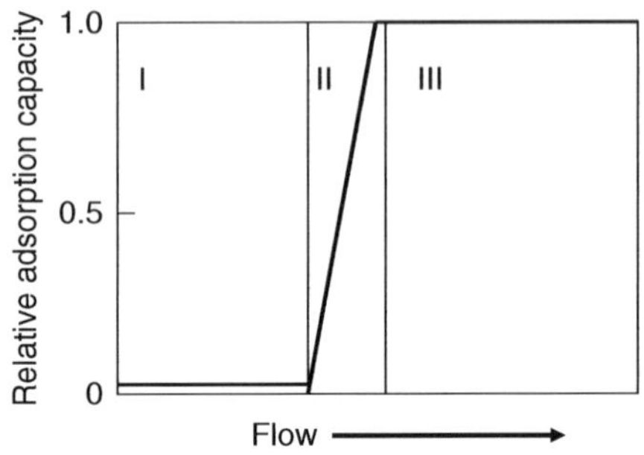

Figure 8.13. Adsorption capacity in relation to bed length (Rodríguez-Reinoso and Sepúlveda-Escribano, 2001).

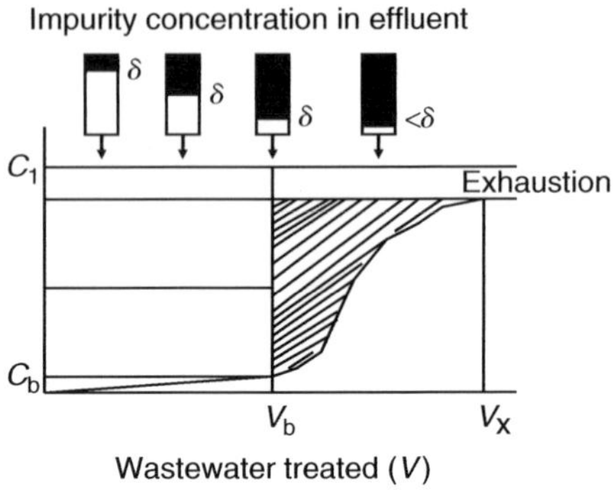

Figure 8.14. Relation between the MTZ position and the breakthrough curve (Rodríguez-Reinoso and Sepúlveda-Escribano, 2001).

high concentration at the exit of the bed. This parameter is determined, not only by the physical shape and adsorptive properties of the adsorbent, but also by the adsorbate to be retained and by process conditions. The adsorbent will be better utilized when the MTZ is as short as possible.

The humidity of the gas to be treated is an important parameter in gas-phase applications of AC. Although the interaction between the polar water molecule and the relatively non-polar carbon surface is very weak, AC can adsorb appreciable amounts of water when the relative humidity of the gas stream is high. The organic molecules that are usually to be retained interact more strongly than water with the carbon surface and thus can easily displace the water molecules form the surface. However, if the water content of the carbon adsorbent is high, the organic molecules to be adsorbed will have to diffuse slowly through adsorbed water to reach the carbon surface, and this leads to a longer MTZ. Consequently, it is highly recommended to avoid relative humidities higher than 70%, and this can be easily achieved in most cases by increasing the gas stream temperature (see also Section 4.5).

When an AC has to be selected (from several potential candidates) for gas-phase applications, certain criteria have to be met:

(a) *Particle size.* The pressure drop in the gas stream strongly increases as the adsorbent particle size decreases. Normally, GAC or extruded AC with particle sizes between 0.8 and 4 mm are used.

(b) *Gas stream velocity.* Generally, the MTZ is shorter for low gas flow rates. However, the adsorption becomes very slow for low rates, and larger facilities are then needed to achieve the treatment capacities required. Gas flow rates normally used range from 5 to 40 $cm^3 s^{-1}$.

(c) *PSD.* A large micropore content enhances the adsorption capacity of the AC, but lack of well-developed macroporosity leads to a longer MTZ, because the adsorbate diffusion to the internal microporosity becomes slower.

(d) Determination of adsorption capacity using equilibrium isotherms.

(e) Kinetics of adsorption.

(f) Comparison of several carbons to choose the best one.

8.2.2 Gas Purification

AC in different forms (granular, extruded, fiber or cloth) is used for production of pure gases in the chemical industry, to reduce pollutant gases to very low concentrations in a single stage, in protection against poison gases, in air conditioning, for removal of oil from compressed air, etc. The actual processes, using AC, include hydrogen sulfide removal from sour natural gas, removal of sulfur dioxide and nitrous oxides from flue gas, gasoline vapor recovery in gasoline loading facilities, chlorofluorocarbon recovery in foam blowing, removal of mercury vapor from air, hydrogen and other gas streams, cigarette filters, military uses, nuclear uses and automotive evaporation control systems.

The use of carbonaceous adsorbents for gas purification, by H_2S removal, benefits from two important properties of these materials: (i) AC is an effective catalyst for the direct oxidation of H_2S with air and (ii) it is a powerful adsorbent for sulfur oxides. The combination of these two properties facilitates further steps in the industrial process, such as the regeneration of the adsorbent and the extraction of the sulfur compounds.

Carbons impregnated with caustic materials such as KOH and NaOH are widely used to control odors of H_2S and organic mercaptans in sewage treatment plants. However, there is a risk of bed self-ignition, likely due to the high heats of reaction. Alternatively, the adsorption capacity for H_2S can be improved by impregnating the AC with compounds such as KI or $KMnO_4$, which promote the oxidation to elemental sulfur. In both cases, the impregnation process decreases the adsorption capacity, as the microporosity may be blocked. Therefore, the development of a non-impregnated carbon adsorbent for H_2S removal would be of interest. It has been reported that the combination of a suitable PSD and the presence of polar groups on the carbon surface (containing oxygen and phosphorous) contributes to the process of H_2S retention, Bandosz (1999). These results have been explained on the basis of the adsorption of water on surface groups, which enhances the oxidation of H_2S to sulfur or even to sulfuric acid.

Boudou *et al.* (2003) introduced nitrogen groups into a viscose-based ACC by reaction with ammonia/air at 300 °C and by reaction with ammonia/steam at 800 °C. Extensive surface characterizations were carried out. Ammonia/steam treatment was more effective for the adsorption of H_2S or SO_2 by enhancing the microporosity and by modifying the distribution of the surface oxygen complexes. A series of successive adsorption–regeneration cycles showed important differences between the oxidation retention of H_2S and SO_2.

The first process for the simultaneous dry removal of SO_2 and nitrogen oxides (NO_x) was developed in the 1960s by the Bergbau Forschung Co. (now DMT) in Germany, by using a two-stage moving bed with activated coke (Knoblauch *et al.* 1981). The flue gas is denitrified in the first bed by reduction of NO_x with ammonia selective catalytic reduction (SCR) and then de-sulfurized in the second bed. Sulfur dioxide is retained as H_2SO_4 on the activated coke, and it is recovered as SO_2 by heating at 400–550 °C. Simultaneously, the activated coke is regenerated and recycled to the first bed.

Mochida *et al.* (2000) reviewed the applicability of ACFs for the removal of SO_x and NO_y. The highest activity for SO_2 removal in the presence of water, at 25 °C, was achieved with pitch-based ACFs, and this activity could also be improved by heat treatment of the ACFs in nitrogen at temperatures ranging from 600 to 900 °C. For NO_x removal by reaction with ammonia, it was shown that the activity at room temperature of the ACFs was markedly increased if the ACFs were previously oxidized to form surface oxygen complexes. A mechanism was proposed on the basis of NO adsorption on the defect sites formed upon decomposition of surface oxygen groups, and subsequent oxidation to NO_2, which reacts with adsorbed ammonia.

Volatile organic compounds (VOCs) are a group of low-boiling-point compounds, with different chemical properties and are generally toxic, which are produced in the petrochemical industry, food processing, wastewater treatment and electronic industries, etc. Benzene, toluene, xylenes, hexane, cyclohexene, thiophene, diethylamine, acetone, acetaldehyde and methyl ethyl ketone are examples of VOCs.

They can be removed by catalytic combustion, but the low concentrations in which they are present in the gas streams to be treated make it necessary to add energy to keep the combustor at the operating temperature. To overcome this, a process by which the VOCs are concentrated by adsorption, and then desorbed and burnt by catalytic combustion, has been proposed (Meeyoo *et al.* 1998).

VOCs can also be removed by adsorption processes (vapor recovery). In this case, ACCs are used in preference over GAC because they are more easily contained, have faster adsorption kinetics and higher adsorption capacities, and can be regenerated, *in situ*, by electro-thermal methods (resistive heating). ACCs have been chemically modified by treatment with ammonia (to introduce basic nitrogen complexes), chlorine (to introduce polar —Cl groups) and nitric acid (to introduce acidic oxygen complexes). In this way, the water vapor adsorption capacity can be tailored to obtain ACCs with enhanced adsorption of individual VOCs in the presence of humidity.

Some VOCs, as well as other air pollutants, are emitted from automotive vehicles. They can occur in the vehicle exhaust emissions, but also can be produced by evaporative emission from the fuel system. These evaporative emissions are fuel vapors, generated and released from the vehicle's fuel system, which depend on the interactions of the specific fuel in use, the fuel system characteristics and environmental factors.

The AC required for use in automotive applications must adsorb gasoline vapors efficiently, and also must be able to release them during the regeneration cycle. The regeneration conditions are mild, and so the adsorbent–adsorbate interaction must not be too strong. The most effective pores for this purpose lie in the mesopore range. Some of the automotive-grade AC produced by Westvaco are WV-A 900 and WV-A 1100 (granular) and BAX 950, BAX 1100 and BAX 1500 (pelleted). They have BET surface areas from 1400 to 2000 $m^2 g^{-1}$ and apparent densities of 0.2–0.35 $g cm^{-3}$.

The fuel vapors evaporated from the carburettor or from the fuel tank are adsorbed on to AC (0.5–2.0 L) contained in a canister. Recovery of the fuel is carried out by drawing an air stream through the canister. The fuel desorbed is thus taken to the engine and mixed with air, to be burned during normal operation. Other systems have been designed to capture fuel vapors emitted during vehicle refueling and when tank trucks are filled and drained.

Lillo-Ródenas *et al.* (2005) studied the adsorption of benzene and toluene (VOC) on 10 AC, 2 commercial from wood, 1 from bituminous coal using steam activation, 5 from anthracite by chemical activation using NaOH and KOH and 2 from bituminous coal by chemical activation using NaOH and KOH. An extensive review of relevant literature is provided. It is concluded that the most important property controlling adsorbate uptake is the narrow microporosity, <0.7 nm. Removal of oxygen surface functionality enhances the adsorption of the benzene and toluene which can reach as high 34 and 64 $g\,g^{-1}$ for benzene and toluene, respectively, from low concentrations of 200 ppmv.

8.2.3 Separation of Gas Mixtures: CMS

CMS, as their name suggests, are designed to separate gases in mixtures by sieving, a process controlled by PSD. CMS are widely used for the separation of air into nitrogen and oxygen. CMS can be prepared from a number of carbonaceous precursors and by several preparation methods, although those to be used industrially for the separation of air are generally prepared from AC, with a post-treatment (chemical vapor deposition) to narrow their porosity. Taking into account that the molecules to be separated have very similar sizes (kinetic diameters 0.346 nm for O_2 and 0.364 nm for N_2), the PSD in these materials has to be precisely tailored in the 0.3–0.4 nm range to within 0.02 nm.

Whereas zeolites adsorb nitrogen preferentially and produce an O_2-rich stream at the exit of the separation device, CMSs separate the components of air on a kinetic basis. Although the adsorption equilibrium isotherms for the two gases are nearly the same, oxygen is adsorbed faster than nitrogen.

The industrial practice of *pressure swing adsorption* (PSA) has become one of the most-applied techniques for the separation of gas mixtures in general, and separation of nitrogen from oxygen in particular (Yang, 1987). A PSA unit generally runs a series of pressurization–depressurization (regeneration) steps in a cyclic fashion. Compressed air is introduced into a column containing the CMS, and a N_2-rich stream is produced. When the adsorbent is saturated with oxygen, the air feed is stopped, and the column is regenerated by depressurization to remove the adsorbed oxygen. Then, the air stream is again fed and a nitrogen-rich product is obtained. In practice, two adsorbing beds are used, in such a way that air is fed to one bed while the other is regenerated. The two beds are linked by valves, and the feed is diverted back and forth between beds to produce a constant flow of nitrogen-rich product. The better understanding of the relationships between the PSD of the CMS and the performance of the PSA process has favored the preparation of CMSs with higher adsorption capacities, higher N_2–O_2 selectivities and higher adsorption rates, and this has increased the purity of nitrogen obtained from PSA processes.

Methane–carbon dioxide mixtures can also be separated using AC (Kyotani, 2000). There are three main applications where this process is of interest, including treatment of landfill

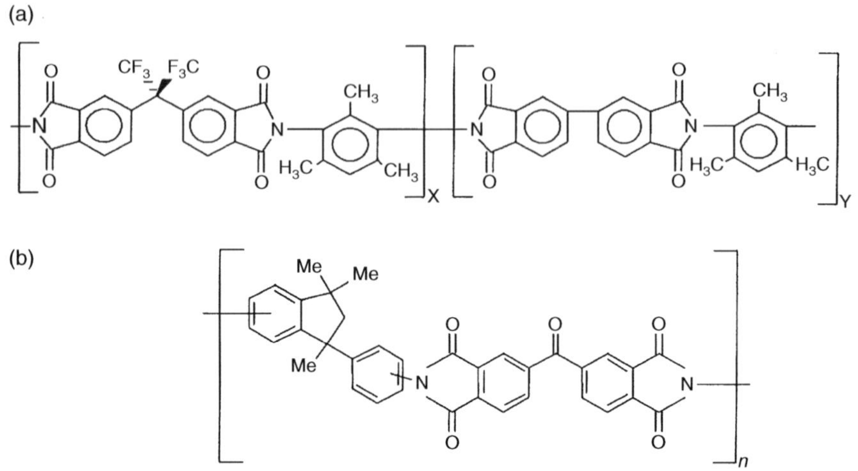

Figure 8.15. Structures of precursors for CMS developed by Steel and Koros (2005). (a) 6FDA/BPDA-DAM and (b) Matrimid (a commercial polymer).

gas, which contains about 50% CO_2 as an impurity, purification of natural gas, which can contain up to 12% CO_2, and tertiary oil recovery, where the effluent gas contains variable CO_2/CH_4 ratios.

The process for obtaining clean natural gas from landfill includes a first purification treatment with AC, followed by separation with CMS in a PSA unit. As for air, separation of methane from carbon dioxide is based on adsorption kinetics, the smaller CO_2 molecule (0.33 nm) being adsorbed faster than the slightly larger CH_4 molecule (0.40 nm).

Steel and Koros (2005), using two polymeric precursors, as in Figure 8.15, describe how the control of HTT and soak time could be used to tailor-make carbons with desired permeabilities and selectivities for CMS. Carbons were prepared for the optimized separations of mixtures of O_2/N_2, CO_2/CH_4 and C_3H_6/C_3H_8. Adsorption isotherms of carbon dioxide at 273 K, in the range to 0.1 MPa pressure, were analyzed to understand how the PSD related to experimental conditions. Increasing HTT tends to close the smallest of the micropores once the maximum in capacity has been reached. Steel and Koros (2005) published a hypothetical (calculated using density functional theory) PSD curve for micropores, as in Figure 8.16. Here, a selected molecule is envisaged as being able to enter all porosity (interconnected) to the right of the vertical line.

8.2.4 Methane Storage

The instability in oil markets and the increase in environmental concerns have stimulated the research for alternative fuels for transportation. Among them, natural gas has become a promising choice from both the economic and environmental points of view. On the one hand, it is a fuel of considerable natural abundance and therefore, commercially attractive.

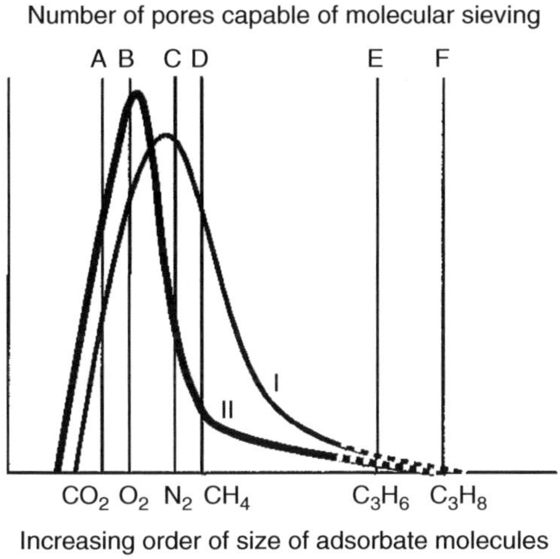

Figure 8.16. Hypothetical semi-quantitative micropore size distributions based on carbons from polymer systems (Steel and Koros, 2005).

On the other hand, it consists mainly of methane (85–95%), higher hydrocarbons, nitrogen and carbon dioxide, and thus it has inherent clean burning characteristics and a less adverse effect on the environment, with the emission of lower levels of pollutants (hydrocarbons, sulfur and nitrogen oxides, etc.), per unit of energy provided.

The main problem arising when natural gas is used as a fuel in mobile vehicles is its storage. Although it has higher hydrogen-to-carbon ratio than other fuels and consequently a greater energy per unit mass, it cannot be stored at the same density (Cook *et al.*, 1999). In a first approach, natural gas can be conveniently stored on a large scale in liquid form (liquefied natural gas (LNG)) at its normal boiling point (112 K) at atmospheric pressure. Under these conditions, the system will deliver 600 volumes of gas per volume of storage (transportation V/V). Although the delivery is very high, the energy density (defined as the heat of combustion per unit volume) is 23 MJ L^{-1}, which is low compared to the values for diesel fuel (37 MJ L^{-1}) or gasoline (32 MJ L^{-1}). Additionally, the cost of liquefaction, the special insulated vessels required and the potential fire hazard are such as to make it unsuitable for use on a small scale.

A second approach is the use of compressed natural gas (CNG). There is a possible market worldwide for vehicles fueled by CNG at pressures up to 20 MPa and room temperature. The delivery is 220 V/V (220 volume units of gas at atmospheric pressure are compressed to one volume of storage container) (Parkyns *et al.*, 1995) and has an energy density of less than 10 MJ L^{-1}. For the same driving range, a CNG vehicle using these pressures requires a storage vessel at least three times the volume of a gasoline tank. Higher pressures would help to increase the energy density, but technical difficulties concerning the storage container make this approach infeasible. Anyway, the size and cost of the cylinders needed for storage are considerable deterrents to the use of these systems on a wide scale, and the cost of providing the high-pressure facilities is also relatively high.

Natural gas can also be stored if adsorbed on a suitable material such as AC, at relatively low pressure (Cook *et al.*, 1999) (see Section 6.1.4). Adsorbed natural gas (ANG) is stored at pressures ranging from 3.5 to 4.0 MPa, with the help of an adsorbent. The density of the natural gas in the adsorbed phase is much higher than in the gaseous phase, and this compensates for the space in the storage container occupied by the solid. If an appropriate adsorbent is used, it is even possible to reach the same delivery (per unit volume of storage) with natural gas adsorbed at relatively low pressures (3.5 MPa) as with natural gas compressed at much higher pressures (20 MPa). It is evident that the cost of the storage system and the potential risks of use will be considerably reduced, thus making more competitive the use of natural gas for transportation.

Efforts have therefore been addressed to the search for a suitable porous material to further improve ANG storage volumetric energy density. The requirements for an adsorbent to have optimal storage capacity are well established, the most important being high microporosity and high density (Menon and Komarneni, 1998).

In this application methane, the main component of natural gas, is adsorbed at room temperature, higher than its critical temperature (191 K). Adsorption of supercritical gases takes place predominantly in pores that are less than two or three molecular diameters in width (Chen *et al.*, 1997). On increasing the pore width, the forces responsible for the adsorption decrease rapidly. Thus, pores larger than 2 nm (meso- and macropores) are not useful for the enhancement of methane storage, although they may be necessary for transport into and out of the micropores. To enhance the adsorption storage of methane, the fraction of micropores should be maximized, with no voids or macropore volume. However, this contribution has to come from pores with a width of around 0.76 nm (larger than the thickness of two methane molecules), in order to maximize the deliverability at ambient pressures. Macropore volume and void volume in a storage system (adsorbent-packed storage vessel) should be minimized.

There are many adsorbents that could fit these requirements (high microporosity, high density and minimum interparticle space). AC, by virtue of their very high micropore volumes, have shown the greatest promise. Furthermore, they have other properties that could enhance methane storage. AC have low heats of adsorption that prevent excessive heating or cooling of the storage vessel during the adsorption–desorption cycles. Also, they are mainly hydrophobic, so the competition between water and methane for the adsorption sites is minimized. Moreover, the costs of highly AC are of the same order as those of the cheapest zeolites. And, finally, AC have slit-shaped pores, whose importance is clearly shown in Figure 8.17, which displays the effect of pore size and shape on the packing density of spherical molecules such as those of methane. The difference in the packing density in slits and cylinders is of importance with higher packing density in slit-shaped pores.

Grand canonical Monte Carlo simulations performed for natural gas adsorbed on carbon have demonstrated that the optimum pore size is 0.76 nm, as mentioned earlier (methane adsorption (capillary condensation) does not takes place at room temperature); thus, a further increase of the slit width would lower the particle density without a significant increase in amounts adsorbed. These calculations predict that the theoretical maximum methane storage capacity of carbon at 3.5 MPa is 209 V/V for a monolithic carbon that fills the vessel and contains a minimum amount of macropore volume and no external

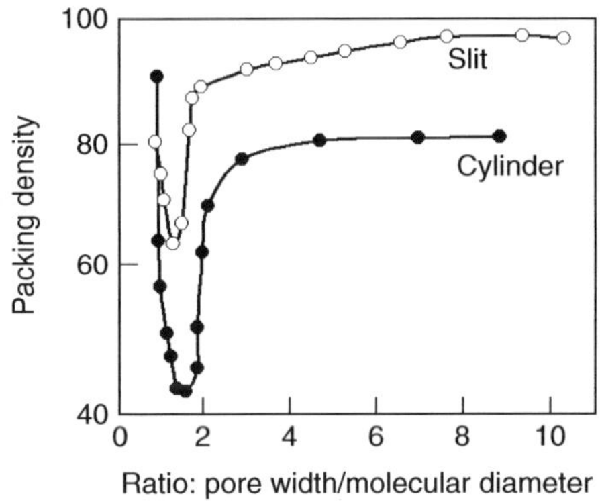

Figure 8.17. Packing of spherical molecules in narrow cylinders and slits (Rouquerol *et al.*, 1999; Rodríguez-Reinoso and Sepúlveda-Escribano, 2001).

void space. Similar calculations applied to a pelletized carbon predict a value of 146 V/V, which is considerably lower than that of the monolithic carbon. The difference is due to the different volume fraction distributions in the two physical forms of AC. The volume of an AC can be divided into three main parts which include the carbon atoms (skeleton), the micropores and the voids (the meso- and macropores and the interparticle space). The carbon atoms do not contribute to the adsorption – on the contrary, they diminish the volume available for the gas – but they determine the density of the carbon. The meso- and macropores and the interparticle space can be grouped in one fraction, the void, because the density of the gas there is the same as in the gaseous phase; therefore, this fraction has to be minimized. Finally, the micropore fraction is where the methane adsorption takes place, and thus this fraction has to be maximized.

The volume fractions in a GAC, such as Calgon BLP, are made up of void 60%, carbon 23% and micropores 17%. In a monolithic AC, such as Saran B, the volume fraction is: void 30%, carbon 38% and micropores 32%. It is then clear that the AC has to be used in the monolithic form to achieve a reasonable level of natural gas storage.

In the early 1990s, the US Department of Energy set a target figure of 150 V/V deliverable for an operational ANG vehicle system, working at 3.5 MPa and 25 °C. This value was considered demanding, but realistic and achievable. In fact, some research groups have already attained it. Different methods have been used to prepare the AC, but all of them agree on the need for a final monolithic form of the carbon. Burchell *et al.* (2005) used a composite of carbon fibre based molecular sieves for air separations. They obtained a storage capacity around 150 V/V at 23 °C. Rodríguez-Reinoso *et al.* (2003) reported an experimental storage of 150 V/V at 25 °C with AC prepared from lignocellulosic precursors by chemical activation, without any binder.

Rather than prepare monoliths, directly, for methane storage, Celzard *et al.* (2005) prepared chemically AC (using NaOH) from an anthracite from a German (Sophia) coal mine, by compacting granular size fractions at pressures up to 220 MPa. It was found that although compaction increased the bulk density of the material, there was no effect on adsorption capacities. Compaction decreased the inter-particle void space but could not affect the microporosity which is responsible for the adsorption of the methane. At the optimal compaction pressure of ~100 MPa, a maximum storage capacity of 193 V/V was obtained, being 163 V/V at 3.5 MPa gas pressure at 30 °C.

Another approach for methane storage is to use the natural gas hydrate (NGH) as the adsorbate in AC. Practical difficulties of gas release precluded any commercial applications of adsorbed natural gas (ANG). However, there appear to be some advantages in adsorbing methane into AC which are pre-loaded with water (Zhou *et al.*, 2005). These authors showed that methane hydrate formed quickly within the carbon at 273 K and >4.12 MPa pressure. A packing density of 0.6 $g\,cm^{-3}$ was obtained, yielding 152 V/V of released methane at a charging pressure of 8 MPa. During the charging/discharging procedures, thermal effects were low and did not affect the effective storage capacity. An AC from coconut shells was used, possessing a surface area of 2585 $m^2\,g^{-1}$ with a pore volume of 1.13 $cm^3\,g^{-1}$ almost entirely made up of pores 2.0 nm and 2.6 nm width.

8.2.5 Solvent Recovery

The main industrial method of recovering solvent vapors from air streams is adsorption by AC. This method has the advantage over water scrubbing that it can be applied to water-insoluble solvents; it has the advantage over condensation that the solvent content of the air can be reduced as much as desired. AC recovery systems are used in the paint, petrochemical, printing, rubber, synthetic fiber, paper, adhesives, metal fabrication, solvent manufacturing and coatings industries. Typical solvents recovered by AC are acetone, benzene, ethanol, ethyl ether, pentane, methylene chloride, tetrahydrofuran, toluene, xylene, chlorinated hydrocarbons and aromatics. An important use lies in the recovery of acetone from cellulose acetate spinning and film production. For every kilogram of acetate or triacetate produced, 3 or 4 kg each of acetic acid and solvent must be recovered to make the process economically viable.

8.3 Liquid-Phase Applications

8.3.1 Introduction

It is impossible, in a single text, to describe adequately the large number of liquid-phase applications allocated to AC. The following list indicates the wide-ranging scenarios for AC. Selected topics are dealt with more fully below:

(a) Drinking water availability, to improve taste, smell and color including removal of chlorinated compounds and other VOCs.

(b) Improvements to ground water purity, contaminants coming from disused sites of heavy industries.

(c) Treatments of both industrial and municipal wastewater.

(d) Mining operations require feed water treatment, metallic ion adsorption (gold and other metals), adsorption of excess flotation reagents and adsorption of NOM.

(e) Pharmaceutical processes, including purification of process water, use with fermentation broths and purification of many products.

(f) The food, beverage and oil industries for removal of small, color and unacceptable tastes.

(g) The dry-cleaning industries require purification of solvents.

(h) The electroplating industries require purification of wastewaters containing Pb, Cr, etc.

(i) Household water purification, cleaning of aquaria and use in oven-extract hoods.

(j) The sugar and sweetener industries need decolorization agents for the production of white sugar, etc.

Both GAC and powdered AC are used in liquid-phase applications, the former being used in continuous processes because they are capable of regeneration, whereas the latter are generally used in batch processes (after completion, the carbon is separated from the liquid and discarded or eluted). Almost 60 wt% of the AC used in the USA for liquid-phase applications is in powdered form. Liquid-phase applications require AC with a larger pore size than gas-phase ones, because of the need for rapid diffusion of the liquid to the interior of the carbon particles and because of the large size of many dissolved molecules to be retained. It is usual for AC for liquid-phase applications to be prepared by chemical activation of wood, peat, lignite, etc. There are two types of liquid-phase applications for AC: odor, color, or taste removal from a solution, and concentration or recovery of a solute from solution including water purification and sugar and sweetener decolorization as principal applications.

8.3.2 Water Treatment

The application of AC in water treatment is mainly centered in the removal of pollutant organic compounds. These compounds can be classified in three different categories: (1) NOM, (2) synthetic organic compounds and (3) by-products of chemical water treatment. NOM is mainly composed of residues of the metabolism of living things. These compounds produce bad tastes and odors, and also may constitute a source of infection. Among the synthetic organic compounds that can be present in water, one can find oil, benzene and toluene, phenols and chlorophenols, trichloromethane and carbon tetrachloride, detergents, pesticides, dyes, surfactants and so on. Finally, trihalomethanes are the most important group of compounds that are to be found in water as a product of chemical treatments to disinfect water with chlorine: $CHCl_3$, $CHBrCl_2$, $CHBr_2Cl$ and $CHBr_3$. They are very strongly adsorbed on AC, and this is the reason for the increasing number of potable-water plants using AC as the polishing step.

The removal of water contaminants by active carbon is the major market (55% in the USA) for liquid-phase applications, which in turn constitute ~80% of total AC demand in the USA. AC is used both as a primary treatment, to render the effluent more amenable to other purification processes, and as the final tertiary stage in the purification of the effluent. Of the total US water treatment market, about 50% is in drinking water, 40% in wastewater and the rest in ground water markets. Both powdered and GAC are used in water treatment, the tendency being toward use of the granular type because of its regeneration capability (Rodríguez-Reinoso, 2002).

When the powdered form of AC is used, it is added to the water as slurry with automatic feeders. After a suitable contact time, it is removed by clarification or filtration. Dosage rates of AC in taste and odor control depend on the type of carbon and the level of impurities in the water, but in general terms the dosage is low, and the carbon can last for up to one year. As a result, it is not usually economic to regenerate the carbon, and spent carbon is generally discarded.

GAC is preferred when there is a persistent problem with taste and odor control, and it is also used in special filters and disposable cartridges in industrial, commercial and residential installations. GAC is used in gravity columns, through which water flows continuously for a set contact time. Contacting systems can be of the up-flow or down-flow type, the former adsorbing organic compounds, whereas the latter filters suspended solids in addition. In an up-flow system, replacement of the spent carbon is carried out from the bottom of the column, with addition of new carbon at the top, while the unit remains in operation. In a down-flow system that does not have pre-filtration, suspended solids may accumulate at the top of the bed, requiring periodic back-washing of the bed to relieve the pressure drop caused by the accumulated solids. This type of bed is operated in series or in parallel. As the carbon will be exhausted first at the top of the bed, it is necessary to remove the entire bed in order to replace the carbon. The world's largest municipal GAC potable-water treatment system was installed in Cincinnati, Ohio in 1989 with a potential capacity of over 830 million litres per day.

Wastewater is treated in four stages.

1. The first stage, or pre-treatment, is carried out when the wastewater contains toxic or non-biodegradable compounds that can affect the subsequent biological treatments. Redox reactions, followed by precipitation and filtration are used to separate the metals contained in the wastewater. Ozone treatment followed by adsorption into GAC is used to eliminate high molecular weight organics, whereas light organics and ammonia are eliminated by air-stripping.
2. The secondary treatment involves the removal of suspended solids, non-soluble oil and floating material by treatment with lime or other chemicals, followed by nitration, as well as neutralization by addition of acids or bases.
3. Dissolved and colloidal organic compounds like proteins, sugars, starches and phenols are removed in the secondary treatment by biological oxidation. These two processes remove about 85% of the suspended solids and of the biological oxygen demand (BOD) of wastewater. For many purposes, however, further purification is needed to satisfy stringent effluent regulations.
4. This treatment involves the removal of inorganic and organic compounds by adsorption. It results in an extremely high-purity effluent, where the BOD can be reduced by over 99%, to 1 mg L^{-1}.

There are three possible locations in a wastewater treatment plant for treatments with AC. This can be used as an adsorbent after primary and secondary biological processes, it can be used as an independent physico-chemical treatment, or it can be added to biological aeration tanks and used as part of the secondary biological treatment. This last choice has been used effectively to obtain a high-quality effluent. The selection of an appropriate treatment depends on the nature and contaminant loading of the wastewater, the scale of operation, specific requirements for effluent purity and the cost of carbon regeneration compared with alternative available treatments.

The use of tertiary AC processes on wastewater that has already undergone conventional secondary biological treatment results in a very high-purity effluent. These processes consist of packed beds of granular material arranged either for down-flow in series, for down-flow in parallel, for up-flow in series, or in moving beds. The efficiency of a tertiary treatment depends on the consistent and efficient operation of the previous treatment. Changes in wastewater composition, large variations in flow and the presence of toxic material can all disrupt the biological oxidation process. One method of improving efficiency is to treat the influent waste streams with ozone prior to treatment with AC, as is widely done in Western Europe.

AC is also used in the treatment of industrial wastewater to upgrade the water for reuse or to pre-treat effluents prior to discharge into municipal treatment plants, rivers and streams. Adsorption by AC may be used as the only treatment before biological treatment or as a tertiary process after biological treatment. AC is used to purify industrial wastewater, as it removes not only biodegradable organic compounds, but also chemicals that are not responsive to, or are toxic to, conventional biological treatments. These include pesticides, phenols, organic dyes, detergents and polyols. AC is used to treat effluent wastes from chemical factories, rubber tread factories, fabric dyeing, fertilizer plants, pulp and paper mills, etc.

AC systems are more flexible than biological ones as they can handle sudden fluctuations in the concentration of impurities, and the water purity can be controlled to meet specific requirements.

Another important use for AC is the removal of oil from effluent water in petroleum refining, petrochemicals, metal extraction, detergent, margarine and soft fat manufacture and mineral extraction. The presence of oil in the effluent inhibits the biological treatment in sewage works. AC is also used to remove oil and organic material from recycled condensate in boilers, because water contaminated with oil would cause foaming or priming even in low-pressure boilers. In high-pressure boilers, where the feed water is de-ionized, the oil causes serious fouling in the ion-exchange resins.

The selection of the AC for a given application (solutes to be retained, concentration, stream flow, etc.) is made by dynamic tests in pilot adsorption units. However, it is more effectively done, with less time consumption, if the adsorption properties of the candidates are established. Properties such as the surface area and PSD, iodine number, bulk density, particle size, hardness and mechanical strength give an idea of the suitability of the AC. The determination of the adsorption isotherm of a given solute over a wide range of concentrations is the method to obtain the equilibrium capacity of the adsorbent. In this way, the amount of solute it can remove from the solution can be estimated and consequently the suitability of the AC for achieving the required removal of the solute.

The analysis of the adsorption isotherm gives the adsorption capacity at equilibrium, but it says nothing about the time required to reach equilibrium, and this is a very important parameter in the design of an adsorbed bed. Once a group of carbons has been selected, it is necessary to carry out dynamic tests to determine the breakthrough curve to determine the volume of water that can be treated by the carbon bed before it becomes saturated, or before the concentration of the solute in the effluent (at the exit of the adsorbent bed) becomes higher than that required (the breakthrough point).

The breakthrough curve in an adsorption system depends on several parameters. On the one hand, it depends on the space velocity of the stream (the ratio between the volumetric flow and the volume of the bed); it is clear that the adsorbent bed can be used for a larger period of time with low space velocities. But, the breakthrough curve also depends on the characteristics of the stream to be treated (composition, concentration of impurities, temperature, pH) and on the adsorbent (surface characteristics, particle size). As an example, the adsorption kinetics is highly favored by decreasing the particle size of the carbon. It results in a shorter MTZ, and thus, the carbon adsorbent is better exploited.

After the AC becomes saturated, it is sometimes necessary to regenerate it in order to recover its adsorption capacity. Physical adsorption is a reversible process, and desorption can be more or less easily achieved. Regeneration is usually carried out by means of a thermal process. The exhausted carbon is taken off the adsorption column as slurry. It is dewatered and passed to a rotary kiln, where it is heat treated under controlled conditions and with a limited oxygen content to avoid carbon combustion. This treatment removes residual water and volatilizes organic compounds, which are also oxidized. Then the carbon is quenched with water, washed and recycled. The carbon is regenerated either on or off site. Some carbon losses (2–10 wt%) are unavoidable during the regeneration step, and fresh carbon has to be supplied. The regeneration process takes about 30 min.

8.3.3 Food and Beverage Processing

AC is used in food and beverage industries to remove color or odor from products. Applications include the processing of fruit juices, honey, sugar, sweeteners, vegetable oils and fats, alcoholic beverages, soft drinks, yeast, maple sirup, etc. The use of AC for these applications is either declining or growing at a low rate, due to new practices in advanced countries or to competition from cheaper materials like kieselguhr and bentonite.

8.3.4 Chemicals and Pharmaceutical

By removing impurities in chemical processes, AC helps to control product quality; it is also used for the removal of toxic chemicals. AC is used to extract pharmaceuticals in processes involving fermentation. Antibiotics, vitamins and steroids are adsorbed into the AC and recovered by solvent extraction followed by distillation. Other uses include filters for the dialysis of poisons and drugs, felts for wounds, etc.

8.3.5 Adsorption of Dyes

It is estimated that about 40,000 tonnes of dyes are not used but are discharged (lost) into wastewaters. This is out of a total production of about 450,000 tonnes. A large variety of dyestuffs is available with such names as acid, basic, reactive, direct, disperse, sulfur and metallic dyes. Losses are at a minimum (5%) in the dyeing of acrylic fibers but can be as high as 50% for the use of reactive dyes with cottons. Pereira *et al.* (2003) review much of the earlier literature of adsorption of dyes by AC. This appears to have been confined to studies of adsorption behavior in terms of surface areas and pore sizes. As a result, little progress was made in understanding of adsorption behavior by dyes. Pereira *et al.* (2003), being aware of the importance of surface functionality of carbons in adsorptions from solution (e.g. phenols from aqueous systems), as well as in catalysis on carbon surfaces, explored this aspect using dyes and AC with surfaces modified by oxidation and reduction procedures.

These authors were able to conclude that surface functionality has a dominant role in the adsorption of dyes, with significantly different acidities and basicities. Thermal treatments of the AC offered a means of controlling surface functionality, with reaction with hydrogen at 700 °C being the best for almost all of the dyes used. For anionic dyes, surface basicity related directly to dye adsorption. This basicity arose from oxygen-free Lewis base sites related to the delocalized π-electrons of the defective graphene layers. As to be expected, for the basic dyes (cationic) the carboxylic acid groups of the carbon surface were effective. Nevertheless, the interactions with the π-electrons of the graphene layers still played a dominant role.

These conclusions are supported by Valix *et al.* (2004) for the adsorption of Acid Blue 80 dye, using carbonized sugar cane (Bagasse) as the adsorbent.

The estimated molecular dimension of this dye is 1.65×0.626 nm, with a molecular formula as in Figure 8.18. Despite the clear indications of the importance of surface functionality of the AC, access to these surfaces is still of paramount importance and that means access into the porosity, and this means that wider microporosity and smaller mesoporosity must also be a necessary characteristic of the AC.

8.3.6 Other Applications

AC has many smaller-scale applications such as industrial dry cleaning, as well as in coin-operated dry-cleaning machines, cleaning of electroplating solutions, public and private

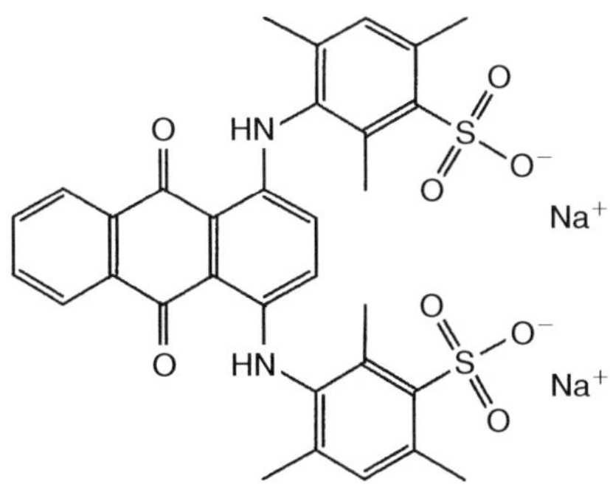

Figure 8.18. The structural formula of the dye Acid Blue 80, manufactured by CIBA Speciality Chemical (Valix *et al.*, 2004).

aquariums, decaffeination, etc. Medical applications include oral ingestion into the stomach to remove poisons or toxic materials.

8.4 Porous Carbons in Catalytic Processes

8.4.1 Introduction

Much that has been written earlier, about the properties of AC, finds relevance to the understanding of the use of carbon as a catalyst support and as a catalyst in its own right (Rodríguez-Reinoso, 1998). Section 4.2 characterizes porosities in carbons, with special reference to the dominant importance of microporosity. The chemistry of functionality attached to the graphene layer, both in-plane and at edges, is immediately relevant to the use of carbons in catalytic processes (Section 4.3). The amphoteric nature of the graphene layer, for example, as discussed with reference to adsorption of anions from aqueous solution, has a role in catalytic processes (Section 8.1.3).

Catalytic processes account for >90 of the chemical manufacturing processes in use throughout the world, the catalyst increasing the rate of a reaction and controlling its selectivity. Carbon materials have been used for some time in heterogeneous catalysis, acting as direct catalysts or as a catalyst support. The Catalytic Reaction Guide, published by Johnson Matthey, one of the world's leading suppliers of catalysts, includes a list of 69 organic reactions catalyzed by precious metals, nine of which use carbon as the only support material. The number of industrial processes using carbon as catalyst is however, very small, although there is a clear and increasing interest in this area, as denoted by the rapid growth in the literature. It is now recognized that carbon as a catalyst support, as well as a catalyst in its own right, offers unparalleled flexibility in tailoring catalyst properties to specific needs. However, the future growth of the use of carbon materials in catalysis depends on the better understanding (and subsequent control) of the chemistry of carbon surfaces.

The presence of oxygen and hydrogen in surface groups (values of up to 30 mol% hydrogen and 15 mol% oxygen are not unusual) has a major effect on the adsorptive properties of the AC. The origin of these surface groups may be the original raw material, the activation process or introduction after preparation via post-treatment. Thus, although PVDC, phenol–formaldehyde and other polymers produce carbons free of oxygen, and with only traces of hydrogen, these carbons always chemisorb atmospheric oxygen, producing oxygen surface groups. Oxygen surface groups are by far the most important factor in influencing the surface characteristics and adsorption behavior of AC, and as such they have attracted much attention in the last few years.

The precise nature of carbon–oxygen structures is that there are several types of oxygen functionality (Section 4.3). The presence of these surface groups gives to AC an acid–base character. Yet, the acid–base adsorption properties of carbons can be varied by different pre-treatment conditions.

Carbon structures offer an unparalleled flexibility for tailoring catalyst properties to specific needs, and this flexibility is a consequence of several characteristics such as (a) porous structure which determines the active phase loading, (b) a chemical structure which influences the interaction with molecules of different nature and (c) a wide range of active sites, as summarized below:

1. the carbon structure is resistant to acidic or basic media.
2. the structure is stable at high temperatures (even above 1000 K), in the absence of air.
3. the pore structure (constituted by slit-shaped pores, something very important when shape selectivity is a must in catalysis) may be tailored to obtain the PSD needed for a given reaction.
4. porous carbons can be prepared with different physical forms (granules, pellets, extrudates, cloth, fibers, etc.).
5. although carbon is usually hydrophobic, the chemical nature of the surface can be modified to increase the hydrophilicity, and even carbons with ion-exchange properties can be prepared.
6. the active phase can be easily recovered from spent catalysts by burning away the carbon support (very important when the active phase is a precious metal).
7. the cost of carbon supports is usually lower than conventional supports such as alumina and silica.

For a long time, a drawback of AC, as catalyst supports, was the presence of mineral matter, arising from either the precursor or introduced during the preparation process. It is true that many AC have ash contents as high as 10–15 wt%, especially if they are prepared from coal, but it is today possible to prepare porous carbons with very low ash content by selecting an appropriate precursor.

The preparation of carbon-supported catalysts is mainly carried out by contacting the support with a solution containing the precursor of the active phase (excess solution of incipient-wetness). In this process, different types of interactions between the carbon surfaces and the active phase precursor can arise, ranging from weak dispersion forces to electrostatic interactions, ion-exchange or chemical reactions. Thus, adsorption capacity is determined by both the porous structure and the chemical nature of the surface. In this way, whereas adsorption on a clean graphite surface is determined by dispersion forces, the random ordering of crystallites and the presence of heteroatoms in AC results in the creation of active sites which influence the adsorption behavior. As a consequence of this, carbons having the same surface area but prepared by different methods may show markedly different adsorption characteristics. Fortunately, the last few years have seen a considerable increase in the understanding of relationships between the physical and chemical properties of carbon materials and their catalytic effectiveness, thus paving the way toward their industrial use.

8.4.2 Carbon-Supported Catalysts

Many commercial catalysts consist of metals (e.g. platinum and palladium) supported on a porous solid, sometimes carbon. High surface areas and a well-developed porosity are essential for high metal dispersions which are very often necessary for high catalytic activity, more so if a promoter is used with the catalyst. The surface area and porosity available in AC are usually much larger than those available in alumina or silica, with a large proportion of the surface area being contained within micropores.

There are many examples of the effects of porosity and surface area on catalytic activity. Thus, in a study with graphitized carbon black (V3G), subject to different degrees of gasification in air to increase the surface area, it was found that as the surface area of the original sample increased from 62 to $136\,m^2\,g^{-1}$, the dispersion of the Pt/C catalysts (dispersion is the fraction of metal atoms that are on the surface of the particle in respect to the metal loading) increased from 0.17 to 0.35. For a completely different metal, molybdenum, supported on AC (prepared from the same precursor but with increasing porosity and surface area), the activity for thiophene hydro-desulfurization (HDS) increased with surface area up to *ca.* $1200\,m^2\,g^{-1}$, remaining constant thereafter (metal loading was constant for all catalysts).

There exists a correlation between support affinity for sulfur and HDS catalytic activity and it is suggested that, in contrast to conventional alumina supports, the narrow slit-shaped micropores in AC, while inaccessible to bulky reactant molecules, lower the vapor pressure of sulfur to such an extent as to create a driving force for sulfur transfer from the active compound to the micropores (sink effect), thus generating active vacancies. It is also suggested that the pore shape is a necessary condition for the sink effect, probably because the slit-shaped pore allows the sulfur accumulated between the pore walls to react with hydrogen from any side of the pore, to desorb as H_2S. This is not possible in cylindrical micropores of comparable dimensions; in fact, the sink effect was not observed for microporous silica.

That in many cases the catalyst properties are a function of porosity and surface area is relatively straightforward, because the microporosity responsible for the high surface area of AC may act as a physical barrier to the deposited metal particles so facilitating dispersion and preventing sintering.

There are, however, many examples showing that neither dispersion nor catalytic activity are a function of the porosity and surface area of the carbon support. Thus, Rodríguez-Reinoso *et al.* (1990), in an attempt to clarify the relationship between the carbon support porosity and catalytic activity, used a series of AC of increasing porosity and surface area to prepare Fe/C catalysts using two different metal precursors: iron nitrate and iron pentacarbonyl, the first one dissolved in water and the second in n-pentane. Results indicated that there is an increase in iron dispersion for the nitrate series, while high and unaffected dispersion was found for the carbonyl series. However, iron-loading on catalysts is a function of the metal precursor used, thus showing that the access of the metal solution is related not only to the increasing porosity of the carbons, but also to the chemical character of its surface and its interaction with the different solutions.

On the other hand, the catalytic activity for CO hydrogenation did not differ by much for both series of catalysts, thus showing that, in this case, there is no correlation with the porosity or surface area of the supports.

These and many other examples of carbon-supported catalysts show that an accessible and large surface area coupled with the right PSD is a necessary but not a sufficient condition for the preparation of adequate catalysts with maximum dispersion. They are not sufficient to explain many of the properties of the carbon-supported catalysts.

It has not been until the last few years that the importance of the role of the chemical nature of the carbon surface has become widely recognized.

8.4.3 Effect of Carbon Surface Chemistry

According to the above comments, it is important to recognize the role of the chemical nature of the carbon surface in the preparation of optimized catalysts. Some examples will be used to illustrate this important effect of the carbon surface chemistry.

It was realized, as from the mid-1970s, that neither the support surface area nor its pore structure were sufficient to explain many of the properties of carbon-supported catalysts, and it was in the late 1980s when the neglected importance of carbon surface chemistry started to be analyzed in depth. Although carbon is considered to be an inert material in comparison with other catalyst supports such as alumina and silica, its surface has a proportion of active sites, made up of unsaturated valences at the edges and defects within the constituent graphene layers. The proportion of these active sites increases as porosity and surface area increase. The presence of heteroatoms (mainly oxygen, hydrogen and nitrogen) also introduces active sites on the carbon surface, and consequently the carbon surface is not as inert as first thought. From the point of view of catalyst support, the presence of oxygen functional groups on the carbon surface is now well understood.

The distribution of a metal such as Pt on a carbon support surface depends on the solvent used for impregnation. Since carbon is essentially non-hydrophilic in nature, it has a very low affinity for solvents of polar character such as water and a high affinity for non-polar solvents such as acetone. The metal precursor, chloroplatinic acid, is mostly located at the external surface of the carbon particle when using water, but it will penetrate to the interior of the porosity when using acetone, thus leading to a more uniform distribution throughout the particle.

Derbyshire *et al.* (1986) and Prado-Burguete *et al.* (1989, 1991) examined and clarified the effect of oxygen surface groups on the dispersion and sintering of Pt/C catalysts. The latter selected a high surface area carbon black, which was heat treated in hydrogen at 950 °C to reduce the oxygen surface groups to a minimum, and then oxidized with solutions of H_2O_2. In this way, several supports with the same porosity but different proportion of oxygen surface groups were available. These authors concluded that the acidic groups (evolving as CO_2 upon TPD at relatively low temperature) introduced by treatment with H_2O_2 decreased the hydrophobicity of the carbon thus making the surface more accessible to the aqueous solution of the metal precursor during impregnation. Platinum dispersion increased with increasing amount of oxygen surface groups and this in turn indicates that the oxygen surface groups have relevance, other factors being constant. Furthermore, the less acidic oxygen groups introduced (evolving as CO at high temperature using TPD) increased the interaction of the metal precursor or the metal particle with the support and thus minimized the sintering propensity of the Pt/C catalysts. The more stable oxygen surface groups are more effective for the anchorage of platinum particles (reduction under hydrogen was carried out at 350 °C, a temperature at which the less stable oxygen groups decomposed), thus enhancing the thermal stability of the platinum particles supported.

As shown already, carbon surfaces may have different amounts and types of oxygen surface groups and consequently both negatively and positively charged surface sites exist in aqueous solution, depending on the pH. At the IEP, pH_{IEP}, the net overall surface charge is zero; at $pH > pH_{IEP}$ the carbon surface, covered by de-protonated carboxyl groups, will attract cations from solution; at $pH < pH_{IEP}$ it will attract anions. This means that when preparing a carbon-supported catalyst, not only are physically accessible sites needed, but also chemical sites have to be accessible to the catalyst precursor. Thus, the electrostatic repulsion forces between anionic catalyst precursor species in solution and a negatively charged surface may be stronger than the non-specific dispersion forces of attraction. In such a case, when the catalyst precursor is dried, the benefits of a high support surface are lost, and the dispersion of the catalyst cannot be maintained.

Leon and Radovic (1994) and Wunder *et al.* (1993) carried out extensive work to understand the amphoteric character of the carbon surface and to use such surfaces to optimize the preparation of Mo/C and Pd/C catalysts. In their study of Pd/C catalysts (Leon y Leon *et al.*, 1992) they concluded that "maximum catalyst dispersion is favored when the entire carbon surface is chemically accessible that is when there is electrostatic attraction between the positively charged surface (below pH_{IEP}) and the catalyst precursor anions or between the negatively charged surface (above pH_{IEP}) and the catalyst precursor cations".

Oxygen surface groups are not the only centers conditioning the catalytic behavior of carbon-supported catalysts. Thus, when a high surface area carbon black is subjected to heat treatment in an inert atmosphere at temperatures ranging from 1600 to 2200 °C there is not only a decrease in surface area, but also an increase in crystalline ordering, associated with an increase in the basicity of the carbon, which cannot be explained by basic groups. The basicity of the carbon surface is explained in terms of the (π) sites of the carbon basal plane, which upon interaction with water lead to the equation:

$$C_\pi + 2H_2O \rightarrow C_nH_3O^+ + OH^- \qquad \text{(Reaction 8.6)}$$

Heat-treated carbon blacks have been used as catalyst supports for platinum using hexachloroplatinic acid in water, Coloma *et al.* (1994). It was found that the greater the difference between the pH of the impregnating solution (2.06 in this case) and the IEP of the carbon, the better will be the interaction between the surface and the anionic precursor. A low dispersion for the carbon black (HTT 1600 °C) is the result of a larger amount of oxygen surface groups formed on the carbon surface upon impregnation with the aqueous solution of hexachloroplatinic acid and the subsequent modification of its initial surface characteristics. Furthermore, sintering tests carried out in hydrogen, using different temperatures and periods of treatment, show that there is an increase in resistance to sintering of the platinum crystallites with increasing degree of pre-graphitization of the supports because the interaction increases between the platinum particles and the carbon support. This increasing metal-support interaction is related to the increasing strength of the π-sites on the support upon pre-graphitization, which act as anchoring centers for platinum.

An important point to be taken into account is that some oxygen surface groups may not be stable under the reduction conditions to which the catalysts are subjected to obtain the active phase, and their decomposition would favor the sintering of the metal species. Also, oxygen surface groups may influence other kinds of centers on the carbon surface. In this way the strength of the π-sites may be diminished by the presence of oxygen groups, due to the electron-withdrawing effect exerted by the oxygen atom. In order to see this effect, the carbon black mentioned above, heat treated at 2000 °C, was oxidized with hydrogen peroxide (to introduce oxygen groups) and the resulting sample was heat treated under a flow of helium at 500 °C (to eliminate the less stable oxygen groups).

The three resulting samples (C2, $C2_{OX}$, $C2_{OXT}$) were used to prepare platinum catalysts by impregnation with an aqueous solution of hexachloroplatinic acid. The dispersion was very high, especially for catalyst C2, among the highest reported to date for platinum catalysts. Interaction of a basic carbon like C2 (the carbon black heated at 2000 °C) with an anion in acidic solution would be more favored than if a more acidic carbon was used. It is known that oxidation by hydrogen peroxide creates acidic oxygen surface groups that decrease the IEP of the carbon. However, modifications in the surface characteristics of the support upon interaction with the precursor modify expected behavior. Thus, a better platinum dispersion has been found, by XPS, in the fresh oxidized catalyst with $C2_{OX}$ than in the one prepared with support C2, also in the fresh state. This is explained by assuming a better wettability of the more hydrophilic $C2_{OX}$ carbon support and a greater amount of oxygen surface groups acting as anchoring centers for the platinum precursor. However, reduced catalysts show a completely different tendency, the platinum dispersion increasing as the oxygen content in the support decreases. The more stable oxygen surface groups (evolving as CO at high temperature) remaining on the support after catalyst reduction in hydrogen play an adverse effect on the metal-support interaction. In a further attempt to clarify the role of oxygen surface groups in this type of catalysts, sintering tests were carried out with catalysts C2 and $C2_{OX}$. Sintering is more important in catalyst $C2_{OX}$ even though the initial dispersion is much higher than in catalyst C2, thus showing that the presence of oxygen near the π-sites decreases the basicity by an electron withdrawal effect, which destroys the π-electron delocalization.

In the impregnation step, the carbon–platinum precursor interaction is favored by the presence of oxygen surface groups, which make the carbon surface more hydrophilic, and thus, a better dispersion of the platinum precursor is achieved. However, after reduction in hydrogen at 350 °C the main part of the less stable oxygen surface groups is decomposed, with the remaining groups affecting the electron delocalization in the *n* sites, so weakening their capability to act as anchoring centers. *As a result of this a better platinum dispersion and an improved resistance to sintering are obtained when the pre-graphitized carbon black support is free of oxygen surface groups.*

As a summary of the possible electrostatic interactions taking place between the sites on the carbon surface and the metal cation or anion, maximum catalyst dispersion and resistance to sintering can be obtained by:

1. oxidation of the carbon typically renders the carbon surface more acidic and thus negatively charged over a wide range of pH, this in turn resulting in electrostatic repulsion of the $PtCl_6^{2-}$ anions, and favoring electrostatic attraction of the $[Pt(NH_3)_4]^{2+}$

cations, and so maximizing catalyst dispersion, when this compound is used as a platinum precursor.

2. increasing the basic C_π sites on the basal plane surface of oxygen-free carbon maximizes the electrostatic attraction with the metal anion (e.g. $C_n—H_3O^+ \cdots PtCl_6^{2-}$), also minimizing the electrostatic repulsion (e.g. $COO^-–PtCl_6^{2-}$), thus increasing the dispersion.

3. C=O groups acting as anchoring centers hinder agglomeration and surface diffusion of catalyst particles across the graphene layers.

Another example of the role of oxygen surface groups on the behavior of carbon-supported catalysts is provided by Pt/AC catalysts as used for the selective hydrogenation of crotonaldehyde, Coloma *et al.* (1996, 1997). The catalytic hydrogenation of α,β-unsaturated aldehydes to give unsaturated alcohols is not easy because the olefinic double bond is preferentially reduced. A de-mineralized AC prepared by steam activation of an olive stone char was oxidized with hydrogen peroxide to introduce oxygen surface groups (carbon AC_{OX}) and later heat treated at 500 °C under a flow of helium to eliminate the less stable oxygen groups (carbon AC_{OXT}). The three carbons were impregnated with a basic (pH = 9.5) aqueous solution of $[Pt(NH_3)_4]Cl_2$ to achieve 1 wt% of platinum. Oxidation of the carbon creates oxygen surface groups that decrease the IEP of the carbon, and the adsorption of a cationic species should be favored in basic media, leading to high dispersion. However, the amount of acidic surface groups increases in the order $AC < AC_{OXT} < AC_{OX}$ whereas platinum dispersion increases in the opposite order $AC_{OX}(0.09) < AC_{OXT}(0.20) < AC(0.24)$. One has to bear in mind that the thermal stability of acidic groups is relatively low, and that these are decomposed at the temperature at which the catalyst is reduced under a flow of hydrogen at 350 °C (after a heat treatment in helium at 400 °C). This reduction in acid groups permits mobility of platinum species and, consequently, this leads to lower metal dispersion (more agglomeration).

The specific activity for the hydrogenation of crotonaldehyde is high and follows the order $AC << AC_{OXT} < AC_{OX}$, but is closer for catalysts AC_{OXT} and AC_{OX}, in spite of their different dispersions, than for AC and AC_{OXT}, with similar dispersions. This suggests that platinum particle size is not the only factor related to catalytic activity. The selectivity to crotyl alcohol is low for catalyst AC, and larger for the other catalysts, especially for AC_{OX}.

In order to find out why the three catalysts behave differently, their catalytic properties were reassessed after reduction in hydrogen at temperatures of 350, 450 and 500 °C. There is an increase in activity and selectivity with increasing reduction temperature, being marginal for carbon AC, but important for the other two catalysts, especially for AC_{OX} after reduction at 500 °C.

Although oxygen surface groups are important from the point of view of carbon as a catalyst support, the effect of other heteroatoms has also been studied, especially nitrogen. Effects of nitrogen are mainly studied by treating carbon with ammonia in the temperature range 400–800 °C, Boehm *et al.* (1984) and Derbyshire *et al.* (1986). It was found that pre-nitrided carbons enhanced the activity of molybdenum catalysts. Guerrero-Ruiz *et al.* (1986) found a similar effect for Fe/C and Ru/C catalysts. Similar results are described for other carbon-supported catalysts, Radovic *et al.* (1997).

8.4.4 *Influence of Carbon Inertness*

It is demonstrated above that the presence of oxygen surface groups is beneficial in many aspects of the preparation, dispersion and activity of carbon-supported catalysts. However, if one considers that one of the main advantages of carbon supports with respect to conventional oxidic supports, like alumina and silica, is the ease of reducibility of the metal on the support, the presence of oxygen groups could be detrimental. However, carbon is an ideal support for this type of study because, in contrast to alumina and silica, the carbon-active phase interaction is weak and the behavior of the catalyst is basically governed by the chemical nature of the active phase. Thus, the effect of adding other metals or promoters may be studied without the problem of undesirable effects with the support.

It has been shown above that monometallic platinum catalysts used in the hydrogenation of crotonaldehyde usually produce the saturated aldehyde (Coloma *et al.*, 1996). It may be possible to improve the selectivity of the process toward the hydrogenation of the carbonyl bond, by addition of a second metal. When the second metal is tin, there are three aspects of interest:

1. the oxidation state of tin after reduction in hydrogen;
2. the possibility of platinum–tin alloy formation;
3. the extent of the tin-support interaction.

All three of which are strongly affected by the characteristics of the support. The use of a relatively inert support such as carbon reduces strong tin-support interactions, thus facilitating the tin–platinum interaction and leading to an easy formation of Pt–Sn alloy phases. On the other hand, reduction of tin species is easier on a carbon support than on typical oxidic supports.

Figure 8.19 shows the evolution of catalytic activity in the vapor-phase hydrogenation of crotonaldehyde as a function of the Sn/Pt atomic ratio for two series of catalysts supported on a pre-graphitized (2000 °C) carbon black, prepared by co-impregnation with ethanolic solutions of $H_2PtCl_6{\cdot}6H_2O$ and $SnCl_2$ (series A) or by successive impregnation with first $H_2PtCl_6{\cdot}6H_2O$ and then $SnCl_2$ (series B) or with first $SnCl_2$ and then $H_2PtCl_6{\cdot}6H_2O$ (series C); the tin contents were 0.25, 0.50 and 0.75 wt% and the platinum content was 1 wt% in all cases. There is a sharp increase in specific activity with increasing tin loading when the Sn/Pt atomic ratio is kept <1. Thus, although the amount of surface platinum atoms decreases by surface enrichment with tin species, the catalytic activity of the remaining platinum atoms is enhanced. XPS measurements reveal the presence of tin in both zero and oxidized states in the reduced catalysts, along with reduced platinum. The selectivity toward crotyl alcohol as a function of conversion is given in Figure 8.20 which shows that the monometallic catalyst promotes, almost exclusively, the hydrogenation of the olefinic bond, but the addition of tin noticeably increases the selectivity to the hydrogenation of the carbonyl bond (Coloma *et al.*, 1996). The XPS results suggest that tin, in an oxidized state, promotes the hydrogenation of the carbonyl bond on platinum atoms placed in its vicinity whereas the formation of Pt–Sn alloys or the dilution of platinum by tin species would inhibit the hydrogenation of the olefinic bond. The best results are obtained when the catalyst is prepared by successive impregnations, loading the tin precursor first.

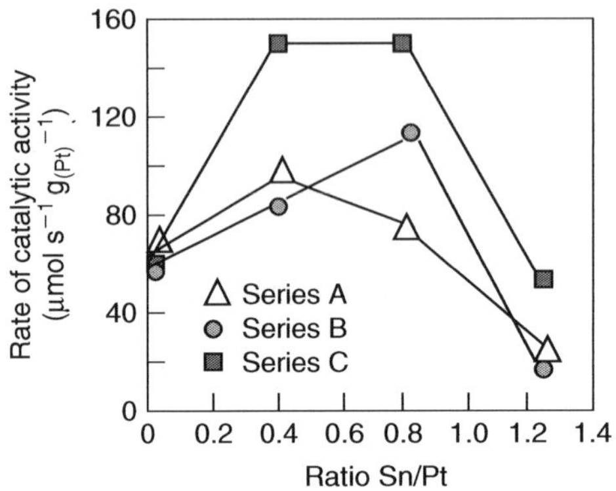

Figure 8.19. Evolution of catalytic activity in the vapor-phase hydrogenation of crotonaldehyde as a function of the Sn/Pt atomic ratio (Coloma *et al.*, 1996).

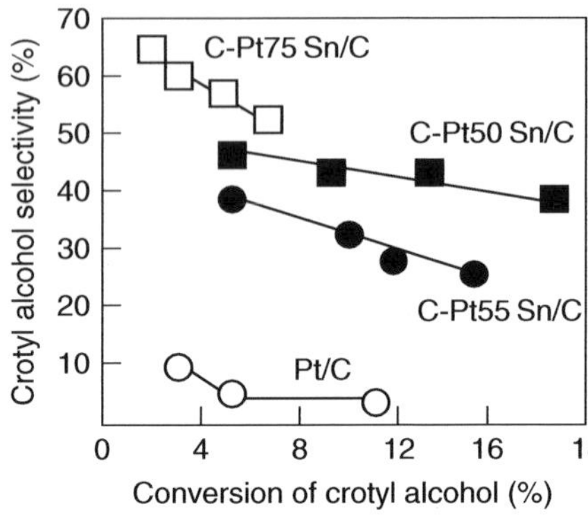

Figure 8.20. Selectivity toward crotyl alcohol as a function of conversion degree for different catalysts (Coloma *et al.*, 1996).

A good example for the exploitation of the *relative inertness of the carbon surface* on catalysis is the case of Fe/C catalysts for CO hydrogenation. High surface area carbon, as a support for iron, has certain advantages over typical oxide supports because it is inert in the syngas reaction, it provides high dispersion and inhibits sintering within its small pore structure and its inertness facilitates the presence of zero-valent iron in the catalyst. On a support such as alumina, iron cations strongly interact with the support surface, resulting in the oxide precursor being difficult to reduce to iron in the zero-valent state and in not being possible to obtain small particles of iron. Chen *et al.* (1986) carried out an extensive study of carbon-supported iron catalysts using different carbons and preparation methods and found that highly dispersed Fe/C catalysts could be prepared on high surface area carbons as a consequence of the weak chemical interaction between iron and the carbon surface.

One of the reasons for the interest in Fe/C catalysts is the possibility of obtaining a high selectivity to light olefins in the hydrogenation of CO, since Fe/C catalysts are more selective to olefines than Fe/Al_2O_3 and Fe/SiO_2 catalysts. The approach, followed by Venter and Vannice (1989) and Sepúlveda-Escribano and Rodríguez-Reinoso (1994) was to use iron carbonyls supported on high surface area carbon supports, to facilitate the access of the catalyst precursor to the internal porosity and the corresponding high iron dispersion, and especially because zero-valent Fe is initially present in the metal precursor. It is mentioned above that Fe/C catalysts were prepared using iron nitrate in aqueous solution and iron pentacarbonyl in *n*-pentane, and that iron dispersion was high and uniform in all AC when the precursor was iron pentacarbonyl, indicating that the porosity and chemical composition of the carbon support affects the iron dispersion for the nitrate precursor but not for the pentacarbonyl. Furthermore, Chen *et al.* (1986) showed that if a high surface area carbon is treated under hydrogen for 12 h at 1225 K and is never exposed to air, most of the Fe particles are very small and resist sintering, whereas if the hydrogen treated carbon is exposed to air before catalyst preparation, iron particles are larger and sinter more rapidly. This is another example of the negative influence of surface groups of carbon in the preparation of some catalysts.

In addition, to facilitating the preparation of highly dispersed iron catalysts, the use of iron carbonyls and carbon supports facilitates the preparation of promoted catalysts to increase the selectivity to olefines. Thus, the use of mixed-metal carbonyl clusters as metal precursors allows the preparation of a variety of stoichiometric metal compositions, something difficult to reach by co-impregnation techniques. Furthermore, the mixed-metal carbonyl cluster should be activated by heating just to that minimum temperature which would decompose the cluster to yield reduced metal and CO, the temperature being <475 K. Hence, it is possible to obtain reduced metals under much less severe conditions than those used for conventional metal salt precursors.

It is known that the addition to iron of promoters such as alkali metals and manganese enhances the selectivity to low molecular olefines in the CO hydrogenation reaction. Studies of carbon-supported Mn–Fe catalysts prepared from mixed-metal carbonyl clusters showed them to be unique in that they had exceptionally high selectivities to olefines and were also very active, with an addition of one potassium atom to a cluster improving the performance further. Further, iron catalysts were prepared by the exchange of the appropriate alkali metal cation with the tetra-ethylammonium cation in the $N(C_2H_5)_4[Fe_2Mn(CO)_{12}]$, Martín-Martínez *et al.* (1989). The supports were AC prepared from olive stones. The utilization of such a cluster as a catalyst precursor has distinct advantages over metal salts because:

1. stoichiometric mixed-metal compounds can be prepared;
2. different promoters can be identically added;
3. high dispersions of iron can be obtained;
4. zero-valent iron (and manganese) initially exists (this implies that no reduction step is required to produce active CO hydrogenation catalysts).

The activity for hydrocarbon formation at 548 K, 0.1 MPa and $H_2/CO = 3$, is highest for the un-promoted Fe crystallites, the addition of Mn decreasing them and the further addition of the alkali metals, with the exception of Rb, reducing the values even more. However, the addition of Mn increases the olefin/paraffin ratio, the increase being greater with the further addition of alkali metals (except Cs), especially K (olefin/paraffin ratio above 50).

Carbonaceous supports have also been used to study the metal-oxide promoter interaction in catalysts for the Fischer–Tropsch reaction. Again, the properties of the support avoided the promoter-support interaction which is common for other supports such as alumina or silica. The promoting effects of magnesium, vanadium and cerium oxides on AC-supported cobalt and ruthenium were compared, the metal-promoter interaction being monitored by temperature-programmed reduction measurements and studies on their catalytic behavior.

Another example of the effective role of carbon inertness is the use of carbon-supported catalysts for petroleum hydroprocessing, since in comparison to conventional alumina supports, carbon is exceedingly inert. Thus, when the conversion of a catalyst precursor to the catalytically active phase involves reduction or sulfidation, this is easier and more complete when carbon is the support. The very complete study of De Beer *et al.* (1984), with a long series of important contributions, initiated by Duchet *et al.* (1983) provides a clear example of this advantage of a carbon black support, which "can be considered as an almost inert support material". In studies on catalysts such as Mo/C, Co/C, Co–Mo/C, Ni–Mo/C, etc., the thiophene HDS was found to be larger when the active phase was supported on a carbon surface.

A further important aspect of carbon inertness in catalysis is its much lower coking propensity in comparison with alumina or silica supports. Coke deposition on the surface of the catalyst reduces the life of the catalyst. De Beer *et al.* (1984) studied this effect and found that the extent of carbon deposition on the blank supports is higher for carbons than for alumina and it increases with increasing surface area. In the absence of a metallic component the cracking appears to be related more to the accessible surface area than to any other particular surface property. However, the addition of metals to the supports causes an increase in the rates and amounts of carbon deposition, but the increase is much higher for the alumina-supported catalysts.

Thus, Figure 8.21 shows the coke deposition as a function of Mo loading for an alumina and carbon black composite having about the same porosity. The latter induces a much smaller carbon deposition and, in addition, is essentially insensitive to metal concentration, Scaroni *et al.* (1985). The larger activity for hydro-treating reactions such as HDS and the lower coking propensity observed in carbon-supported mono- and bimetallic-sulfide catalysts, as compared with conventional alumina-supported catalysts, was originally attributed to the almost inert character of the carbon support, where the conversion of oxidic metal species into the sulfide phase is easier, and also to the slit-shaped pores of AC. It is hoped that a better understanding of the role of the support in HDS and the advantages of carbon mentioned above (in addition to the easy recovery of catalysts by burning off the carbon, important from the environmental point of view) will help to change the situation in the near future. A problem remaining to be solved is the fact that the large surface area of porous carbons is mainly contained in micropores and these will be inaccessible to the large sulfur-containing molecules of industrial feedstocks. For this reason, carbon-covered alumina is of special interest because it combines three virtues of carbon and/or alumina catalysts support materials: the mechanical strength of alumina, the

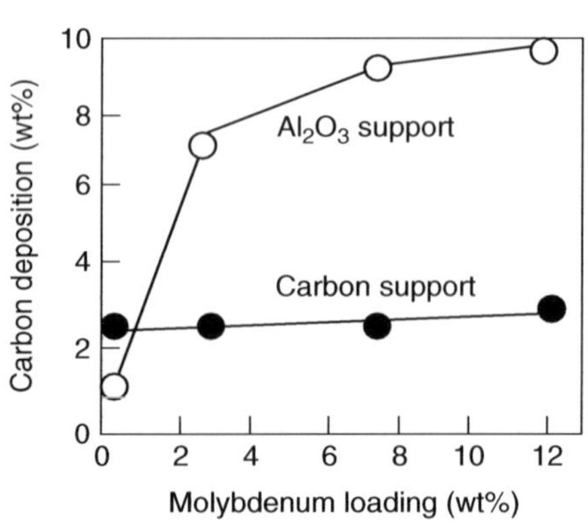

Figure 8.21. Effect of molybdenum loading on carbon deposition for carbon and alumina supports (Scaroni *et al.*, 1985).

flexibility of carbon in tailoring the degree of catalyst-support interaction and the absence of microporosity in alumina.

8.4.5 Catalysis by Carbon Surfaces

Although carbon is much more frequently used as a catalyst support than a catalyst in its own right, there are several industrial reactions using carbon catalysts, the most important one being the formation of phosgene (carbonyl chloride) by combination of carbon monoxide and chlorine over carbon (production of phosgene amounted 1.42×10^9 lb in 1983).

In a brief overview of reactions catalyzed by both carbon materials and by carbon deposits, Trimm (1981) presented a list with reactions catalyzed by carbon materials ranging from hydrogenation through oxidation and reduction to polymerization and chlorination, a summarizing version of which is given in Table 8.7.

For carbon supports, there are reports in which authors suggest a good correlation between the activity of carbon materials (mostly AC) and apparent surface area of the carbon, but in most cases, such correlation is not found when using AC with different surface areas and PSD for a given catalytic reaction, a more detailed knowledge of the carbon's surface chemistry being needed to explain the experimental results.

Thus, when studying the catalysis on both carbonaceous deposits and high surface area carbons, it was found that the surface oxygen complexes have an important role in the reaction, Lisovskii and Ahorone (1994), Petrosius *et al.* (1993). Szymanski and Rychlicki (1991) found that a high surface area carbon annealed in hydrogen at 950 °C exhibited essentially a low activity for the dehydration and dehydrogenation of alcohols while the same carbon oxidized with nitric acid considerably increased its activity by two orders of magnitude. A treatment with nitric acid resulted in the formation of carboxylic groups on the carbon surface whereas treatment in hydrogen produced a basic carbon.

Table 8.7. Some reactions catalyzed by carbon (Trimm, 1981; Rodríguez-Reinoso, 1998).

General type	Examples
Reactions involving hydrogen	$RX + H_2 \rightarrow RX$ (X = Cl, Br)
	$HCOOH \rightarrow CO_2 + H_2$
	$CH_3CHOHCH_3 \rightarrow CH_3COCH_3 + H_2$
Reactions involving oxygen	$SO_2 + (1/2)O_2 \rightarrow SO_3$
	$NO + (1/2)O_2 \rightarrow NO_2$
	$2H_2S + O_2 \rightarrow S_2 + H_2O$
	Catalytic oxidation of inorganic anions
Reactions involving halogen	$CO + Cl_2 \rightarrow COCl_2$
	$C_2H_4 + 5Cl_2 \rightarrow C_2Cl_6 + 4HCl$
	$SO_2 + Cl_2 \rightarrow SO_2Cl_2$
	$C_6H_5CH_3 \rightarrow C_6H_5CH_2Cl + HCl$

A reaction attracting special attention in the last few years has been the dehydrogenation of hydrocarbons (especially alkyl aromatics) to unsaturated hydrocarbons. Since the former are more stable, dehydrogenation has to be carried out at high temperature, and an interesting alternative is to use oxidative dehydrogenation, in which the reaction products are the unsaturated hydrocarbon and water. The introduction of oxygen makes the reaction irreversible and exothermic, and the operation temperature can be much lower than in direct dehydrogenation. Many catalysts, including semi-conducting oxides have been tested for the oxidative dehydrogenation of alkyl aromatics, and it was found that the activity gradually increases as coke deposition increases.

The acid centers of the catalyst facilitate the formation of coke deposits, which in the presence of oxygen lead to oxygen surface groups. The reactant, for instance, ethylbenzene, is adsorbed on the coke surface where a complex is formed by reaction of oxygen surface groups with hydrogen atoms of the alkyl radical of ethylbenzene, with decomposition of this complex leading to styrene and water. Conversions of *ca.* 60–70% and selectivities of 70–90% are possible as a consequence of the formation of coke deposits on the acid catalysts, thus suggesting that carbon catalysts can be used for the formation of styrene from ethylbenzene. Furthermore, the review by Lisovskii and Ahorone (1994) indicates that it is the oxygen of the surface groups of the carbon which is responsible for the catalytic oxidative dehydrogenation, because the role of oxygen gas is to re-oxidize the surface. More specifically, Pereira *et al.* (1999) studied the catalytic dehydrogenation of ethylbenzene using AC for which the type and amount of functional groups were modified by oxidative treatments followed by heat treatment to selectively remove some of the functional groups. The styrene yield and selectivity are associated with an increase in the amounts of carbonyl/quinone groups on the surface, and were identified as active sites for the reaction.

8.4.6 Additional Investigations

Kang *et al.* (2005) studied the effects of heat treating an AC (as distinct from pre-graphitized carbon blacks mentioned above) on the loading and activity of platinum as a catalyst. A first effect was that concentrations of Pt^0 increased with decreases in concentrations of PtO and PtO_2. This may be the result of more π-sites being created as the perfection of constituent graphene layers increases. It was also found that different degrees of heat treatment caused different surface Pt^0. An optimum carbon was prepared at 1673 K. Increasing HTT of these AC increases their resistance to air oxidation.

Santiago *et al.* (2005) studied the oxidation of phenols in aqueous solution, by oxygen at 140 °C and 0.2 MPa pressure, over the surface of two AC, one (CI) was prepared as a charcoal (Merck #2514), the other (CII) was prepared from coconut shell. Both carbons were initially oxidized with nitric acid, hydrogen peroxide and $(NH_4)_2S_2O_8$. Two other carbons were prepared by acid extraction (HCl) of the parent material. It was unexpectedly observed that the various oxidation treatments only marginally affected the rates of phenol depletion, with small decreases being attributed to loss of basic surface groups. But, there were major differences in the behavior of the two carbons. Carbon CI was much more effective than CII. The formation of stable phenolic polymers appears to block access to the catalytic sites of the carbon.

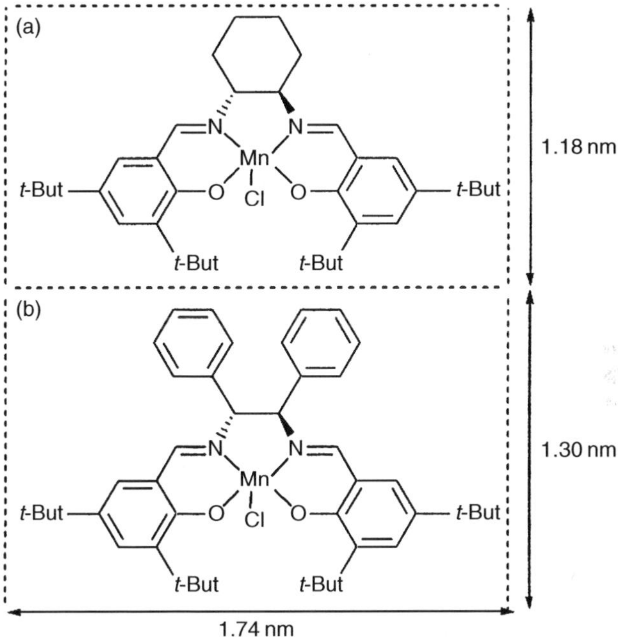

Figure 8.22. Molecular structure and dimensions of two Mn(III) *salen* complexes as attached to carbon surfaces (Silva *et al*., 2005).

Metals deposited on to carbon surfaces are effective catalysts, where catalytic efficiency and selectivity are functions of carbon structure and surface functionality. Here, the functionality has a secondary role in the way it influences the characteristics of the deposited metal. However, surface functionality can have a direct positive role when the catalyst is anchored directly to the carbon surface via surface oxygen complexes, Silva *et al*. (2004, 2005). Modification of the surface of an AC (NORIT ROX 0.8) was brought about by reaction with NaOH to form phenolate groups ($C_{OX}ONa$). Attachment of manganese(III) *salen* catalysts (see Figure 8.22) is via these phenolate groups through axial coordination of the metal center to these groups. These materials were active and enantioselective in the epoxidation of styrene and α-methylstyrene in dichloromethane at 0 °C. For the former, reuse did not lead to loss of activity while for the α-methylstyrene a significant loss was observed. This method of anchoring the catalyst to the surface is effective against active phase leaching.

8.4.7 Aerogel Carbons for Catalysis

The aerogel carbons are already mentioned in Section 3.4.18 and have seen development studies in terms of their abilities to act as catalyst support materials. Since these carbons are synthesized from monomeric materials, catalytic elements can be introduced *within* resultant carbons as distinct from methods of deposition from the liquid phase or by vapor deposition on to an already prepared carbon. Moreno-Castilla and Maldonado-Hódar (2005) published an overview which summarizes recent literature relevant to this topic. Aerogels are prepared from the sol-gel polycondensation of, for example, resorcinol and formaldehyde. This process facilitates some sensitivity in the formulation of surface areas, pore volumes and PSD together with surface functionality within resultant carbons. The form adopted for the carbons can be as monoliths, beads, powders or thin films. Metal-doped monolithic organic aerogels, promising applications for adsorption and catalysis, can be prepared by (a) addition of the metal precursor to the original mixture, (b) ion-exchange with the functionality of the gels and (c) the traditional method of deposition. Moreno-Castilla and Maldonado-Hódar (2005) report that the distribution/dispersion of metal atoms within the aerogel carbons can be very uniform and that the dispersion is stable on heat treatment of the carbon, suggesting some anchoring of the metal particle to the surface. Reactions which have been studied include the isomerization of 1-butene, toluene combustion, synthesis of methyl-*tert*-butyl ether, the formation of nano-filaments from CO decomposition and cathode catalysis in proton exchange membrane fuel cells (PEMFC), at atmospheric pressure.

8.4.8 Summary of Carbon in Catalysis

This overview of carbon materials and their use in heterogeneous catalysis shows their unparalleled flexibility in tailoring their properties (physical and chemical) to specific needs, thus illustrating the remarkably wide range of potential applications. Carbon materials are currently used in some industrial catalytic processes, but their full potential has not been exploited yet because the fundamental issues have become clear only recently.

It has been shown that although extents of surface area and the shape of porosity are critical in the preparation and properties of the catalysts, the role of carbon surface chemistry is equally important. It is these two aspects what makes carbon supports so attractive for heterogeneous catalysis, because carbons can be tailored to satisfy any specific need.

The role of surface carboxyl groups in the catalyst loading stage and of surface carbonyl groups in anchoring the active phase has been discussed, and, at the same time, the importance has been demonstrated of utility of an inert carbon surface.

Although the state-of-the-art for carbon as a catalyst in its own right is not as well developed as for carbon-supported catalysts, it is clear that the future is promising once it is understood that not only the surface area and porosity of carbon materials are the key for their use, and that both physical and chemical surface properties of carbon have to be taken into account when designing a catalyst.

8.5 Impregnated Carbons

There is no doubt of the aptitude and versatility of microporous AC to adsorb selected materials from both the gas and liquid phases. Despite this aptitude, there are limitations as to what AC can do. For example, the adsorption of toxic gases with low molecular weight, low-boiling point and appreciable polarity is not possible in conventional AC because adsorption capacity is too low and retentivity of the adsorbate is too limited.

To overcome these limitations, recourse is made to the use of impregnation of AC, as already exemplified in Section 8.4 for catalysis. Impregnated carbons offer additional capabilities through (a) optimization of existing properties of an AC, for example, that of oxidation, where, for example, impregnation with potassium iodide promotes the oxidation of hydrogen sulfide to sulfur, (b) synergism between AC and the impregnating agent, for example, the reaction between mercury and sulfur occurs at room temperature, (c) the availability of a large internal surface area, where the carbon surface is replaced by the surface of another material (as in a coating), for example, impregnation with phosphoric acid is used for the removal of ammonia vapors from the atmosphere.

There is a growing market for AC that have been impregnated with appropriate chemicals, organic and inorganic, these being added either by soaking impregnation or by spray impregnation, followed by drying and thermal after treatment at 150–200 °C. Care has to be taken in these processes not to block the functional porosity, and this can be checked-out using adsorption or calorimetric (heat of immersion) techniques.

Such impregnated carbons have to function at *room temperature, often in moist air*, unlike impregnation with catalysts which operate at higher temperatures.

Carbons for protection against toxic gases such as hydrogen cyanide, cyanogen and cyanogen chloride (typical examples of potential chemical warfare agents), are manufactured by impregnating AC with solutions of chromium and copper salts, sometimes coupled with silver salts, Molina-Sabio *et al.* (1994). Due to the aging problems associated with these impregnated carbons, mainly due to the change in oxidation state of chromium with age, other impregnation processes have been introduced, in which the impregnant may be, for instance, a combination of molybdenum and triethylenediamine, Tolles (1989).

The removal of ammonia and amines can be achieved by impregnating the AC with zinc acetate or with phosphoric acid. Carbons impregnated with iodine compounds can be used to remove mercury from waste air, although there are reports of the use of sulfur compounds for the same application. Iodine compounds can also be used to produce activated compounds for the extraction of radioactive methyl iodide and other gaseous compounds produced in nuclear reactors. Formaldehyde can be oxidized to non-hazardous compounds using AC impregnated with manganese dioxide. The use of impregnation by silver is reported for domestic purifiers for drinking water. A comprehensive list of applications of impregnated carbons is in Table 8.8.

The possibility of gas warfare necessitated the governments of the USA, the UK and of the USSR to undertake significant research programs for impregnated carbons to protect against the low-boiling point nerve gases. Most of this work never reached the open literature but an insight is possible via a few publications. In the USA, the late Victor R. Deitz made significant contributions into the manufacture and use of *Whetlerite*. This material is an AC supporting oxidizable metal ions, adsorbed from ammoniacal solutions, with co-adsorbed ligand structures. Reference should be made to Deitz and Karwacki (1994).

Table 8.8. Column 1: gases to be removed from atmosphere; Column 2: impregnating agents; Column 3: materials for gas protection; Column 4: impregnating agents.

Gas purification	Impregnated material	Civil and military gas protection	Impregnated material
Hydrogen sulfide	Iron oxide	Sulfur dioxide	Potassium carbonate
Mercaptans	Iron oxide	Hydrogen chloride	Potassium carbonate
Mercury		Hydrogen fluoride	Potassium carbonate
Ammonia	Sulfuric acid Phosphoric acid	Nitrogen oxide	Potassium carbonate
Amine	Sulfuric acid Phosphoric acid	Amines	Sulfuric acid Phosphoric acid
Acid gases (HCl, SO_2, HF, HCN)	Potassium carbonate	Hydrogen sulfide	Potassium permanganate
Arsine	Potassium iodide	Mercury	Sulfuric acid sulfur
Phosphine	Potassium iodide	Radioactive iodine and methyl chloride	Potassium iodide
Aldehyde	Manganese oxide(IV)	Phosgene	Cr/Cu/Ag salts
Radioactive iodine	Triethylene diamine	Hydrogen cyanide	Cr/Cu/Ag salts Cu/Zn/Mo salts
Radioactive methyl iodide	Triethylene diamine	Chlorine	Cr/Cu/Ag salts
Nitrogen oxide	Potassium carbonate	Arsine	Cr/Cu/Ag salts
Carbon disulfide	Potassium carbonate	Sarin and other nerve gases	Cr/Cu/Ag salts

Figure 8.23. An advertisement for shoe insoles which contain AC should the feet be odoros.

8.6 AC: Worldwide

The roles of AC extend, ultimately, to the entire atmosphere of this plant as well as to its rivers, seas and ocean, Derbyshire *et al.* (2001). The role also extends to the individual person, who, by bad luck, is blessed with odoros feet. The shoe insole, fitted with an AC, comes to the rescue, Figure 8.23.

8.7 The Outlook for AC

The AC industry, in a competitive market, is subject to changes which call for the use of both less AC and for more AC. Alternative systems of adsorption may be more cost-effective in some environmentally driven applications than those using AC. Continued emphasis by the EPA and industry toward the prevention of generated waste, rather than "end-of-pipe" solutions, is one limiting concern. AC markets are growing at less than the gross domestic product (GDP), in the range of 0–3%. However, AC is expected to grow in the USA at approximately 3% annually, mostly driven by environmental considerations. Large under-utilized capacity outside the USA will provide a ready supply to potential imports which will tend to limit price increases for most grades. Manufacturers, meanwhile, are developing specialized products, usually by impregnations, which offer higher user value and sell for a premium price.

References

Alfarra A, Frackowiak E, Béguin F. The HSAB concept as a means to interpret the adsorption of metal ions onto activated carbons. *Appl Surf Sci* 2004;228:84–92.

Babel S, Kurniawan TA. Cr(VI) removal from synthetic wastewater using coconut shell charcoal and commercial activated carbon modified with oxidizing agents and chitosan. *Chemosphere* 2004;54:951–967.

Bandosz TJ. Effect of pore structure and surface chemistry on virgin activated carbons on removal of hydrogen sulphide. *Carbon* 1999;36(3):483–491.

Biniak S, Pakula M, Szymanski GS, Swiatkowski A. Effect of activated carbon surface oxygen and/or nitrogen containing groups on adsorption of copper(II) ions from aqueous solution. *Langmuir* 1999;15:6117–6122.

Boehm HP, Mair G, Stoehr T, de Rincon AR, Tereczki B. Carbon as a catalyst in oxidation reactions and hydrogen halide elimination reactions. *Fuel* 1984;63(8):1061–1063.

Boudou JP, Chehimi M, Broniek E, Siemieniewska T, Bimer J. Adsorption of H_2S or SO_2 on an activated carbon cloth modified by ammonia treatment. *Carbon* 2003;41(10):1999–2007.

Burchell TD, Omatete OO, Gallego NC, Baker FS. Use of carbon fibre composite molecular sieves for air separations. *Adsorption Science and Technology* 2005;23(3):175–194.

Carrott PJM, Ribeiro-Carrott MML, Nabais MV, Prates-Ramalho JP. Influence of surface ionization on the adsorption of aqueous zinc chloride species by activated carbons. *Carbon* 1997;35(3):403–410.

Celzard A, Albiniak A, Jasienko-Halat M, Marêché JF, Furdin G. Methane storage capacities and pore textures of active carbons undergoing mechanical densification. *Carbon* 2005;43(9):1990–1999.

Chen AA, Kaminsky M, Geoffroy GL, Vannice MA. Carbon monoxide hydrogenation over carbon-supported iron–cobalt and potassium–ironicobalt carbonyl cluster-derived catalysts. *J Phys Chem* 1986;90:4810–4819.

Chen XS, McEnaney B, Mays TJ, Alcañiz-Monge J, Cazorla-Amorós D, Linares-Solano A. Theoretical and experimental studies of methane adsorption on microporous carbon. *Carbon* 1997;35(9):1251–1258.

Coloma F, Sepúlveda-Escribano A, Fierro JLG, Rodríguez-Reinoso F. Preparation of platinum supported on pre-graphitized carbon blacks. *Langmuir* 1994;10(3):750–755.

Coloma F, Sepúlveda-Escribano A, Fierro JLG, Rodríguez-Reinoso F. Crotonaldehyde hydrogenation over bimetallic Pt–Sn catalysts on pre-graphitized carbon black. Effect of the Sn/Pt atomic ratio. *Appl Catal A Gen* 1996;136:231–248.

Coloma F, Sepúlveda-Escribano A, Rodríguez-Reinoso F. Gas phase hydrogenation of crotonaldehyde over Pt/activated carbon catalyst. Influence of oxygen surface groups of the support. *Appl Catal A Gen* 1997;150:165–183.

Cook TL, Komodromos C, Quinn DF, Ragan S. Adsorbent storage for natural gas vehicles. In: *Carbon Materials for Advanced Technologies*, Ed. Burchell TD. Pergamon Press, Amsterdam, 1999, pp. 269–302.

Coughlin R, Ezra FS. Role of surface acidity in the adsorption of organic pollutants on the surface of carbon. *Environ Sci Technol* 1968;2:291–297.

Dąbrowski A, Podkościelny P, Hubicki Z, Barczak M. Adsorption of phenolic compounds by activated carbon – a critical review. *Chemosphere* 2004;58(8):1049–1070.

Dastgheib SA, Rockstraw DA. A model for the adsorption of single metal ion solutes in aqueous solution onto activated carbon produced from pecan shells. *Carbon* 2002;40(11):1843–1851.

de Beer VHJ, Derbyshire FJ, Groot CK, Prins R, Scaroni AW, Solar JM. Hydrodesulphurization activity and coking propensity of carbon and alumina supported catalysts. *Fuel* 1984;63(8):1095–1100.

Deitz VR, Karwacki CJ. Chemisorption of cyano-containing vapors by metal-ligand structures adsorbed by activated carbon. *Carbon* 1994;32(4):703–707.

Derbyshire F, de Beer VHJ, Abotsi GMK, Scaroni AW, Solar JM, Skrovanek DJ. The influence of surface functionality on the activity of carbon-supported catalysts. *Appl Catal* 1986;27(1):117–131.

Derbyshire F, Jagtoyen M, Andrews R, Rao A, Martin-Gullón I, Gruike E. Carbon materials in environmental applications. In: *Chemistry and Physics of Carbon*, Vol. 27, Ed. Radovic LR. Marcel Dekker Inc, New York, 2001, pp. 21–66.

Duchet JC, van Oers EM, de Beer VHJ, Prins R. Carbon-supported sulfide catalysts. *J Catal* 1983;80(2):386–402.

Franz M, Arafat HA, Pinto NG. Effect of chemical surface heterogeneity on the adsorption mechanism on an activated carbon with different oxidations. *Carbon* 2000;38(13):1807–1819.

Garten CH, Weiss DE. Adsorption from solution by carbon. *Rev Pure Appl Chem* 1957;7:69–122.

Giles CH, Smith D, Huitson A. A general treatment and classification of the solute adsorption isotherm. I. *Theor J Coll Interf Sci* 1974a;47:755–765.

Giles CH, D'Silva AP, Easton IA. A general treatment and classification of the solute adsorption isotherm. II. Experimental interpretation. *J Coll Interf Sci* 1974b;47:766–778.

Goel J, Kadirvelu K, Rajagopal C, Garg VK. Removal of mercury(II) from aqueous solution by adsorption on carbon aerogel: response surface methodological approach. *Carbon* 2005;43(1):195–213.

González-García CM, González-Martín ML, Gómez-Serrano V, Bruque JM, Labajos-Broncano L. Analysis of the adsorption isotherms of a non-ionic surfactant from aqueous solution onto activated carbons. *Carbon* 2001;39(6):849–855.

Grant TM, King CJ. Mechanisms of irreversible adsorption of phenolic compounds by activated carbons. *Ind Eng Chem Res* 1990;29:264–271.

Guerrero-Ruiz A, López-González JD, Rodríguez-Ramos I, Rodríguez-Reinoso F. Pt catalysts supported on activated carbons. I. Preparation and characterization. *React Kinet Catal Lett* 1986;31:349–354.

Haydar S, Ferro-García MA, Rivera-Utrilla J, Joly JP. Adsorption of *p*-nitrophenol on an activated carbon with different oxidations. *Carbon* 2003;41(3):387–395.

Hill A, Marsh H. A study of the adsorption of iodine and acetic acid from aqueous solutions on characterized porous carbons. *Carbon* 1968;6(1):31–39.

Ismadji S, Bhatia SK. Characterization of activated carbons using liquid-phase adsorption. *Carbon* 2001;39(8):1237–1250.

Jia YF, Thomas KM. Adsorption of cadmium ions on oxygen surface sites in activated carbon. *Langmuir* 2000;16:1114–1122.

Jia YF, Steele CJ, Hayward JP, Thomas KM. Mechanism of adsorption of gold and silver species on activated carbons. *Carbon* 1998;36(9):1299–1308.

Kang M, Bae Y-S, Lee C-H. Effect of heat treatment of activated carbon supports on the loading and activity of Pt catalyst. *Carbon* 2005;43(7):1512–1516.

Knoblauch K, Richiter E, Juntgen H. Application of active coke in processes of SO_2 and NO_x removal from flue gases. *Fuel* 1981;60(9):832–838.

Kononova ON, Kholmogorov AG, Lukianov AN, Kachin SV, Pashkov GL, Kononov YS. Sorption of Zn(II), Cu(II), Fe(III) on carbon adsorbents from manganese sulfate solutions. *Carbon* 2001;39(3):383–387.

Kononova ON, Kholmogorov AG, Danilenko NV, Kachin SV, Kononov YS, Dmitrieva ZhV. Sorption of gold and silver on carbon adsorbents from thiocyanate solutions. *Carbon* 2005;43(1):17–22.

Kotaś J, Stasicka Z. Chromium occurrence in the environment and methods of its speciation. *Environ Pollut* 2000;107:263–283.

Kwon S, Borguet E, Vidic RD. Impact of surface heterogeneity on mercury uptake by carbonaceous sorbents under UHV and atmospheric pressure. *Environ Sci Technol* 2002;36:4162–4169.

Kyotani T. Control of pore structure in carbon. *Carbon* 2000;38(2):269–286.

Leon CA, Radovic LR. Interfacial chemistry and electrochemistry of carbon surfaces. In: *Chemistry and Physics of Carbon*, Vol. 24, Ed. Thrower PA. Marcel Dekker Inc, New York, 1994, pp. 213–310.

Leon y Leon CA, Solar JM, Calemma V, Radovic LR. Evidence for the protonation of basal plane sites on carbon. *Carbon* 1992;30(5):797–811.

Li YH, Lee CW, Gullet BK. Importance of activated carbon's oxygen surface functional groups on elemental mercury adsorption. *Fuel* 2003;82(4):451–457.

Lillo-Ródenas MA, Cazorla-Amorós D, Linares-Solanao A. Behaviour of activated carbons with different pore size distributions and surface oxygen groups for benzene and toluene adsorption at low concentrations. *Carbon* 2005;43(8):1758–1767.

Lisovskii AE, Ahorone C. Carbonaceous deposits as catalysts for oxydehydrogenation of alkylbenzenes. *Catal Rev Sci Eng* 1994;36(1):25–74.

López-González JD, Moreno-Castilla C, Guerrero-Ruiz, Rodríguez-Reinoso F. Effect of carbon–oxygen and carbon–sulphur surface complexes on the adsorption of mercuric chloride in aqueous solution by activated carbons. *J Chem Technol Biotechnol* 1982;32:575–579.

López-Ramón MV, Stoeckli F, Moreno-Castilla C, Carrasco-Marín F. Specific and non-specific interactions of water molecules with carbon surfaces from immersion calorimetry. *Carbon* 2000;38(6):825–829.

Lu Q, Sorial GA. The role of adsorbent pore size distribution in multi-component adsorption on activated carbon. *Carbon* 2004;42(15):3113–3142.

MacDonald JAF, Evans MJB. Adsorption and enthalpy of phenol on BLP carbon. *Carbon* 2002;40(5):703–707.

Magne P, Walker Jr PL. Phenol adsorption on activated carbons: application to the regeneration of activated carbons polluted with phenol. *Carbon* 1986;24(2):101–107.

Mahajan OP, Moreno-Castilla C, Walker Jr PL. Surface-treated activated carbon for removal of phenol from water. *Separ Sci Technol* 1980;15:1733–1752.

Maroto-Valer MM, Zhang Y, Granite EJ, Tang Z, Pennline HW. Effect of pore structure and surface functionality on the mercury capacity of a fly ash carbon and its activated sample. *Fuel* 2005;84(1):105–108.

Martín-Martínez JM, Rodríguez-Reinoso F, Vannice MA. Carbon-supported, (Álkali Metal) ($Fe_2Mn(CO)_{12}$)-derived catalysts. Adsorption properties and catalytic behavior in carbon monoxide hydrogenation. *Appl Catal* 1989;51:93–112.

Mattson JS, Mark Jr HB, Malbin MD, Weber Jr WJ, Crittenden JV. Surface chemistry of active carbon: specific adsorption of phenols. *J Coll Interf Sci* 1969;31:116–130.

Meeyoo V, Lee JH, Trimm DL, Cant NW. Hydrogen sulphide emission control by combined adsorption and catalytic combustion. *Catal Today* 1998;44(1–4):67–72.

Menon VC, Komarneni S. Porous adsorbents for vehicular natural gas storage: a review. *J Porous Mater* 1998;5:43–48.

Mochida I, Korai Y, Shirahama M, Kawano S, Hada T, Seo Y, Yoshikawa M, Yasutake A. Removal of SO_x and NO_x over activated carbon fibres. *Carbon* 2000;38(2):227–239.

Molina-Sabio M, Pérez V, Rodríguez-Reinoso F. Impregnation of activated carbon with chromium and copper salts. Effect of porosity and metal content. *Carbon* 1994;32(7):1259–1265.

Moreno-Castilla C. Adsorption of organic molecules from aqueous solutions on carbon materials. *Carbon* 2004;42(1):83–94.

Moreno-Castilla C, Maldonado-Hódar FJ. Carbon aerogels for catalysis applications: an overview. *Carbon* 2005;43(3):455–465.

Moreno-Castilla C, Rivera-Utrilla J, López-Ramón MV, Carrasco-Marín F. Adsorption of some substituted phenols on activated carbons from a bituminous coal. *Carbon* 1995;33(6):845–851.

Nagoka T, Yoshino T. Surface properties of electrochemically pre-treated glassy carbon. *Anal Chem* 1986;58:1037–1042.

Newcombe G, Drikas M. Adsorption of NOM onto activated carbon: electrostatic and non-electrostatic effects. *Carbon* 1997:35(9):1239–1250.

Parkyns ND, Quinn DF. Natural gas adsorbed on carbon. In: *Porosity in Carbons*, Ed. Patrick JW. Edwards Arnold, London, 1995, pp. 291–325.

Pavlish JH, Sondreal EA, Mann MD, Olson ES, Galbreath KC, Laudal DL, Benson SA. Status review of mercury control options for coal-fired power plants. *Fuel Proc Tech* 2003;82(2–3):89–165.

Pendleton P, Wong SH, Schumann R, Levay G, Denoyel R, Rouquerol J. Properties of activated carbon controlling 2-methylisoborneol adsorption. *Carbon* 1997;35(8):1141–1149.

Pendleton P, Wu SH, Badalyan A. Activated carbon oxygen content influence on water and surfactant adsorption. *J Coll Interf Sci* 2002;246:235–240.

Pereira MFR, Órfão JJM, Figueiredo JL. Oxidative dehydrogenation of ethylbenzene on activated carbon catalysts. I. Influence of surface chemical groups. *Appl Catal A* 1999;184:153–160.

Pereira MFR, Soares SF, Órfão JJM, Figueiredo JL. Adsorption of dyes on activated carbons: influence of surface chemical groups. *Carbon* 2003;41(4):811–821.

Petrosius GC, Drago RS, Young V, Grunewald GC. Low-temperature decomposition of some halogenated hydrocarbons using metal oxide/porous carbon catalysts. *J Am Chem Soc* 1993;115(14):6131–6137.

Prado-Burguete C, Linares-Solano A, Rodríguez-Reinoso F, Salinas-Martínez de Lecea C. The effect of oxygen surface groups of the support on platinum dispersion in Pt/Carbon catalysts. *J Catal* 1989;115:98–106.

Prado-Burguete C, Linares-Solano A, Rodríguez-Reinoso F, Salinas-Martínez de Lecea C. The effect of carbon support and mean Pt particle size on hydrogen chemisorption by carbon supported Pt catalysts. *J Catal* 1991;128:397–404.

Radovic LR, Silva IF, Ume JI, Menéndez JA, Léon y Léon C, Scaroni AW. An experimental and theoretical study of the adsorption of aromatics possessing electron-withdrawing and electron-donating functional groups by chemically modified activated carbons. *Carbon* 1997;35(9):1339–1348.

Radovic LR, Moreno-Castilla C, Rivera-Utrilla J. Carbon materials as adsorbents in aqueous solutions. In: *Chemistry and Physics of Carbon*, Vol. 27, Ed. Radovic LR. Marcel Dekker Inc, New York, 2000, pp. 227–405.

Rivera-Utrilla J, Utrera-Hidalgo E, Ferro-Garcia MA, Moreno-Castilla C. Comparison of activated carbons prepared from agricultural raw materials and Spanish lignites when removing chlorophenols from aqueous solutions. *Carbon* 1991;29(4–5):613–619.

Rodríguez-Reinoso F. Activated carbon: structure, characterization, preparation and applications. In: *Introduction to Carbon Technologies*, Eds. Marsh H, Heintz EA, Rodríguez-Reinoso F. Secretariado de Publicaciones, Universidad de Alicante, 1997, Chapter 2, pp. 35–95.

Rodríguez-Reinoso F. The role of carbon catalysts in heterogeneous catalysis. *Carbon* 1998;36(3):159–175.

Rodríguez-Reinoso F. Production and applications of activated carbons. In: *Handbook of Porous Solids*, Eds. Schüth F, Sing KSW, Weitkamp J. Wiley-VCH Verlag GmmH, Weiheim, Germany, 2002, pp. 1766–1827.

Rodríguez-Reinoso F, Sepúlveda-Escribano A. Porous carbons in adsorption and catalysis. In: *Handbook of Surfaces and Interfaces of Materials*, Vol. 5, Ed. Nalwa HS. Academic Press, San Diego, 2001, pp. 309–355.

Rodríguez-Reinoso F, Salinas-Martínez de Lecea C, Sepúlveda-Escribano A, López-Gonzalez JD. Effect of support porosity in the preparation and catalytic activity for CO hydrogenation of carbon-supported Fe catalysts. *Catal Today* 1990;7:287–298.

Rodríguez-Reinoso F, Molina-Sabio M, Almansa C. Process for the manufacture of activated carbon monoliths and their use. Spanish Patent P9902459, 2003.

Rouquerol F, Rouquerol J, Sing K. *Adsorption by Powders and Porous Solids*. Academic Press, San Diego, 1999.

Santiago M, Stüber F, Fortuny A, Fabregat A, Font J. Modified activated carbons for catalytic wet oxidation of phenol. *Carbon* 2005;43(10):2134–2145.

Scaroni AW, Jenkins RG, Walker Jr PL. Coke deposition on Co–Mo/Al_2O_3 and Co–Mo/C catalysts. *Appl Catal* 1985;14:173–183.

Sepúlveda-Escribano A, Rodríguez-Reinoso F. Mo-promoted Fe/activated carbon catalysts for CO hydrogenation. *J Mol Catal* 1994;90:291–301.

Silva AR, Freire C, Castro B de. Jacobsen catalysts anchored onto an activated carbon as an enantioselective heterogeneous catalyst for the epoxidation of alkenes. *Carbon* 2004:42(14):3027–3030.

Silva AR, Budarin V, Clark JH, Castro B de, Freire C. Chiral manganese(III) schiff base complexes anchored onto activated carbon as enantioselective heterogeneous catalysts for alkene epoxidation. *Carbon* 2005;43(10):2096–2105.

Steel K.M, Koros WJ. An investigation of the effects of pyrolysis parameters on gas separation properties of carbon materials. *Carbon* 2005:43(9):1843–1857.

Stoeckli F, López-Ramón MV, Moreno-Castillo C. The adsorption of phenolic compounds from aqueous solution by activated carbons described by the Dubinin–Astakhov equation. *Langmuir* 2001;17:3301–3306.

Szymanski GS, Rychlicki G. Influence of oxygen surface groups in catalytic dehydration and dehydrogenation of butan-2-ol promoted by carbon catalysts. *Carbon* 1991;29(4–5):489–498.

Tamon H, Okazaki M. Desorption characteristics of aromatic compounds in aqueous solution on solid adsorbents. *J Coll Interf Sci* 1996;179:181–187.

Tanada S, Uchida M, Nakamura T, Kawasaki N, Doi H, Takebe Y. *J Environ Sci Health* 1997;A32:1451–1458.

Terzyk AP. Further insights into the role of carbon surface functionalities in the mechanism of phenol adsorption. *J Coll Interf Sci* 2003;268(2):301–329.

Terzyk AP, Rychilicki G, Biniak S, Lukaszewicz JP. New correlations between the composition of the surface layer of carbon and its physicochemical properties exposed while paracetamol is adsorbed at different temperatures and pH. *J Coll Interf Sci* 2003;257(1):13–30.

Tolles ED. Method and apparatus for removing hydrogen cyanide, cyanogen a. US Patent 4,801,311, 1989.

Trimm DL. *Catalysis*, Vol. 4. The Royal Society of Chemistry, London, 1981, p. 210.

Valix M, Cheung WH, McKay G. Preparation of activated carbon using low temperature carbonization and physical activation of high ash raw bagasse for acid dye adsorption. *Carbon Chemosphere* 2004;56:493–501.

Venter JJ, Vannice MA. Diffuse reflectance infra-red Fourier transform spectroscopic investigation of the decomposition of carbon supported iron carbonyl clusters. *J Phys Chem* 1989;93 (10):4158–4167.

Wunder RW, Cobes JW, Phillips J, Radovic LR, López-Peinado AJ, Carrasco-Marín F. Microcalorimetric study of the adsorption of hydrogen by palladium powders and carbon-supported palladium particles. *Langmuir* 1993;9:984–992.

Xiao J-X, Zhang Y, Wang C, Zhang J, Wang C-M, Bao Y-X, Zhao Z-G. Adsorption of cationic–anionic surfactant mixtures on activated carbon. *Carbon* 2005;41(5):1032–1038.

Yang TR. *Gas Separation by Adsorption Processes*. Butterworths, Boston, 1987.

Yardim MF, Budinova T, Ekinci E, Petrov N, Razigorova M, Minkova V. Removal of mercury(II) from aqueous solution by activated carbon obtained from furfural. *Chemosphere* 2003;52:835–841.

Zhou Y, Wang Y, Chen H, Zhou L. Methane storage in wet activated carbon: studies on the charging/discharging process. *Carbon* 2005:43(9):2007–2012.

CHAPTER 9

Production and Reference Material

9.1 Production

9.1.1 Introduction

This concluding chapter summarizes some of the industrial aspects associated with the production and use of activated carbon. Industrial experience is vast with most being summarized in the books listed in Section 2.1. Although this book is dominantly concerned with the structure of activated carbons and explanations of how they work as well as they do, some indication of how they are produced and are utilized in industrial environments is essential complimentary material.

9.1.2 Manufacturing Methods

Activated carbon is produced from nearly all carbon-containing organic materials, mainly wood, sawdust, nutshells, fruit stones, peat, lignite, coal, petroleum coke, etc. The use of a suitable precursor is mainly conditioned by its availability and cost, although it also depends on the main applications of the manufactured carbon and the type of installation available (Jankowska, 1991; Leon and Radovic, 1994; Rodríguez-Reinoso, 2002). A general flow sheet for the manufacture of activated carbon is as Figure 9.1.

9.1.3 Precursors

Activated carbon is a non-graphitizable carbon, meaning that it cannot be transformed into graphite solely by high heat treatment. In this type of carbon the structure of the precursor remains during carbonization, because fusion cannot take place (most graphitizable carbons pass through a fluid stage during carbonization, thus facilitating the formation of a graphitic structure). This means that the precursor, upon carbonization, must produce a non-graphitizable char (Section 2.6.2).

As mentioned already, most organic materials rich in carbon, that do not fuse when carbonized, produce a porous carbon (sometimes called a *char*). The selection of the parent

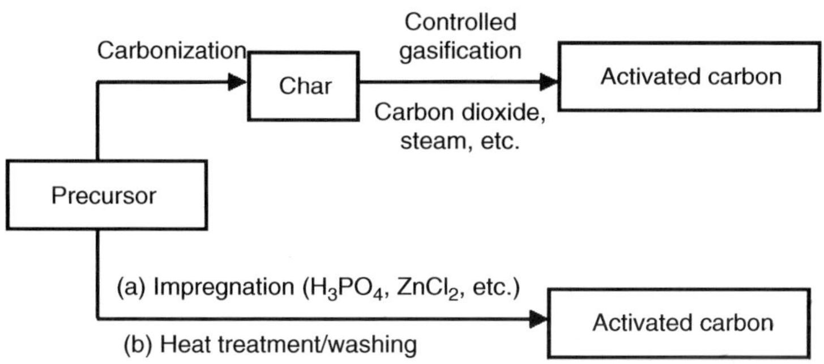

Figure 9.1. General flow sheet for the manufacture of activated carbon. (Rodríguez-Reinoso, 2002).

Table 9.1. Properties of materials used in the manufacture of activated carbons.

Raw material	Carbon (wt%)	Volatiles (wt%)	Density ($cm^3 g^{-1}$)	Ash (wt%)	Texture of activated carbon
Soft wood	40–50	55–60	0.4–0.5	0.3–1.1	Soft, large pore volume
Hard wood	40–42	55–60	0.55–0.80	0.3–1.2	Soft, large pore volume
Lignin	35–40	58–60	0.3–0.4		Soft, large pore volume
Nutshells	30–45	55–60	1.4		Hard, large micropore volume
Lignite	55–70	25–40	1.0–1.35	5–15	Hard, small pore volume
Soft coal	65–80	20–30	1.25–1.50	2–12	Medium hard. Medium pore volume
Petroleum coke	70–85	15–20	1.35	0.5–0.7	Medium hard Medium pore volume
Hard coal	70–75	10–15	1.45	5–15	Hard, large pore volume
Anthracite	85–95	5–10	1.5–1.8	2–15	Hard, large pore volume

Adapted from Bansal *et al.* (1988).

material for activated carbon production, however, is based mainly on the following criteria: (i) low in inorganic matter; (ii) availability and cost; (iii) low degradation upon storage; and (iv) ease of activation (calcined coke is a difficult material while wood char is easily activated). Some data for typical precursors used in the manufacture of activated carbon are listed in Table 9.1, this table including information on typical precursor properties and the type of activated carbon expected for each precursor. Cellulose-type materials such as wood, sawdust, nutshells, and fruit stones are mainly used for chemical activation. Although some of these precursors (especially nutshells and fruit stones) are also used for thermal activation, the general tendency is to use peat, lignite and several types of coal (mainly subbituminous and bituminous), according to availability and cost.

Lignocellulosic materials constitute the more commonly used precursor and account for around 45 wt% of the total raw materials used for the manufacture of activated carbon. Low contents of inorganic materials are important to produce activated carbons with low ash content, but relatively high volatile content is also needed for the control of the manufacturing process. Both characteristics are common to most lignocellulosic materials used for the production of activated carbon. Although pine wood is the more commonly used (especially in the form of sawdust), precursors such as coconut and fruit stones are popular for many types of activated carbon, because their relatively high density, hardness and volatile content are ideal for the manufacture of hard granular activated carbon. Coconut shells, together with peach and olive stones are used commercially for the production of microporous activated carbons, useful for a very wide range of applications.

All these lignocellulosic precursors produce a char upon carbonization with a yield ranging from 20 to 30 wt%, this meaning that after thermal activation the overall yield may be around 10 wt%. Woods, as sawdust, are favored for the manufacture of powdered activated carbons by the chemical activation method (addition of H_3PO_4). The precursor is crushed and sieved to a coarse powder before impregnation, and the mixture is then subjected to heat treatment.

Hard coals (high-rank bituminous coals) are the preferred precursor in many countries, because they can be used both for production of inexpensive activated carbons, and also for the more expensive granular, hard carbons with well-developed porous structures. As-received coals have some porosity (which decreases with increasing rank of coal), and consequently further treatments are needed to increase the porosity. Low rank coals (peat, lignite, brown coal), which do not fuse on carbonization, are used to produce activated carbons with a wide pore-size distribution. The yield of activated carbons from coal is generally larger than for lignocellulosic materials, above 30 wt%.

As many bituminous coals have the problem of coking (softening) upon heat treatment, pre-oxidation in air is needed to prevent coking. This is not a simple process, because the exothermicity makes the control of temperature difficult; the temperature must be maintained in the 150–350 °C range, depending on the type of coal. Pre-oxidation is followed by conventional carbonization. Coals of lower rank, with larger volatile content and lower coking propensity, can be used (without pre-oxidation) to prepare extruded or granular activated carbons with lower hardness. Anthracites, which exhibit a significant microporosity, are also used for the production of activated carbon after devolatization. The reactivity of the resulting coke is very low, and activation is usually preceded by oxidation with air, or chemical treatment with acid or inorganic compounds. The resulting activated carbons are intrinsically microporous, this being the reason why anthracite has also been used to prepare carbon molecular sieves (Section 8.2.3).

Brown coal is a popular precursor in Eastern Europe (because of availability), and because the semi-cokes and cokes obtained from it exhibit a well-developed porous structure (from micro-to macropores). They are even used as direct adsorbents (without activation) for operations in which the adsorbent is discarded after operation. Thermal activation by steam produces activated carbons with a wide range of pore size, useful in adsorption from solutions, although the granules obtained do not have good mechanical strength. An additional problem with brown coal is the high ash and sulfur content, this making pre-treatment necessary to reduce them to reasonable levels (Figure 9.2).

Peat is another popular precursor in Central Europe, and lignite in both Europe and USA, both materials being mainly used for the production of activated carbon for water treatment.

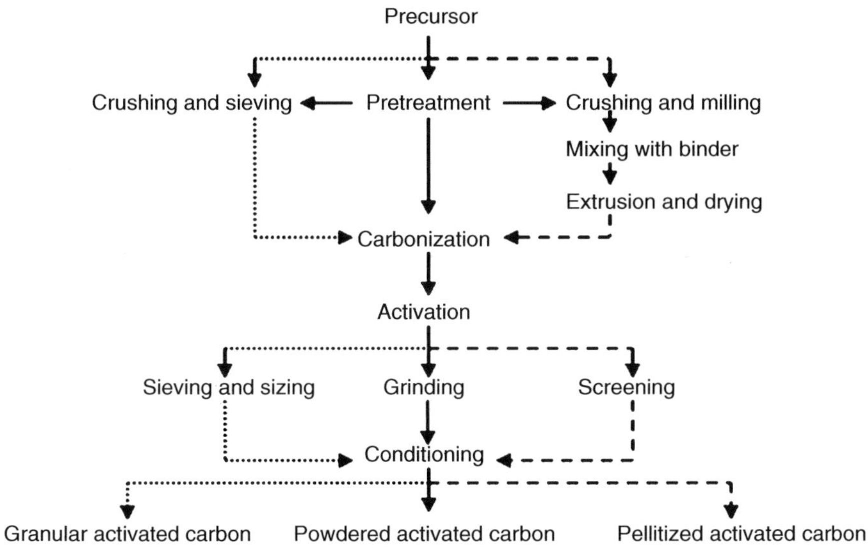

Figure 9.2. General flow sheet for thermal (physical) activation (Rodríguez-Reinoso, 2002).

In addition to the precursors included in Table 9.1, there are other precursors which are used either for the manufacture of specialty carbons or for scientific investigations. Thus, polymers such as Saran and phenol-formaldehyde yield activated carbons with large extents of microporosity and high surface area. Polyacrylonitrile (PAN) and acrylonitrile textile wastes can be used to produce activated carbons with a relatively high content in nitrogen, this being useful for the adsorption of sulfur species. There are many reports on the possible use of a wide range of precursors for the preparation of activated carbons, including mineral oil products, asphalt and tar. Petroleum residues are used by Witco Chemical (USA) and green petroleum coke is used by the Kansai Coke & Chemicals (Japan). In the latter case, the use is limited to green coke, since calcined coke cannot be activated because the structure is too ordered. Green coke is either pre-treated before activation or chemically activated with potassium hydroxide (KOH) to produce activated carbons with extremely high surface areas.

Activated carbon fibers and cloths are commercially prepared from rayon, PAN, phenolic resins, acetate, petroleum and coal tar pitch, etc. Most commercial activated carbon cloths are produced from viscous rayon impregnated with a solution of a Lewis acid before the heat treatment, although some other precursors such as Kynol, polyamide, etc., are also used.

The series of precursors included in this survey cover a wide range of properties, but only a few of them are really appropriate for the production of commercial activated carbon. Many of them do not comply with all of the selection criteria given at the beginning of this section, and their study has been used to gain an in-depth knowledge of the pyrolysis and gasification processes, to open the way to future types of activated carbon. It is quite clear that the manufacture of a product with well-defined and reproducible properties requires a most careful and adequate selection of the precursor material, and this limits, to a few of the precursors, what can be used at an industrial level.

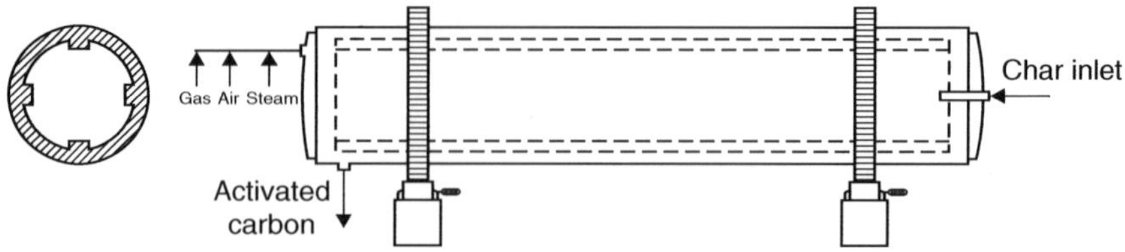

Figure 9.3. Schematic representation of a rotary kiln (Rodríguez-Reinoso, 2002).

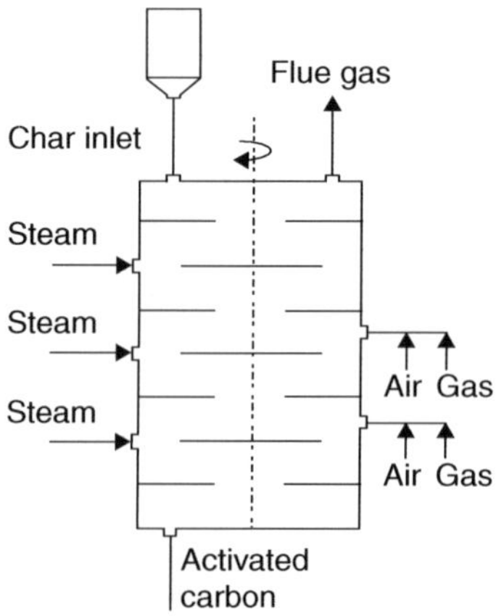

Figure 9.4. Schematic representation of a multiple-hearth furnace (Rodríguez-Reinoso, 2002).

9.1.4 Furnaces

Although several types of furnaces are described in the literature, three are the most commonly used by producers of activated carbon, that is rotary kilns, multiple hearth furnaces and, to a lesser extent, fluidized bed furnaces. It is to be understood that carbonization can be carried out in any of these furnaces. Rotary kilns are the most popular among the producers because, in general terms, they produce activated carbons with a more controlled porous structure. The residence time is greater than for other furnaces, and because their length is large, several burners and gas supply lines (natural gas, propane, or the gases from the carbonization unit) are distributed along the kiln in order to improve the control of the temperature and the temperature profile (Figure 9.3). The lifters placed along the interior of the kiln improve the mixing of the material thus providing a more homogeneous gasification. The rate of production is defined by the rotation speed, the angle of the kiln, the steam injection, and the rate of carbon dosage and withdrawal. It is generally admitted that more microporous carbons, with lower proportion of mesopores, can be manufactured using rotary kilns.

Multiple hearth furnaces are used as activating furnaces (Figure 9.4). There is a central vertical axis moving the rotary arms of the stationary floor on each stage, and the material is transported downwards falling alternatively near to the axis on one stage and near to the

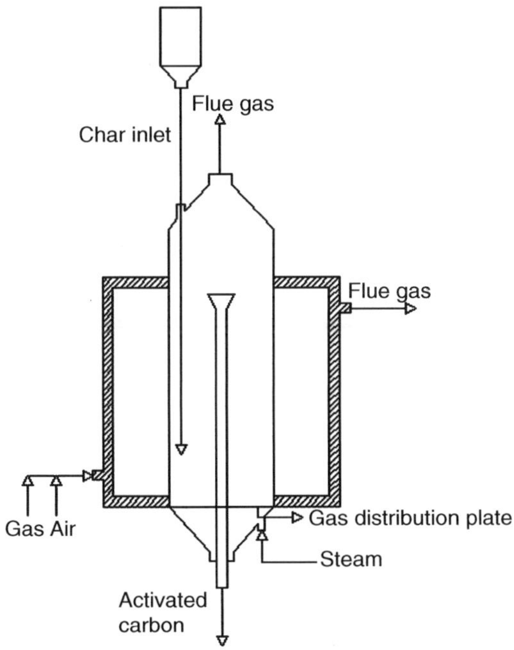

Figure 9.5. Schematic representation of a fluidized-bed furnace (Rodríguez-Reinoso, 2002).

wall at the next stage. As there are several supplies of heat and steam at different levels (in each stage in some furnaces), the temperature regulation is relatively simple. However, the relatively low carbon bed height makes the residence time shorter than in rotary kilns and the quality of the carbon may be slightly lower, with more development of mesoporosity. This type of furnace is also used for regeneration of spent activated carbon.

Fluidized bed furnaces (Figure 9.5) are characterized by a good heat and mass transfer, the activating gases being in excellent contact with the carbon particles. This means that the residence time is the shortest of all activating furnaces. Industry uses fluidized furnaces that are operated continuously, with several fluidized beds running in series. In this way, the experimental conditions may be different in each bed, to favor a more controlled activation. The main problem with this type of furnace is that reaction is extensive and there is much external burning of the particles, with the end result of activated carbons of relatively poor quality, mainly used in powdered form for decolorizing purposes. However, there have been some recent developments in the design of fluidized bed furnaces, which are now more commonly used by producers.

9.1.5 Quality Control: Testing

The application of activated carbon in many industrial processes requires a precise knowledge of the physical and adsorptive properties. Most of the testing methods have been developed and approved by the American Society for Testing Materials (ASTM) (Annual book, 1992), the American Water Works Association (AWWA, 1990), the International Organization for Standardization (ISO, 1999) or the Deutsches Institut für Normung e.V (DIN, 1986). In many manufacturing companies the tests are slightly modified, but only in non-essential parts.

From the point of view of the producer and the user many properties of activated carbon need to be known. Relevant physical tests are summarized as follows:

Bulk density: it is defined as the mass of a unit volume of the sample in air, including both the pore system and the voids among the particles. It is very useful for the estimation of the packing volume or to determine the grade of carbon needed for an existing system. Powdered carbons used for decolorization usually have a bulk density in the range 0.25–0.75 g cm^{-3}, while granular grades used in gas adsorption have a bulk density of around 0.40–0.50 g cm^{-3}. The ASTM D2854-70 is the base for this test.

Real density: it is also called absolute and helium density and is defined as the mass of a unit volume of the solid carbon skeleton, inaccessible to helium. In this method it is assumed that helium enters the smallest pores present, without being adsorbed. However, it is well documented that helium may be adsorbed in microporous carbons and, consequently, this possibility has to be checked and taken into consideration to give a realistic value. This is not a routine test; the main problem being that commercially available equipment shows a poor reproducibility.

Apparent density: also known as particle or mercury density, is defined as the mass of unit volume of the carbon particle, including its pore system. From this value it is possible to obtain the bed porosity. This bed porosity, or void volume is an important value with respect to the flow characteristics of the activated carbon bed. The method uses a pycnometer. If both apparent and real densities are known, the difference of their inverses will provide the total pore volume of the activated carbon.

Particle size: it is an important property because it influences the flow characteristics, filterability, adsorption kinetics, etc. The rate of adsorption of activated carbon depends inversely on particle size: small particles have the fastest rate of adsorption. The separation between powdered and granular activated carbon is established by the AST D2652-74 definition as 80 mesh or approximately 0.18 mm.

Mechanical strength: is an important factor for most technical applications of granular activated carbon and it is determined to simulate the resistance to abrasion or attrition under practical conditions. Activated carbon needs a high mechanical strength and attrition resistance where pressure drop and carbon losses may be significant. The determination of the ball-abrasion hardness is carried out following the AWWA B604-74 (Ro-Tap abrasion test) test and the determination of the attrition hardness according to the AWWA B694-74 (Stirring abrasion test) test.

Other relevant chemical and physico-chemical tests include:

Moisture content: it is determined in the conventional way by weighing before and after heating to 100 °C (ASTM D 2867099).

Ash content: This test provides the total ash content although in some cases a more detailed analysis of the ash is needed. The ash content increases in direct proportion to the degree of activation and it can be used to determine the raw material used to produce an activated carbon. For instance, coconut shell carbons contain 1–3 wt% ash, whereas coal-based carbons have 6–20 wt% ash content. The determination follows the ASTM 2866-70 test.

Ignition temperature (kindling point): it can be critical in solvent recovery or other applications in which hot air is involved; it must be high enough to prevent excessive carbon oxidation in gas-phase adsorption, where high heat evolution is involved. It is strongly dependent on the activation procedure used and thus high ignition temperatures are usually encountered in steam activated carbons. The test most commonly used is ANSI/ASTM D3466.

Self-ignition test: since self-ignition (especially in powdered activated carbon) is a potential hazard on transportation in elevated ambient temperatures, this test is used to determine whether a carbon sample ignites at 140 °C under specified conditions.

pH value: it takes into consideration that activated carbons carry inorganic and surface groups that may alter the pH of liquids to which they are added (Section 8.1.4.6).

Water-soluble content: it gives an estimation of purity in relation to water extractable substances.

Of all adsorptive tests needed to characterize the adsorptive properties of activated carbon, the most commonly used by both producer and user are as follows:

Carbon tetrachloride activity: it is intended to determine the activation level of activated carbon. The activity is defined as the ratio (in percent) of the weight of carbon tetrachloride adsorbed by an activated carbon sample to the weight of the sample when the carbon is saturated with CCl_4. The standard used is ASTM D3467-88.

Benzene adsorption: it is a test mainly used for granular activated carbons. It provides very relevant information about the adsorptive capacity in gaseous phase. In many cases the adsorption–desorption isotherm is determined in order to obtain the pore-size distribution of the carbon. There is not a reference standard test, and some industries only use it for information to users.

Iodine adsorption: it is a simple and quick test, giving an indication of the internal surface area of the carbon; in many activated carbons the iodine number (expressed as milligrams of iodine per gram of carbon) is close to the Brunauer–Emmett–Teller (BET) surface area. In this method, the activated carbon is boiled with 5% HCl and, after cooling, a 0.1 N iodine solution is added and shaken for 30 s; after filtration, the filtrate is titrated with 0.1 N sodium thiosulfate solution, with starch as indicator. The standard used is AWWA B600-78 (see Section 8.1.2).

Methylene blue adsorption: it gives an indication of the adsorption capacity for large molecules having similar dimensions to methylene blue; it is a quick test for medicinal and other carbons prepared to adsorb large molecules. The methylene blue value is defined as the number of millilitres standard methylene blue solution decolorized by 0.1 g of activated carbon (dry basis).

Phenol adsorption: it provides information on the adsorptive properties of activated carbon in applications such as drinking water treatment. This test measures the adsorption of phenol in percentage by weight on activated carbon, and is used when it is necessary to reduce the original concentration from $10\,mg\,L^{-1}$ to $1\,mg\,L^{-1}$ phenol. The ASTM D1738B and AWWA B600-66 are followed for this test.

Molasses decolorization: it is a test specifically directed towards the use of activated carbon in sugar and glucose industries, and it covers predominantly the medium size pore range. In this test a weighed amount of carbon is shaken in a water bath for 30 min with a molasses solution and filtered under strictly controlled conditions; optical transmission is measured and compared to filtrates of a standard carbon treated in the same way. Both transmissions have to be matched within certain limits by changing the amount of carbon under test. The results are expressed as the number of milligrams giving the same decolorizing effect as the standard weight of standard carbon.

Butane adsorption: it gives an indication of the applicability in solvent recovery and other gas-phase applications. The usual test consists in the determination of the adsorption of *n*-butane at near ambient temperature, and for some applications such as gasoline vapor recovery in automotive vehicles the adsorption and desorption capacity are measured in a specific number of cycles.

Phenazone adsorption: this is a test to estimate the adsorption capacity of activated carbon for pharmaceutical purposes.

There are additional data that some producers supply to the users to provide a maximum of information about their products. The most common is the surface area, as deduced from the application of the BET equation to the adsorption data of nitrogen at 77 K (it can also be applied to other gases and vapors at different temperatures). Although the BET surface area has not much physical meaning in microporous solids (Section 4.1.3) it is widely used to give an idea of the adsorptive capacity of activated carbon. The adsorption capacity as measured by adsorption of different gases and vapors is also frequently given by some producers.

More detailed information on these and other commercial tests for activated carbon can be obtained from sources such as ASTM, AWWA, DIN and ISO.

9.1.6 Washed Carbons

Some applications need very pure activated carbons, and to comply with this need many manufacturers remove most of the ash components by washing the carbon with water or acids such as hydrochloric or nitric acid. In some cases a more exhaustive washing requires the use of hydrochloric and hydrofluoric acids to eliminate the aluminosilicate components of the ash. The washing must be particularly thorough when the carbon is to be used in pharmaceutical preparations, food industries or in catalysis (when the carbon is used either as a catalyst or a catalyst support).

9.1.7 Regeneration

The main mechanism by which activated carbon removes impurities is one of physical adsorption, this being a reversible process. Consequently one can expect that desorption of the impurities will render the carbon surface available again for adsorption. Regeneration of spent activated carbon is not only important from the point of view of restoring the adsorption capacity of the carbon, but also because in many cases the recovery of the adsorbed species is important. If the adsorption is of chemical type (chemisorption), the formation of a bond between the carbon and the adsorbate makes the process non-reversible, and even if desorption is possible the desorbed species will be different to those originally adsorbed. Additionally, adsorption (especially in liquid phase) is often accompanied by precipitation of species which cannot be removed by simple desorption.

Regeneration of activated carbon is based on the principle that carbon is a stable material that can withstand changes in temperature and it is resistant to acidic and basic media. Regeneration of activated carbon is normally carried out only on granular activated carbon because it is uneconomical for powdered activated carbon, which is usually discarded after use. The more common regeneration process is by passing a flow of superheated steam through the carbon bed at a temperature lower than the one used during activation. As the temperature increases during regeneration there is a desorption of the more volatile adsorbed compounds; above 400 °C there is also decomposition of adsorbed organic material, thus leaving some less-organized carbon, which is eliminated by reaction with steam at around 600 °C. Of course, although the less-organized carbon is more reactive, a portion of the activated carbon is also gasified and regeneration usually means a loss of 8–15 wt% of the original activated carbon. For this reason, when regeneration is carried out under these conditions it is also called reactivation. It is important to note that the off-gases must be properly handled, passing them through a post-burner and a scrubber, where they are washed properly.

As regeneration implies manipulation of the activated carbon, the harder it is the lower will be the amount lost during regeneration. If regeneration is properly carried out the bulk density and the surface area of the resulting carbon will be very similar to that of the original carbon. If the surface area is larger after regeneration this means that a proper reactivation has taken place, with the subsequent reduction in bulk density.

Additional regeneration procedures include desorption by hot inert gases, desorption under vacuum, and more important, use of conventional liquid solvents. The main problem for the latter is the slow desorption and the difficult solvent regeneration, and the last few years have seen the development of regeneration with supercritical solvents, mainly carbon dioxide and water (Modell, 1997). Reports on the use of supercritical water seem to indicate that this is an interesting approach to the regeneration of activated carbon used in water treatments (Salvador *et al.*, 1999).

A problem exists with the use of regenerated activated carbon which contains calcium (Mazyck *et al.*, 2005). The introduction of such carbons to water and wastewater utilities brings about significant and unacceptable excursions of pH, sometimes >pH 11. This is attributed to hydrolysis reactions of the CaO to form $Ca(OH)_2$ and OH^-. Washing of the regenerated (calcium-containing) activated carbon with HCl and NaOH removed the calcium. Such a carbon, when used to treat to wastewater with a pH of 5–6, changed the pH to 8.5, compared with a pH of 11.5 for the carbon containing the calcium.

9.1.8 Industrial Production

The world production capacity (excluding China and other Eastern countries, for which figures are not accurately known) is estimated to be around 400,000 tonnes per year. The largest producer is the USA (accounting for more than 40 wt% of the total production capacity), followed by Europe, Japan and the Pacific Rim countries.

A comparison of production capacity, actual production and apparent consumption in 1994 for the three major regions (USA, Western Europe and Japan) is given in Figure 9.6 (Waterhouse *et al.*, 1996). The balance between actual production and apparent consumption is the exports–imports made in 1994.

The largest producer in the USA is Calgon Carbon Corporation, with plants in several locations. Its annual capacity represents around 42% of the total production capacity of the country; the main activation process is thermal and the precursors used are bituminous coal, coconut shell and charcoal. The second largest producer is Norit Americas, Inc., with around 23% of total capacity. Thermal activation is carried out with lignite and bituminous coal, and chemical activation (with phosphoric acid) is carried out with wood and peat. The third producer is Westvaco Corporation, with around 12%, sawdust being the main precursor for both thermal and chemical (with phosphoric acid) activation. Regeneration capacity of spent activated carbon has considerably increased in the last few years, this capacity being estimated to be now over 50,000 tonnes per year.

The largest producer in Western Europe is Norit NV (plants in the Netherlands and UK), with an annual capacity share of almost 50% of the total capacity. Peat for thermal activation and sawdust for chemical (with phosphoric acid) activation are the main precursors used. Other producers are Chemviron (around 20% of total capacity), CECA (around 15%) and Pica (around 12%). A new company, ICASA, has started production in Spain of activated carbon from olive stones. Reactivation of spent activated carbon is important

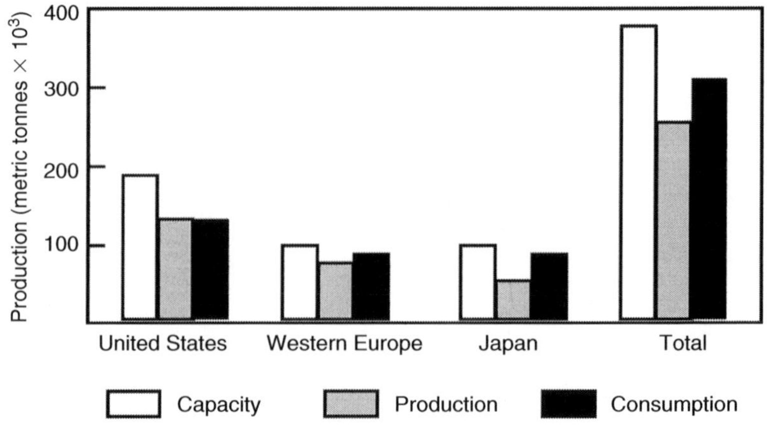

Figure 9.6. Supply/demand for activated carbon in the three major regions. Data from Waterhouse *et al.* (1996).

in Europe. Reactivation is normally carried out on-site for carbons used in gas-phase applications, but it must be carried out off-site (mostly by the same suppliers) for carbons used in liquid-phase applications like water treatment.

Production capacity of activated carbon in Japan is widely distributed among many companies, the largest producers being: Takeda Chemicals Industries (23%), Kuraray Chemicals (18%), Futamura Chemical Industries (18%) and Mitsubishi Chemical Corporation (11%). Lignocellulosic precursors such as coconut shell, wood and sawdust are the more frequently used, with smaller proportion of coal, resinous pitch, etc. Regeneration capacity is estimated to be around 30,000 tonnes per year. Some of the activated carbon is produced outside Japan, mainly in Southeast Asia, where joint venture companies have been established.

9.1.9 Market Consumption

As mentioned above, the major markets for activated carbon in decreasing order of importance are: water treatment, decolorizing, chemical and pharmaceutical processing, food processing, air and gas purification and solvent vapor recovery. The market has been increasing constantly as a consequence of environmental issues, especially fore water and air purification. Furthermore, as more and more countries are becoming industrialized the need for activated carbon to comply with environmental regulation will grow at a faster rate. Thus, while the activated carbon consumption is growing at a rate between 1% and 5% per year in areas like USA, Europe and Japan, the growth is much faster in developing countries.

The market of activated carbon is not uniform throughout the world, but there are many common characteristics. Information up to 1994 is discussed below for the three most industrialized regions in the world: USA, Western Europe and Japan. Figure 9.7 includes

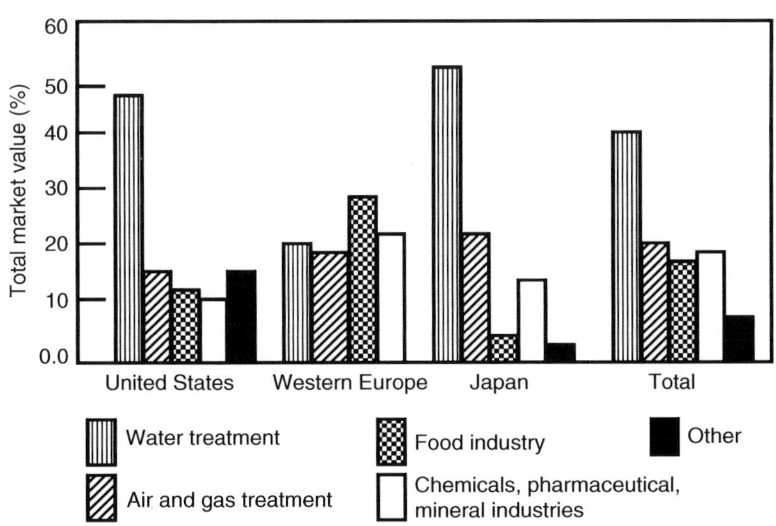

Figure 9.7. Activated carbon consumption in three major regions. Data from Waterhouse *et al.* (1996).

the different percentages of activated carbon consumption as a function of end use for the three regions (Waterhouse *et al.*, 1996).

Consumption in the USA has grown an average of around 5% per year up to 1994, when the consumption was estimated to be around 143,000 tonnes. More than 80% of that amount was used in liquid-phase applications. Sweetener decolorization used to be the main end-use of activated carbon, but it has been displaced by water treatment as a consequence of the increasing severity of government legislation on pollution control. Potable water is now the largest end-use market for activated carbon, around 26% of total consumption in 1994, 47% of which is powdered and 53% granular. The last years have seen a considerable increase in the use of granular activated carbon for potable water treatment as the definition of potable water standards becomes more restrictive.

Wastewater (industrial and municipal) treatment is the second largest market for activated carbon in the USA, accounting for 22% of the total in 1994. Here, powdered activated carbon is more popular, accounting for around two thirds of the total. Groundwater constitutes a relatively smaller market, but it is growing at a relatively high rate with increasing legislation on the cleaning of contaminated groundwater.

Use of activated carbon for sweetener (corn syrup, cane and beet sugars) decolorizing has considerably decreased from 26% of total USA consumption in 1980 to around 10% in 1994. The market seems to be now stabilized.

Other liquid-phase applications of activated carbon with lower consumption, but accounting for around 20% of the total liquid-phase consumption, include: mining (for the usage of low grade ores of silver and gold), household uses (home water taps for purification, oven hoods and aquaria), food, beverages and oils, pharmaceuticals, dry cleaning, electroplating, chemical processes and others.

Gas-phase applications of activated carbon in the USA are mainly concentrated towards granular carbons. Total consumption in 1994 accounted for somewhat <20% of the total market. The main applications are air purification (it includes industrial gas purification), solvent vapor recovery, automotive evaporation control systems, and others, of which the two first constitute the main share of the total.

The USA demand for activated carbon will very probably reach 200,000 tonnes per year by 2005, and the market will very probably be negatively affected by imports from the Asia-Pacific region. The trend is for more rapid growth in gas-phase applications, mainly air purification, solvent recovery and motor vehicle emission canisters. This means that granular activated carbon will have a better growth potential than powdered activated carbon.

As shown in Figure 9.7, the consumption structure of Western Europe is rather different to that of USA, because the most important use in 1995 was the food industry, with a share of 32%, followed by chemical–pharmaceutical industry (24%), water treatment (23%) and gas–air purification (21%). Main reasons for this difference are that the number of plants for water purification using activated carbon is much smaller and that wastewater treatment is not as common in Europe as it is in the USA. The market for powdered activated carbon in Europe seems to be stabilized, but that of granular, extruded and pelletized activated carbon is continuously growing. The trend for the market in the next years is somewhat paralleled to that of the USA, with a larger growth for granular carbons in gas-phase applications.

The structure of consumption of activated carbon in Japan is more similar to that of the USA, with the two main end-uses being water treatment (52%) and air–gas purification (24%). The main reason for the large demand of activated carbon for water treatment was the revision of water quality standard in 1993, to include a much larger number of compounds which had to be eliminated from drinking water. The demand of granular activated carbon is being pushed by the increasing number of advanced water purification systems being installed in and around large cities. There is also an increasing demand for domestic compact units to improve the taste and odor of potable water (antibacterial, silver-impregnated granular activated carbon is being introduced into this new market). Powdered activated carbon is used mainly for water treatment, production of industrial chemicals and beverage clarification. The use in gas-phase applications is mainly centred around granular activated carbon, 40–45% of which is used in cigarette filters. Other gas-phase applications are automobile canisters, deodorants in wastewater treatment plants, refrigerator deodorizing, etc.

There is in Japan a large interest in the development of new types of activated carbon, the research and development being carried out at universities and research centres of companies such as Takeda and Kuraray. Activated carbon fibers are produced by several companies, with a total of around 200 tonnes per year. This market is not growing because the price is very high, up to 20–50 times higher than granular activated carbon. Carbon molecular sieves are mainly produced by Takeda and Kuraray, and they are used for gas separation processes using pressure swing adsorption (PSA) technology.

9.1.10 Price

The market price of activated carbon varies widely because it is a function of demand, quality of the product, cost of production, etc. Oversupply, as in the USA in the early 1980s, led to a decrease in price, but the subsequent increasing demand improved the price in the early 1990s and has increased substantially since then.

In the USA, the price for granular activated carbon ranged in 1995 from \$1.65 to \$9.90 kg^{-1}, while most powdered activated carbon was priced in the range of \$0.80–2.00 kg^{-1}. The largest difference in price is for carbons used in water treatment, with granular carbons being up to three times more expensive. However, in such applications as gas and air purification the costs of granular and powdered activated carbon are similar.

The range of prices of activated carbon in Europe is wide because of the large number of grades and qualities available. Fluctuation of price in the last 20 years has been much smaller than in the USA and, in general terms, costs are higher in Europe. As a typical example, prices of powdered activated carbon for water treatment ranged, in 1995, from \$0.70 to \$1.50 kg^{-1}, whereas for food and pharmaceutical applications the range is \$1.40–5.00 kg^{-1}. Granular activated carbon is more expensive, \$1.60–1.75 kg^{-1} for water treatment and \$2.50–5.00 for gas-phase applications. Extruded activated carbon was in the price range of \$4.00–6.00. Average total regeneration costs were around \$0.70–0.85 kg^{-1}.

The price of activated carbon in Japan has not changed much in the last 10 years, very probably because of the imports of cheaper carbons from the Asia-Pacific region. However, the price of activated carbon in Japan is in general higher than in the USA and Europe.

9.1.11 Conclusions

The increasing demand for activated carbon consumption is largely based on the growing concern for pollution control and for increasing the quality of life. The growth of the activated carbon market in the last two decades in the most industrialized regions will continue in the near future as more developing industrial areas of the world realize the importance of controlling water and air pollution (this, independently of the fact that mankind has to try to reduce pollution by not creating it). This demand can be satisfied by considering the large number of precursors available for the production of activated carbon, the variety of activation processes described, and the available forms of activated carbon. This review shows that the porosity of activated carbon can be tailored to cover a wide range of applications thus opening the possibility of satisfying the demand for future needs. However, researchers and producers have to be aware of these new needs and, consequently, special attention has to be given to aspects such as the development of new activated carbons, specific for a given need. This implies not only looking for new precursors but also for adequate modifications in the activation procedures.

Areas in which further developments are expected are related to the optimization of the solution of air and water pollution, gas purification (removal of oxides of sulfur and nitrogen, of hydrogen sulfide, motor vehicle emissions, etc.), gas separation, mineral industries, regeneration, etc. Many of these areas will require the use of new forms of activated carbon such as cloth, felts, fibers, monoliths, etc., and consequently a search for the appropriate precursor and preparation mode is essential. Other areas in continuous progress will be gas storage, carbon molecular sieves and heterogeneous catalysis, all of them requiring considerable research efforts in the next few years.

9.2 Reference Information

9.2.1 Introduction

The world of carbon is not small and is multi-discipline. The structural range of carbon materials is wide with applications to be found almost everywhere. Although this book is devoted to activated carbons, it is appropriate to have access, for easy reference within the same covers, to the vocabulary of carbon materials. This is done in two ways; first, Section 9.2.2 is a list of the keywords contained within the journal *carbon*, and used to characterize (a) types of carbon known to exist, (b) methods of preparation and processing, (c) experimental techniques used for their characterization and (d) properties and phenomena.

In 1995, IUPAC published a list of definitions of carbon materials in an attempt to promote better understandings in the current literature (Fitzer *et al.* 1995). This attempt has met with success, but it is an ongoing process and so this list is included between these covers in an attempt to maintain progress in the way that we talk about carbon material (Section 9.2.3).

Surface chemistry is dominated by the names of Langmuir, BET and Dubinin with Radushkevich (DR). These scientists changed the world of surface science for ever and therefore it is appropriate to have, easily available, a brief biography of each scientist, to establish, if nothing more, how the Langmuir, BET and DR equations came about. The personal lives of these men also make interesting reading (Section 9.2.4).

9.2.2 Keywords from the Journal Carbon

Types of carbon

Activated carbon
Battery carbon
Carbon aerogel
Carbon anions
Carbon beads
Carbon black
Carbon cloth
Carbon clusters
Carbon composites
Carbon fibers
Carbon filaments
Carbon films
Carbon microbeads
Carbon microcoils
Carbon nanofibers
Carbon nanoparticles
Carbon nanotubes
Carbon precursor
Carbon xerogels
Carbon/carbon composites
Carbyne
Catalytically grown carbon
Char
Charcoal
Chemically modified carbons
Coal
Coal-tar pitch
Coke
Diamond
Diamond-like carbon
Doped carbons
Electrodes
Exfoliated graphite
Fullerene
Glass-like carbon
Graphite
Graphite oxide
Graphitic carbon
Highly oriented graphite
Intercalation compounds
Isotropic carbon
Mesophase
Mesophase pitch (MPP)
Molecular sieves
Natural graphite
Needle coke
Non-graphitic carbon
Nuclear graphite
Petroleum coke
Petroleum pitch
Pitch
Porous carbon
Pyrolytic carbon
Resins
Shungite
Single crystals
Soot
Synthetic graphite
Tars
Vapor-grown carbon

Preparation and processing

Activation
Annealing
Arc discharge
Baking
Calcination
Carbonization
Catalyst
Catalyst support

Chemical treatment
Chemical vapor deposition (CVD)
Chemical vapor infiltration
Coating
Coking
Combustion
Cracking
Doping
Electrochemical treatment
Etching
Gasification
Graphitization
Grinding
Heat treatment
High pressure
Hydrothermal treatment
Implantation
Impregnation
Intercalation
Laser irradiation
Mixing
Oxidation
Plasma deposition
Plasma reactions
Plasma spluttering
Pyrolysis
Sintering
Stabilization

Techniques

Adsorption
Atomic force microscopy
BET surface area
Chemisorption
Chromatography
Computational chemistry
Differential scanning calorimetry
Dynamical mechanical thermal analysis
Electron diffraction
Electron energy loss spectroscopy
Electron microscopy
Electron spin resonance
Ellipsometry
Image analysis
Infrared spectroscopy
Ion scattering spectroscopy
Light scattering
Mass spectrometry
Mercury porosimetry
Microcalorimetry
Modeling
Molecular simulation
Mossbauer spectroscopy
Neutron scattering
Nuclear magnetic resonance
Optical microscopy
Paramagnetic resonance
Photoelectron spectroscopy
Raman spectroscopy
Rheology
Scanning electron microscopy
Scanning tunneling microscopy
Small angle X-ray scattering
Spectrophotometry
Sputtering
Temperature programmed desorption
Thermal analysis
Thermodynamic analysis
Transmission electron microscopy
ultrasonic measurements
X-ray diffraction
X-ray photoelectron spectroscopy
X-ray scattering

Properties and phenomena

Absorption
Activation energy
Adsorption properties
Aggregation
Bioactivity
Biocompatibility
Carbon yield
Catalytic properties
Chemical structure
Crystal structure

Crystallite size
Defects
Density
Dielectric properties
Diffusion
Elastic properties
Electrical (electronic) properties
Electrochemical properties
Electronic properties
Field emission
Fracture
Frictional properties
Functional groups
Galvanomagnetic properties
Gas storage
Heat of adsorption
Immersion enthalpy
Intercalation reactions
Interfacial properties
Lattice constant
Lattice dynamics
Luminescence
Magnetic properties
Mechanical properties
Microporosity
Microstructure
Mössbauer effect
Optical properties
Particle size
Phase equilibria
Phase transitions
Phonons
Photoconductivity
Porosity
Radiation damage
Reaction kinetics
Reactivity
Rheology
Specific heat
Superconductivity
Surface areas
Surface oxygen complexes
Surface properties
Texture
Thermal conductivity
Thermal diffusivity
Thermal expansion
Thermodynamics properties
Transport properties
Viscoelasticity

9.2.3 IUPAC Definitions (Fitzer et al., *1995)*

ACETYLENE BLACK
Description:
Acetylene black is a special type of *carbon black* formed by an exothermic decomposition of acetylene. It is characterized by the highest degree of aggregation and crystalline orientation when compared with all types of *carbon black.*
See *carbon black*

Note:
Acetylene black *must not be confused with the* carbon black *produced as a by-product during the production of acetylene in the electric arc process.*
See *carbon black*

ACHESON GRAPHITE
Description:
Acheson graphite is a *synthetic graphite* made by the Acheson process.
See *synthetic graphite*

Note:
Reference to Acheson in combination with synthetic graphite *honors the inventor of the first technical* graphitization. *Today the term* Acheson graphite, *however, is of historical interest only because it no longer covers the plurality of* synthetic graphite.
See *synthetic graphite*

ACTIVATED CARBON
Description:
Activated carbon is a porous *carbon material*, a *char* which has been subjected to reaction with gases, sometimes with the addition of chemicals (e.g. $ZnCl_2$) before, during or after *carbonization* in order to increase its adsorptive properties.
See *carbon material, carbonization, char*

Notes:
Activated carbons *have a large adsorption capacity, preferably for small molecules, and are used for purification of liquids and gases. By controlling the process of* Carbonization *and activation, a variety of active carbons having different porosity can be obtained.* Activated carbons *are used mainly in granular and powdered forms, but can also be produced in textile form by controlled* carbonization *and activation of textile fibers. Other terms used in the literature: active carbons, active charcoals.*
See *carbonization*

ACTIVATED CHARCOAL
Description:
Activated charcoal is a traditional term for *activated carbon.*
See *activated carbon*

AGRANULAR CARBON
Description:
A*granular carbon* is a mono-granular or monolithic *carbon material* with homogeneous microstructure which does not exhibit any structural components distinguishable by optical microscopy.
See *carbon material*

Notes:
The above definition of a homogeneous microstructure does not pertain to pores and structural components which may be visible by contrast differences in optical microscopy with polarized light. *As a consequence*, glass-like carbon *with visible pores is still an* agranular carbon. *The same is true, for instance, for* pyrolytic carbon *with preferred orientation, such as conical or lamellar structures, visible in optical microscopy with polarized light. The use of the term* agranular carbon *is not restricted to bulk materials of a minimum size. Only* particulate carbon *should be excluded even if the isolated particles exhibit a homogeneous microstructure.*
See *glass-like carbon, particulate carbon, pyrolytic carbon*

AMORPHOUS CARBON

Description:

Amorphous carbon is a *carbon material* without long-range crystalline order. Short-range order exists, but with deviations of the inter-atomic distances and/or inter-bonding angles with respect to the graphite lattice as well as to the diamond lattice.

See *carbon material, DLC*

Notes:

The term amorphous carbon *is restricted to the description of* carbon materials *with localized π-electrons as described by PW Anderson (Phys.Rev. 109, 1492 (1958)). Deviations in the C–C distances >5 (i.e.* $\pm\Delta x/X_0 > 0.05$*, where* X_0 *is the inter-atomic distance in the crystal lattice for the* sp^2 *as well as for the* sp^3 *configuration) occur in such materials, as well as deviations in the bond angles because of the presence of "dangling bonds".*

Above description of amorphous carbon *is not applicable to* carbon materials *with two-dimensional structural elements present in all pyrolysis residues of carbon compounds as polyaromatic layers with a nearly ideal inter-atomic distance of* $a = 142$ *pm and an extension >1000 pm.*

See *carbon material*

ARTIFICIAL GRAPHITE

Description:

The term *artificial graphite* is often used in place of *synthetic graphite*.

See *synthetic graphite*

Note: *This term is not recommended.*

BAKING

Description:

The process in which the carbonaceous *binder*, usually *coal tar pitch* or *petroleum pitch*, as part of a shaped *carbon mix* is converted to *carbon* yielding a rigid carbon body by the slow application of heat. The process can take as little as 14 days in coarse-grained, electrothermic grades (low binder level) and as long as 36 days in ultra-fine-grained, specialty grades (high binder level). The final baking temperature can be in the range of 1100 K–1500 K, depending on the grade.

See *binder, carbon, coal-tar pitch, petroleum pitch*

BINDER

Description:

A *binder* is usually a *coal tar pitch* or *petroleum pitch* (but may include thermosetting resins or *MPP* powders) which, when mixed with a *binder coke* or a *filler*, constitutes a *carbon mix*. This is used in preparation of the formation of shaped green bodies and subsequently *carbon artifacts*.

See *binder coke, carbon artifact, carbon mix, coal-tar pitch, MPP, petroleum pitch, pitch*

BINDER COKE
Description:
Binder coke is a constituent of a carbon (or ceramic) artifact resulting from *carbonization* of the binder during baking.
See *carbonization*

Note:
Pitches *are mainly used as binders (i.e. as precursors for* binder cokes*) but the term* binder *should include any carbonaceous binder material, for example thermosetting resins such as poly(furfuryl alcohol) or phenolics and similar compounds which may form a* char *during* carbonization.
See *carbonization, char*

BROOKS AND TAYLOR STRUCTURE IN THE CARBONACEOUS MESOPHASE
Description:
The *Brooks and Taylor structure in the carbonaceous mesophase* refers to the structure of the anisotropic spheres which precipitate from isotropic *pitch* during pyrolysis. The structure of the spheres consists of a lamellar arrangement of aromatic molecules in parallel layers which are perpendicular to the polar axis of the sphere and which are perpendicular to the mesophase–isotropic phase interface.

Note:
The term Brooks and Taylor structure *is recommended to describe the particular lamellar morphology of the spherules most commonly precipitated from pyrolyzed PITCH. The term honors the workers who first recognized the significance of* carbonaceous mesophase *to carbon science and technology and who first defined this spherical morphology. The term* Brooks and Taylor structure *does not cover all structures found in the spherical mesophase, because other lamellar arrangements have been observed.*
See *carbonaceous mesophase, pitch*

BULK MESOPHASE
Description:
Bulk mesophase is a continuous anisotropic phase formed by coalescence of mesophase spheres. *Bulk mesophase* retains fluidity and is deformable in the temperature range up to about 770 K, and transforms into *green coke* by further loss of hydrogen or low-molecular-weight species.
See *green coke*

Note:
This bulk mesophase *can sometimes be formed directly from the isotropic* pitch *without observation of intermediate spheres.*
See *pitch*

CALCINED COKE
Description:
Calcined coke is a *petroleum coke* or *coal-derived pitch coke* obtained by heat treatment of *green coke* to about 1600 K. It will normally have a hydrogen content of <0.1 wt%.
See *Coal-derived pitch coke, green coke, petroleum coke*

Note:
Calcined coke *is the main raw material for the manufacture of* polygranular, carbon *and* polygranular graphite *products (e.g.* carbon *and* graphite, electrodes*).*
See *Carbon electrode polygranular carbon, polygranular graphite*

CARBON
Description:
Carbon is the element number 6 of the Periodic table of elements (electronic ground state $1s^2\ 2s^2\ 2p^2$).

Note:
For description of the various types of carbon as a solid *the term* carbon *should be used only in combination with an additional noun or a clarifying adjective.*
See *amorphous carbon, carbon fibers, carbon material, glass-like carbon, graphitic carbon, non-graphitic carbon, pyrolytic carbon*

CARBON ARTIFACT
Description:
Carbon artifact means an "artificially produced" solid body which consists mainly of carbonaceous material in a distinct shape.

Note:
Sometimes this term is also used for artificially (in the sense of technically) produced non-shaped carbon materials *such as* coke, *blacks*, etc. *This application of the term* carbon artifact *is not recommended. Synonyms to the term* carbon artifact *are*: *"artificial carbon article" or "artificial carbon body".*
See *carbon material, coke*

CARBON BLACK
Description:
Carbon black is an industrially manufactured colloidal carbon material in the form of spheres and of their fused aggregates with sizes below 1000 nm.
See *colloidal carbon*

Notes:
Carbon black *is a commercial product manufactured by thermal decomposition, including detonation, or by incomplete combustion of carbon hydrogen compounds and has a well-defined morphology with a minimum content of tars or other extraneous materials.*

For historical reasons, however, carbon black *is popularly but incorrectly regarded as a form of* soot*. In fact, in many languages, the same word is used to designate both materials.*

CARBON FELT
Description:
Carbon felt is a textile material consisting of, in approximation, randomly oriented and intertwined *carbon fibers*.
See *carbon fibers*

Note:
Carbon felts *are usually fabricated by* carbonization *of organic felts but they can also be produced from short* carbon fibers.
See *carbon fibers, carbonization*

CARBON FIBER
Description:
Carbon fibers are fibers (filaments, tows, yams, rovings) consisting of at least 92 wt% (mass fraction) *carbon*, usually in the *non-graphitic* state.
See *carbon, non-graphitic carbon*

Notes:
Carbon fibers *are fabricated by pyrolysis of organic precursor fibers or by growth from gaseous hydrocarbons. The use of the term* graphite fibers *instead of* carbon fibers *as often observed in the literature is incorrect and should be avoided. The term* graphite fibers *is justified only if three-dimensional crystalline order is confirmed, for example, by X-ray diffraction measurements.*
See *graphite fibers*

CARBON FIBER FABRICS
Description:
Carbon fiber fabrics are woven textile materials made of *carbon fibers*.
See *carbon fibers*

CARBON FIBERS TYPE HM
Description:
Carbon fibers type HM (high modulus) are *carbon fibers* with a value of Young's modulus (tensile modulus) larger than 300 GPa (nearly 30 of the C_{II} elastic constant of a graphite single crystal).
See *carbon fibers*

Notes:
The level of the tensile modulus of carbon fibers *is controlled by the degree of preferred orientation of the layer planes in the direction parallel to the fiber axis.* C_{II}*, the elastic constant of graphite single crystals in the direction of the layer planes is 1060* $\pm$ *20 GPa. In general, the ratio of tensile strength to tensile modulus is smaller than* $1\text{–}10^{22}$ *for carbon fibers type HM (but the tensile strength is influenced by flaws in the fibers and may be improved in the future).*

Carbon fibers type UHM (ultra-high modulus) *have moduli of elasticity in excess of 600 GPa, surpassing 50 of the theoretical C_{II}* number. Such high values of Young's modulus can be achieved most readily in MPP-based carbon fibers.
See *carbon fibers*

CARBON FIBERS TYPE HT
Description:
Carbon fibers type HT are carbon fibers with values of Young's modulus between 150 and 275 to 300 GPa. The term HT, referring to high tensile strength, was early applied because fibers of this type display the highest tensile strengths.
See *carbon fibers*

Notes:
The disposition of boundaries between the fiber types is somewhat arbitrary. For carbon fibers type HT, *the values of the strength-to-stiffness ratio are typically larger than 1.5-10-2. The tensile strength of* carbon fibers *is flaw controlled, however, and therefore the measured values increase strongly as the diameter of the filaments is decreased.*
See *carbon fibers*

CARBON FIBERS TYPE IM CARBON FIBERS TYPE IM
Description:
The *carbon fibers type IM* (*intermediate modulus*) are related to *carbon fibers type HT* because of the comparable values of tensile strength, but are characterized by greater stiffness (Young's modulus up to approximately 35% of the theoretical C_{II} value).
See *carbon fibers, carbon fibers type HT*

Notes:
The tensile modulus (Young's modulus) varies between ca. 275 and 350 GPa, but the disposition of the boundaries is somewhat arbitrary. The relatively high ratio of tensile strength to tensile modulus, typically above $1–10^{-2}$, in carbon fibers type IM, *in spite of an increase of Young's modulus, requires a further increase of strength, which is achievable by a significant reduction of the monofilament diameter down to about 5 μm. Such small filament diameters are typical of* carbon fibers type IM.

CARBON FIBERS TYPE LM
Description:
Carbon fibers type LM (*low modulus*) are *carbon fibers* with isotropic structure, tensile modulus values as low as 10% of the C_{II} values of the graphite single crystal, and low strength values.
See *carbon fibers*

Notes:
The term carbon fibers type LM *is sometimes used for various types of isotropic* carbon fibers *known as* pitch-based or rayon-based carbon fibers *that have not been subjected to*

hot-stretching. Such fibers are not used for reinforcement purposes in high-performance composites.
See *carbon fibers, pitch-based carbon fibers, rayon-based carbon fibers*

CARBON FIBERS TYPE UHM
Description:
Carbon fibers type UHM designates a class of *carbon fibers* having very high values of Young's modulus larger than 600 GPa (i.e. >55% of the theoretical C_{II} value of *graphite*).
See *carbon fiber, carbon fiber type HM, graphite*

CARBON MATERIAL
Description:
Carbon material is a solid high in content of the element *carbon* and structurally in a *non-graphitic* state.
See *carbon, non-graphitic carbon*

Notes:
The use of the term carbon *as a short term for a material consisting of* non-graphitic carbon *is incorrect. The use of the term* carbon *without a second noun or a clarifying adjective should be restricted to the chemical element carbon. The term* carbon *can be used in combination with other nouns or clarifying adjectives for special types of* carbon materials (carbon electrode, carbon fibers, pyrolytic carbon, glass-like carbon *and others*).
See *Carbon, carbon electrode, carbon fibers, glass-like carbon,non-graphitic carbon, pyrolytic carbon*

CARBON MIX
Description:
Carbon mix is a mixture of *filler coke*, for example grains and/or powders of solid *carbon materials*, and a carbonaceous *binder* and selected additives, prepared in heated mixers at temperatures in the range of 410 K–445 K as a preliminary step for the formation of shaped green bodies.
See *binder, carbon material, filler coke*

CARBON WHISKERS
Description:
See *graphite whiskers*

CARBONACEOUS MESOPHASE
Description:
Carbonaceous mesophase is a liquid-crystalline state of *pitch* which shows the optical birefringence of disc-like (discotic) nematic liquid crystals. It can be formed as an intermediate phase during thermolysis (pyrolysis) of an isotropic molten PITCH or by development from *pitch* fractions prepared by selective extraction. Generally, the spherical mesophase developed from a pyrolyzing *pitch* has the *Brooks and Taylor structure*. With continuous heat treatment the *carbonaceous mesophase* coalesces to a state of *bulk mesophase* before solidification to *green coke* with further loss of hydrogen or low-molecular-weight compounds.
See *Brooks and Taylor structure in the carbonaceous, mesophase, bulk mesophase, green coke, pitch*

Notes:
In the formation of carbonaceous mesophase *by thermolysis (pyrolysis) of isotropic molten* pitch, *the development of a liquid-crystalline phase is accompanied by simultaneous aromatic polymerization reactions. The reactivity of* pitch *with increasing heat treatment temperature and its thermosetting nature are responsible for the lack of a true reversible thermotropic phase transition for the* bulk mesophase *in most* pitches. *Due to its glass-like nature most of the liquid-crystalline characteristics are retained in the super-cooled solid state.*
See *Brooks and Taylor structure in the carbonaceous, mesophase, bulk mesophase, pitch*

CARBONIZATION
Description:
Carbonization is a process by which solid residues with increasing content of the element carbon are formed from organic material usually by pyrolysis in an inert atmosphere.

Notes:
As with all pyrolytic reactions, carbonization *is a complex process in which many reactions take place concurrently such as dehydrogenation, condensation, hydrogen transfer and isomerization. It differs from* coalification *in that its reaction rate is faster by many orders of magnitude. The final pyrolysis temperature applied controls the degree of* carbonization *and the residual content of foreign elements, for example, at* $T \sim 1200$ *K the carbon content of the residue exceeds a mass fraction of 90 wt%, whereas at* $T \sim 1600$ *K more than 99 wt% carbon is found.*
See *calcined coke, coalification*

CATALYTIC GRAPHITIZATION
Catalytic graphitization refers to a transformation of *non-graphitic carbon* into *graphite* by heat treatment in the presence of certain metals or minerals.
See *graphite, graphitization, non-graphitic carbon*

Notes:
Catalytic graphitization *gives a fixed degree of* graphitization *at lower temperature and/or for a shorter heat treatment time than in the absence of the catalytic additives (or a higher degree of* graphitization *at fixed heat treatment conditions). Often it involves dissolution of* carbon *and precipitation of* graphite *at the catalyst particles so that* non-graphitizing carbons *can be graphitized by this procedure.*
See *carbon, graphite, graphitization, non-graphitizing carbon*

CHAR
Description:
Char is a solid decomposition product of a natural or synthetic organic material.

Notes:
If the precursor has not passed through a fluid stage, char *will retain the characteristic shape of the precursor (although becoming of smaller size). For such materials the term "pseudomorphous" has been used. Some simple organic compounds (e.g. sugar) melt at an early stage of decomposition and then polymerize during* carbonization *to produce* chars.
See *carbonization*

CHARCOAL
Description:
Charcoal is a traditional term for a *char* obtained from wood, peat, coal or some related natural organic materials.
See *char*

Note:
Charcoal *has highly reactive inner surfaces and a low sulfur content. It has or has had, therefore, a variety of uses, for example, in ferrous metallurgy and for gunpowder (minor uses: medical purpose and paint materials).*

COAL-DERIVED PITCH COKE
Description:
Coal-derived pitch coke is the primary industrial solid *carbonization* product obtained from *coal tar pitch*, and is mainly produced in chamber or *delayed coking processes.*
See *carbonization, coal-tar pitch, delayed coking process*

Note:
Coal-derived pitch coke*, although it exhibits a pre-graphitic microstructure, has often a lower graphitizability than* petroleum coke. *Fractions of* coal tar pitches *(obtained by extraction or filtration) may form* cokes *with needle-like structures and have an improved graphitizability. The usually lower graphitizability compared to* petroleum coke *is due to an inhibition of mesophase growth because of chemical and physical differences of the cokes.*
See *coal tar pitches, coke, petroleum coke*

COAL TAR PITCH
Description:
Coal tar pitch is a residue produced by distillation or heat treatment of coal tar. It is a solid at room temperature, consists of a complex mixture of numerous predominantly aromatic hydrocarbons and heterocyclic compounds, and exhibits a broad softening range instead of a defined melting temperature.

Note:
The hydrogen aromaticity in coal tar pitch *(ratio of aromatic to total content of hydrogen atoms) varies from 0.7 to 0.9.*

COALIFICATION
Description:
Coalification is a geological process of formation of materials with increasing content of the element carbon from organic materials that occurs in a first, biological stage into peats, followed by a gradual transformation into coal by action of moderate temperature (about 500 K) and high pressure in a geochemical stage.

Notes:
Coalification *is a dehydrogenation process with a reaction rate slower by many orders of magnitude than that of* carbonization. *Some specific reactions approach completion before others have started. The dehydrogenation remains incomplete. The degree of* coalification *reached by an organic material in the process of* coalification *increases progressively and*

can be defined by means of the measured C/H ratio and of the residual contents of oxygen, sulfur and nitrogen.
See *carbonization*

COKE
Description:
Coke is a solid high in content of the element carbon and structurally in the *non-graphitic* state. It is produced by pyrolysis of organic material which has passed, at least in part, through a liquid or liquid-crystalline state during the *carbonization* process. *Coke* can contain mineral matter.
See *carbonization, non-graphitic carbon*

Notes:
As some parts, at least, of the carbonization *product have passed through a liquid or liquid-crystalline state, the resulting* non-graphitic carbon *is of the graphitizable variety. From a structural viewpoint, the term* coke *characterizes the state of* graphitizable carbon *before the beginning of* graphitization.
See *carbonization, graphitizable carbon, graphitization, non-graphitic carbon*

COKE BREEZE
Description:
Coke breeze is a by-product of *coke* manufacture. It is the residue from the screening of heat-treated *coke*; the particle size is <10 mm. Generally *coke breeze* has a volatile matter content of <3 wt%.
See *coke*

COLLOIDAL CARBON
Description:
Colloidal carbon is a *particulate carbon* with particle sizes below ca 1000 nm in at least one dimension.

Note:
Colloidal carbon *exists in several morphologically distinct forms.*
See *particulate carbon, carbon black*

DELAYED COKE
Description:
Delayed coke is a commonly used term for a primary *carbonization* product (*green* or *raw coke*) from high-boiling hydrocarbon fractions (heavy residues of petroleum or coal processing) produced by the *delayed coking process.*
See *carbonization, delayed coking process, green* or *raw coke*

Notes:
Delayed coke *has, with only a few exceptions, a better graphitizability than* cokes *produced by other coking processes even if the same feedstock is used.* Delayed coke *contains a mass fraction of volatile matter between 4 and 15 wt% which can be released during heat treatment.*
See *coke, delayed coking process*

DELAYED COKING PROCESS
Description:
Delayed coking process is a thermal process which increases the molecular aggregation or association in petroleum-based residues or *coal tar pitches* leading to extended mesophase domains. This is achieved by holding them at an elevated temperature (usually 750 K–765 K) over a period of time (12 to 36 h). It is performed in a coking drum and is designed to ultimately produce *delayed coke*. The feed is rapidly pre-heated in a tubular furnace to about 760 K.
See *coal tar pitch, delayed coke*

Notes:
Needle coke *is the premium product of the* delayed coking process. *It is generally produced from highly aromatic residues from, for instance, the steam cracking of gas oil. Its appearance and preferred orientation of the* graphene layers *is the consequence of the evolved gaseous products percolating through the mesophase which must not have too high a viscosity. A close control of temperature, time and feedstock is essential. Lower grades, for instance* isotropic cokes, *are used for* carbon electrodes *applied, for example, in the production of aluminum.*
See *isotropic carbon, needle coke*

DIAMOND
Description:
Diamond is an allotropic form of the element carbon with cubic structure (space group $0\,h^7$–$Fd3\,m$) which is thermodynamically stable at pressures above 6 GPa at room temperature and metastable at atmospheric pressure. At low pressures *diamond* converts rapidly to *graphite* at temperatures above 1900 K in an inert atmosphere. The chemical bonding between the carbon atoms is covalent with sp^3 hybridization.
See *carbon, graphite*

Note:
There is also a hexagonal diamond-like structure of the element carbon *(Lonsdaleite).*
See *carbon*

DIAMOND BY CVD
Description:
Diamond by CVD is formed as crystals or as films from various gaseous hydrocarbons or other organic molecules in the presence of activated, atomic hydrogen. It consists of sp^3-hybridized carbon atoms with the three-dimensional crystalline structure of the diamond lattice.
See *DLC films*

Notes:
"CVD diamond" or "low-pressure diamond" are synonyms of the term diamond by CVD. Diamond by CVD *can be prepared in a variety of ways. Deposition parameters are: total (low) pressure, partial hydrogen pressure, precursor molecules in the gas phase, temperature for activation of the hydrogen and that of the surface of the underlying substrate. The energy supply for the hydrogen activation may be, for instance: heat, radio frequency, microwave excitation (plasma deposition) or accelerated ions (e.g.* Ar^+ ions). *CVD diamond has also been obtained at atmospheric pressure from oxyacetylene torches*

and by other flame-based methods. Often CVD carbon films consist of a mixture of sp^2*- and* sp^3*-hybridized carbon atoms and do not have the three-dimensional structure of the DIAMOND lattice. In this case they should be called* hard amorphous carbon *or DLC films.*
See *DLC films*

DLC FILMS
Description:
Diamond-like carbon (DLC) films are hard, amorphous films with a significant fraction of sp^3-hybridized carbon atoms and which can contain a significant amount of hydrogen. Depending on the deposition conditions, these films can be fully amorphous or contain *diamond* crystallites. These materials are not called *diamond* unless a full three-dimensional crystalline lattice of *diamond* is proven.
See *diamond*

Notes:
Diamond-like films without hydrogen can be prepared by carbon ion-beam decomposition ion-assisted sputtering from graphite *or by laser ablation of* graphite. DLC films *containing significant contents of hydrogen are prepared by CVD. The hydrogen content is usually over 25 at.%. The deposition parameters are (low) total pressure, hydrogen partial pressure, precursor molecules, and plasma ionization. The plasma activation can be radio frequency, microwave, or* Ar^+ ions. *High ionization favors amorphous films while high atomic hydrogen contents favor* diamond *crystallite formation. Due to the confusion about structure engendered by the term* DLC films, *the term* hard amorphous carbon *has been suggested as a synonym.*

ELECTROGRAPHITE
Electrographite is a *synthetic graphite* made by electrical heating of *graphitizable carbon*.
See *synthetic graphite*

EXFOLIATED GRAPHITE
Description:
Exfoliated graphite is the product of very rapid heating (or flash heating) of graphite intercalation compounds, such as graphite hydrogen sulfate of relatively large particle diameter (flakes). The vaporizing intercalated substances force the graphite layers apart. The *exfoliated graphite* assumes an accordion-like shape with an apparent volume often hundreds of times that of the original graphite flakes.

Notes:
Exfoliated graphite is usually prepared from well-crystallized natural *flake* graphite *It is used for the production of graphite foils.* Exfoliated graphite *is different from the deflagration product of graphite oxide (graphitic acid).*
See *natural graphite*

FIBROUS ACTIVATED CARBON
Description:
Fibrous activated carbon is an *activated carbon* in the form of fibers, filaments yarns or rovings and fabrics or felts. Such fibers differ from *carbon fibers* used for reinforcement

purposes in composites by their high surface area, high porosity and low mechanical strength.
See *activated carbon; carbon fibers*

Note:
Sometimes fabrics of fibrous activated carbon *are named* charcoal *cloths; a more precise term is "activated carbon cloth".*
See *activated charcoal, charcoal*

Fibrous carbon
Description:
See *filamentous carbon*

FILAMENTOUS CARBON
Description:
Filamentous carbon is a carbonaceous deposit from gaseous carbon compounds, consisting of filaments grown by the catalytic action of metal particles.

Notes:
In general, such deposits are obtained at pressures of <100 kPa in the temperature region 600 K to 1300 K on metals such as iron, cobalt or nickel. Typical filaments consist of a duplex structure, a relatively oxidation-resistant skin surrounding a more easily oxidizable core, with a metal particle located at the growing end of the filament. They generally range from 0.01 to 0.5 pm in diameter and up to 10 pm in length.

In some systems, the metal particles are located in the middle of the filaments, and there are also examples where several filaments originate from a single particle.

The filaments may be produced in different conformations, such as helical, twisted and straight.
See RTK Baker and PS Harris. In: *Chemistry and Physics of Carbon*, Vol. 14, Eds. PL Walker Jr., PA Thrower. Marcel Dekker, New York, 1978, pp. 83–165.

Filler
Description:
Filler (also called GRIST) is a petroleum- or coal-based coke fraction of a green, carbon mix or formulation. Coarse particles, >0.425 mm, are sometimes referred to as tailings; fine particles, <0.074 mm, are referred to as flour. *graphite* flour is also used as a filler.

FILLER COKE
Description:
Filler coke is the main constituent of a carbon artifact, introduced as solid component (predominantly in the form of *particulate carbon*) into the "*carbon mix*" from which *polygranular carbon* and *graphite* materials are obtained by heat treatment.
See *carbon material, carbon mix, graphite material, particulate carbon, polygranular carbon, polygranular graphite*

Note:
Filler coke *is not necessarily the only, but it is commonly the most important* filler *material used in a* "carbon mix" *which consists of* filler *and binder.*
See *carbon mix, filler*

FLUID COKE
Description:
Fluid coke is the *carbonization* product of high-boiling hydrocarbon fractions (heavy residues of petroleum or coal processing) produced by the fluid coking process.
See *carbonization*

Notes:
Fluid coke *consists of spherulitic grains with a spherical layer structure and is generally less graphitizable than* delayed coke. *Therefore, it is not suitable as* filler coke *for* polygranular graphite *products and is also less suitable for* polycrystalline carbon *products. Due to its isotropy it is less suitable to produce an anisotropic* synthetic graphite. All cokes *contain a fraction of matter that can be released as volatiles during heat treatment. This mass fraction, the so-called volatile matter, is in the case of* fluid coke *about 6 wt%.*
See *coke, delayed coke, filler coke, polygranular graphite polycrystalline carbon, synthetic graphite*

FULLERENES
See *note to the preface*

FURNACE BLACK
Description:
Furnace black is a type of *carbon* that is produced industrially in a furnace by incomplete combustion in an adjustable and controllable process that yields a wide variety of properties within the product.
See *carbon black*

Note:
The most widely employed industrial process for carbon black *production is the furnace process.*
See *carbon black*

GAS-PHASE-GROWN CARBON FIBERS
Description:
Gas-phase-grown carbon fibers are *carbon fibers* grown in an atmosphere of hydrocarbons with the aid of fine particulate solid catalysts such as iron or other transition metals and consisting of *graphitizable carbon.*
See *carbon fibers, graphitizable carbon*

Notes:
Gas-phase-grown carbon fibers *transform during* graphitization heat treatment *into* graphite fibers. *These show a very high degree of preferred orientation and are particularly suitable for intercalation treatments. The term "vapor-grown carbon fibers" alternatively used in the literature is acceptable. The use of the term "CVD fibers" is not recommended as an alternative for* gas-phase-grown carbon fibers *since the term "CVD fibers" also describes fibers grown by a CVD process on substrate fibers.*
See *graphite fibers, graphitization heat treatment*

GLASS-LIKE CARBON

Description:

Glass-like carbon is an *agranular non-graphitizable carbon* with a very high isotropy of its structural and physical properties and with a very low permeability for liquids and gases. The original surfaces and the fracture surfaces have a pseudo-glassy appearance.
See *agranular carbon, non-graphitizable carbon*

Note:

The often used synonyms "Glassy Carbon" and "Vitreous Carbon" have been introduced as trademarks and should not be used as terms. From a scientific viewpoint, all synonymous terms suggest a similarity with the structure of silicate glasses which does not exist in glass-like carbon, *except for the pseudo-glassy appearance of the surface.*

Glass-like carbon cannot be described as *amorphous carbon* because it consists of two-dimensional structural elements and does not exhibit "dangling" bonds.
See *amorphous carbon*

GRANULAR CARBON

Description:

The term *granular carbon* is equivalent to coarse *particulate carbon*. This is a *carbon material* consisting of separate particles or grains which are monolithic, on the average larger than about 100 urn in diameter, but smaller than about 1 cm.
See *carbon material, particulate carbon*

Notes:

Although limits of size cannot be exactly defined, coke grains obtained by grinding belong to coarse particulate carbon *for grain sizes above ca. 100 urn, or to fine* particulate carbon *for grain sizes below ca. 100 urn.* Colloidal graphite *obtained by grinding of* natural graphite *is a typical extra fine* particulate carbon. *Industrial* carbon materials *(such as electrodes) are made with* fillers *composed of coarse* particulate carbon *(coke grains) and fine* particulate carbon *(flour), and sometimes even* colloidal carbon (carbon blacks *or* soot). *They are therefore polygranular materials.*
See *carbon black, carbon material,colloidal carbon, filler, natural graphite, particulate carbon, soot*

GRAPHENE LAYER

Description:

Graphene is a single carbon layer of the graphite structure, describing its nature by analogy to a polycyclic aromatic hydrocarbon of quasi-infinite size.

Notes:

Previously, descriptions such as graphite layers, carbon layers or carbon sheets have been used for the term graphene.

As graphite *designates that modification of the chemical element* carbon, *in which planar sheets of carbon atoms, each atom bound to three neighbors in a honeycomb-like structure, are stacked in a three-dimensional regular order, it is not correct to use for a single layer a term which includes the term* graphite, *which would imply a three-dimensional structure. The term* graphene *should be used only when the reactions, structural relations or other properties of individual layers are discussed.*
See *carbon, graphite*

GRAPHITE

Description:

Graphite is an allotropic form of the element carbon consisting of layers of hexagonally arranged carbon atoms in a planar condensed ring system (*graphene layers*). The layers are stacked parallel to each other in a three-dimensional crystalline long-range order. There are two allotropic forms with different stacking arrangements, hexagonal and rhombohedral. The chemical bonds within the layers are covalent with sp^2 hybridization and with a C–C distance of 141.7 pm. The weak bonds between the layers are metallic with a strength comparable to *van der Waals* bonding only.

See *carbon, hexagonal graphite, rhombohedral graphite*

Note:

The term graphite *is also used often but incorrectly to describe* graphite materials, that is *materials consisting of* graphitic carbon *made from* carbon materials *by processing to temperatures* >2500 *K, even though no perfect graphite structure is present.*

See *graphitic carbon, carbon material, graphite material*

GRAPHITE ELECTRODE

Description:

See *carbon electrode*

GRAPHITE FIBERS

Description:

Graphite fibers are *carbon fibers* consisting mostly of *synthetic graphite* for which three-dimensional crystalline order is confirmed by X-ray diffraction.

See *carbon fibers, synthetic graphite*

Note:

Graphite fibers *can be obtained by* graphitization heat treatment *of* carbon fibers *if these consist mostly of* graphitizable carbon. *If the h,k,l diffraction lines are difficult to recognize because they are of minor intensity, the mean interlayer spacing c/2 can be used as indication for the presence of a graphitic structure. The c/2 value of 0.34 nm is generally considered as an upper limit for* synthetic graphite.

See *carbon fibers, graphitizable carbon, graphitization heat treatment, synthetic graphite*

GRAPHITE MATERIAL

Description:

Graphite material is a material consisting essentially of *graphitic carbon*.

See *graphite, graphitic carbon*

Note:

The use of the term graphite *as a short term for material consisting of* graphitic carbon *is incorrect. The term* graphite *can only be used in combination with other nouns or clarifying adjectives for special types of* graphite materials *(graphite electrodes,* natural graphite *and others). The use of the term* graphite *without a noun or clarifying adjective should be restricted to the allotropic form of the element* carbon.

See *carbon, graphite, graphitic carbon, natural graphite*

GRAPHITE WHISKER

Description:

Graphite whiskers consist of thin, approximately cylindrical filaments in which *graphene layers* are arranged in a scroll-like manner. There is, at least in part, a regular stacking of the layers as in the *graphite* lattice, giving rise to *h,k,l,* X-ray reflections. The physical properties of *graphite whiskers* approach, along the cylinder axis, those of *graphite*.

Note:

If there is, due to misalignment of the layers caused by their bending, no three-dimensional stacking order as in graphite, *the term* carbon whiskers *should be used*. Graphite whiskers *and* carbon whiskers *should be distinguished from more disordered* filamentous carbon.
See *filamentous carbon, graphitic carbon*

GRAPHITIC CARBON

Description:

Graphitic carbons are all varieties of substances consisting of the element carbon in the allotropic form of *graphite* irrespective of the presence of structural defects.
See *graphite*

Note:

The use of the term graphitic carbon *is justified if three-dimensional hexagonal crystalline long-range order can be detected in the material by diffraction methods, independent of the volume fraction and the homogeneity of distribution of such crystalline domains. Otherwise, the term* non-graphitic carbon *should be used.*
See *non-graphitic carbon*

GRAPHITIZABLE CARBON

Description:

Graphitizable carbon is a *non-graphitic carbon* which upon *graphitization heat treatment* converts into *graphitic carbon*.
See *graphitic carbon, graphitization heat treatment, non-graphitic carbon*

Note:

If it is preferred to define the characterizable state of material instead of its behavior during subsequent treatment, the term "pre-graphitic carbon" could be considered.

GRAPHITIZATION

Description:

Graphitization is a solid-state transformation of thermodynamically unstable *non-graphitic carbon* into *graphite* by means of heat treatment.
See *graphite, non-graphitic carbon*

Note:

Graphitization *is also used for the transformation of metastable* diamond *into* graphite *by heat treatment, as well as in metallurgy for the formation of* graphite *from thermodynamically unstable carbides by thermal decomposition at high temperatures. Such uses of the term* graphitization *are in line with the above definition. The use of the term* graphitization

to indicate a process of thermal treatment of carbon materials *at* $T > 2500\,K$ *regardless of any resultant crystallinity is incorrect.*
See *carbon materials, diamond, graphite, graphitization, graphitization heat treatment*

GRAPHITIZATION HEAT TREATMENT
Description:
Graphitization heat treatment is a process of heat treatment of a *non-graphitic carbon*, industrially performed at temperatures in the range between 2500 and 3300 K, to achieve transformation into *graphitic carbon.*
See *graphitic carbon graphitization, non-graphitic carbon*

Note:
The term graphitization heat treatment *does not include information as to the crystallinity achieved by the heat treatment, that is the extent of transformation into* graphitic carbon *or the degree of* graphitization. *Only for such a transformation into* graphitic carbon *should the term* graphitization *be used. Consequently the common use of the term* graphitization *for the heat treatment process only, regardless of the resultant crystallinity, is incorrect and should be avoided*
See *graphitic carbon, graphitization*

GRAPHITIZED CARBON
Description:
Graphitized carbon is a *graphitic carbon* with more or less perfect three-dimensional hexagonal crystalline order prepared from *non-graphitic carbon* by *graphitization heat treatment.*
See *graphitic carbon, graphitization heat treatment, non-graphitic carbon*

Note:
Non-graphitizable carbons *do not transform into* graphitic carbon *on heat treatment at temperatures above 2500 K and therefore are not* graphitized carbons.
See *graphitizable carbon, non-graphitizable carbon.*

GREEN COKE
Description:
Green coke (*raw coke*) is the primary solid *carbonization* product from high-boiling hydrocarbon fractions obtained at temperatures below 900 K. It contains a fraction of matter that can be released as volatiles during subsequent heat treatment at temperatures up to approximately 1600 K. This mass fraction, the so-called volatile matter, is in the case of *green coke* between 4 and 15 wt%, but it depends also on the heating rate.
See *carbonization, raw coke*

Note:
Raw coke *is an equivalent term to* green coke *although it is now less frequently used.*

The so-called volatile matter of green coke *depends on temperature and time of coking, but also on the method for its determination.*
See *raw coke*

HARD AMORPHOUS CARBON FILMS

Description:

Hard amorphous carbon films is a synonym for *DLC films*

See *DLC films*

HEXAGONAL GRAPHITE

Description:

Hexagonal graphite is the thermodynamically stable form of *graphite* with an ABAB stacking sequence of the *graphene layers*. The exact crystallographic description of this allotropic form is given by the space group $D6h^4$ – PG_3/mmc (unit cell constants – $a = 245.6\,pm$, $c = 670.8\,pm$). *Hexagonal graphite* is thermodynamically stable below approximately 2600 K and 6 GPa.

See *graphite*

Note:

The use of the term graphite *instead of the more exact term* hexagonal graphite *may be tolerated in view of the minor importance of* rhombohedral graphite, *the other allotropic form.*

See *graphite, rhombohedral graphite*

HIGH-PRESSURE GRAPHITIZATION

Description:

High-pressure graphitization refers to a solid-state transformation of *non-graphitic carbon* into *graphite* by heat treatment under elevated pressure (e.g. 100 to 1000 MPa) so that a definitely higher degree of *graphitization* is achieved at lower temperature and/or for a shorter heat treatment time than in heat treatment of the same.

See *graphite, graphitization, non-graphitic carbon*

HIGHLY ORIENTED PYROLYTIC GRAPHITE

Description:

Highly oriented pyrolytic graphite (*HOPG*) is a *pyrolytic graphite* with an angular spread of the *c*-axes of the crystallites of <1 degree.

See *graphite, pyrolytic graphite*

Note:

Commercial HOPG *is usually produced by stress annealing at approximately 3300 K.*

ISOTROPIC CARBON

Description:

Isotropic carbon is a monolithic *carbon material* without preferred crystallographic orientation of the microstructure.

See *carbon material*

Note:

Isotropic carbon *can also be a* graphite material. *The isotropy can be gross (bulk), macroscopic, or microscopic, depending on the structural level at which isotropy is obtained. This word is widely used today and its meaning covers all the above levels. For example, the aerospace graphites have isotropy built in by random grain orientation. Some* nuclear graphites *are isotropic at the crystalline (sub-grain) level.*

See *graphite material, nuclear graphite*

ISOTROPIC PITCH-BASED CARBON FIBERS
Description:
Isotropic pitch-based carbon fibers are *carbon fibers* obtained by *carbonization* of isotropic pitch fibers after these have been stabilized (i.e. made non-fusible).
See *carbon fibers, carbonization, pitch-based carbon fibers, stabilization treatment*

Notes:
During fabrication of isotropic pitch-based carbon fibers *no means (neither mechanical nor chemical) are applied to achieve preferred orientation of the polyaromatic molecules in the fiber direction. They belong to the* carbon fibers type *LM, and because of the relatively low values of strength and Young's modulus this* pitch-based carbon fiber *type is not used for high-performance reinforcement purposes.*
See *carbon fibers type LM, pitch-based carbon fibers*

LAMP BLACK
Description:
Lamp black is a special type of *carbon black* produced by incomplete combustion of a fuel rich in aromatics that is burned in flat pans. *Lamp black* is characterized by a relatively broad particle size distribution.
See *carbon black*

MESOGENIC PITCH
Description:
Mesogenic pitch is a *pitch* with a complex mixture of numerous essentially aromatic hydrocarbons. It does not contain anisotropic particles detectable by optical microscopy. *Mesogenic pitch* is low in quinoline-insoluble fractions and capable of transforming into *MPP* during continuous heat treatment above 750 K by the formation of optically detectable *carbonaceous mesophase.*
See *carbonaceous mesophase, MPP, pitch*

MPP
Description:
MPP is a *pitch* with a complex mixture of numerous essentially aromatic hydrocarbons containing anisotropic liquid-crystalline particles (*carbonaceous mesophase*) detectable by optical microscopy and capable of coalescence into the *bulk mesophase.*
See *bulk mesophase, carbonaceous mesophase, mesogenic pitch, pitch*

Notes:
The carbonaceous mesophase *particles are formed from the aromatics of high molecular mass in* mesogenic pitch, *which have not yet been aggregated to particles detectable by optical microscopy within the apparently isotropic* pitch *matrix. The* carbonaceous mesophase *is insoluble in quinoline and pyridine, but the amount of I mesophase measured from microscopical observation appears somewhat higher because parts of the* carbonaceous mesophase *can be extracted by the solvents.*
See *carbonaceous mesophase, mesogenic pitch, pitch*

MPP-BASED CARBON FIBERS

Description:

MPP-based carbon fibers are *carbon fibers* obtained from *mesogenic pitch* after it has been transformed into MPP at least during the process of spinning, and after the spun *MPP* fibers have been made non-fusible (stabilized) and carbonized.

See *carbon fibers, mesogenic pitch, MPP, pitch, pitch-based carbon fibers*

METALLURGICAL COKE

Description:

Metallurgical coke is produced by *carbonization* of coals or coal blends at temperatures up to 1400 K to produce a macroporous *carbon material* of high strength and relatively large lump size.

See *carbon material, carbonization*

Notes:

Metallurgical cokes *must have a high strength to support heavy loads in the blast furnace without disintegration.* Metallurgical coke *is also used as* filler coke *for* polygranular carbon *products.*

See *filler coke, polygranular carbon*

MICROPOROUS CARBON

Description:

Microporous carbon is a porous *carbon material*, usually a *char* or *carbon fibers*, which may or may not have been subjected to an activation process to increase its adsorptive properties. A *microporous carbon* is considered to have a major part of its porosity in pores of <2 nm width and to exhibit apparent surface areas usually higher than $500\,m^2\,g^{-1}$.

See *activated carbon, fibrous activated carbon*

Notes:

For definition of micropores see: IUPAC Manual of Symbols and Terminology, Appendix 2, Pt.1, Colloid and Surface Chemistry. Pure Appl. Chem. 31, 518 (1972).

The surface areas determined by the BET method are apparent surface areas only since the BET adsorption equation is, in principle, not valid when micropore filling occurs. The determination of the true surface area in the micropores depends on the method used for the evaluation of the adsorption isotherms and on the model used for the shape of the micropores (cylindrical, slit-shaped or other).

MPP-BASED CARBON FIBERS

Description:

See *MPP-based carbon fibers*

NATURAL GRAPHITE

Description:

Natural graphite is a mineral found in nature. It consists of *graphitic carbon* regardless of its crystalline perfection.

See *graphitic carbon*

Notes:
Some natural graphites, *often in the form of large flakes, show very high crystalline perfection. Occasionally, they occur as single crystals of* graphite. *The use of the term* natural graphite *as a synonym for the term "graphite single crystal" is incorrect and should be avoided. Varieties of* natural graphite *with lower structural perfection are classified as "Microcrystalline* natural graphite*". Commercial* natural graphite *is often contaminated with other minerals (e.g. silicates) and may contain* rhombohedral graphite *due to intensive milling.*
See *graphite, rhombohedral graphite*

NEEDLE COKE
Description:
Needle coke is the commonly used term for a special type of *coke* with extremely high graphitizability resulting from a strong preferred parallel orientation of its turbostratic layer structure and a particular physical shape of the grains.
See *coke*

Notes:
Needle coke *is derived mainly from clean (i.e. lacking hetero atoms and solids) and highly aromatic (i.e. several condensed rings per cluster) feedstocks with a very low concentration of insolubles. Upon solidification a material with a distinctive streaked or flow-like macroscopic appearance is produced. Upon grinding the* coke *breaks up first into macroscopic needles and then, after further grinding, into micro-platelets. Sometimes the word "acicular" is used as a synonym for needle-like.*
See *coke, delayed coking process*

NON-GRAPHITIC CARBON
Description:
Non-graphitic carbons are all varieties of solids consisting mainly of the element carbon with two-dimensional long-range order of the carbon atoms in planar hexagonal networks, but without any measurable crystallographic order in the third direction (c-direction) apart from more or less parallel stacking.
See *amorphous carbon*

Note:
Some varieties of non-graphitic carbon *convert on heat treatment to* graphitic carbon (graphitizable carbon) *but some others do not* (non-graphitizable carbon).
See *graphitic carbon, graphitizable carbon, non-graphitic carbon, non-graphitizable carbon*

NON-GRAPHITIZABLE CARBON
Description:
Non-graphitizable carbon is a *non-graphitic carbon* which cannot be transformed into *graphitic carbon* solely by high-temperature treatment up to 3300 K under atmospheric pressure or lower pressure.
See *graphitic carbon, graphitization heat treatment, non-graphitic carbon*

Note:
The term non-graphitizable *is limited to the result of heat treatment without additional influence of foreign matter or neutron radiation.* Non-graphitizable carbon *can be transformed*

into graphitic carbon *by a high-temperature process via intermediate dissolution in foreign matter and precipitation under high pressure or by radiation damage.*
See *graphitic carbon*

NUCLEAR GRAPHITE
Description:
Nuclear graphite is a *polygranular graphite* material for use in nuclear reactor cores consisting of *graphitic carbon* of very high chemical purity. High purity is needed to avoid absorption of low-energy neutrons and the production of undesirable radioactive species.
See *graphite material, graphitic carbon, polygranular graphite*

Notes:
Apart from the absence of neutron-absorbing impurities, modern reactor graphites are also characterized by a high degree of graphitization *and no preferred bulk orientation. Such properties increase the dimensional stability of the* nuclear graphite *at high temperatures and in a high flux of neutrons. The term* nuclear graphite *is often, but incorrectly, used for any* graphite material *in a nuclear reactor, even if it serves only for structural purposes.*
See *graphite material, graphitization*

PAN-BASED CARBON FIBERS
Description:
Pan-pased carbon fibers are *carbon fibers* obtained from PAN precursor fibers by *stabilization treatment, carbonization*, and final heat treatment.
See *carbon fibers, carbonization, stabilization treatment*

PARTICIPATE CARBON
Description:
Particulate carbon is a *carbon material* consisting of separated monolithic particles.
See *carbon material*

Note:
Distinctions should be made between coarse particulate carbon *or* granular carbon *(larger than about 100 urn, but smaller than about 1 cm in average size), fine* particulate carbon *or powder or flour (between 1 urn and 100 urn in average size) and* colloidal carbon *(below approximately 1 urn in size in at least one direction), for example,* carbon blacks *and* colloidal carbon.
See *carbon black, colloidal carbon, granular carbon, particulate carbon*

PETROLEUM COKE
Description:
Petroleum coke is a *carbonization* product of high-boiling hydrocarbon fractions obtained in petroleum processing (heavy residues). It is the general term for all special *petroleum coke* products such as *green, calcined* and *needle* petroleum *coke.*
See *calcined coke, carbonization, green or raw coke, needle coke*

Notes:
High-boiling hydrocarbon fractions (heavy residues) used as feedstock for petroleum coke *are residues from distillation (atmospheric pressure, vacuum) or cracking (e.g. thermal,*

catalytic, steam-based) processes. The nature of feedstock has a decisive influence on the graphitizability of the calcined coke.
See *calcined coke*

PETROLEUM PITCH
Description:
Petroleum pitch is a residue from heat treatment and distillation of petroleum fractions. It is solid at room temperature, consists of a complex mixture of numerous predominantly aromatic and alkyl-substituted aromatic hydrocarbons, and exhibits a broad softening range instead of a defined melting temperature.

Note:
The hydrogen aromaticity (ratio of aromatic to total hydrogen atoms) varies between 0.3 and 0.6. The aliphatic hydrogen atoms are typically present in alkyl groups substituted on aromatic rings or as naphthenic hydrogen.

PITCH
Description:
Pitch is a residue from pyrolysis of organic material or tar distillation which is solid at room temperature, consisting of a complex mixture of numerous, essentially aromatic hydrocarbons and heterocyclic compounds. It exhibits a broad softening range instead of a defined melting temperature. When cooled from the melt, pitches solidify without crystallization.

Notes:
The ratio of aromatic to aliphatic hydrogen depends mainly on the source of the starting material. The hydrogen aromaticity (ratio of aromatic to total hydrogen atoms) varies between 0.3 and 0.9.

The aliphatic hydrogen in pitch is largely associated with alkyl side chains substituted on aromatic rings. The content of heterocyclic compounds in pitches varies depending on their origins. Also the softening temperature can vary in a broad range between about 320 and 570 K depending on the molecular weight (relative molecular mass) and composition of the constituents.

PITCH-BASED CARBON FIBERS
Description:
Pitch-based carbon fibers are *carbon fibers* obtained from *pitch* precursor fibers after *stabilization treatment, carbonization*, and final heat treatment.
See *carbon fibers, pitch, stabilization treatment*

Notes:
The term pitch-based carbon fibers *comprises the* isotropic pitch-based carbon fibers *as well as the anisotropic* MPP-based carbon fibers . *The isotropic type belongs to the* carbon fibers type LM *and is mainly used as filler in polymers and insulation materials and for similar applications. The anisotropic type* (MPP-based carbon fibers) *belongs to the* carbon fibers type *HM and is used mainly for reinforcement purposes due to its high Young's modulus value.*
See *carbon fibers type HM, carbon fibers type LM, isotropic pitch-based carbon fibers, MPP-based carbon fibers*

POLYCRYSTALLINE GRAPHITE

Description:

Polycrystalline graphite is a graphite material with coherent crystallographic domains of limited size regardless of the perfection and preferred orientation (texture) of their crystalline structure.

See *graphite material*

Notes:

The common use of the term polycrystalline graphite *for* polygranular graphite *is in line with this definition but may be inexact because usually all grains of* polygranular graphite *are polycrystalline themselves.* Polycrystalline graphite *can exhibit a random orientation, more or less preferred orientation, or a highly oriented texture as in some* pyrolytic graphites. *There is no sharp transition, however, between the typical polycrystalline texture and the "single crystal-like" texture of* HOPG.

See *HOPG, polygranular graphite, pyrolytic graphite*

POLYGRANULAR CARBON

Description:

Polygranular carbon is a *carbon material* composed of grains, which can be clearly distinguished by means of optical microscopy.

See *carbon material*

Notes:

Industrial carbon materials *(such as electrodes) are mostly polygranular, but special grades are agranular materials, such as* glass-like carbon, carbon fibers *or* pyrolytic carbon. *Such materials are covered by the term* agranular carbon.

See *agranular carbon, carbon fibers, carbon material, glass-like carbon, pyrolytic carbon*

POLYGRANULAR GRAPHITE

Description:

Polygranular graphite is a *graphite material* composed of grains which can be clearly distinguished by means of optical microscopy.

See *graphite material*

Note:

From the viewpoint of crystal Unity, a polygranular graphite *is always a* polycrystalline graphite, *but not vice versa. Most industrial* graphite materials *are polygranular. Monogranular materials consist mostly of* non-graphitic carbon, *such materials are called monolithic or* agranular carbons.

See *agranular carbon, graphite material, non-graphitic carbon, polycrystalline graphite*

PREMIUM COKE

Description:

Premium coke is an extremely well graphitizing carbon with a high degree of optical anisotropy (isochromatic areas of optical texture above about 100 pm) and is characterized by a combination of the following properties which differ significantly from those of *regular coke*: high real density, low reversible thermal expansion, and low ash content combined, in most cases, with low sulfur content.

See *regular coke*

Note:
Premium coke *is mainly produced from tars or residues from petrochemistry by the* delayed coking process. *Also refined* coal tar pitches *are used as precursors for* premium coke *production.*
See *coal tar pitch, delayed coking process, regular coke*

PUFFING
Description:
The term *puffing* describes an irreversible expansion of some *carbon artifacts* during *graphitizaion heat treatment* between 1650 and 2700 K.
See *carbon artifact, coke, graphitization heat treatment*

Notes:
Puffing *is caused by the release of heteroatoms, for instance sulfur atoms, from the* coke *in association with specific microstructural rearrangements.*
See *coke, polygranular carbon, puffing inhibitor*

PUFFING INHIBITOR
Description:
Puffing inhibitors are metals or metal compounds with a high chemical affinity for the heteroatoms in the carbons. They are distributed as fine particles within the *carbon materials* to be graphitized.
See *carbon material*

Notes:
Iron and iron compounds are most frequently used as puffing inhibitors *when* puffing *is related to sulfur.*
See *puffing*

PYROLYTIC CARBON
Description:
Pyrolytic carbon is a *carbon material* deposited from gaseous hydrocarbon compounds on suitable underlying substrates (*carbon materials*, metals, ceramics) at temperatures ranging from 1000 to 2500 K (CVD).
See *carbon material*

Notes:
A wide range of microstructures, for example isotropic, lamellar, substrate-nucleated and a varied content of remaining hydrogen, can occur in pyrolytic carbons, *depending on the deposition conditions (temperature, type, concentration and flow rate of the source gas, surface area of the underlying substrate, etc.).*

"Pyrocarbon" *which is synonymous with* pyrolytic carbon *was introduced as a trademark and should not be used as a term.*

The term pyrolytic carbon *does not describe the large range of* carbon materials *obtained by thermal degradation (thermolysis, pyrolysis) of organic compounds when they are not*

formed by CVD. Also carbon materials, *obtained by physical vapor deposition (PVD) are not covered by the term* pyrolytic carbon.
See *carbon material*

PYROLYTIC GRAPHITE
Description:
Pyrolytic graphite is a *graphite material* with a high degree of preferred crystallographic orientation of the *c*-axes perpendicular to the surface of the substrate, obtained by *graphitization heat treatment* of *pyrolytic carbon* or by CVD at temperatures above 2500 K.
See *graphite material, graphitization heat treatment, pyrolytic carbon*

Notes:
"Pyrographite", a synonym for pyrolytic graphite, *was introduced as a trademark and should not be used as term.*

Hot working of pyrolytic graphite *(by heat treatment under compressive stress at temperatures above 3000 K) results in* HOPG.
See *HOPG*

RAW COKE
Description:
See *green coke*

Note:
The term raw coke *is equivalent to* green coke *although it now used less frequently.*

RAYON-BASED CARBON FIBERS
Description:
Rayon-based carbon fibers are *carbon fibers* made from rayon (cellulose) precursor fibers.
See *carbon fibers*

Notes:
Rayon-based carbon fibers *have a more isotropic structure than similarly heat-treated PAN or* MPP-based carbon fibers. *Their Young's modulus values are therefore drastically lower ($E^a < 100\,GPa$, $\sigma^b < 100\,MPa$).* Rayon-based carbon fibers *can be transformed into anisotropic* carbon fibers *with high strength and Young's modulus values by hot-stretching treatment at temperatures of approximately 2800 K.*
See *carbon fibers, pan-based carbon fibers, MPP-based carbon fibers*

[a]E, Young's modulus; [b]σ, tensile strength

REGULAR COKE
Description:
Regular coke is a *petroleum coke* with good graphitizability and is characterized by a combination of properties which differ significantly from those of *metallurgical coke* but do not reach the quality level of *premium coke*: These properties are: optical anisotropy, medium reversible thermal expansion, and low ash content.
See *metallurgical coke, petroleum coke, premium coke*

Notes:
Typical characteristics for regular coke *in comparison with those of* metallurgical coke *and of* premium coke *calcined at 1620 K are*:

		Regular coke	Premium coke	Metallurgical coke
Real Density[a]	($g\,cm^{-3}$)	2.07–2.09	2.12–2.14	1.95–2.02
CTE[b] (293–277 K)	(K^{-1})	2.0×10^{-6}	1.1×10^{-6}	3.0×10^{-6}
CTE[b] (293–773 K)	(K^{-1})	1.0×10^{-6}	0.5×10^{-6}	2.0×10^{-6}
Ash	(wt%)	0.4	0.05	8–12
Sulfur	(wt%)	1.0–1.5	0.6	0.6–5.0

[a] measured with *m*-zylene, [b] coefficient of thermal expansion

regular coke *is mainly used for the production of* synthetic carbon *and* graphite materials.
See *carbon material, graphite material, metallurgical coke, premium coke, synthetic graphite*

RHOMBOHEDRAL GRAPHITE
Description:
Rhombohedral graphite is a thermodynamically unstable allotropic form of *graphite* with an ABCABC stacking sequence of the layers. The exact crystallographic description of this allotropic form is given by the space group D_3d^5–R^3m (unit cell constants: $a = 256.6$ pm, $c = 1006.2$ pm).
See *graphite*

Notes:
The structure of rhombohedral graphite *can be best considered as an extended stacking fault in* hexagonal graphite. Rhombohedral graphite *cannot be isolated in pure form (*natural graphite *and laboratory preparations contain* <40% *of* rhombohedral graphite *in combination with* hexagonal graphite*) It is produced by shear deformation of* hexagonal graphite *and transforms progressively to the hexagonal (ABAB) modification on heating above 1600 K.*
See *hexagonal graphite, natural graphite*

SEMI-COKE
Description:
Semi-coke is a carbonaceous material intermediate between a fusible MPP and a non-deformable *green coke* produced by incomplete *carbonization* at temperatures between the onset of fusion (of coal, ca. 620 K), and complete devolatilization. *Semi-coke* still contains volatile matter, therefore:
See *carbonization, coal tar pitch, green coke*

Note:
Semi-coke *may be conceived as covering a continuous range from coal that has not yet been fused to* coke breeze. Semi-coke *can also be used as a* filler *in carbon mixtures.*
See *coke breeze, filler*

SOOT

Description:

Soot is a randomly formed *particulate carbon* material and may be coarse, fine, and/or colloidal in proportions dependent on its origin. *Soot* consists of variable quantities of carbonaceous and inorganic solids together with absorbed and occluded tars and resins.
See *particulate carbon*

Notes:

Soot *is generally formed as an unwanted by-product of incomplete combustion or pyrolysis.* Soot *generated within flames consists essentially of aggregates of spheres of carbon.* Soot *found in domestic fireplace chimneys contains few aggregates but may contain substantial amounts of particulate fragments of* coke *or* char. Soot *from diesel engines consists essentially of aggregates together with tars and resins. For historical reasons, the term* soot *is sometimes incorrectly used for* carbon black. *This misleading use should be avoided.*
See *carbon black, char, coke*

SPHERICAL CARBONACEOUS MESOPHASE

Description:

The term *spherical carbonaceous mesophase* describes the morphology of *carbonaceous mesophase* which is formed in the isotropic *pitch* matrix. The *spherical carbonaceous mesophase* usually has a lamellar structure consisting of flat aromatic molecules arranged in parallel layers which are perpendicular to the sphere/isotropic phase interface as described by *Brooks and Taylor*. On coalescence this spherical mesophase loses its characteristic morphology and is converted to the *bulk mesophase*.
See *Brooks and Taylor structure, bulk mesophase, carbonaceous mesophase, pitch*

STABILIZATION TREATMENT OF THERMOPLASTIC PRECURSOR FIBERS FOR CARBON FIBERS

Description:

Stabilization treatment is a process applied to fusible organic precursor fibers for *carbon fibers* with the aim of obtaining non-fusible polymer fibers suitable for subsequent *carbonization*. The original fiber shape is maintained.
See *carbon fibers, carbonization*

Notes:

The stabilization treatment *of thermoplastic precursor fibers for* carbon fibers *is usually a heat treatment process performed in an oxidizing atmosphere above 470 K. For* stabilization treatment *of PAN fibers, 600 K is the highest temperature up to which cyclization, dehydrogenation and oxidation processes prevail.*
See *carbon fibers*

STRESS GRAPHITIZATION

Description:

Stress graphitization refers to the solid-state transformation of *non-graphitic carbon* into *graphite* by heat treatment combined with application of mechanical stress, resulting in a defined degree of *graphitization* being obtained at a lower temperature and/or after a shorter time of heat treatment than in the absence of applied stress.
See *graphite, graphitization, non-graphitic carbon*

Note:
Stress graphitization *may also occur in volume elements of a carbon body in the process of heat treatment as a result of the action of internal residual or thermal stresses.*

SYNTHETIC GRAPHITE
Synthetic graphite is a material consisting of *graphitic carbon* which has been obtained by graphitizing of *non-graphitic carbon*, by CVD from hydrocarbons at temperatures above 2500 K, by decomposition of thermally unstable carbides or by crystallizing from metal melts supersaturated with carbon.
See *graphitic carbon, graphitization, non-graphitic carbon*

Notes:
The term artificial graphite *is often used as a synonym for* synthetic graphite. *The term* synthetic graphite *is preferred, however, since graphite crystals can be considered to consist of carbon macromolecules. Although the term* synthetic graphite *also covers the CVD product* pyrolytic graphite *as well as the residues of carbide decomposition, it is predominantly used for* graphitized carbon. *Such common uses are in line with the above definition. Synonyms for this most important type of* synthetic graphite *are* Acheson graphite and electrographite.
See *Acheson graphite, artificial graphite, electrographite, graphitized carbon, pyrolytic graphite*

THERMAL BLACK
Description:
Thermal black is a special type of *carbon black* produced by pyrolysis of gaseous hydrocarbons in a preheated chamber in the absence of air. *Thermal black* consists of relatively large individual spheres (100–500 nm diameter) and aggregates of a small number of pseudo-spherical particles. The preferred alignment of the layer planes is parallel to the surface of the spheres.
See *carbon black*

9.3 Surface Chemists: Biographies

9.3.1 Irving Langmuir (1881–1957)

Irving Langmuir was born in Brooklyn, New York, on January 31, 1881, as the third of four sons of Charles Langmuir and Sadie, née Comings. His early education was obtained in various schools and institutes in the USA, and in Paris (1892–1895). He graduated as a metallurgical engineer from the School of Mines at Columbia University in 1903. Postgraduate work in Physical Chemistry, under Nernst in Gottingen, earned him the degrees of M.A. and Ph.D. in 1906.

Returning to America, Dr Langmuir became Instructor in Chemistry at Stevens Institute of Technology, Hoboken, New Jersey, where he taught until July 1909. He then entered the Research Laboratory of the General Electric Company at Schenectady where he eventually became Associate Director. Langmuir's studies embraced chemistry, physics, and engineering and were largely the outgrowth of studies of vacuum phenomena. In seeking the atomic and molecular mechanisms of these he investigated the properties of adsorbed

films and the nature of electric discharges in high vacuum and in certain gases at low pressures. His work on filaments in gases led directly to the invention of the gas-filled incandescent lamp and to the discovery of atomic hydrogen. He later used the latter in the development of the atomic hydrogen welding process.

The Langmuir equation of adsorption related directly to monolayer chemisorption processes. There is no indication that porous solids were studied.

He was the first to observe the very stable adsorbed monatomic films on tungsten and platinum filaments, and was able, after experiments with oil films on water, to formulate a general theory of adsorbed films. He also studied the catalytic properties of such films. Langmuir's work on space charge effects and related phenomena led to many important technical developments which have had a profound effect on later technology. In chemistry, his interest in reaction mechanism caused him to study structure and valence, and he contributed to the development of the Lewis theory of shared electrons.

Among the awards made to him were: Nichols Medal (1915 and 1920); Hughes Medal (1918); Rumford Medal (1921); Cannizzaro Prize (1925); Perkin Medal (1928); School of Mines Medal (Columbia University, 1929); Chardler Medal (1929); Willard Gibbs Medal (1930); Popular Science Monthly Award (1932); Nobel Prize in Chemistry (1932); Franklin Medal and Holly Medal (1934); John Scott Award (1937); "Modern Pioneer of Industry" (1940); Faraday Medal (1944); Mascart Medal (1950). In addition, he was a Foreign Member of the Royal Society of London, Fellow of the American Physical Society, Honorary Member of the British Institute of Metals, and of the Chemical Society (London). He had served as President of the American Chemical Society and as President of the American Association for the Advancement of Science. Honorary degrees were bestowed upon Langmuir by the following colleges and universities: Northwestern, Union, Edinburgh

(Scotland), Columbia, Kenyon, Princeton, Lehigh, Harvard, Oxford, Johns Hopkins, Rutgers, Queens (Canada), and Stevens Institute of Technology. Dr Langmuir's hobbies were mountaineering, skiing, flying, and, most of all, to understand the mechanism of simple and familiar natural phenomena. He married Marion Mersereau in 1912. They had a son, Kenneth, and a daughter, Barbara. After a short illness, he died on August 16, 1957.

9.3.2 Brunauer, Emmett and Teller

Although these three scientists have become bonded together, probably for ever, by reason of the BET equation (1938) their working relationship in Washington was almost casual. Brunauer and Teller came to the USA from Hungary to escape the political situation that was developing there in the 1930s. Emmett was a native born American. All three came together in Washington, to the fixed nitrogen laboratory (FNL), independently with quite different scientific interests. Their brief interactive encounter resulted in the BET equation which has dominated adsorption studies for almost 70 years. It was an equation initially designed to describe multilayer adsorption on non-porous surfaces and was quickly applied to porous materials, in particular activated carbons with a broad range of pore sizes. Although the equation was an important development it was not always reliable to investigate adsorption within of well-defined size and shape. This problem persists to today. Of the three scientists, it was Brunauer who was the closest to being a surface scientist. His book *The Adsorption of Gases and Vapors*, published in 1943, was, and still is, a masterpiece of the subject.

9.3.3 Stephen Brunauer (1903–1986)

Even before the BET paper was published, the three scientists were going separate ways. Brunauer, following the publication of the BET paper, found himself a well-known person within the FNL that wanted to de-emphasize theoretical studies of fertilizers, and to focus on other research areas. Thus, Brunauer was faced with saving his job in a time when his organization wanted to eliminate it. In his effort, Brunauer obtained contracts from the military, and defence work began to become more important to him. With the outbreak of World War II, it became apparent to Brunauer that he would not be able to continue much longer at the FNL. He tried to join the military's Naval research effort. In his first attempt, he was turned down. Undaunted, Brunauer called upon his friends as well as those of his wife, to write letters in support of his second application. This resulted in a flood of letters from senators, ambassadors, etc., and in 1942 he became a Lt, Junior Grade, in the Naval Reserve. He was appointed to head a high explosives research group in the Bureau of Ordnance which, at the time, included only three people. Eventually this group expanded to more than 100 people, and included both Naval officers and civilian employees. By the end of the war Brunauer had become a Lt Commander, and had led a research effort that made considerable progress in improving the surface and underwater explosives used by the Navy. Brunauer drew many well-known scientists into the war effort; these included Albert Einstein and John von Neumann, the father of the computer. Brunauer remained in the same position following the war, but as a civilian employee.

Shortly after the end of World War II, Brunauer went to Hungary to develop a roster of Hungarian scientists. He was instrumental in helping many scientists come to the USA.

Drs John Gregg (left), Stephen Brunauer and Ken Sing (right) (1969). Note: John Gregg was a pioneer of adsorption studies (Exeter University, UK). Ken Sing, from Brunel University, London, UK, of international reputation, interacted with Brunauer and Dubinin. (Photograph by courtesy of Professor Sing).

By 1950 he had been appointed to the highest grade level for civilian employees. In 1944, Brunauer's wife Esther joined the State Department and rose quickly, in spite of the State Department not having females in high positions at the time. Esther was appointed by President Truman to the rank of Minister to represent the USA at the preparatory commission for the formation of UNESCO. In 1950, Senator Joseph McCarthy launched his infamous charge that there were 205 rank-and-file Communists in the State Department, and he waved a piece of paper that he claimed contained their identities. The Brunauers were on such a list. McCarthy listed Esther as the first person from the State Department who should be investigated. McCarthy turned to JB Matthews, a former investigator for the House Un-American Activities Committee, and was provided material about the Brunauer couple. Matthews had been a successful organizer of communist front organizations in the 1930s. He claimed to have met the Brunauers during the 1930, and that they were Communists. For some time thereafter the Brunauers were the subject of front page headlines. Esther was called to testify at the hearings of the Tydings Committee, and even pair was incorrect. The result was this was terminated. The loyalty committee voted 2–1 that, while Esther was completely loyal, she was married to Stephen, a potential security risk because he had belonged to the Young Workers League. Brunauer then joined the Portland Cement Association in Chicago and his work made him an internationally known expert in this area. In Chicago, he became active in the Baptist Church, and soon became a Deacon. From 1952 until her death in 1958, Esther and Stephen remained together and were both active in church activities.

9.3.4 Paul Emmett (1900–1985)

Paul Emmett arrived in Washington in 1926 to work at the FNL because he missed the research days of graduate school. Emmett was born in Portland, Oregon. His father, a railroad worker, was frequently absent from home and Emmett was raised primarily by his mother. When he went to Cal Tech to obtain his Ph.D., his mother accompanied him to Pasadena to establish a home there. Nobel Prize winner Linus Pauling was a fellow student with Emmett in both high school and undergraduate days, and lived with the Emmett family for a year. Cal Tech was a small school just starting graduate studies. Only four Ph.D.s had been awarded prior to the class of four that included Linus Pauling and Paul Emmett. These were not two ordinary graduate students. They published a paper in the prestigious Journal of the American Chemical Society while still studying as graduate students, and without including a faculty member. Emmett returned to Portland to teach at Oregon State University, but missed his research work. He arrived at the FNL, Washington, just as the direction of the laboratory was turning from development work on the synthetic ammonia catalyst to that of understanding the mechanism of the catalytic reaction. Emmett was a brilliant organizer and, despite his low-key personality, an intense competitor. He was an outstanding writer and a dynamic speaker who was an effective salesman of the BET method. In addition to the work at the FNL that led to the BET equation, Emmett led a team consisting of Brunauer and five other professionals, in defining the kinetics and mechanism of the ammonia synthesis. In science where change is the final objective, tile understanding of ammonia synthesis developed by Emmett and his co-workers is just as valid today as it was when the work was carried out.

9.3.5 Edward Teller (1908–2003)

Teller was a nuclear physicist, and has been described as one of the giants of the nuclear age. He was a formidable mathematician who became impaled by the nuclear arms race, leading to the Star Wars program of the US President Ragan (Goodchild, 2004).

Edward Teller, like Brunauer, was born in Hungary. He immigrated to the USA at the inducement of a George Gamow. The President of George Washington University (USA) wanted to develop an internationally known physics department and, seeking advise on how to do this, was told, "Develop a department emphasizing theoretical physics; it is much cheaper than experimental physics." Gamow accepted the position only if he could bring Teller along as an assistant. Teller obtained his Ph.D. working with Professor Werner Heisenberg, famous for his uncertainty principle. He was an excellent table tennis player, beating everyone in the school except Professor Heisenberg, his mentor. Teller was one of the few people who had a background in quantum physics, which was in an early stage of development at the time. Thus, many of the most well-know scientists in the Washington area made their way to Teller's home to discuss research topics, and to obtain help from Teller.

It was when in Washington that Teller worked with Brunauer, a student of his, and assisted in the formulation of what has now become known as the BET equation. To Teller, it appears that this was a mathematical exercise.

Emmett left Washington in 1937 to become chairman of the Chemical Engineering Department at Johns Hopkins University. Here he was charged with obtaining professional

Picture taken in 1969 at Clarkson University (USA) when (from left) Brunauer, Emmett and Teller received honorary Ph.D. degrees. Photograph courtesy of Andrea Master, Director, Publications Office, Clarkson University.

accreditation for the newly formed small department. This he accomplished in 3 years. During World War II Emmett became involved with defence research. Initially this involved the development and evaluation of adsorbents for gas masks. It is here that he could have drawn upon his experiences with Brunauer in studies of adsorption processes. He was soon called to the Manhattan Project at Columbia University. Here he reported to Harold Urey, winner of a Nobel Prize for separating the heavy hydrogen isotope, deuterium. Emmett's work at Columbia University involved the development of fluorocarbons to serve as barrier material and pump fluids for use in separating uranium isotopes using the diffusion process.

Teller had turned to the development of the H-Bomb, based on nuclear fusion and not the fission process used to construct the bombs dropped on Japan during the war. Teller is credited with developing the science and pushing the political process that led to the H-Bomb.

After the FNL years, Brunauer, Emmett and Teller came together only twice. They were awarded honorary Doctor of Science degrees from Clarkson University in 1969 (photograph) and 10 years later, the three, along with their wives, had a quiet dinner together in San Francisco to celebrate their paper. The three individuals involved with developing the BET method do not fit the public's image of scientists. In the scientific area they rank at the top of their profession. They exhibited personalities to be found in the more dynamic people of any profession. Politically they represented various views, ranging from very liberal to very conservative. They came together for a short while to contribute a significant advancement to science and lived the rest of their lives following other pursuits.

9.3.6 Mikhail M. Dubinin (1901–2001)

It is impossible to overestimate the contributions made to chemical science, and carbon science in particular, by Mikhail Mikhailovich Dubinin. For about 70 years, Mikhail

Dubinin resolutely and successfully developed the basics of adsorption science which became applied to both military and domestic applications, from lethal war gases to the enhancement of the quality of potable water. He created the Department of Sorption Studies in Moscow where he excelled as a researcher and as a teacher. In 1947, working with the mathematician LV Radushkevich, he suggested a general equation for adsorption on microporous materials from which was developed the theory of volume filling of micropores leading to the DR equation. Currently, this approach is used worldwide in assessments of the characteristics of activated carbons and other such materials, including transport phenomena, sorption dynamics and diffusion coefficients. Mikhail Dubinin was a pioneer worker in studies of gases at high temperatures and in the preparation of activated carbons with tailor-made properties, including the use of oxygen.

Professor H. Akamatsu (Japan) (left) and Professor M.M. Dubinin (USSR) (right) in the USA, June 1983.

Mikhail Dubinin was educated in the Moscow Higher Technical School, 1921–1932, and was appointed, as a Professor, to the Academy of Chemical Defence, 1921–1945. He then became Head of the Department of Sorption Sciences, Institute of Physical Chemistry, USSR, Academy of Sciences, 1946 until his death. He was an Honorary Member of the Hungarian Academy of Sciences, Foreign Member of the Academy of the German Democratic Republic (1982). In 1972 he became an Honorary Member of the DS Slovak Technical Institute, and in 1981 a Member of Leipzig University. In the period 1946–1950, he acted in the capacity of President of the Mendeleyev Chemical Society. He also acted as Editor-in-Chief of the Bulletin of the USSR Academy of Sciences, Chemical Series.

Mikhail Dubinin will always be remembered as an outstanding scientist, a most stimulating teacher, a brilliant organizer, a man of tireless energy, very modest and shy, charming and most supportive. He was able to travel in Western Europe and the USA.

References

Annual Book of ASTM Standards, Section 15, Vol. 15.01, American Society for Testing Materials, Easton, MD, 1992.

AWWA Standard for Powdered Activated Carbon, B600, American Water Works Association, Denver, CO, 1990.

Bansal RC, Donnet J-B, Stoeckli F. *Active Carbon.* Marcel Dekker Inc., New York, 1988, p. 482.

DIN Standards for Carbon Materials, DIN Deutsches Institut für Normung e.V. Beut Verlag GMBH, Berlin, 1986.

Fitzer E, Köchling KH, Boehm HP, Marsh H. Recommended terminology for the description of carbon as a solid. *Pure Appl Chem*, 1995;67(3):473–506.

Goodchild P. Edward Teller. Weidenfeld & Nicolson, London, 2004.

ISO Catalogue of Standards, International Organization for Standardization, Geneva, 1990.

Jankowska H. *Active Carbon*. Ellis Horwood Limited, Chichester, England, 1991.

Leon CA, Radovic LR. Interfacial chemistry and electrochemistry of carbon surfaces. In: *Chemistry and Physics of Carbon*, Vol. 24, Ed. Thrower PA. Marcel Dekker Inc., New York, 1994, pp. 213–310.

Mazyck DW, Cannon FS, Bach MT, Radovic LR. The role of calcium in high pH excursions for reactivated GAC. *Carbon* 2005;43(3):511–518.

Modell M. US Patent 4,061,566 (1977).

Rodríguez-Reinoso F. Production and applications of activated carbons. In: *Handbook of Porous Solids*. Eds. Schüth F, Sing KSW, Weitkamp J. Wiley-VCH Verlag GmmH, Weinheim, Germany, 2002, pp. 1766–1827.

Salvador F, Sánchez-Jiménez C. Effect of regeneration treatment with liquid water at high pressure and temperature on the characteristics of three commercial activated carbons. *Carbon* 1999;37:577–583.

Waterhouse BR, Jaeckel M, Ishikawa M. *Chemical Economics Handbook,* SRI International, Menlo Park, California, 1996, p. 731.2000E.

Author Index

Subject Index

Note: Entries taken from the figures and tables are denoted by *italicized* page numbers.